# THEORY AND DESIGN FOR MECHANICAL MEASUREMENTS

# THEORY AND DESIGN FOR MECHANICAL MEASUREMENTS

SEVENTH EDITION

**Richard S. Figliola**
*Clemson University*

**Donald E. Beasley**
*Clemson University*

# WILEY

| VP AND DIRECTOR | Laurie Rosatone |
| EDITORIAL DIRECTOR | Don Fowley |
| EDITOR | Jennifer Brady |
| EDITORIAL MANAGER | Judy Howarth |
| CONTENT MANAGEMENT DIRECTOR | Lisa Wojcik |
| CONTENT MANAGER | Nichole Urban |
| SENIOR CONTENT SPECIALIST | Nicole Repasky |
| PRODUCTION EDITOR | Kavitha Balasundaram |
| COVER PHOTO CREDIT | © Richard Figliola |

This book was set in 10/12 pt TimesLTStd-Roman by SPi Global and printed and bound by Quad/Graphics.

Founded in 1807, John Wiley & Sons, Inc. has been a valued source of knowledge and understanding for more than 200 years, helping people around the world meet their needs and fulfill their aspirations. Our company is built on a foundation of principles that include responsibility to the communities we serve and where we live and work. In 2008, we launched a Corporate Citizenship Initiative, a global effort to address the environmental, social, economic, and ethical challenges we face in our business. Among the issues we are addressing are carbon impact, paper specifications and procurement, ethical conduct within our business and among our vendors, and community and charitable support. For more information, please visit our website: www.wiley.com/go/citizenship.

ISBN: 978-1-119- 47566-8 (PBK)

ISBN: 978-1-119- 49522-2 (EVALC)

*Library of Congress Cataloging-in-Publication Data*

Names: Figliola, R. S., author. | Beasley, Donald E., author.
Title: Theory and design for mechanical measurements / Richard S. Figliola,
    Clemson University, Donald E. Beasley, Clemson University.
Description: Seventh edition. | Hoboken, NJ : Wiley, 2019. | Includes
    bibliographical references and index. |
Identifiers: LCCN 2018056461 (print) | LCCN 2018056669 (ebook) | ISBN
    9781119495192 (Adobe PDF) | ISBN 9781119475651 (ePub) | ISBN 9781119475668
    (PBK) | ISBN 9781119495222 (EVALC)
Subjects: LCSH: Measurement. | Measuring instruments.
Classification: LCC T50 (ebook) | LCC T50 .F54 2019 (print) | DDC
    530.8—dc23
LC record available at https://lccn.loc.gov/2018056461

The inside back cover will contain printing identification and country of origin if omitted from this page. In addition, if the ISBN on the back cover differs from the ISBN on this page, the one on the back cover is correct.

V10008047_020819

# PREFACE

The revised edition of *Theory and Design for Mechanical Measurements* continues to provide a well-founded, fundamental background in the theory and practice of engineering measurements. This edition takes advantage of web-based and interactive technology to improve the online and e-format access to materials for improved pedagogy while retaining archival material in the published edition. This format minimizes print materials in our effort to reduce text costs.

As with each other edition, we have integrated the necessary elements to conduct engineering measurements through the design of measurement systems and measurement test plans, with an emphasis on the role of statistics and uncertainty analyses in that process. The text is appropriate for undergraduate-level and graduate-level study in engineering. The text is suitably advanced and oriented to serve as a reference source for professional practitioners. Rather than a topical source of information, we have provided sufficient background material to enable intelligent selection of instruments and design of measurement systems. The pedagogical approach invites independent study or use in related fields that require an understanding of instrumentation and measurements.

The organization used in each edition of this text develops from our view that certain aspects of measurements can be generalized, such as test plan design, signal analysis and reconstruction, and measurement system response. Topics such as statistics and uncertainty analysis require a basic development of principles but are then best illustrated by integrating these topics throughout the text material. Other aspects are better treated in the context of the measurement of a specific physical quantity, such as strain or temperature.

## Pedagogical Tools to Aid Learning

We have retained certain successful elements from prior editions in this textbook:

- Each chapter begins by defining a set of **learning outcomes**.

- The text develops an **intuitive understanding** of measurement concepts with its focus on test system modeling, test plan design, and uncertainty analysis.

- Each chapter includes carefully constructed **example problems** that illustrate new material and problems that build on prior material. Where appropriate case studies and vignettes illustrate current applications of measurement theory or design.

- Example problems make use of a **KNOWN, FIND, SOLVE** approach as an organizational aid toward solution. This methodology for problem solutions helps new users to link words and concepts with symbols and equations. Many problems contain **COMMENTS** that expand on the solution, provide a proper context for application of the principle, or offer design application insight.

- **End-of-chapter practice problems** are included for each chapter to exercise new concepts.
  - Practice problems range from those focused on concept development to building of advanced skills to open-ended design applications. Some problems require independent research for specific applications or measurement techniques.
  - Within each chapter, we have added new practice problems and removed some old problems from prior editions.

- We provide online access to a solutions manual for instructors who have adopted the book and registered with Wiley. Materials are posted on the textbook website at www.wiley.com/go/figliola/Theory&designformechanicalmeasurements7e.

- Use of the software in problem solving allows in-depth exploration of key concepts that would be prohibitively time consuming otherwise. The text includes online access to **interactive software** of focused examples based on software using National Instruments Labview® or Matlab®. The Labview programs are available as executables so that they can be run directly without a Labview license. The software is available on both the Wiley student and instructor websites.

## New to This Revised Edition

With this revised edition, we have new or expanded material on a number of topics:

- We have updated the end-of-chapter practice problems by removing some older problems and adding new problems. We have replaced more than 25% of the problems in the text.

- We have added a dedicated set of Student Problems with "click-to-access" to view solutions. These are roughly 25% of the end-of-chapter problems. This capability is accessible to readers who have purchased the capability through Wiley.

- Given the new e-format, we have added some data files that can be accessed electronically and used in practice problems.

- We have set aside a dedicated set of Instructor Problems with solutions to assist in lectures or assessment. These are visible only to instructors who have registered with Wiley.

## Suggested Course Coverage (from Sixth Edition)

To aid in course preparation, Chapters 1–5 provide an introduction to measurement theory with statistics and uncertainty analysis, Chapters 6 and 7 provide a broad treatment of analog and digital sampling methods, and Chapters 8–12 focus on instrumentation.

Many users report to us that they use different course structures—so many that it makes a preferred order of topical presentation difficult to anticipate. To accommodate this, we have written the text in a manner that allows any instructor to customize the order of material presentation. Although the material of Chapters 4 and 5 is integrated throughout the text and should be taught in sequence, the other chapters can stand on their own. The text is flexible and can be used in a variety of course structures at both the undergraduate and graduate levels.

For a complete measurements course, we recommend the study of Chapters 1–7 with use of the remaining chapters as appropriate. For a lab course sequence, we recommend using chapters as they best illustrate the course exercises while building complete coverage over the several lab courses normally within a curriculum. The manner of the text allows it to be a resource for a lab-only course with minimal lecture. Over the years we have used it in several forums, as well as in professional development courses, and we simply rearrange material and emphasis to suit our audience and objective.

We express our sincerest appreciation to the students, teachers, and engineers who have used our earlier editions. We are indebted to the many who have written us with their constructive comments and encouragement. Special thanks to Kathy Hays-Stang, University of Texas-Arlington, and Todd Schweisinger, Clemson University, for their detailed feedback and suggestions.

RICHARD S. FIGLIOLA
DONALD E. BEASLEY

*Clemson, South Carolina*

# CONTENTS

# Basic Concepts of Measurement Methods

## 1.1 Introduction

We take measurements every day. We routinely read the temperature of an outdoor thermometer to choose appropriate clothing for the day. We add exactly 5.3 gallons (about 20 liters) of fuel to our car fuel tank. We use a tire gauge to set the correct car tire pressures. We monitor our body weight. And we put little thought into the selection of instruments for these measurements. After all, the instruments and techniques are routine or provided, the direct use of the information is clear to us, and the measured values are assumed to be good enough. But as the importance and complexity increases, the selection of equipment and techniques and the quality of the results can demand considerable attention. Just contemplate the various types of measurements and tests needed to certify that an engine meets its stated design specifications.

The objective in any measurement is to assign numbers to variables so as to answer a question. The information acquired is based on the output of some measurement device or system. We might use that information to ensure that a manufacturing process is executing correctly, to diagnose a defective part, to provide values needed for a calculation or a decision, or to adjust a process variable. There are important issues to be addressed to ensure that the output of the measurement device is a reliable indication of the true value of the measured variable. In addition, we must address some important questions:

1. How can a measurement or test plan be devised so that the measurement provides the unambiguous information we seek?

2. How can a measurement system be used so that the engineer can easily interpret the measured data and be confident in their meaning?

There are procedures that address these measurement questions.

At the onset, we want to stress that the subject of this text is real-life–oriented. Specifying a measurement system and measurement procedures represents an open-ended design problem. That means there may be several approaches to meeting a measurements challenge, and some will be better than others. This text emphasizes accepted procedures for analyzing a measurement challenge to aid selection of equipment, methodology, and data analysis.

Upon completion of this chapter, the reader will be able to:

- Identify the major components of a general measurement system and state the function of each

- Develop an experimental test plan

- Distinguish between random and systematic errors

- Become familiar with the hierarchy of units standards, and with the existence and use of test standards and codes

- Understand the international system of units and other unit systems often found in practice

- Understand and work with significant digits

## 1.2   General Measurement System

*Measurement*[1] is the act of assigning a specific value to a physical variable. That physical variable is the *measured value*. A measurement system is a tool used to quantify the measured variable. Thus, a measurement system is used to extend the abilities of the human senses, which, although they can detect and recognize different degrees of roughness, length, sound, color, and smell, are limited and relative: They are not very adept at assigning specific values to sensed variables.

A system is composed of components that work together to accomplish a specific objective. We begin by describing the components that make up a measurement system, using specific examples. Then we will generalize to a model of the generic measurement system.

### Sensor and Transducer

A *sensor* is a physical element that employs some natural phenomenon to sense the variable being measured. To illustrate this concept, suppose we want to measure the profile of a surface at the nanometer scale. We discover that a very small cantilever beam placed near the surface is deflected by atomic forces. Let's assume for now that they are repulsive forces. As this cantilever is translated over the surface, the cantilever will deflect, responding to the varying height of the surface. This concept is illustrated in Figure 1.1; the device is called an atomic-force microscope. The cantilever beam is a sensor. In this case, the cantilever deflects under the action of a force in responding to changes in the height of the surface.

A *transducer* converts the sensed information into a detectable signal. This signal might be mechanical, electrical, optical, or it may take any other form that can be meaningfully quantified. Continuing with our example, we will need a means to change the sensor motion into something that we can quantify. Suppose that the upper surface of the cantilever is reflective, and we shine a laser onto the upper surface, as shown in Figure 1.2. The movement of the cantilever will deflect the laser. Employing a number of light sensors, also shown in Figure 1.2, the changing deflection of the light can be measured as a time-varying current signal with the magnitude corresponding to the height of the surface. Together the laser and the light sensors (photodiodes) form the transducer component of the measurement system.

A familiar example of a complete measurement system is the bulb thermometer. The liquid contained within the bulb of the common bulb thermometer shown in Figure 1.3 exchanges energy with its surroundings until the two are in thermal equilibrium. At that point they are at the same temperature. This energy exchange is the input signal to this measurement system. The phenomenon of thermal expansion of the liquid results in its movement up and down the stem, forming an output signal from which we determine temperature. The liquid in the bulb acts as

**FIGURE 1.1**   Sensor stage of an atomic-force microscope.

---

[1]A glossary of the italicized terms is located in the back of the text for your reference.

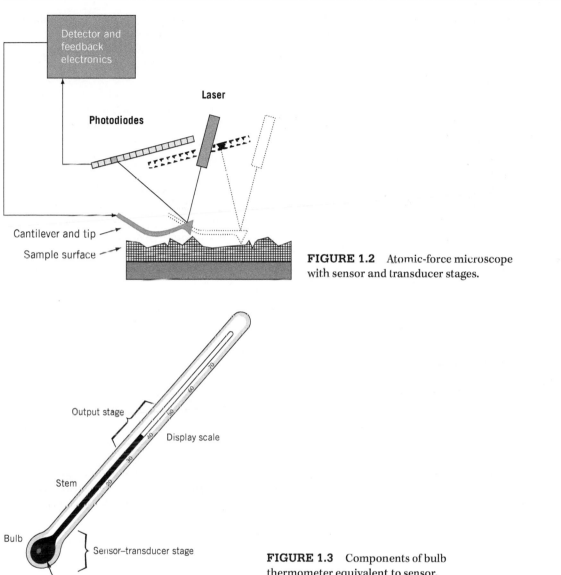

**FIGURE 1.2** Atomic-force microscope with sensor and transducer stages.

**FIGURE 1.3** Components of bulb thermometer equivalent to sensor, transducer, and output stages.

the sensor. By forcing the expanding liquid into a narrow capillary, this measurement system transforms thermal information into a mechanical displacement. Hence the bulb's internal capillary acts as a transducer.

Sensor selection, placement, and installation are particularly important to ensure that the sensor output signal accurately reflects the measurement objective. After all, the interpretation of the information indicated by the system relies on what is actually sensed by the sensor.

## Signal-Conditioning Stage

The *signal-conditioning stage* takes the transducer signal and modifies it to a desired magnitude or form. It might be used to increase the magnitude of the signal by amplification, remove portions of the signal through some filtering technique, or provide mechanical or optical linkage between the transducer and the output stage. The diameter of the thermometer capillary relative to the bulb volume (Figure 1.3) determines how far up the stem the liquid moves with increasing temperature.

## Output Stage

The goal of a measurement system is to convert the sensed information into a form that can be easily quantified. The *output stage* indicates or records the value measured. This might be by a simple readout display, a marked scale, or even a recording device such as computer memory from which it can later be accessed and analyzed. In Figure 1.3, the readout scale of the bulb thermometer serves as its output stage.

As an endnote, the term "transducer" is often used in reference to a packaged measuring device that contains the sensor and transducer elements described above, and even some signal-conditioning elements, in one unit. This terminology differs from the term "transducer" when describing the function served by an individual stage of a measurement system. Both uses are correct, and we use both: one to refer to how a sensed signal is changed into another form and the other to refer to a packaged device. The context in which the term is used should prevent any ambiguity.

## General Template for a Measurement System

A general template for a measurement system is illustrated in Figure 1.4. Essentially such a system consists of part or all of the previously described stages: (1) sensor–transducer stages, (2) signal-conditioning stage, and (3) output stage. These stages form the bridge between the input to the measurement system and the output, a quantity that is used to infer the value of the physical variable measured. We discuss later how the relationship between the input information acquired by the sensor and the indicated output signal is established by a calibration.

Some systems may use an additional stage, the feedback-control stage shown in Figure 1.4. Typical to measurement systems used for process control, the *feedback-control stage* contains a controller that compares the measured signal with some reference value and makes a decision regarding actions required in the control of the process. In simple controllers, this decision is based

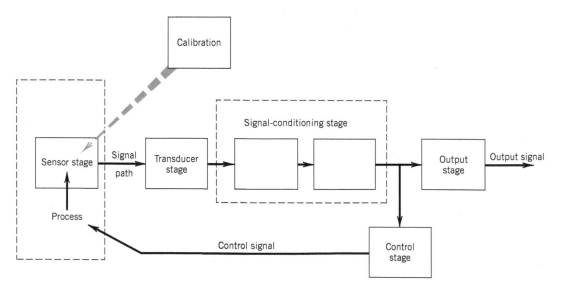

**FIGURE 1.4**    Components of a general measurement system.

on whether the magnitude of the signal exceeds some set point, whether high or low—a value set by the system operator. For example, a household furnace thermostat activates the furnace as the local temperature at the thermostat, as determined by a sensor within the device, rises above or falls below the thermostat set point. The cruise speed controller in an automobile works in much the same way. A programmable logic controller (PLC) is a robust industrial-grade computer and data-acquiring device used to measure many variables simultaneously and to take appropriate corrective action per programmed instructions. We discuss some features of control systems in detail in Chapter 12.

## 1.3 Experimental Test Plan

An experimental test is designed to make measurements so as to answer a question. The test should be designed and executed to answer that question and that question alone. Consider an example.

Suppose you want to design a test to answer the question "What fuel use can my family expect from my new car?" In a test plan, you identify the variables that you will measure, but you also need to look closely at other variables that will influence the result. Two important measured variables here will be distance and fuel volume consumption. The accuracies of the odometer and station fuel pump affect these two measurements. Achieving a consistent final volume level when you fill your tank affects the estimate of the fuel volume used. In fact, this effect, which can be classified as a zero error, could be significant. From this example, we can surmise that a consistent measurement technique must be part of the test plan.

What other variables might influence your results? The driving route, whether highway, city, rural, or a mix, affects the results and is an independent variable. Different drivers drive differently, so the driver becomes an independent variable. So how we interpret the measured data is affected by variables in addition to the primary ones measured. Imagine how your test would change if the objective were to establish the fleet average fuel use expected from a car model used by a rental company.

In any test plan, you need to consider just how accurate an answer you need. Is an accuracy within 2 liters per 100 kilometers (or, in the United States, 1 mile per gallon) good enough? If you cannot achieve such accuracy, then the test may require a different strategy. Last, as a concomitant check, is there a way to check whether your test results are reasonable, a sanity check to avoid subtle mistakes? Interestingly, this one example contains all the same elements of any sophisticated test. If you can conceptualize the factors influencing this test and how you will plan around them, then you are on track to handle almost any test.

Experimental design involves itself with developing a measurement test plan. A test plan draws from the following three steps:[2]

1. ***Parameter design plan.*** Determine the test objective and identify the process variables and parameters and a means for their control. Ask: "What question am I trying to answer? What needs to be measured?" "What variables will affect my results?"

2. ***System and tolerance design plan.*** Select a measurement technique, equipment, and test procedure based on some preconceived tolerance limits for error.[3] Ask: "In what ways can I do the measurement? How good do the results need to be to answer my question?"

3. ***Data reduction design plan.*** Plan how to analyze, present, and use the data. Ask: "How will I interpret the resulting data? How will I apply the data to answer my question? How good is my answer?

---

[2]These three strategies are similar to the bases for certain design methods used in engineering system design [1].

[3]The tolerance design plan strategy used in this text draws on uncertainty analysis, an extension of sensitivity analysis. Sensitivity methods are common in design optimization.

Going through all three steps in the test plan before any measurements are taken is a useful habit of the successful engineer. Often step 3 will force you to reconsider steps 1 and 2. We will develop methods to help address each step as we proceed.

## Variables

A *variable* is a physical quantity whose value influences the test result. Variables can be continuous or discrete by nature. A *continuous variable* can assume any value within its range. A *discrete variable* can assume only certain values.

Functionally, variables can be classed as being dependent, independent, or extraneous. A variable whose value is affected by changes in the value of one or more other variables is known as a *dependent variable*. Variables that do not influence each other are *independent variables*. Dependent variables are functions of independent variables. The test result is a dependent variable whose value depends on the values of the independent variables. For example, the value of strain measured in a loaded cantilever beam (dependent variable) results from the value of the loading applied (independent variable).

The values of independent variables are either purposely changed or held fixed throughout a test to study the effect on the dependent variable. A *controlled variable* is a variable whose value is held fixed. A test *parameter* is a controlled variable or grouping of variables whose fixed value sets the behavior of the process being measured. The dimensions and material properties of the beam in the loaded cantilever beam are parameters. The value of the parameter is changed when you want to change the process behavior.

In engineering, groupings of variables, such as moment of inertia or Reynolds number, are also referred to as parameters when these relate to certain system behaviors. For example, the damping ratio of an automotive suspension system affects how the displacement velocity will change with differing input conditions; this parameter affects the variables that measure vehicle behavior, such as handling and ride quality. In our treatment of statistics, we use the term parameter to refer to quantities that describe the behavior of the population of a measured variable, such as the true mean value or variance.

Variables that are not purposely manipulated or controlled during measurement but that do affect the test result are called *extraneous variables*. Dependent variables are affected by extraneous variables. If not treated properly within a test plan, extraneous variables can impose a false trend or impose variation onto the behavior of the measured variable. This influence can confuse the clear relation between cause and effect.

There are other uses of the term "control." Experimental control describes how well we can maintain the prescribed value of an independent variable within some level of confidence during a measurement. For example, if we set the loading applied in a bending beam test to 103 kN, does it stay exactly fixed during the entire test, or does it vary by some amount on repeated tries? Different fields use variations on this term. In statistics, a control group is one held apart from the treatment under study. But the nuance of holding something fixed is retained.

### EXAMPLE 1.1

Consider an introductory thermodynamics class experiment used to demonstrate phase change phenomena. A beaker of water is heated and its temperature measured over time to establish the temperature at which boiling occurs. The results, shown in Figure 1.5, are for three separate tests conducted on different days by different student groups using the same equipment and method. Why might the data from three seemingly identical tests show different results?

**FIGURE 1.5**   Results of a boiling point test for water.

**KNOWN**  Temperature-time measurements for three tests conducted on three different days

**FIND**  Boiling point temperature

**SOLUTION**
Each group of students anticipated a value of exactly 100.0°C. Individually, each test result is close, but when compared, there is a clear offset between any two of the three test results. Suppose we determine that the measurement system accuracy and natural chance account for only 0.1°C of the disagreement between tests—then something else is happening. A plausible contributing factor is an effect due to an extraneous variable.

Fortunately, an assistant recorded the local barometric pressure during each test. The boiling point (saturation) temperature is a function of pressure. Each test shows a different outcome, in part, because the local barometric pressure was not controlled (i.e., it was not held fixed between the tests).

**COMMENT**  One way to control the effects of pressure here might be to conduct the tests inside a barometric pressure chamber. Direct control of all variables is not often possible in a test. So another way to handle the extraneous variable applies a special strategy: Consider each test as a single block of data taken at the existing barometric pressure. Then consolidate the three blocks of test data. In that way, the measured barometric pressure becomes treated as if it were an independent variable, with its influence integrated into the entire data set. That is, we can actually take advantage of the differences in pressure so as to study its effect on saturation temperature. This is a valuable control treatment called randomization, a procedure we discuss later in this chapter. *Identify and control important variables, or be prepared to solve a puzzle!*

---

**EXAMPLE 1.2   Case Study**

The golf cart industry within the United States records nearly $1 billion in sales annually. The manufacturer of a particular model of golf cart sought a solution to reduce undesirable vibrations in the cart steering column (Figure 1.6). Engineers identified the source of the vibrations as the

**FIGURE 1.6**

Measured time-based *z*-acceleration signal at the steering wheel during an engine run-up test: baseline and the attenuated prototype (proposed solution) signals.

gas-powered engine and consistent with operating the engine at certain speeds (revolutions/minute or rpm). The company management ruled out expensive suspension improvements, so the engineers looked at cheaper options to attenuate the vibrations at the steering wheel.

Using test standard ANSI S2.70-2006 [19] as a guide for placing sensors and interpreting results, accelerometers (discussed in Chapter 12) were placed on the steering column, engine, suspension, and frame. The measurement chain followed Figure 1.4. The signal from each accelerometer sensor–transducer was passed through a signal-conditioning charge amplifier (Chapter 6) and on to a laptop-based data acquisition system (DAS; Chapter 7). The amplifier converted the transducer signal to a 0–5 V signal that matched the input range of the DAS. The DAS served as a voltmeter, output display, and time-based recorder.

Measurements showed that the offensive vibration at the steering wheel was at the natural frequency of the steering column ($\sim$20 Hz). This vibration was excited at forcing function frequencies between 17 and 37 Hz, corresponding to normal operating speeds of the engine (1,000–2,200 rpm). From a lumped parameter model (Chapter 3) of the steering assembly, they determined that a slightly heavier steering column would reduce its natural frequency to below the engine idle speed and thereby attenuate the amplitude of the vibrations (Chapter 3) excited at higher engine speeds. The analysis was verified and the fix was refined based on the tests. The measured vibration signals during engine runup, both before (baseline) and with the proposed fix (prototype), are compared in Figure 1.6. Peak vibration amplitudes were shown to reduce to comfortable levels with the proposed and validated solution. The inexpensive fix was implemented by the manufacturer.

**COMMENT** This example illustrates how measurements applied to a mechanical test provided information to diagnose a problem and then formulate and validate a proposed solution.

## Noise and Interference

Just how extraneous variables affect measured data can be described in terms of noise and interference. *Noise* is a random variation in the value of the measured signal. Noise increases data scatter. Building vibrations, variations in ambient conditions, and random thermal noise of electrons passing through a conductor are examples of common extraneous sources for the random variations found in a measured signal.

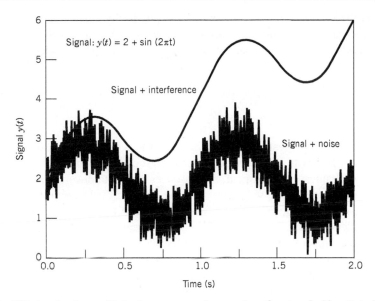

**FIGURE 1.7**   Effects of noise and interference superimposed on the signal $y(t) = 2 + \sin 2\pi t$.

*Interference* imposes undesirable deterministic trends on the measured signal. A common interference in electrical instruments comes from an AC power source and is seen as a sinusoidal wave superimposed onto the measured signal. Hum and acoustic feedback in public address systems are ready examples of interference effects superimposed onto a desirable signal. Sometimes the interference is obvious. But if the period of the interference is longer than the period over which the measurement is made, the false trend imposed may go unnoticed. A goal should be either to control the source of interference or to break up its trend.

Consider the effects of noise and interference on the signal, $y(t) = 2 + \sin 2\pi t$. As shown in Figure 1.7, noise adds to the scatter of the signal. Through statistical techniques and other means, we can sift through the fuzziness to extract desirable signal information. Interference imposes a false trend onto the signal. The test plan should be devised to break up such trends so that they become random variations in the data set. Although this increases the scatter in the measured values of a data set, noise may mask, but does not change, the deterministic aspects of a signal. It is far more important to eliminate false trends in the data set.

In Example 1.1, by combining the three tests at three random but measured values of barometric pressure into a single dataset, we incorporated the uncontrolled pressure effects into the results for boiling temperature while breaking up any offset in results found in any one test. This approach imparts some control over an otherwise uncontrolled variable.

## Randomization

An engineering test purposely changes the values of one or more independent variables to determine the effect on a dependent variable. *Randomization* refers to test strategies that apply the changes in the independent variables in some random order. The intent is to neutralize effects that may not be accounted for in the test plan.

In general, consider the situation in which the dependent variable, $y$, is a function of several independent variables, $x_a, x_b, \ldots$. To find the dependence of $y$ on the independent variables, they are varied in a prescribed manner. However, the measurement of $y$ can also be influenced by extraneous variables, $z_j$, where $j = 1, 2, \ldots$, such that $y = f(x_a, x_b, \ldots; z_j)$. Although the influence of the

$z_j$ variables on these tests cannot be eliminated, the possibility of their introducing a false trend on $y$ can be minimized by using randomization strategies. One approach is to add randomness to the order in which an independent variable is manipulated. This works well for continuous extraneous variables.

## EXAMPLE 1.3

In the pressure calibration system shown in Figure 1.8, a pressure transducer is exposed to a known pressure, $p$. Increasing pressure deforms the elastic diaphragm sensor of the device, which is sensed as a mechanical displacement. A transducer element converts the displacement into a voltage that is measured by the voltmeter. The test approach is to control the applied pressure through the measured displacement of a piston that is used to compress a gas contained within the piston–cylinder chamber. The gas closely obeys the ideal gas law. Hence, piston displacement, $x$, which sets the chamber volume, $\forall = (x \times \text{area})$, is easily related to chamber pressure. Accordingly, the known chamber pressure can be associated with the transducer output signal. Identify the independent and dependent variables in the calibration and possible extraneous variables.

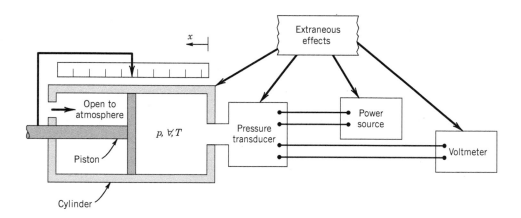

**FIGURE 1.8**
Pressure calibration system.

**KNOWN** Pressure calibration system of Figure 1.8.

**FIND** Independent, dependent, and extraneous variables.

**SOLUTION**
The sensor is exposed to the chamber gas pressure. The pressure transducer output signal will vary with this chamber pressure. So the dependent variable is the chamber gas pressure, which is also the pressure applied to the transducer. An independent variable in the calibration is the piston displacement, which sets the volume. A parameter for this problem is formed from the ideal gas law: $p\forall/T = \text{constant}$, where $T$ is the gas temperature and $\forall = (x \times \text{area})$. This parameter shows that the gas pressure also depends on gas temperature, so gas temperature is an independent variable. Volume is to be varied through variation of piston displacement, but $T$ can be controlled provided that a mechanism is incorporated into the scheme to maintain a constant chamber gas temperature. However, $T$ and $\forall$ are not in themselves independent. The chamber cross-sectional area is a function of temperature and $p = f(\forall, T)$, where $\forall = f_1(x, T)$. If chamber temperature is held constant, the applied variations in $\forall$ will be the only variable with an effect on pressure, as desired $p = f(\forall, T)$.

If gas temperature is not controlled, or if environmental effects allow uncontrolled variations in chamber and transducer temperature, then temperature behaves as an extraneous variable, $z_1$. Hence

$$p = f(\forall, z_1), \text{ where } \forall = f_1(x, T)$$

**COMMENT** To a lesser extent, line voltage variations and electrical interference, $z_2$, affect the excitation voltage from the power supply to the transducer or affect the performance of the voltmeter. This list is not exhaustive.

---

## EXAMPLE 1.4

Develop a test matrix that could minimize interference owing to any uncontrolled temperature effects discussed in Example 1.3.

**KNOWN** Control variable $\forall$ is changed through displacement $x$. Dependent variable $p$ is measured.

**FIND** Randomize possible effects on the measured pressure resulting from uncontrolled thermal expansion of the cylinder volume.

**SOLUTION**
Part of our test strategy is to vary volume and measure pressure. If the uncontrolled temperature variations are related to the value of the chamber volume, for example, $\forall = f(T)$, this would be an interference effect. These effects can be randomized using a random test that shuffles the order by which $\forall$ is applied. Say that we pick six values of volume, $\forall_1, \forall_2, \forall_3, \forall_4, \forall_5,$ and $\forall_6$, such that the subscripts correspond to an increasing sequential order of the respective values of volume. The volume is set by the displacement and area. Any order will do fine. One possibility is

$$\forall_2 \forall_5 \forall_1 \forall_4 \forall_6 \forall_3$$

From the test parameter, we expect pressure to vary linearly with $\forall$ for a fixed gas temperature. If we perform our measurements in a random order, interference trends due to $z_1$ will be broken up. This approach should serve also to break up any interference trends resulting from environmental effects that might influence the transducer, excitation voltage, and voltmeter.

---

The use of different test operators, different instruments, and different test-operating conditions are examples of discrete extraneous variables that can affect the outcome of a measurement. Randomizing a test matrix to minimize discrete influences can be done efficiently through the use of experimental design methods using random blocks. A block consists of a data set comprised of the measured (dependent) variable in which the extraneous variable is fixed while the independent variable is varied. The extraneous variable will vary between blocks. This enables some amount of local control over the discrete extraneous variable through randomization.

In the fuel-usage discussion earlier, we might consider several blocks, each comprising a different driver (an extraneous variable) driving similar routes, and averaging the fuel consumption results for our car. In Example 1.1, we imposed the strategy of using several tests (blocks) under different values of barometric pressure to break up the interference effect found in a single test. Many strategies for randomized blocks exist, as do advanced statistical methods for data analysis [2–5].

## EXAMPLE 1.5

The manufacture of a particular composite material requires mixing a percentage by weight of binder with resin to produce a gel. The gel is used to impregnate a fiber to produce the composite material in a manual process called the lay-up. The strength, $\sigma$, of the finished material depends on the percent binder in the gel. However, the strength may also be lay-up operator–dependent. Formulate a test matrix by which the strength to percent binder–gel ratio under production conditions can be established.

**KNOWN** $\sigma = f$ (binder; operator)

**ASSUMPTION** Strength is affected only by binder and operator.

**FIND** Test matrix to randomize effects of operator.

**SOLUTION**

The dependent variable, $\sigma$, is to be tested against the independent variable, percent binder–gel ratio. The operator is an extraneous variable in actual production. As a simple test, we could test the relationship between three binder–gel ratios, $A, B,$ and $C$ and measure strength. We could also choose three typical operators ($z_1$, $z_2$, and $z_3$) to produce $N$ separate composite test samples for each of the three binder–gel ratios. This gives the three-block test pattern:

|  Block |         |     |     |     |
|--------|---------|-----|-----|-----|
| 1      | $z_1$:  | $A$ | $B$ | $C$ |
| 2      | $z_2$:  | $A$ | $B$ | $C$ |
| 3      | $z_3$:  | $A$ | $B$ | $C$ |

In the analysis of the test, all these data can be combined. The results of each block will include each operator's influence as a variation. We can assume that the order used within each block is unimportant. But if only the data from one operator are considered, the results may show a trend consistent with the lay-up technique of that operator. The test matrix above will randomize the influence of any one operator on the strength test results by introducing the influence of several operators.

## EXAMPLE 1.6

Suppose that after lay-up, the composite material of Example 1.5 is allowed to cure at a controlled but elevated temperature. We wish to develop a relationship between the binder–gel ratio and the cure temperature and strength. Develop a suitable test matrix.

**KNOWN** $\sigma = f$ (binder, temperature, operator)

**ASSUMPTION** Strength is affected only by binder, temperature, and operator.

**FIND** Test matrix to randomize effect of operator.

**SOLUTION**

We develop a simple matrix to test for the dependence of composite strength on the independent variables of binder–gel ratio and cure temperature. We could proceed as in Example 1.5 and set up three randomized blocks for ratio and three for temperature for a total of 18 separate tests.

Suppose instead we choose three temperatures $T_1$, $T_2$, and $T_3$, along with three binder–gel ratios $A$, $B$, and $C$ and three operators $z_1$, $z_2$, and $z_3$ and set up a $3 \times 3$ test matrix representing a single randomized block. If we organize the block such that no operator runs the same test combination more than once, we randomize the influence of any one operator on a particular binder–gel ratio, temperature test.

|   | $z_1$ | $z_2$ | $z_3$ |
|---|-------|-------|-------|
| $A$ | $T_1$ | $T_2$ | $T_3$ |
| $B$ | $T_2$ | $T_3$ | $T_1$ |
| $C$ | $T_3$ | $T_1$ | $T_2$ |

**COMMENT** The suggested test matrix not only randomizes the extraneous variable, but also has reduced the number of tests by half over the direct use of three blocks for ratio and for temperature. However, either approach is fine. The above matrix is referred to as a Latin square [2–5].

## Replication and Repetition

In general, the estimated value of a measured variable improves with the number of measurements. Repeated measurements of the same variable made during any single test run or on a single batch are called *repetitions*. Repetition helps quantify the variation in a measured variable under a set of test conditions. For example, a bearing manufacturer would obtain a better estimate of the mean diameter and the variation in the diameter within a new batch containing thousands of bearings by measuring many bearings rather than just a few.

An independent duplication of a set of measurements or test is referred to as a *replication*. Replication allows for quantifying the variation in a measured variable as it occurs between duplicate tests conducted under similar conditions. If the bearing manufacturer was interested in how closely the bearing mean diameter was controlled in day-in and day-out operations with a particular machine or test operator, duplicate tests run on different days would be needed.

If the bearing manufacturer were interested in how closely bearing mean diameter was controlled when using different machines or different machine operators, duplicate tests using these different configurations holds the answer. Here, replication provides a means to randomize the interference effects of the different bearing machines or operators.

**EXAMPLE 1.7**

Consider a room furnace thermostat. Set to some temperature, we can make repeated measurements (repetition) of room temperature and come to a conclusion about the average value and the fluctuations in room temperature achieved at that particular thermostat setting. Repetition permits an assessment of the set condition and how well we can maintain it.

Now suppose we change the set temperature to another arbitrary value but some time later return it to the original setting and duplicate the measurements. The two sets of test data are replications of each other. We might find that the average temperature in the second test differs from the first. The different averages suggest something about our ability to set and control the temperature in the room. Replication provides information to assess how well we can duplicate a set of conditions.

## Concomitant Methods

Is my test working? What value of result should I expect from my measurements? To help answer these, a good strategy is to incorporate *concomitant methods* in a measurement plan. The goal is to obtain two or more estimates for the result, each estimate based on a different method, which can be compared as a check for agreement. This may affect the experimental design in that additional variables may need to be measured. Or the different method could be an analysis that estimates the expected value for the measurement. The two need not be of same accuracy as one method serves to check the reasonableness of the other. For example, suppose we want to establish the volume of a cylindrical rod of known material. We could measure the diameter and length of the rod to compute this. Alternatively, we could measure the weight of the rod and compute volume based on the specific weight of the material. The second method complements the first and provides an important check on the adequacy of the first estimate.

# 1.4 Calibration

A *calibration* applies a known input value to a measurement system for the purpose of observing the system output value. It establishes the relationship between the input and output values. The known value used for the calibration is called the *standard*.

## Static Calibration

The most common type of calibration is known as a static calibration. In this procedure, a value is applied (input) to the system under calibration and the system output is recorded. The value of the applied standard is known to some acceptable level. The term "static" means that the input value does not vary with time or space or that average values are used.

By applying a range of known input values and by observing the system output values, a direct calibration curve can be developed for the measurement system. On such a curve the applied input, $x$, is plotted on the abscissa against the measured output, $y$, on the ordinate, as indicated in Figure 1.9. In a calibration the input value is an independent variable, while the measured output value is the dependent variable of the calibration.

The static calibration curve describes the static input–output relationship for a measurement system and forms the logic by which the indicated output can be interpreted during an actual measurement. For example, the calibration curve is the basis for fixing the output display scale on a measurement system, such as that of Figure 1.3. Alternatively, a calibration curve can be used as part of developing a functional relationship, an equation known as a correlation, between input and output. A correlation will have the form $y = f(x)$ and is determined by applying physical reasoning and curve fitting techniques to the calibration curve. The correlation can then be used in later measurements to ascertain an unknown input value based on the measured output value, the value indicated by the measurement system.

## Dynamic Calibration

When the variables of interest are time-dependent (or space-dependent) and such varying information is sought, we need dynamic information. In a broad sense, dynamic variables are time (or space) dependent in both their magnitude and amplitude and frequency content. A *dynamic calibration* determines the relationship between an input of known dynamic content and the measurement system output. For example, does the output signal track the input value exactly in time

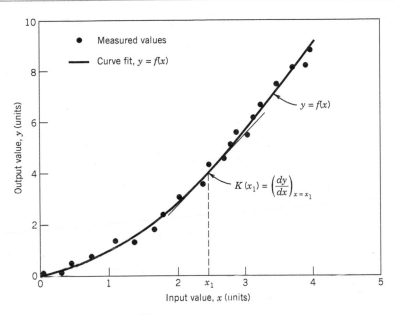

**FIGURE 1.9**   Representative static calibration curve.

and space, is there lag, or is the output value input frequency dependent? Usually, such calibrations involve applying either a sinusoidal signal of known amplitude and frequency or a step change as the input signal. The dynamic response of measurement systems is explored fully in Chapter 3.

## Static Sensitivity

The slope of a static calibration curve provides the static sensitivity[4] of the measurement system. As depicted graphically in the calibration curve of Figure 1.9, the static sensitivity, $K$, at any particular static input value, say $x_1$, is evaluated by

$$K = K(x_1) = \left(\frac{dy}{dx}\right)_{x=x_1} \tag{1.1}$$

where $K$ can be a function of the applied input value $x$. The static sensitivity is a measure relating the change in the indicated output associated with a given change in a static input.

## Range and Span

A calibration applies known inputs ranging from the minimum to the maximum values for which the measurement system is to be used. These limits define the operating *range* of the system. The input full scale operating range (FSO) is defined as extending from $x_{min}$ to $x_{max}$. The *span* is the value difference of the range limits. For example, a transducer with a range of 0 to 100 N has a span of 100 N. The input span may be expressed as

$$r_i = x_{max} - x_{min} \tag{1.2}$$

---

[4]Some texts refer to this as the *static gain*.

The output full scale operating range extends from $y_{min}$ to $y_{max}$. The output span for the FSO is expressed as

$$r_o = y_{max} - y_{min} \tag{1.3}$$

## Resolution

The *resolution* represents the smallest increment in the measured value that can be discerned. In terms of a measurement system, it is quantified by the smallest scale increment or the *least count* (least significant digit) of the output readout indicator.

## Accuracy and Error

The exact value of a variable is called the *true value*. The value of a variable as indicated by a measurement system is called the *measured value*. The *accuracy* of a measurement refers to the closeness of agreement between the measured value and the true value. Unfortunately, the true value is rarely known exactly in engineering practice, and various influences, called errors, have an effect on both of these values. This means that the concept of the accuracy of a measurement is a qualitative one.

Instead, an appropriate approach to stating this closeness of agreement is to identify the measurement errors and to quantify them by the value of their associated uncertainties, where an uncertainty is the estimated range of the value of an error. We define an *error, e*, as the difference between the measured value and the true value—that is,

$$e = \text{Measured value} - \text{True value} \tag{1.4}$$

The exact value for error is usually not known. So Equation 1.4 serves only as a reference definition. Errors exist, and they have a magnitude, as given by Equation 1.4. Sometimes we can correct a reading to account for an estimated amount of error, even if that correction is approximate. We'll discuss this concept next and then develop it extensively in Chapter 5.

Often an estimate for the value of error is based on a reference value used during the instrument's calibration, which becomes a surrogate for the true value. A relative error based on this reference value is estimated by

$$A = \frac{|e|}{\text{Reference value}} \times 100 \tag{1.5}$$

Some vendors may refer to this term as the "relative accuracy."

## Random and Systematic Errors and Uncertainty

Errors cause a measured value to differ from its true value. *Random error* causes a random variation in measured values found during repeated measurements of the variable. *Systematic error* causes an offset between the mean value of the data set and its true value. Both random and systematic errors affect a system's accuracy.

The concept of accuracy and the effects of systematic and random errors in instruments and measurement systems can be illustrated by dart throws. Consider the dart boards in Figure 1.10; the goal is to throw the darts into the bullseye. For this analogy, the bullseye can represent the true

**(a)** High repeatability gives low random error but no direct indication of accuracy.

**(b)** High accuracy means low random and systematic errors.

**(c)** Systematic and random errors lead to poor accuracy.

**FIGURE 1.10**   Throws of a dart: illustration of random and systematic errors and accuracy.

value, and each throw can represent a measured value. The error in each throw can be calculated as the distance between the dart and the bullseye. In Figure 1.10a, the thrower displays good repeatability (i.e., small random error) in that each throw repeatedly hits nearly the same spot on the board, but the thrower is not accurate in that the dart misses the bullseye each time. The average value of the error gives an estimate of the systematic error in the throws. The random error is some average of the variation between each throw, which is near zero in Figure 1.10a. We see that a small amount of random error is not a complete measure of the accuracy of this thrower. This thrower has an offset to the left of the target, a systematic error. If the effect of this systematic error could be reduced, then this thrower's accuracy would improve. In Figure 1.10b, the thrower displays a high accuracy, hitting within the bull's-eye on each throw. Both scatter and offset are near zero. High accuracy must be associated with small values of random and systematic errors as shown. In Figure 1.10c, the thrower does not show good accuracy, with errant throws scattered around the board. Each throw contains a different amount of error. While an estimate of the systematic error is the average of the errors in the throws, the estimate of the random error is related to the varying amount of error in the throws, a value that can be estimated using statistical methods. The estimates in the random and systematic errors of the thrower can be computed using the statistical methods that are discussed in Chapter 4 or the methods of comparison discussed in Chapter 5.

Suppose we used a measurement system to measure a variable whose value was kept constant and known almost exactly, as in a calibration. For example, 10 independent measurements are made with the results, as shown in Figure 1.11. The variations in the measurements, the observed scatter in the data, would be related to the random error associated with the measurement of the variable. This scatter is mainly owing to [1] the measurement system and the measurement method, [2] any natural variation in the variable measured, and [3] any uncontrolled variations in the variable. However, the offset between the apparent average of the readings and the true value would provide a measure of the systematic error to be expected from this measurement system.

## Uncertainty

The *uncertainty* is the numerical estimate of the possible range of the error in the stated value of a variable. In any measurement, the error is not known exactly since the true value is rarely known exactly. But based on available information, the operator might feel confident that the error is within certain bounds, a plus or minus range of the indicated reading. This is the assigned uncertainty. Uncertainty is brought about by each of the errors that are present in a measurement—its system calibration, the data set statistics, and the measurement technique. Individual errors are properties of the instruments, the test method, the analysis, and the measurement system.

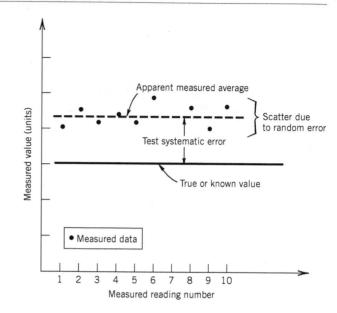

**FIGURE 1.11** Effects of random and systematic errors on calibration readings.

Uncertainty is a property of the test result. In Figure 1.11, we see that we might assign an estimate to the random error, that is, the random uncertainty, based on the data scatter. The systematic uncertainty might be an estimate of the offset based on a comparison against a value found by a concomitant method. An uncertainty based on the estimates of all known errors would then be assigned to the stated result. A method of estimating the overall uncertainty in the test result is treated in detail in Chapter 5.

The uncertainty values assigned to an instrument or measurement system specification are usually the result of several interacting random and systematic errors inherent to the measurement system, the calibration procedure, and the standard used to provide the known value. An example of some known calibration errors affecting a typical pressure transducer is given in Table 1.1. The value assigned to each stated error is its uncertainty.

**Table 1.1    Manufacturer's Specifications: Typical Pressure Transducer**

| Operation | |
|---|---|
| Input range | 0–1000 cm $H_2O$ |
| Excitation | ±15 V DC |
| Output range | 0–5 V |
| Temperature range | 0–50°C |
| **Performance** | |
| Linearity error | ±0.5% FSO |
| Hysteresis error | Less than ± 0.15% FSO |
| Sensitivity error | ±0.25% of reading |
| Thermal sensitivity error | ±0.02%/°C of reading |
| Thermal zero drift | ±0.02%/°C FSO |

FSO, full-scale operating range.

## Sequential Test

A *sequential test* varies the input value sequentially over the desired input range. This may be accomplished by increasing the input value (upscale direction) or by decreasing the input value (downscale direction) over the full input range. The output value is the measured value.

## Hysteresis

*Hysteresis error* refers to differences in the measured value between an upscale sequential test and a downscale sequential test. The sequential test is an effective diagnostic technique for identifying and quantifying hysteresis error in a measurement system. The effect of hysteresis in a sequential test calibration curve is illustrated in Figure 1.12a. The hysteresis error of the system is estimated by its uncertainty, $u_h = (y)_{upscale} - (y)_{downscale}$. Hysteresis is usually specified for a measurement system in terms of an uncertainty based on the maximum hysteresis error as a percentage of the full-scale output, $r_o$,

$$\%u_{h_{max}} = \frac{u_{h_{max}}}{r_o} \times 100 \tag{1.6}$$

such as the value indicated in Table 1.1. Hysteresis occurs when the output of a measurement system is dependent on the previous value indicated by the system. Such dependencies can be brought about through system limitations such as friction or viscous damping in moving parts or residual charge in electrical components. Some hysteresis is normal for any system and affects the repeatability of the system.

## Random Test

A *random test* applies the input value in a random order over the intended calibration range. The random application of input tends to reduce the effects of interference. It breaks up hysteresis effects and observation errors. It ensures that each application of input value is independent of the previous. Thus it reduces calibration systematic error, converting it to random error. Generally, such a random variation in input value will more closely simulate the actual measurement situation.

A random test provides an important diagnostic for the delineation of several measurement system performance characteristics based on a set of random calibration test data. In particular, linearity error, sensitivity error, zero error, and instrument repeatability error, as illustrated in Figure 1.12b–e, can be quantified from a static random test calibration.

## Linearity Error

Many instruments are designed to achieve a linear relationship between the applied static input and indicated output values. Such a linear static calibration curve would have the general form

$$y_L(x) = a_0 + a_1 x \tag{1.7}$$

where the curve fit $y_L(x)$ provides a predicted output value based on a linear relation between $x$ and $y$. As a measure of how well a linear relation is actually achieved, measurement device specifications usually provide a statement as to the expected linearity of the static calibration

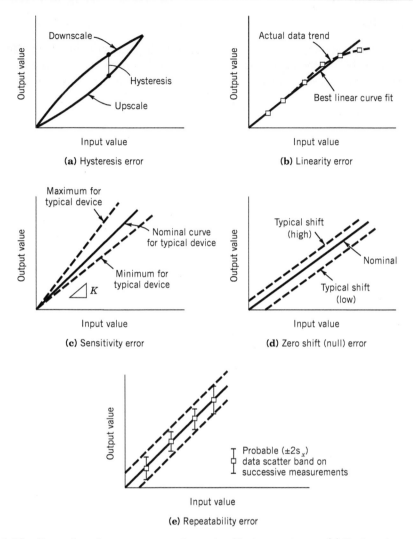

**FIGURE 1.12**   Examples of some common elements of instrument error. (a) Hysteresis error. (b) Linearity error. (c) Sensitivity error. (d) Zero shift (null) error. (e) Repeatability error.

curve for the device. The relationship between $y_L(x)$ and measured value $y(x)$ is a measure of the nonlinear behavior of a system:

$$u_L(x) = y(x) - y_L(x) \tag{1.8}$$

where the uncertainty $u_L(x)$ is a measure of the *linearity error* that arises in describing the actual system behavior by Equation 1.7. Such behavior is illustrated in Figure 1.12b, in which a linear curve has been fit through a calibration data set. For a measurement system that is essentially linear in behavior, the extent of possible nonlinearity in a measurement device is often specified in terms of the maximum expected linearity error as a percentage of full-scale output, $r_o$,

$$\%u_{L_{max}} = \frac{u_{L_{max}}}{r_o} \times 100 \tag{1.9}$$

This is how the linearity error for the pressure transducer in Table 1.1 was estimated. Statistical methods of quantifying data scatter about a curve fit are discussed in Chapter 4.

## Sensitivity and Zero Errors

The scatter in the data measured during a calibration affects the precision in predicting the slope of the calibration curve. As shown for the linear calibration curve in Figure 1.12c, in which the zero intercept is fixed, the scatter in the data about the curve fit indicates random errors. The *sensitivity error* as described by its uncertainty, $u_K$, is a statistical measure of the random error in the estimate of the slope of the calibration curve (we discuss the statistical estimate further in Chapter 4). In Table 1.1, the sensitivity error reflects calibration results at a constant reference ambient temperature, whereas the thermal sensitivity error was found by calibration at different temperatures.

A drift in the zero intercept introduces a vertical shift of the calibration curve, as shown in Figure 1.12d. This shift is known as the *zero error* with uncertainty, $u_z$. Zero error can usually be reduced by periodically nulling the output from the measurement system under a zero input condition. However, some random variation in the zero intercept is common, particularly with electronic and digital equipment subjected to external noise and temperature variations (e.g., thermal zero drift in Table 1.1). Zero error can be a major source of uncertainty in many measurements.

## Instrument Repeatability

The ability of a measurement system to indicate the same value on repeated but independent application of the same input provides a measure of the instrument *repeatability*. Specific claims of repeatability are based on multiple calibration tests (replication) performed within a given lab on the particular unit. Repeatability, as shown in Figure 1.12e, is an uncertainty based on a statistical measure (developed in Chapter 4) called the standard deviation, $s_x$, a measure of the variation in the output for a given input. The value claimed is usually in terms of the maximum expected error as a percentage of full-scale output:

$$\%u_{R_{\max}} = \frac{2s_x}{r_o} \times 100 \tag{1.10}$$

The instrument repeatability reflects only the variations found under controlled calibration conditions.

## Reproducibility

The term "reproducibility," when reported in instrument specifications, refers to the closeness of agreement in results obtained from duplicate tests carried out under similar conditions of measurement. As with repeatability, the uncertainty is based on statistical measures. Manufacturer claims of instrument reproducibility must be based on multiple tests (replication) performed in different labs on a single unit or model of instrument.

## Instrument Precision

The term "instrument precision," when reported in instrument specifications, refers to a random uncertainty based on the results of separate repeatability tests. Manufacturer claims of instrument precision must be based on multiple tests (replication) performed on different units of the same model, either performed in the same lab (same-lab precision) or, preferably, performed in different labs (between-lab precision).

## Overall Instrument Error and Instrument Uncertainty

An estimate of the *overall instrument error* is made by combining the uncertainty estimates of all identified instrument errors into a term called the *instrument uncertainty*. The estimate is computed from the square root of the sum of the squares of all known uncertainty values. For $M$ known errors, the overall instrument uncertainty, $u_c$, is calculated as

$$u_c = [u_1^2 + u_2^2 + \cdots + u_M^2]^{1/2} \tag{1.11}$$

For example, for an instrument having known hysteresis, linearity, and sensitivity errors, the instrument uncertainty is estimated by

$$u_c = [u_h^2 + u_L^2 + u_K^2]^{1/2} \tag{1.12}$$

An assumption in Eq. 1.11 is that each error contributes equally to $u_c$. If not, then sensitivity factors are used as described in Chapter 5.

## Verification and Validation

*Verification* refers to executing the intended model correctly. Within a test or measurement system, this means verifying that the measurement system, along with each of its components, is functioning correctly. One approach is to place the measurement system into a state in which its output can be anticipated, a systems-level calibration. Unacceptable differences suggest a problem somewhere within the system. This is a debugging tool. A close analogy is checking a computer program for programming errors.

*Validation* refers to ensuring that the experimental model used is itself correct. To do this, we compare a set of test results to results obtained using another reliable method for the same conditions. For example, we can compare the results from a wind tunnel test of a vehicle to the results obtained from a road test. Gross differences indicate shortcomings in one or both models. A close analogy is determining whether the algorithm programmed within a computer model actually simulates the intended physical process. Verification and validation should always be a part of a test plan to ensure meaningful results.

# 1.5 Standards

During calibration, the indicated value from the measurement system is compared directly with a reference or known value, which is applied as the input signal. This reference value forms the basis of the comparison and is known as the *standard*. Primary standards, as we shall discuss next, are exact but are not available for routine calibrations. So instead the standard used could be a close copy with direct traceability to the primary standard, or it could be based on the output from a trusted piece of equipment or even from a well-accepted technique known to produce a reliable value. Of course, the accuracy of the calibration is limited to the accuracy of the standard used. The following sections explore how certain standards are the foundation of all measurements and how other standards are derived from them.

## Primary Unit Standards

A *dimension* defines a physical variable that is used to describe some aspect of a physical system. A *unit* defines a quantitative measure of a dimension. For example, mass, length, and time describe base dimensions with which we associate the units of kilogram, meter, and second.

A *primary standard* defines the exact value of a unit. It provides the means to describe the unit with a unique number that can be understood throughout the world. In 1960, the General Conference on Weights and Measures (CGPM), the international agency responsible for maintaining exact uniform standards of measurements, formally adopted the International System of Units (SI) as the international standard of units [6]. The SI system has been adopted worldwide and contains seven base units. As of 2019, these base units are defined around exact definitions of fixed natural physical constants, such as the speed of light, Planck's constant and Avogadro's number. All other units are derived from these seven.

Other unit systems are commonly used in the consumer market and so deserve mention. Examples of these include the inch-pound (I-P) unit system found in the United States and the gravitational mks (meter-kilogram-second or metric) unit system common in much of the world. The units used in these systems are directly related, through conversion factors, to SI units. Although SI defines our unit system exactly, practical units sometimes better fit our customs. For example, we use temperature units of Celsius and Fahrenheit in daily life rather than kelvin.

Primary unit standards are necessary because the value assigned to a unit is actually quite arbitrary. For example, over 4,500 years ago, the Egyptian cubit was used as a standard of length and was based on the length from outstretched fingertips to the elbow. It was later codified with a master of marble, a stick about 52 cm in length, on which scratches were etched to define subunits of length. This standard served well for centuries.

So whether today's standard unit of length, the meter, is the length of a king's forearm or the distance light travels in a fraction of a second really only depends on how we want to define it. To avoid contradiction, the units of primary standards are defined by international agreement. Once agreed upon, a primary standard forms the exact definition of the unit until it is changed by some later agreement. Important features sought in any standard should include global availability, continued reliability, and stability with minimal sensitivity to external environmental sources. Next we examine dimensions and the primary standards that form the definition of the units used to describe them [6, 7].

## Base Dimensions and Their Units

### Mass

The kilogram (kg) is the base unit of mass. The value of the kilogram is defined by taking the value of Planck's constant $h$ to be exactly $6.62607015 \times 10^{-34}$ kg-m$^2$/s. Standard mass can be realized using a Kibble balance [9]. This balance scale generates an electromagnetic force by electric power, which is used as reference to compare against a tested mass. An added advantage is that masses of well less than 1 kg can also be realized. By this definition, mass can be determined to an uncertainty of about 20 µg/kg, perhaps improving in the future [10]. Prior to 1889, the kilogram was defined by the mass of 1 liter of water at room temperature. Between 1889 and 2018, one kilogram was defined as being exactly equal to the mass of a particular platinum-iridium cylindrical bar maintained at the International Bureau of Weights and Measures (BIPM) [6–10].

In the United States, the I-P unit system (also referred to as the U.S. customary units) remains widely used. In the I-P system, *mass* is defined by the pound-mass, lb$_m$, which is derived directly from the definition of the kilogram:

$$1 \text{ lb}_m = 0.4535924 \text{ kg} \tag{1.13}$$

Equivalent standards for the kilogram and other standards units are maintained by national labs around the globe. In the United States, this role is assigned to the National Institute of Standards and Technology (NIST) in Gaithersburg, Maryland.

**EXAMPLE 1.8**    **The Changing Kilogram**

Until 2019, the kilogram remained as the only basic unit still defined in terms of a unique physical object, known as the International Prototype Kilogram mass (IPK) and defined in 1889. A problem with using a physical object is that it can change over time. For example, atmospheric pollutants added mass and periodic cleanings removed metal molecules, reducing its mass. The standard IPK had been removed from protected storage several times in the past century to compare with exact official copies [8]. By 1989, there were differences detected between all of the official masses. This difference was about 50 micrograms and anticipated to drift larger with time. This instability was a concern for a primary standard. In response, the CGPM voted in November 2018 to change the standard to one based around an exact definition of Planck's constant [9, 10]. By using Planck's constant, the constants that define the units of meter and second are also needed but the method is more accurate than the IPK standard and will not drift with time. This change is consistent with the international movement to define basic units in terms of fixed constants of nature.

### Time and Frequency

The second (s) is the base unit of time. One second (s) is defined as the time elapsed during 9,192,631,770 periods of the radiation emitted between two excitation levels of the fundamental state of cesium-133. Despite this seemingly unusual definition, this primary standard can be reliably reproduced at suitably equipped laboratories throughout the world to an uncertainty of within 2 parts in 10 trillion.

The Bureau International de l'Heure (BIH) in Paris maintains the primary standard for clock time. Periodically, adjustments to clocks around the world are made relative to the BIH clock so as to keep time synchronous.

The standard for cyclical frequency is a derived unit based on the time standard (s). The standard unit is the hertz (1 Hz = 1 cycle/s). The cyclical frequency is related to the circular frequency (radians/s) by

$$1 \text{ Hz} = 1 \text{ cycle/s} = \frac{2\pi \text{ rad}}{1 \text{ s}} \tag{1.14}$$

### Length

The meter (m) is the base unit for length. In 1982, a new primary standard was adopted by the CGPM to define the unit of a meter. One meter (m) is defined as the length traveled by light in 1/299,792,458 of a second, a number derived from the velocity of light in a vacuum (defined exactly to be 299,792,458 m/s).

The I-P system unit of the inch and the related unit of the foot are derived exactly from the meter.

$$1 \text{ ft} = 0.3048 \text{ m}$$
$$1 \text{ in.} = 0.0254 \text{ m} \tag{1.15}$$

### Temperature

The kelvin (K) is the base unit of temperature. The magnitude of the kelvin is determined from the Boltzmann constant $k$ defined exactly as $1.380649 \times 10^{-23}$ J/K. This fixed constant forms the fundamental relationship between energy and temperature. Prior to 2019, the kelvin was defined as the fraction 1/273.16 of the thermodynamic temperature of the triple point of water.

A temperature scale was devised by William Thomson, Lord Kelvin (1824–1907) that is based on polynomial interpolation between the equilibrium phase change points of a number of common pure substances from the triple point of equilibrium hydrogen (13.81 K) to the freezing point of pure gold (1337.58 K). Above 1337.58 K, the scale is based on Planck's law of radiant emissions. The details of the standard scale have been modified over the years but are governed by the International Temperature Scale–1990 [11].

The I-P unit system uses the absolute scale of Rankine (R). This and the common scales of Celsius (°C), used in the metric system, and Fahrenheit (°F) are related to the Kelvin scale by the following:

$$(°C) = (K) - 273.15$$
$$(°F) = (R) - 459.67 \tag{1.16}$$
$$(°F) = 1.8 \times (°C) + 32.0$$

## Current

The ampere (A) is the base unit for electric current. The value of the ampere is arrived at by assigning the value of the elementary charge of an electron e to be exactly $1.602176634 \times 10^{-19}$ ampere-seconds.

## Measure of Substance

The mole (mol) is the base unit defining the quantity of a substance. One mole contains exactly $6.02214076 \times 10^{23}$ elementary entities. This number is the Avogadro constant $N$ expressed in the unit of $mol^{-1}$. An elementary entity may be an atom, molecule, ion, electron, or some specified particle. Prior to 2019, the mole was defined as the number of atoms in 0.012 kg of carbon-12. The current definition eliminates this link to measuring mass.

## Luminous Intensity

The candela (cd) is the base unit of the intensity of light. One candela is defined by taking the value of the luminous efficacy of monochromatic radiation of frequency $540 \times 10^{12}$ Hz to be 683 lumens/watt. A watt is a derived unit of power.

# Derived Units

Other dimensions and their associated units are defined in terms of and derived from the base dimensions and units [7].

## Force

From Newton's law, force equals mass times acceleration. Force is defined by the unit of the newton (N), which is derived from the base units for mass, length, and time:

$$1 \text{ N} = 1\frac{\text{kg-m}}{s^2} \tag{1.17}$$

One Newton is defined as the force applied to a 1 kg mass to accelerate it at $1 \text{ m/s}^2$.

When working within and between unit systems, it can be helpful to view force in terms of

$$\text{Force} = \frac{\text{Mass} \times \text{Acceleration}}{g_c}$$

where $g_c$ is a proportionality constant used to maintain consistency between the units of force and mass. So for the SI system, the value of $g_c$ must be 1.0 $\text{kg-m/s}^2$-N. The term $g_c$ does not need to be explicitly stated, but the relationship between force and mass units holds.

However, in I-P units, the units of force and mass are related through the definition: One pound-mass ($lb_m$) exerts a force of 1 pound (lb) in a standard earth gravitational field. With this definition,

$$1 \text{ lb} = \frac{(1 \text{ lb}_m)(32.1740 \text{ ft/s}^2)}{g_c} \tag{1.18}$$

and $g_c$ must take on the value of 32.1740 $lb_m$-ft/$s^2$-lb.

Richard Figliola

**FIGURE 1.13** Bicycle tire pressure expressed in three different unit systems to meet different consumer preferences within a global market.

Similarly, in the gravitational mks (metric) system, which uses the kilogram-force ($kg_f$),

$$1 \; kg_f = \frac{(1 \; kg)(9.80665 \; m/s^2)}{g_c} \tag{1.19}$$

Here the value for $g_c$ takes on a value of exactly $9.80665 \; kg\text{-}m/s^2\text{-}kg_f$.

Many engineers have some difficulty distinguishing between units of mass and force in the non-SI unit systems. Actually, whenever force and mass appear in the same expression, just relate them using $g_c$ through Newton's law:

$$g_c = \frac{mg}{F} = 1\frac{kg\text{-}m/s^2}{N} = 32.1740\frac{lb_m\text{-}ft/s^2}{lb} = 9.80665\frac{kg\text{-}m/s^2}{kg_f} \tag{1.20}$$

Shown in Figure 1.13 is a common example of multiple unit systems being cited on a consumer product marketed for use in different parts of the world: a bicycle tire with recommended inflation pressures shown in I-P, gravitational metric, and SI ($1 \; kPa = 1000 \; N/m^2$) units.

### Other Derived Dimensions and Units

Energy is defined as force times length and uses the unit of the joule (J), which is derived from base units as

$$1 \; J = 1\frac{kg\text{-}m^2}{s^2} = 1 \; N - m \tag{1.21}$$

Power is defined as energy per unit time in terms of the unit of the watt (W), which is derived from base units as

$$1 \; W = 1\frac{kg\text{-}m^2}{s^3} = 1\frac{J}{s} \tag{1.22}$$

Stress and pressure are defined as force per unit area in terms of the pascal (Pa), which is derived from base units as

$$1 \; Pa = 1\frac{kg}{m - s^2} = 1 \; N/m^2 \tag{1.23}$$

### Electrical Dimensions

The units for the dimensions of electrical potential, resistance, charge, and capacitance are based on the definitions of the absolute volt (V), ohm ($\Omega$), coulomb (C), and farad (F), respectively.

Derived from the ampere, 1 ohm is defined as 0.9995 times the resistance to current flow of a column of mercury that is 1.063 m in length and that has a mass of 0.0144521 kg at 273.15 K. The volt is derived from the units for power and current, $1\,V = 1\,\text{N-m/s-A} = 1\,\text{W/A}$. The ohm is derived from the units for electrical potential and current, $1\,\Omega = 1\,\text{kg-m}^2/\text{s}^3\text{-A}^2 = 1\,\text{V/A}$. The coulomb is derived from the units for current and time, $1\,C = 1\,\text{A-s}$. One volt is the difference of potential between two points of an electrical conductor when a current of 1 ampere flowing between those points dissipates a power of 1 watt. The farad (F) is the standard unit for capacitance derived from the units for charge and electric potential, $1\,F = 1\,\text{C/V}$.

On a practical level, working standards for resistance and capacitance take the form of certified standard resistors and capacitors or resistance boxes and are used as standards for comparison in the calibration of resistance measuring devices. The practical potential standard makes use of a standard cell consisting of a saturated solution of cadmium sulfate. The potential difference of two conductors connected across such a solution is set at 1.0183 V and at 293 K. The standard cell maintains constant voltage over very long periods of time, provided that it is not subjected to a current drain exceeding 100 μA for more than a few minutes. The standard cell is typically used as a standard for comparison for voltage measurement devices.

A chart for converting between units is included at the back of the text. Table 1.2 lists the standard base and derived units used in SI and the corresponding units used in the I-P and gravitational metric systems.

**Table 1.2  Dimensions and Units[a]**

| Unit | SI | I-P |
|---|---|---|
| **Primary** | | |
| Length | meter (m) | inch (in.) |
| Mass | kilogram (kg) | pound-mass ($\text{lb}_\text{m}$) |
| Time | second (s) | second (s) |
| Temperature | kelvin (K) | rankine (R) |
| Current | ampere (A) | ampere (A) |
| Substance | mole (mol) | mole (mol) |
| Light intensity | candela (cd) | candela (cd) |
| **Derived** | | |
| Force | newton (N) | pound-force (lb) |
| Voltage | volt (V) | volt (V) |
| Resistance | ohm (Ω) | ohm (Ω) |
| Capacitance | farad (F) | farad (F) |
| Inductance | henry (H) | henry (H) |
| Stress, pressure | pascal (Pa) | pound-force/inch$^2$ (psi) |
| Energy | joule (J) | foot pound-force (ft-lb) |
| Power | watt (W) | foot pound-force/second (ft-lb/s) |

[a]SI dimensions and units are the international standards. I-P units are presented for convenience.

**EXAMPLE 1.9    Units Conversion Issues**

In 1999, communications were lost with the Mars Climate Orbiter during maneuvers intended to place the spacecraft into permanent Mars orbit. The fault was tracked to units conversion incompatibility between the ground-based software and the spacecraft-based software used in tandem to control the spacecraft [12]. At the root of the problem was that the systems and software were each developed by different vendors. In one set of instructions, the thruster pulsation program was written to apply vehicle thrust in units of newtons, whereas the other set was written to apply thrust in units of pounds. Thruster execution placed the spacecraft in a much lower orbit than intended, and (presumably) it burned up in the atmosphere. This example illustrates the value of rigorous systems-level verification.

In 1983, Air Canada Flight 143 ran out of fuel during a flight requiring an emergency landing in Gimli, Manitoba. Although Canada had recently undergone a metrification program and this Boeing 767 was a metric aircraft, many aircraft and operations were still based on Imperial units (pounds and gallons). A sequence of conversion errors by ground fuel crews resulted in 22,300 lbs of fuel being loaded rather than the 22,300 kg specified. A compounding factor was a fuel computer glitch that also rendered the onboard fuel gauges inoperative. Fortunately, the pilots were able to glide the aircraft from 41,000 ft to a complicated but safe landing, thus earning the aircraft the nickname "The Gimli Glider" [13].

## Hierarchy of Standards

The known value or reference value applied during calibration becomes the standard on which the calibration is based. So how do we pick this standard, and how good does it need to be? Obviously, actual primary standards are unavailable for normal calibration use. But they serve as a reference for exactness. So for practical reasons, there exists a hierarchy of secondary standards used to duplicate the primary standards. Just below the primary standard in terms of absolute accuracy are the national reference standards maintained by designated standards laboratories throughout the world. These provide a reasonable duplication of the primary standard while allowing for worldwide access to an extremely accurate standard. Next to these, we develop transfer standards. These are used to calibrate individual laboratory standards that might be used at various calibration facilities within a country. Laboratory standards serve to calibrate working standards. Working standards are used to calibrate everyday devices used in manufacturing and research facilities. In the United States, NIST maintains primary, national reference, and transfer standards and recommends standard procedures for the calibration of measurement systems.

Each subsequent level of the hierarchy is derived by calibration against the standard at the previous higher level. Table 1.3 lists an example of such a lineage for standards from a primary or reference standard maintained at a national standards lab down to a working standard used in a typical laboratory or production facility to calibrate everyday working instruments. If the facility does not maintain a local (laboratory or working) standard, then the instruments must be sent off and calibrated elsewhere. In such a case, a standards traceability certificate will be issued for the instrument that details the lineage of standards used toward its calibration and the uncertainty in the calibration. For measurements that do not require traceable accuracy, a trusted value can be used to calibrate another device.

Moving down through the standards hierarchy, the degree of exactness by which a standard approximates the primary standard deteriorates. That is, increasing amounts of error are introduced into the standard as we move down from one level of hierarchy to the next. As a common example, a company might maintain its own working standard used to calibrate the measurement devices found in the individual laboratories throughout the company. Periodic calibration of

**Table 1.3    Hierarchy of Standards[a]**

| | |
|---|---|
| Primary standard | Maintained as absolute unit standard |
| Reference standard | Duplicate of a primary standard at a national lab |
| Transfer standard | Used to calibrate local standards |
| Local standard | Used to calibrate working standards |
| Working standard | Used to calibrate local instruments |
| Calibration standard | Any reference value used in a calibration |

[a]There may be additional intermediate standards between each hierarchy level.

**Table 1.4    Example of a Temperature Standard Traceability**

| Level | Standard Method | Uncertainty [K][a] |
|---|---|---|
| Primary | Fixed thermodynamic points | 0 |
| Transfer | Platinum resistance thermometer | ±0.005 |
| Working | Platinum resistance thermometer | ±0.05 |
| Local | Thermocouple | ±0.5 |

[a]Typical combined instrument systematic and random uncertainties.

that working standard might be against the company's well-maintained local standard. The local standard would be periodically sent off to be calibrated against the NIST (or appropriate national standards lab) transfer standard (and traceability certificate issued). NIST will periodically calibrate its own transfer standard against its reference or primary standard. This idea is illustrated for a temperature standard traceability hierarchy in Table 1.4. The uncertainty in the approximation of the known value increases as one moves down the hierarchy. It follows, then, that the accuracy of the calibration will be limited by the accuracy of the standard used. But if typical working standards contain errors, how is accuracy ever determined? At best, this closeness of agreement is quantified by the estimates of the known uncertainties in the calibration.

## Test Standards and Codes

The term "standard" is also applied in other ways in engineering. A *test standard* is a document, established by consensus, that provides rules, guidelines, methods or characteristics for activities or their results. They define appropriate use of certain terminology as used in commerce. The goal of a test standard is to provide consistency between different facilities or manufacturers in the conduct and reporting of a certain type of measurement on some test specimen. Similarly, a *test code* is a standard that refers to procedures for the manufacture, installation, calibration, performance specification, and safe operation of equipment.

Diverse examples of test standards and codes are illustrated in readily available documents [14–19] from professional societies, such as the American Society of Mechanical Engineers (ASME), the American Society of Testing and Materials (ASTM), the American National Standards Institute (ANSI), and the International Organization for Standardization (ISO). For example, ASME Power Test Code 19.5 provides detailed designs and operation procedures for flow meters, and ANSI S2.70-2006 details sensor placement and data reduction methods to evaluate and report human exposure to vibrations transmitted to the hand from operating machinery.

Engineering standards and codes are consensus documents agreed on by knowledgeable parties interested in establishing a common basis for comparing equipment performance between manufacturers or for consistent and safe operation. These are not binding legal documents unless specifically adopted and implemented as such by a government agency. Still, they present a convincing argument for best practice.

# 1.6 Presenting Data

Data presentation conveys significant information about the relationship between variables. Software is readily available to assist in providing high-quality plots, or plots can be generated manually using graph paper. Several forms of plotting formats are discussed next.

## Rectangular Coordinate Format

In rectangular grid format, such as Figure 1.14a, both the ordinate and the abscissa have uniformly sized divisions providing a linear scale. This is a common format used for constructing plots and establishing the form of the relationship between the independent and dependent variable.

## Semilog Coordinate Format

In a semilog format, one coordinate has a linear scale and one coordinate has a logarithmic scale. Plotting values on a logarithmic scale performs a logarithmic operation on those values—for example, plotting $y = f(x)$ on a logarithmic $x$-axis is the same as plotting $y = \log f(x)$ on rectangular axes. Logarithmic scales are advantageous when one of the variables spans more than one order of magnitude. A linear curve results when the data follow a trend of the form $y = ae^x$, as in Figure 1.14b, or $y = a10^x$. These two logarithmic forms are related as $\ln y = 2.3 \log y$.

## Full-Log Coordinate Format

The full-log or log-log format (Figure 1.14c) has logarithmic scales for both axes and is equivalent to plotting $\log y$ vs. $\log x$ on rectangular axes. Such a format is preferred when both variables contain data values that span more than one order of magnitude. With data that follow a trend of the form $y = ax^n$ as in Figure 1.14c, a linear curve will result.

## Significant Digits

Just how many digits should you assign to a number when you report it? This number should reflect both your level of confidence and the level of relevance for the information it represents. In this section, we discuss how to identify significant digits and how to round to an appropriate number of significant digits, offering suggestions for assigning significant digits in calculated values. Despite the suggestions here, the number of significant digits reported should also reflect practical sense.

The term *digit* refers to a numerical figure between 0 and 9. The position or *place value* of a digit within a number provides its order of magnitude. A *significant digit* refers to any digit that is used together with its place value to represent a numerical value to a desired approximation. In writing numbers, the leftmost nonzero digit is the *most significant digit*. The rightmost digit

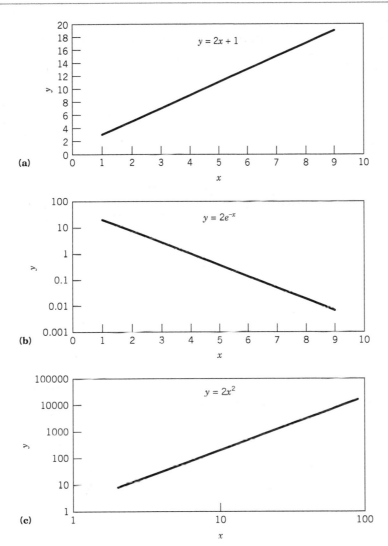

**FIGURE 1.14**   Plotting formats: (a) rectangular, (b) semi-log, (c) full log.

is the *least significant digit*. Leading zeros refer to the leftmost zeros leading any nonzero digits. Trailing zeros are the rightmost zeros following nonzero digits.

All nonzero digits present in a number are significant. Zeros require special attention within the following guidelines: (1) All leading zeros, whether positioned before or after a decimal point, are not significant, as these serve only to set order of magnitude and not value; (2) zeros positioned between nonzero digits are significant; (3) all trailing zeros when situated to either side of a decimal point are significant; and (4) within an exact count, the zeros are significant. These guidelines are consistent with *ASTM E29—Standard Practice for Using Significant Digits* [18].

Writing a number using scientific notation is a convenient way to distinguish between meaningful digits and nonsignificant zero digits used only to hold place value. For example, 0.0042 can be written as $4.2 \times 10^{-3}$. From the guidelines above, this number has two significant digits, which are represented by the two nonzero digits. In this example, the leading zeros serve only to hold place values for the order of magnitude, which is explicit (i.e., $10^{-3}$) when written in scientific notation.

When a number does not have a decimal point cited, trailing zeros can be problematic. If the zeros are intended to be significant, use a decimal point. Otherwise, they will be assumed not significant. To avoid ambiguity, use scientific notation. For example, writing 150,000 Pa as $1.5 \times 10^4$ Pa clearly indicates that two significant digits are intended.

## Rounding

Rounding is a process in which the number of digits is reduced and the remaining least significant digit(s) adjusted appropriately. In reducing the number of digits, (1) if the digits to be discarded begin with a digit less than 5, the preceding digit is not changed; (2) if the digits to be discarded begin with a 5 and at least one of the following digits is greater than 0, the preceding digit is increased by 1; (3) if the digits to be discarded begin with a 5 and all of the following digits are 0, the digit preceding the 5 is unchanged if it is an even digit but increased by 1 if it is an odd digit. When a result must meet a specification limit for conformance, do not round the number to meet the specification.

## Numerical Operations

Here are suggestions for adjusting the significant digits of the result after common numerical operations:

1. In addition or subtraction, the number of digits following the decimal in the reported result should not be greater than the least number of digits found following a decimal in any of the data points used.

2. In other operations, the significant digits in the result should not be greater than the least number of significant digits in any of the data points or the operand used.

3. The number of significant digits in an exact count is not considered when establishing the number of significant digits to be reported.

4. Round your final result, but do not round intermediate calculations.

Computations within software, hand calculators, and spreadsheets carry large numbers of digits, usually many more than can be justified. However, rounding intermediate calculations introduces new truncation errors, so best practice is to round only the final result.

**EXAMPLE 1.10**    **Rounding Decisions**

Rounding is a decision process and practical sense should prevail. As an example, when the most significant digit in the result of a numerical operation is increased in order of magnitude relative to any of the data points used, it may make sense to retain extra significant digit(s). Consider $5 \times 3 = 15$. Would you keep the 15 or round to 20 to maintain the one significant digit suggested in the guidelines? Consider $5.1 \times 2.1 = 10.71$. The rounded result might be better expressed as 10.7 rather than 11. One rationale is that the maximum rounding error in 5.1 is $0.05/5.1 = 0.94\%$, in 2.1 is $0.05/2.1 = 2.4\%$, in 11 is $0.5/11 = 4.5\%$, and in 10.7 is $0.05/10.7 = 0.46\%$. Rounding to 11 discards information present in the data, whereas rounding to 10.7 adds information. In rounding, the engineer must choose between removing known information and adding unwarranted information while keeping the quality of the known information and the intended use of the result in perspective [7]. In short, the correct approach is at the discretion of the engineer.

## Applications

How many significant digits should you carry when recording a measurement? For direct measurements using an analog readout device (e.g., dial indicator), the number of significant digits recorded should include all the digits known exactly, plus up to one more digit from interpolation when possible. This holds the same for data taken from charts. For direct measurements from a digital readout, record all known digits. With tables, interpolate as needed but retain the same number of digits as used in the table. When recording data by computer-based data acquisition methods, carry as many digits as possible and then round a final result to an appropriate number of digits.

### EXAMPLE 1.11

Round the following numbers to three significant digits and then write them in scientific notation.

| | | |
|---|---|---|
| 49.0749 | becomes | 49.1 or $4.91 \times 10^1$ |
| 0.0031351 | becomes | 0.00314 or $3.14 \times 10^{-3}$ |
| 0.0031250 | becomes | 0.00312 or $3.12 \times 10^{-3}$ |

### EXAMPLE 1.12

A handheld appliance consumes 1.41 kW of power. So two identical units (exact count $N = 2$) consume $1.41 \text{ kW} \times 2 = 2.82 \text{ kW}$ of power. The exact count does not affect the significant digits in the result.

### EXAMPLE 1.13

Many conversion factors can be treated as exact counts. Reference 7 provides a list. For example, in converting 1 hour into seconds, the result is exactly 3,600 s. Similarly, 1 ft equals exactly 0.3048 m. Otherwise, apply the rounding suggestions using discretion.

## SUMMARY

During a measurement the input signal is not known but is inferred from the value of the output signal from the measurement system. We discussed the process of calibration as the means to relate the measured input value to the measurement system output value. We discussed the role of standards in that process. An important step in the design of a measurement system is the inclusion of a means for a reproducible calibration that closely simulates the type of signal to be input during actual measurements. A test is the process of "asking a question." The idea of a test plan was developed to allow measurements that answer that question. However, a measured output signal can be affected by many variables that will introduce variation and trends and confuse that answer. Careful test planning is required to reduce such effects. A number of test plan strategies were developed, including randomization. The popular term "accuracy" was explained in terms of the more useful concepts of random error, systematic error, random uncertainty, and systematic uncertainty and their effects on a measured value. We explored the idea of test standards and engineering codes, legal documents that influence practically every manufactured product around us.

# NOMENCLATURE

| | | | | |
|---|---|---|---|---|
| $e$ | absolute error | | $u_R$ | repeatability uncertainty; uncertainty assigned to repeatability error |
| $p$ | pressure $(ml^{-1}t^{-2})$ | | $u_z$ | zero uncertainty; uncertainty assigned to zero error |
| $r_i$ | input span | | $x$ | independent variable; input value; measured variable |
| $r_o$ | output span | | $y$ | dependent variable; output value |
| $s_x$ | standard deviation of $x$ | | $y_L$ | linear polynomial |
| $u_c$ | overall instrument uncertainty | | $A$ | relative error; relative accuracy |
| $u_h$ | hysteresis uncertainty; uncertainty assigned to hysteresis error | | $K$ | static sensitivity |
| $u_K$ | sensitivity uncertainty; uncertainty assigned to sensitivity error | | $T$ | temperature $(°)$ |
| $u_L$ | linearity uncertainty; uncertainty assigned to linearity error | | $\forall$ | volume $(l^3)$ |

# REFERENCES

1.  Peace, G. S., *Taguchi Methods*, Addison-Wesley, Reading, MA, 1993.

2.  Lipsen, C., and N. J. Sheth, *Statistical Design and Analysis of Engineering Experimentation*, McGraw-Hill, New York, 1973.

3.  Peterson, R. G., *Design and Analysis of Experiments*, Marcel-Dekker, New York, 1985.

4.  Mead, R., *The Design of Experiments: Statistical Principles for Practical Application*, Cambridge Press, New York, 1988.

5.  Montgomery, D., *Design and Analysis of Experiments*, 7th ed., Wiley, New York, 2009.

6.  Bureau International des Poids et Mesures, *The International System of Units (SI)*, 8th ed., Organisation Intergouvernementale de la Convention du Mètre, Paris, 2006.

7.  Taylor, B., and A. Thompson, Guide for the Use of the International System of Units, NIST Special Publication 811, 2008. http://physics.nist.gov/cuu/pdf/sp811.pdf

8.  Davis, R., "The SI unit of mass," *Metrologia* 40 299–305, 2003.

9.  Robinson, I.A., Schlamminger, S., "The Watt or Kibble Balance: A Technique for Implementing the New SI Definition of the Unit of Mass," *Metrologia* 53(5): A46, 2016.

10. Richard, P., et al., "Foundation for the redefinition of the kilogram," *Metrologia* 53(5), A6, 2016.

11. Committee Report, International Temperature Scale—1990 *Metrologia* 27(1): 3–10, 1990.

12. Stephenson, A. (Chairman), "Mars Climate Orbiter Mishap Investigation Board Phase I Report," NASA, Washington, DC, November 1999.

13. Nelson, Wade, "The Gimli Glider," Soaring Magazine, October 1997, 19–22.

14. *ASME Power Test Codes*, American Society of Mechanical Engineers, New York.

15. *ASHRAE Handbook of Fundamentals*, American Society of Heating, Refrigeration and Air Conditioning Engineers, New York, 2009.

16. *ANSI Standard*, American National Standards Institute, New York, 2009.

17. *Annual Book of ASTM Standards*, American Society for Testing and Materials, Philadelphia, 2010.

18. *Standard Practice for Using Significant Digits, ASTM E29-13*, American Society for Testing and Materials, Philadelphia, 2013.

19. *Guide for the Measurement and Evaluation of Human Exposure to Vibration Transmitted to the Hand, ANSI S2.70-2006*, American National Standards Institute, New York, 2006.

# Static and Dynamic Characteristics of Signals

2

## 2.1 Introduction

A measurement system takes an *input* quantity and transforms it into an *output* quantity that can be observed or recorded, such as the movement of a pointer on a dial or the magnitude of a digital display. This chapter discusses the characteristics of both the input signals to a measurement system and the resulting output signals. The shape and form of a signal are often referred to as its *waveform*. The waveform contains information about the *magnitude*, which indicates the size of the input quantity, and the *frequency*, which indicates the rate at which the signal changes in time. An understanding of waveforms is required for the selection of measurement systems and the interpretation of measured signals.

## 2.2 Input/Output Signal Concepts

Two important tasks that engineers face in the measurement of physical variables are [1] selecting a measurement system and [2] interpreting the output from a measurement system. A simple example of selecting a measurement system might be the selection of a tire gauge for measuring the air pressure in a bicycle tire or in a car tire, as shown in Figure 2.1. The gauge for the car tire would be required to indicate pressures up to 275 kPa (40 lb/in.$^2$), but the bicycle tire gauge would be required to indicate higher pressures, perhaps up to 700 kPa (100 lb/in.$^2$). This idea of the *range* of an instrument, its lower to upper measurement limits, is fundamental to all measurement systems and demonstrates that some basic understanding of the nature of the input signal, in this case the *magnitude*, is necessary in evaluating or selecting a measurement system for a particular application.

A much more difficult task is the evaluation of the output of a measurement system when the time or spatial behavior of the input is not known. The pressure in a tire does not change while we are trying to measure it, but what if we wanted to measure pressure in a cylinder in an automobile engine? Would the tire gauge or another gauge based on its operating principle work? We know that the pressure in the cylinder varies with time. If our task was to select a measurement system to determine this time-varying pressure, information about the pressure variations in the cylinder would be necessary. From thermodynamics and the speed range of the engine, it may be possible to estimate the magnitude of pressures to be expected and the rate with which they change. From that we can select an appropriate measurement system. But to do this we need to develop a way to express this idea of the magnitude and rate of change of a measured variable.

Upon completion of this chapter, the reader will be able to:

- Define the range of an instrument
- Classify signals as analog, discrete time, or digital
- Compute the mean and rms values for time-varying signals
- Characterize signals in the frequency domain

**FIGURE 2.1**
Measurement system selection based on input signal range.

## Generalized Behavior

Many measurement systems exhibit similar responses under a variety of conditions, which suggests that the performance and capabilities of measuring systems may be described in a generalized way. To examine further the generalized behavior of measurement systems, we first examine the possible forms of the input and output signals. We will associate the term "signal" with the "transmission of information." A *signal* is the physical information about a measured variable being transmitted between a process and the measurement system, between the stages of a measurement system, or as the output from a measurement system.

## Classification of Waveforms

Signals may be classified as analog, discrete time, or digital. *Analog* describes a signal that is continuous in time. Because physical variables tend to be continuous, an analog signal provides a ready representation of their time-dependent behavior. In addition, the magnitude of the signal is continuous and thus can have any value within the operating range. An analog signal is shown in Figure 2.2a; a similar continuous signal would result from a recording of the pointer rotation with time for the output display shown in Figure 2.2b. Contrast this continuous signal with the signal shown in Figure 2.3a. This format represents a *discrete time signal*, for which information about the magnitude of the signal is available only at discrete points in time. A discrete time signal usually results from the sampling of a continuous variable at repeated finite time intervals. Because information in the signal shown in Figure 2.3a is available only at discrete times, some assumption must be made about the behavior of the measured variable during the times when it is not available. One approach is to assume the signal is constant between samples—a

(a) Analog signal representation        (b) Analog display

**FIGURE 2.2**    Analog signal concepts.

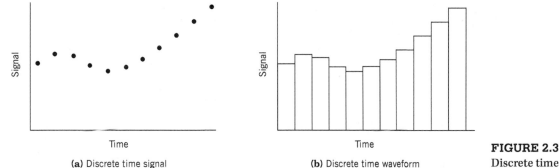

**(a)** Discrete time signal

**(b)** Discrete time waveform

**FIGURE 2.3**
Discrete time signal concepts.

sample and hold method. The resulting waveform is shown in Figure 2.3b. Clearly, as the time between samples is reduced, the difference between the discrete variable and the continuous signal it represents decreases.

*Digital signals* are particularly useful when data acquisition and processing are performed using a computer, a device that handles information in digital form. A digital signal has two important characteristics. First, a digital signal exists at discrete values in time, like a discrete time signal. Second, the magnitude of a digital signal is discrete, determined by a process known as quantization at each discrete point in time. *Quantization* assigns a single number to represent a range of magnitudes of a continuous signal. As an example of quantization, consider a digital clock that displays time in hours and minutes. For the entire duration of 1 minute, a single numerical value is displayed until it is updated at the next discrete time step. As such, the continuous physical variable of time is quantized in its conversion to a digital display.

Figure 2.4a shows digital and analog forms of the same signal where the magnitude of the digital signal can have only certain discrete values. Thus, a digital signal provides a quantized magnitude at discrete times. The waveform that would result from assuming that the signal is constant between sampled points in time is shown in Figure 2.4b.

Sampling of an analog signal to produce a digital signal can be accomplished by using an *analog-to-digital (A/D) converter,*[1] a solid-state device that converts an analog voltage signal into a binary number system representation. The limited resolution of the binary number that corresponds to a range of voltages creates the quantization levels and establishes the range.

For example, digital music recording technology relies on the conversion of a continuously available signal, such as the voltage signal from a microphone, into a digital form such as an MP3 file. However, because headphones and the human ear are analog devices, the digital information

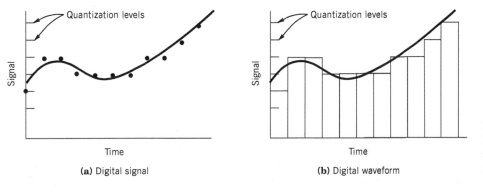

**(a)** Digital signal

**(b)** Digital waveform

**FIGURE 2.4**
Digital signal representation and waveform. Solid line is the analog signal representation.

---

[1]The processes and devices employed in A/D conversion are detailed in Chapter 7.

is converted back into a continuous voltage signal for playback. Broadcast of digital signals makes possible high-definition television service in our homes and satellite radio in our vehicles.

## Signal Waveforms

In addition to classifying signals as analog, discrete time, or digital, some description of the waveform associated with a signal is useful. Signals, represented here by a function $y(t)$, may be characterized as being either static or dynamic. As a note, we will discuss time varying signals but recognize that signals varying in space, as $y(x)$, can be classified similarly. A *static signal* does not vary with time (or space). In addition to signals that are constant in time (or space), many physical variables change slowly enough, compared to the process with which they interact, that for all practical purposes, may be considered static. For example, consider measuring temperature by using an outdoor thermometer. Since the outdoor temperature does not change significantly in a matter of minutes, this input signal might be considered static compared to a longer time period of interest, say over the course of 1 hour A mathematical representation of a static signal is given by a constant, as indicated in Table 2.1. In contrast, often we are interested in how the measured variable changes with time. This leads us to consider time-varying signals further.

A *dynamic signal* is defined as a signal that changes with time, that is, one that is time-dependent. In general, dynamic signal waveforms, $y(t)$, may be classified as shown in Table 2.1. A *deterministic signal* varies in time in a predictable manner, such as a sine wave, a step function, or a ramp function, as shown in Figure 2.5. A signal is *steady periodic* if the variation of the magnitude of the signal repeats at regular intervals in time, such as the sine wave. Examples of steady periodic behaviors are the motion of an ideal pendulum and the temperature variations in the cylinder of an internal combustion engine under steady operating conditions. Periodic waveforms may be classified as simple or complex. A *simple periodic waveform* contains only one frequency, $\omega$ and an *amplitude*, $C$. A more general or complex *periodic waveform* is represented as a superposition of multiple simple periodic waveforms. Periodic signals contain a fundamental frequency (the lowest frequency component) with its equivalent period. Whether simple or complex, a periodic signal repeats itself every period. *Aperiodic* is the term used to describe deterministic signals that do not repeat at regular intervals, such as a once occurring step function or ramp function.

Also described in Figure 2.5 is a *nondeterministic signal* that has no discernible pattern of repetition. A nondeterministic signal cannot be prescribed before it occurs, although certain

**Table 2.1  Classification of Waveforms**

| I. | Static | $y(t) = A_0$ |
|---|---|---|
| II. | Dynamic | |
| | Periodic waveforms | |
| | Simple periodic waveform | $y(t) = A_0 + C\sin(\omega t + \phi)$ |
| | Periodic waveform | $y(t) = A_0 + \sum_{n=1}^{\infty} C_n \sin(n\omega t + \phi_n)$ |
| | Aperiodic waveforms | |
| | Step[a] | $y(t) = A_0 U(t)$ $= A_0$ for $t > 0$ |
| | Ramp | $y(t) = A_0 t$ for $0 < t < t_f$ |
| | Pulse[b] | $y(t) = A_0 U(t) - A_0 U(t - t_1)$ |
| III. | Nondeterministic waveform[c] | |

[a] $U(t)$ represents the unit step function, which is zero for $t < 0$ and 1 for $t \geq 0$.
[b] $t_1$ represents the pulse width.
[c] Nondeterministic waveforms can only be approximated using statistical information.

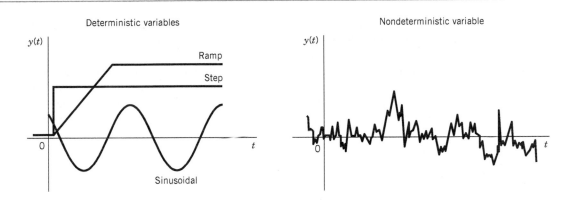

**FIGURE 2.5** Examples of dynamic signals.

characteristics of the signal may be known in advance. As an example, consider the transmission of data files from one computer to another. Signal characteristics such as the rate of data transmission and the possible range of signal magnitude are known for any signal in this system. However, it would not be possible to predict future signal characteristics based on existing information in such a signal. Such a signal is properly characterized as nondeterministic. Nondeterministic signals are generally described by their statistical characteristics or by a model signal that represents the statistics of the actual signal.

## 2.3 Signal Analysis

In this section, we consider concepts related to the characterization of signals. A measurement system produces a signal that may be analog, discrete time, or digital. An analog signal is continuous with time and has a magnitude that is analogous to the magnitude of the physical variable being measured. Consider the analog signal shown in Figure 2.6a, which is continuous over the recorded time period from $t_1$ to $t_2$. The average or mean value[2] of this signal is found by

$$\bar{y} \equiv \frac{\int_{t_1}^{t_2} y(t)dt}{\int_{t_1}^{t_2} dt} \qquad (2.1)$$

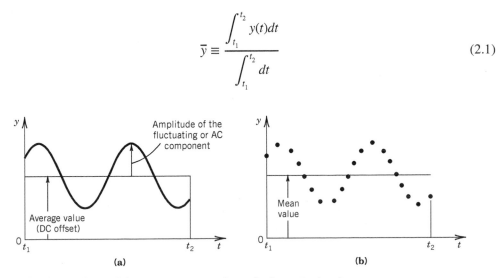

**FIGURE 2.6** Analog and discrete representations of a dynamic signal.

---

[2]Strictly speaking, for a continuous signal, the mean value and the average value are the same. This is not true for discrete-time signals.

The mean value, as defined in Equation 2.1, provides a measure of the static portion of a signal over the time period from $t_1$ to $t_2$. The average value of a fluctuating signal is also called the *DC component* or DC offset of the signal.

The mean value does not provide any indication of the amount of variation in the dynamic portion of the signal. The characterization of the dynamic portion, or *AC component*, of the signal may be illustrated by considering the average power dissipated in an electrical resistor through which a fluctuating current flows. The power dissipated in a resistor due to the flow of a current is

$$P = I^2 R$$

where

$P$ = power dissipated = time rate of energy dissipated

$I$ = current

$R$ = resistance

If the current varies in time, the total electrical energy dissipated in the resistor over the time $t_1$ to $t_2$ would be

$$\int_{t_1}^{t_2} P dt = \int_{t_1}^{t_2} [I(t)]^2 R dt \tag{2.2}$$

The current $I(t)$ would, in general, include both a static DC component and a changing AC component.

## Signal Root-Mean-Square Value

Consider finding the magnitude of a constant effective current, $I_e$, that would produce the same total energy dissipation in the resistor as the time-varying current, $I(t)$, over the time period $t_1$ to $t_2$. Assuming that the resistance, $R$, is constant, this current would be determined by equating $(I_e)^2 R(t_2 - t_1)$ with Equation 2.2 to yield

$$I_e = \sqrt{\frac{1}{t_2 - t_1} \int_{t_1}^{t_2} [I(t)]^2 dt} \tag{2.3}$$

This value of the current is called the root-mean-square (rms) value of the current. Based on this reasoning, the rms value of any continuous analog variable $y(t)$ over the time $t_2 - t_1$ is expressed as

$$y_{\text{rms}} = \sqrt{\frac{1}{t_2 - t_1} \int_{t_1}^{t_2} [y(t)]^2 dt} \tag{2.4}$$

## Discrete Time or Digital Signals

A time-dependent analog signal, $y(t)$, can be represented by a discrete set of $N$ numbers over the time period from $t_1$ to $t_2$ through the conversion

$$y(t) \rightarrow \{y(r\delta t)\} \quad r = 0, 1 \ldots, (N-1)$$

which uses the sampling convolution

$$\{y(r\delta t)\} = y(t)\delta(t - r\delta t) = \{y_i\} \quad i = 0, 1, 2, \ldots, (N-1)$$

Here, $\delta(t - r\delta t)$ is the delayed unit impulse function, $\delta t$ is the sample time increment between each number, and $N\delta t = t_2 - t_1$ gives the total sample period over which the measurement of $y(t)$ takes place. The effect of discrete sampling on the original analog signal is demonstrated in Figure 2.6b in which the analog signal has been replaced by $\{y(r\delta t)\}$, which represents $N$ values of a discrete time signal representing $y(t)$.

For either a discrete time signal or a digital signal, the mean value can be estimated by the discrete equivalent of Equation 2.1 as

$$\bar{y} = \frac{1}{N} \sum_{i=0}^{N-1} y_i \tag{2.5}$$

where each $y_i$ is a discrete number in the data set of $\{y(r\delta t)\}$. The mean approximates the static component of the signal over the time interval $t_1$ to $t_2$. The rms value can be estimated by the discrete equivalent of Equation 2.4 as

$$y_{\mathrm{rms}} = \sqrt{\frac{1}{N} \sum_{i=0}^{N-1} y_i^2} \tag{2.6}$$

The rms value takes on additional physical significance when either the signal contains no DC component or the DC component has been subtracted from the signal. As a special case, the rms value of a signal having a zero mean or one in which the mean has been removed is a statistical measure of the magnitude of the fluctuations in the signal. It is called a standard deviation.

## Direct Current Offset

When the AC component of the signal is of primary interest, the DC component can be removed. This procedure often allows a change of scale in the display or plotting of a signal such that fluctuations that were a small percentage of the DC signal can be more clearly observed without the superposition of the large DC component. The enhancement of fluctuations through the subtraction of the average value of a signal is illustrated in Figure 2.7.

Figure 2.7a contains the entire complex waveform consisting of both static (DC) and dynamic (AC) parts of a voltage signal. When the DC voltage component is removed, the characteristics of the fluctuating component of the signal are more readily observed, as shown in Figure 2.7b. The statistics of the fluctuating component of the signal may contain important information for a particular application.

(a) Signal prior to subtracting the DC voltage offset

(b) Fluctuating component of signal

**FIGURE 2.7**  Effect of subtracting the DC offset from a dynamic signal.

**EXAMPLE 2.1**

Suppose the current passing through a resistor can be described by

$$I(t) = 10 \sin t$$

where $I$ represents the time-dependent current in amperes and $t$ represents time in seconds. Establish the mean and rms values of current over a time from 0 to $t_f$, with $t_f = \pi$ and then with $t_f = 2\pi$. How do the results relate to the power dissipated in the resistor?

**KNOWN** $I(t) = 10 \sin t$ A

**FIND** $\bar{I}$ and $I_{\text{rms}}$ with $t_f = \pi$ and $2\pi$ s

**SOLUTION**
The average value for a time from 0 to $t_f$ is found from Equation 2.1 as

$$\bar{I} = \frac{\int_0^{t_f} I(t)dt}{\int_0^{t_f} dt} = \frac{\int_0^{t_f} 10 \sin t \, dt}{t_f}$$

Evaluation of this integral yields

$$\bar{I} = \frac{1}{t_f}[-10 \cos t]_0^{t_f}$$

With $t_f = \pi$, the average value, $\bar{I}$, is $20/\pi$. For $t_f = 2\pi$, the valuation of the integral yields an average value of zero.

The rms value for the time period 0 to $t_f$ is given by the application of Equation 2.4, which yields

$$I_{\text{rms}} = \sqrt{\frac{1}{t_f} \int_0^{t_f} I(t)^2 dt} = \sqrt{\frac{1}{t_f} \int_0^{t_f} (10 \sin t)^2 dt}$$

This integral is evaluated as

$$I_{\text{rms}} = \sqrt{\frac{100}{t_f} \left( -\frac{1}{2} \cos t \sin t + \frac{t}{2} \right) \Big|_0^{t_f}}$$

For $t_f = \pi$, the rms value is $\sqrt{50}$. Evaluation of the integral for the rms value with $t_f = 2\pi$ also yields $\sqrt{50}$.

**COMMENT** Although the average value over the period $2\pi$ is zero, the power dissipated in the resistor must be the same over both the positive and negative half-cycles of the sine function. Thus, the rms value of current is the same for the time period of $\pi$ and $2\pi$ and is indicative of the rate at which energy is dissipated.

## 2.4 Signal Amplitude and Frequency

A key factor in measurement system behavior is the nature of the input signal to the system. A means is needed to classify waveforms for both the input signal and the resulting output signal relative to their magnitude and frequency. It would be very helpful if the behavior of measurement

**FIGURE 2.8** Separation of white light into its color spectrum. Color corresponds to a particular frequency or wavelength; light intensity corresponds to varying amplitudes.

systems could be defined in terms of their response to a limited number and type of input signals. This is, in fact, exactly the case. The method of expressing a signal as a series of sines and cosines is called *Fourier analysis.*

Nature provides some experiences that support our contention that complex signals can be represented by the addition of a number of simpler periodic functions. For example, combining a number of different pure tones can generate rich musical sound. And an excellent physical analogy for Fourier analysis is provided by the separation of white light through a prism. Figure 2.8 illustrates the transformation of a complex waveform, represented by white light, into its simpler periodic components, represented by the colors in the spectrum. In this example, the colors in the spectrum are represented as simple periodic functions that combine to form white light. Fourier analysis is roughly the mathematical equivalent of a prism and yields a representation of a complex signal in terms of simple periodic functions.

The representation of complex and nondeterministic waveforms by simple periodic functions allows measurement system response to be reasonably well defined by examining the output resulting from a few specific input waveforms, one of which is a simple periodic. As represented in Table 2.1, a simple periodic waveform has a single, well-defined amplitude and a single frequency. Before a generalized response of measurement systems can be determined, an understanding of the method of representing complex signals in terms of simpler functions is necessary.

## Periodic Signals

The fundamental concepts of frequency and amplitude can be understood through the observation and analysis of periodic motions. Although sines and cosines are, by definition, geometric quantities related to the lengths of the sides of a right triangle, for our purposes sines and cosines are best thought of as mathematical functions that describe specific physical behaviors of systems. These behaviors are described by differential equations that have sines and cosines as their solutions. As an example, consider a mechanical vibration of a mass attached to a linear spring, as shown in Figure 2.9. For a linear spring, the spring force $F$ and displacement $y$ are related by $F = ky$, where $k$ is the constant of proportionality, called the spring constant. Application of Newton's second law to this system yields a governing equation for the displacement $y$ as a function of time $t$ as

$$m\frac{d^2y}{dt^2} + ky = 0 \tag{2.7}$$

**FIGURE 2.9**   Spring-mass system.

This linear, second-order differential equation with constant coefficients describes the motion of the idealized spring mass system when there is no external force applied. The general form of the solution to this equation is

$$y = A \cos \omega t + B \sin \omega t \tag{2.8}$$

where the frequency of oscillation is $\omega = \sqrt{k/m}$. Physically we know that if the mass is displaced from the equilibrium point and released, it will oscillate about the equilibrium point. The time required for the mass to finish one complete cycle of the motion is called *period* and is generally represented by the symbol $T$.

Frequency is related to the period and is defined as the number of complete cycles of the motion per unit time. This *frequency*, $f$, is measured in cycles per second (Hz; 1 cycle/s = 1 Hz). The term $\omega$ is also a frequency, but instead of having units of cycles per second it has units of radians per second. This frequency, $\omega$, is called the *circular frequency* because it relates directly to cycles on the unit circle, as illustrated in Figure 2.10.

The relationship among $\omega, f$, and the period, $T$, is

$$T = \frac{2\pi}{\omega} = \frac{1}{f} \tag{2.9}$$

In Equation 2.8, the sine and cosine terms can be combined if a phase angle is introduced such that

$$y = C \cos(\omega t - \phi) \tag{2.10a}$$

or

$$y = C \sin(\omega t + \phi^*) \tag{2.10b}$$

The values of $C$, $\phi$, and $\phi^*$ are found from the following trigonometric identities:

$$
\begin{aligned}
A \cos \omega t + B \sin \omega t &= \sqrt{A^2 + B^2} \cos(\omega t - \phi) \\
A \cos \omega t + B \sin \omega t &= \sqrt{A^2 + B^2} \sin(\omega t + \phi^*) \\
C &= \sqrt{A^2 + B^2} \\
\phi = \tan^{-1}\frac{B}{A} \quad &\phi^* = \tan^{-1}\frac{A}{B} \quad \phi^* = \frac{\pi}{2} - \phi
\end{aligned} \tag{2.11}
$$

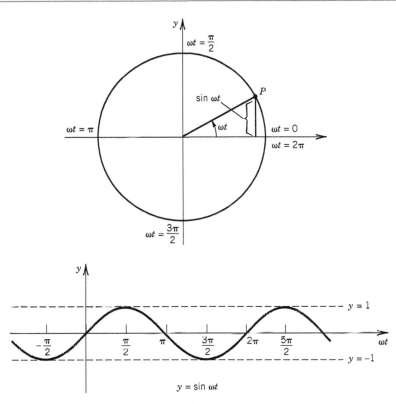

**FIGURE 2.10**  Relationship between cycles on the unit circle and circular frequency.

The size of the maximum and minimum displacements from the equilibrium position, or the value of magnitude $C$, is the *amplitude* of the oscillation. The concepts of amplitude and frequency are essential for the description of time-dependent signals.

## Frequency Analysis

Many signals that result from the measurement of dynamic variables are nondeterministic in nature and have a continuously varying rate of change. These signals, having *complex waveforms*, present difficulties in the selection of a measurement system and in the interpretation of an output signal. However, it is possible to separate a complex signal, or any signal for that matter, into a number of sine and cosine functions. *In other words, any complex signal can be thought of as consisting of sines and cosines of differing periods and amplitudes, which are added together in a trigonometric series.* When a signal is broken down into a series of periodic functions, the importance of each frequency can be easily determined. This information about frequency content allows proper choice of a measurement system and precise interpretation of output signals.

In theory, Fourier analysis allows essentially all mathematical functions of practical interest to be represented by an infinite series of sines and cosines.[3]

---

[3] A periodic function may be represented as a Fourier series if the function is piecewise continuous over the limits of integration and the function has a left- and right-hand derivative at each point in the interval. The sum of the resulting series is equal to the function at each point in the interval except points where the function is discontinuous.

The following definitions relate to Fourier analysis:

1. A function $y(t)$ is *a periodic function* if there is some positive number $T$ such that

$$y(t + T) = y(t)$$

The *period* of $y(t)$ is $T$. If both $y_1(t)$ and $y_2(t)$ have period $T$, then

$$ay_1(t) + by_2(t)$$

also has a period of $T$ ($a$ and $b$ are constants).

2. A *trigonometric series* is given by

$$A_0 + A_1 \cos t + B_1 \sin t + A_2 \cos 2t + B_2 \sin 2t + \dots$$

where $A_n$ and $B_n$ are the coefficients of the series.

## EXAMPLE 2.2

As a physical (instead of mathematical) example of frequency content of a signal, consider stringed musical instruments, such as guitars and violins. When a string is caused to vibrate by strumming, plucking, or bowing, the sound is a result of the motion of the string and the resonance of the instrument. (The concept of resonance is explored in Chapter 3.) The musical pitch for such an instrument is the lowest frequency of the string vibrations. Our ability to recognize differences in musical instruments is primarily a result of the higher frequencies present in the sound, which are usually integer multiples of the fundamental frequency. These higher frequencies are called *harmonics*.

The motion of a vibrating string is best thought of as composed of several basic motions that together create a musical tone. Figure 2.11 illustrates the vibration modes associated with a string plucked at its center. The string vibrates with a fundamental frequency and odd-numbered harmonics, each having a specific phase relationship with the fundamental. The relative strength of each harmonic is graphically illustrated through its amplitude in Figure 2.11. For a string plucked one-fifth of the distance from a fixed end, Figure 2.12 shows the resulting frequencies. Notice that the fifth harmonic is missing from the resulting sound.

Musical sound from a vibrating string is analogous to a measurement system input or output signal, which may contain few or many frequency components. Fourier analysis is a method used to express such a time-based signal in terms of its amplitude, signal power or phase versus its frequency content. A resulting plot of this type of information is referred to broadly as a *frequency spectrum*, but when the ordinate value is amplitude, it is specifically called an *amplitude spectrum*, such as the plot in Figure 2.12. So while a time-based signal superposes this information as a function of time, a frequency spectrum delineates the information as a function of frequency. Fourier analysis provides an insightful and practical means for decomposing signals into a combination of simple waveforms. Next, we explore the frequency and amplitude analysis of signals.

The companion software's Labview[®][4] program *Sound from Microphone* uses your computer's microphone and sound board to sample ambient sounds and to decompose them into their amplitude–frequency components—try humming a tune.

---

[4]LabView[®] is a registered trademark of National Instruments.

**FIGURE 2.11**  Modes of vibration for a string plucked at its center.

**FIGURE 2.12**    Amplitude versus frequency plot (spectrum) for a string plucked one-fifth of the distance from a fixed end.

## Fourier Series and Coefficients

A Fourier series is an expansion of a periodic function in terms of an infinite sum of sines and cosines that are integral multiples of the fundamental frequency. A periodic function $y(t)$ with a period $T$ is to be represented by a trigonometric series, such that for any $t$,

$$y(t) = A_0 + \sum_{n=1}^{\infty} A_n \cos\left(\frac{2\pi nt}{T}\right) + B_n \sin\left(\frac{2\pi nt}{T}\right) \tag{2.12}$$

With the signal $y(t)$ known, the coefficients $A_n$ and $B_n$ are to be determined.

The trigonometric series corresponding to $y(t)$ is called the *Fourier series for $y(t)$*, and the coefficients $A_n$ and $B_n$ are called the *Fourier coefficients* of $y(t)$. When $n = 0$, the coefficient represents the average or mean value of the series. When $n = 1$, the corresponding terms in the Fourier series are called *fundamental* and have the lowest frequency in the series, termed the *fundamental frequency*. Frequencies corresponding to $n = 2, 3, 4, \ldots$ are known as *harmonics*, with, for example, $n = 2$ representing the second harmonic.

## Fourier Coefficients

The coefficients of a trigonometric series representing a function having an arbitrary period $T$ are given by

$$A_0 = \frac{1}{T} \int_{-T/2}^{T/2} y(t)\,dt$$

$$A_n = \frac{2}{T} \int_{-T/2}^{T/2} y(t) \cos\frac{2\pi nt}{T}\,dt \tag{2.13}$$

$$B_n = \frac{2}{T} \int_{-T/2}^{T/2} y(t) \sin\frac{2\pi nt}{T}\,dt$$

where $n = 1, 2, 3, \ldots$, and $T = 2\pi/\omega$ is the period of $y(t)$. The trigonometric series that results from these coefficients is a Fourier series and may be written as

$$y(t) = A_o + \sum_{n=1}^{\infty} A_n \cos\left(\frac{2\pi nt}{T}\right) + B_n \sin\left(\frac{2\pi nt}{T}\right) \tag{2.14}$$

A series of sines and cosines may be written as a series of either sines or cosines through the introduction of a phase angle, so that the Fourier series in Equation 2.14 may be written as

$$y(t) = A_0 + \sum_{n=1}^{\infty} C_n \cos\left(\frac{2\pi nt}{T} - \phi_n\right) \tag{2.15}$$

or

$$y(t) = A_0 + \sum_{n=1}^{\infty} C_n \sin\left(\frac{2\pi nt}{T} + \phi_n^*\right) \tag{2.16}$$

where $C_n$ is an amplitude and $\phi_n$ is a phase angle found by

$$C_n = \sqrt{A_n^2 + B_n^2}$$
$$\tan \phi_n = \frac{B_n}{A_n} \quad \text{and} \quad \tan \phi_n^* = \frac{A_n}{B_n} \tag{2.17}$$

## Special Case: Functions with $T = 2\pi$

A classical mathematical result occurs for a function $y(t)$ with a period $2\pi$. In this case, the coefficients of the trigonometric series representing $y(t)$ are given by the Euler formulas:

$$y(t) = A_0 + \sum_{n=1}^{\infty} A_n \cos nt + B_n \sin nt$$

$$A_0 = \frac{1}{2\pi} \int_{-\pi}^{\pi} y(t)dt$$

$$A_n = \frac{1}{\pi} \int_{-\pi}^{\pi} y(t) \cos nt \, dt \tag{2.18}$$

$$B_n = \frac{1}{\pi} \int_{-\pi}^{\pi} y(t) \sin nt \, dt$$

## Even and Odd Functions

A function $g(t)$ is even if it is symmetric about the vertical axis, which may be stated, for all $t$, as

$$g(-t) = g(t)$$

A function $h(t)$ is odd if, for all $t$,
$$h(-t) = -h(t)$$

For example, $\cos nt$ is even, whereas $\sin nt$ is odd. A particular function or waveform may be even, odd, or neither even nor odd.

### Fourier Cosine Series

If $y(t)$ is even, its Fourier series will contain only cosine terms:

$$y(t) = \sum_{n=0}^{\infty} A_n \cos \frac{2\pi nt}{T} = \sum_{n=0}^{\infty} A_n \cos n\omega t \tag{2.19}$$

### Fourier Sine Series

If $y(t)$ is odd, its Fourier series will contain only sine terms:

$$y(t) = \sum_{n=1}^{\infty} B_n \sin \frac{2\pi nt}{T} = \sum_{n=1}^{\infty} B_n \sin n\omega t \qquad (2.20)$$

*Note: Functions that are neither even nor odd result in Fourier series that contain both sine and cosine terms.*

### EXAMPLE 2.3

Determine the Fourier series that represents the function shown in Figure 2.13.

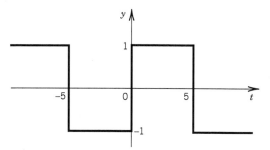

**FIGURE 2.13**   Function represented by a Fourier series in Example 2.3.

**KNOWN**   $T = 10$ (i.e., $-5$ to $+5$)

$\qquad\qquad A_0 = 0$

**FIND**   The Fourier coefficients $A_1, A_2, \ldots$ and $B_1, B_2, \ldots$

### SOLUTION

Since the function shown in Figure 2.13 is odd, the Fourier series will contain only sine terms (see Eq. 2.20):

$$y(t) = \sum_{n=1}^{\infty} B_n \sin \frac{2\pi nt}{T}$$

where

$$B_n = \frac{2}{10} \left[ \int_{-5}^{0} (-1)\sin\left(\frac{2\pi nt}{10}\right) dt + \int_{0}^{5} (1)\sin\left(\frac{2\pi nt}{10}\right) dt \right]$$

$$B_n = \frac{2}{10} \left\{ \left[ \frac{10}{2n\pi} \cos\left(\frac{2\pi nt}{10}\right) \right]_{-5}^{0} + \left[ \frac{-10}{2n\pi} \cos\left(\frac{2\pi nt}{10}\right) \right]_{0}^{5} \right\}$$

$$B_n = \frac{2}{10} \left\{ \frac{10}{2n\pi} [1 - \cos(-n\pi) - \cos(n\pi) + 1] \right\}$$

The reader can verify that all $A_n = 0$ for this function. For even values of $n$, $B_n$ is identically 0, and for odd values of $n$,

$$B_n = \frac{4}{n\pi}$$

The resulting Fourier series is then

$$y(t) = \frac{4}{\pi} \sin \frac{2\pi}{10}t + \frac{4}{3\pi} \sin \frac{6\pi}{10}t + \frac{4}{5\pi} \sin \frac{10\pi}{10}t + \cdots$$

Note that the fundamental frequency is $\omega = \frac{2\pi}{10}\,\mathrm{rad/s}$ and the subsequent terms are the odd-numbered harmonics of $\omega$.

**COMMENT** Consider the function given by

$$y(t) = 1 \quad 0 < t < 5$$

We may represent $y(t)$ by a Fourier series if we extend the function beyond the specified range (0 to 5) either as an even periodic extension or as an odd periodic extension. (Because we are interested only in the Fourier series representing the function over the range $0 < t < 5$, we can impose any behavior outside the domain that helps to generate the Fourier coefficients!) Let's choose an odd periodic extension of $y(t)$; the resulting function remains identical to the function shown in Figure 2.13.

---

**EXAMPLE 2.4**

Find the Fourier coefficients of the periodic function

$$y(t) = -5 \text{ when } -\pi < t < 0$$

$$y(t) = +5 \text{ when } 0 < t < \pi$$

and $y(t + 2\pi) = y(t)$. Plot the resulting first four partial sums for the Fourier series.

**KNOWN** Function $y(t)$ over the range $-\pi$ to $\pi$.

**FIND** Coefficients $A_n$ and $B_n$.

**SOLUTION**
The function as stated is periodic, having a period of $T = 2\pi$ (i.e., $\omega = 1$ rad/s), and is identical in form to the odd function examined in Example 2.3. Since this function is also odd, the Fourier series contains only sine terms, and

$$B_n = \frac{1}{\pi}\int_{-\pi}^{\pi} y(t)\sin n\omega t\, dt = \frac{1}{\pi}\left[\int_{-\pi}^{0}(-5)\sin nt\, dt + \int_{0}^{\pi}(+5)\sin nt\, dt\right]$$

which yields upon integration

$$\frac{1}{\pi}\left\{\left[(5)\frac{\cos nt}{n}\right]_{-\pi}^{0} - \left[(5)\frac{\cos nt}{n}\right]_{0}^{\pi}\right\}$$

Thus,

$$B_n = \frac{10}{n\pi}(1 - \cos n\pi)$$

which is zero for even $n$, and $20/n\pi$ for odd values of $n$. The Fourier series can then be written[5] as

$$y(t) = \frac{20}{\pi}\left(\sin t + \frac{1}{3}\sin 3t + \frac{1}{5}\sin 5t + \cdots\right)$$

---

[5]If we assume that the sum of this series most accurately represents the function $y$ at $t = \pi/2$, then

$$y\left(\frac{\pi}{2}\right) = 5 = \frac{20}{\pi}\left(1 - \frac{1}{3} + \frac{1}{5} - + \cdots\right)$$

or

$$\frac{\pi}{4} = 1 - \frac{1}{3} + \frac{1}{5} - \frac{1}{7} + - \cdots$$

This series approximation of $\pi$ was first obtained by Gottfried Wilhelm Leibniz (1646–1716) in 1673 from geometrical reasoning.

Figure 2.14 shows the first four partial sums of this function, as they compare with the function they represent. Note that at the points of discontinuity of the function, the Fourier series takes on the arithmetic mean of the two function values.

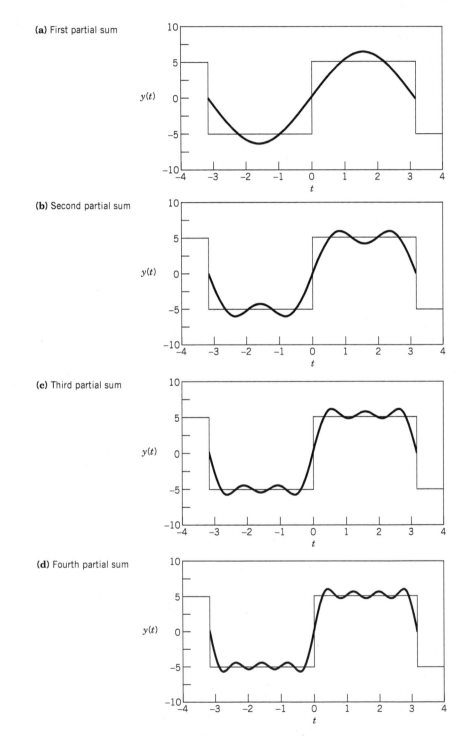

**(a)** First partial sum

**(b)** Second partial sum

**(c)** Third partial sum

**(d)** Fourth partial sum

**FIGURE 2.14**    First four partial sums of the Fourier series $(20/\pi)(\sin t + 1/3 \sin 3t + 1/5 \sin 5t + \cdots)$ in comparison with the exact waveform.

**COMMENT** As the number of terms in the partial sum increases, the Fourier series approximation of the square wave function becomes even better. For each additional term in the series, the number of "humps" in the approximate function in each half-cycle corresponds to the number of terms included in the partial sum.

The Matlab[®6] program file *FourSeries* with the companion software illustrates the behavior of the partial sums for several waveforms. The LabView program *Waveform-Generation* reconstructs signals from several selected trigonometric series.

---

### EXAMPLE 2.5

As an example of interpreting the frequency content of a given signal, consider the output voltage from a rectifier. A rectifier functions to "flip" the negative half of an alternating current (AC) into the positive half plane, resulting in a signal that appears as shown in Figure 2.15. For the AC signal, the voltage is given by

$$E(t) = |120 \sin 120\pi t|$$

where $E(t)$ is in volts and $t$ is in seconds. The period of the signal is 1/60 s, and the frequency is 60 Hz.

**KNOWN** The rectified signal can be expressed as

$$E(t) = |120 \sin 120\pi t|$$

**FIND** The frequency content of this signal is determined from a Fourier series analysis.

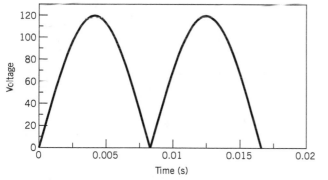

**FIGURE 2.15** Rectified sine wave.

### SOLUTION
The frequency content of this signal can be determined by expanding the function in a Fourier series. We can determine the coefficients using the Euler formulas, keeping in mind that the rectified sine wave is an even function. The coefficient $A_0$ is determined from

$$A_0 = 2\left[\frac{1}{T}\int_0^{T/2} y(t)dt\right] = \frac{2}{1/60}\int_0^{1/120} 120\sin 120\pi t\, dt$$

with the result that

$$A_0 = \frac{2\times 60\times 2}{\pi} = 76.4$$

---

[6]Matlab® is a registered trademark of Mathworks, Inc.

The remaining coefficients in the Fourier series may be expressed as

$$A_n = \frac{4}{T} \int_0^{T/2} y(t) \cos \frac{2n\pi t}{T} dt$$

$$= \frac{4}{1/60} \int_0^{1/120} 120 \sin 120\pi t \cos n\pi t\, dt$$

For values of $n$ that are odd, the coefficient $A_n$ is identically zero. For values of $n$ that are even, the result is

$$A_n = \frac{120}{\pi} \left( \frac{-2}{n-1} + \frac{2}{n+1} \right)$$

The Fourier series for the function $|120 \sin 120\pi t|$ is

$$76.4 - 50.93 \cos 240\pi t - 10.10 \cos 480\pi t - 4.37 \cos 720\pi t \dots$$

Figure 2.16 shows the amplitude versus frequency content of the rectified signal, based on a Fourier series expansion.

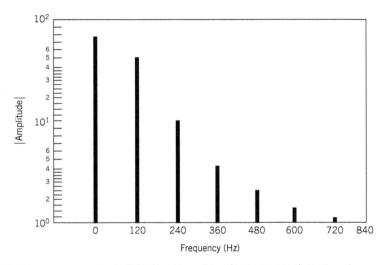

**FIGURE 2.16**    Frequency content of the function $y(t) = |120 \sin 20\pi t|$ displayed as an amplitude spectrum.

**COMMENT** The frequency content of a signal is determined by examining the amplitude of the various frequency components that are present in the signal. For a periodic mathematical function, expanding the function in a Fourier series and plotting the amplitudes of the contributing sine and cosine terms can illustrate these frequency contributions.

The LabView program *WaveformGeneration* follows Examples 2.4 and 2.5 and provides the amplitude spectra for a number of signals. Several Matlab programs are provided (e.g., *FourSeries*, *FunSpect*, *DataSpect*) to explore the concept of superposition of simple periodic signals to create a more complex signal. *DataSpect* allows you to import your own files (or use the provided files) to explore their frequency and amplitude content.

# 2.5 Fourier Transform and the Frequency Spectrum

The previous discussion of Fourier analysis demonstrates that an arbitrary, but known, periodic function can be expressed as a series of sines and cosines known as a Fourier series. The coefficients of the Fourier series specify the amplitudes of the sines and cosines, each having a specific frequency. Unfortunately, in most practical measurement applications, the input signal may not be known in functional form. Furthermore, most measured signals are not exact periodic signals. We now describe a general technique for the decomposition of a measured dynamic signal in terms of its amplitude and frequency content. The technique applies to aperiodic signals. We move from the distinction between periodic signals and aperiodic signals by considering the period to become very large. The method can be applied to any measured signal of finite duration.

The fundamental mathematics required to develop a method for analyzing measured data requires taking the limit of a Fourier series as the period of the signal approaches infinity. In the limit as $T$ approaches infinity, the Fourier series becomes an integral. The spacing between frequency components becomes infinitesimal. This means that the coefficients $A_n$ and $B_n$ become continuous functions of frequency and can be expressed as $A(\omega)$ and $B(\omega)$ where

$$A(\omega) = \int_{-\infty}^{\infty} y(t) \cos \omega t \, dt$$
$$B(\omega) = \int_{-\infty}^{\infty} y(t) \sin \omega t \, dt \tag{2.21}$$

The Fourier coefficients $A(\omega)$ and $B(\omega)$ are known as the components of the *Fourier transform* of $y(t)$.

To further develop the Fourier transform, consider the complex number defined as

$$Y(\omega) \equiv A(\omega) - iB(\omega) \tag{2.22}$$

where $i = \sqrt{-1}$. Then from Equation 2.21 it follows directly that

$$Y(\omega) \equiv \int_{-\infty}^{\infty} y(t)(\cos \omega t - i \sin \omega t) dt \tag{2.23}$$

Introducing the identity

$$e^{-i\theta} = \cos \theta - i \sin \theta$$

leads to

$$Y(\omega) \equiv \int_{-\infty}^{\infty} y(t) e^{i\omega t} dt \tag{2.24}$$

Alternately, recalling from Equation 2.9 that the cyclical frequency $f$, in hertz, is related to the circular frequency and its period by

$$f = \frac{\omega}{2\pi} = \frac{1}{T}$$

Equation 2.24 is rewritten in terms of $f$, with $\omega = 2\pi f$, as

$$Y(f) \equiv \int_{-\infty}^{\infty} y(t) e^{-i2\pi f t} dt \tag{2.25}$$

Equation 2.24 or 2.25 provides the two-sided *Fourier transform* of $y(t)$. If $y(t)$ is known, then its Fourier transform will provide the amplitude–frequency properties of the signal, $y(t)$, which otherwise are not readily apparent in its time-based form. We can think of the Fourier transform as a decomposition of $y(t)$ into amplitude versus frequency information. This property is analogous to the optical properties displayed by the prism in Figure 2.8.

If $Y(f)$ is known or measured, we can recover the signal $y(t)$ from

$$y(t) = \int_{-\infty}^{\infty} Y(f)e^{i2\pi ft}df \qquad (2.26)$$

Here Equation 2.26 describes the *inverse Fourier transform* of $Y(f)$. It suggests that given the amplitude–frequency properties of a signal we can reconstruct the original signal $y(t)$.

The Fourier transform is a complex number having a magnitude and a phase,

$$Y(f) = |Y(f)|e^{i\phi(f)} = A(f) - iB(f) \qquad (2.27)$$

The magnitude of $Y(f)$, also called the modulus, is given by

$$|Y(f)| = \sqrt{\text{Re}[Y(f)]^2 + \text{Im}[Y(f)]^2} \qquad (2.28)$$

and the phase by

$$\phi(f) = \tan^{-1}\frac{\text{Im}[Y(f)]}{\text{Re}[Y(f)]} \qquad (2.29)$$

As noted earlier, the Fourier coefficients are related to cosine and sine terms. Each of these terms assigns amplitude and phase information to its unique frequency. The general form of this information is called a frequency spectrum. When amplitude is expressed in terms of frequency, the frequency spectrum is referred to as the amplitude spectrum. This is given by

$$C(f) = \sqrt{A(f)^2 + B(f)^2} \qquad (2.30)$$

Note that the value of $C(f)$ will be positive regardless of the values of $A(f)$ or $B(f)$. When phase angle is expressed in terms of frequency, the frequency spectrum is referred to as the phase spectrum. This is given by

$$\phi(f) = \tan^{-1}\frac{B(f)}{A(f)} \qquad (2.31)$$

Thus, the introduction of the Fourier transform provides a method to decompose a measured signal $y(t)$ into its amplitude, phase, and frequency components. Later we will see how important this method is, particularly when digital sampling is used to measure and interpret an analog signal.

Another variation of the frequency spectrum is the power spectrum, in which the ordinate is $C^2(f)/2$. Further details concerning the properties of the Fourier transform and spectrum functions can be found in Bracewell [1] and Champeney [2]. An excellent historical account and discussion of the wide-ranging applications are found in Bracewell [3].

## Discrete Fourier Transform

As a practical matter, it is likely that if $y(t)$ is measured and recorded, then it will be stored in the form of a discrete time series or as a digital array. A computer-based data-acquisition system is the most common method for measuring and recording data in this form. These data are acquired over a finite period of time rather than the mathematically convenient infinite period of time. A discrete data set containing $N$ values representing a measured time interval from 0 to $t_f$ will accurately represent the signal provided that the measuring period has been properly chosen and is sufficiently long. We deal with the details for such period selection in a discussion on sampling concepts in Chapter 7. The preceding analysis is now extended to accommodate a discrete series.

Consider the time-dependent portion of the signal $y(t)$, which is measured $N$ times at equally spaced time intervals $\delta t$. In this case, the continuous signal $y(t)$ is replaced by the discrete time signal given by $y(r\delta t)$ for $r = 0, 1, ..., (N - 1)$. In effect, the relationship between $y(t)$ and $\{y(r\delta t)\}$ is described by a set of impulses of an amplitude determined by the value of $y(t)$ at each time step $r\delta t$. This transformation from a continuous to discrete time signal is described by

$$\{y(r\delta t)\} = y(t)\delta(t - r\delta t) \quad r = 0, 1, 2, ..., N - 1 \tag{2.32}$$

where $\delta(t - r\delta t)$ is the delayed unit impulse function and $\{y(r\delta t)\}$ refers to the discrete data set given by $y(r\delta t)$ for $r = 0, 1, 2, ..., N - 1$.

An approximation to the Fourier transform integral of Equation 2.25 for use on a discrete data set is the discrete Fourier transform (DFT). The DFT is given by

$$Y(f_k) = \frac{2}{N} \sum_{r=0}^{N-1} y(r\delta t)e^{-i2\pi rk/N} \quad \text{for } k = 1, 2, ..., \left(\frac{N}{2} - 1\right)$$

$$Y(f_0) = \frac{1}{N} \sum_{r=0}^{N-1} y(r\delta t) \quad \text{for } k = 0 \tag{2.33}$$

$$f_k = k\delta f \quad \text{for } k = 0, 1, 2, ..., \left(\frac{N}{2} - 1\right)$$

$$\delta f = 1/N\delta t$$

Here, $\delta f$ is the frequency resolution of the DFT such that each value of $Y(f_k)$ is spaced at frequency increments of $\delta f$. In developing Equation 2.33 from Equation 2.25, $t$ was replaced by $r\delta t$ and $f$ replaced by $k/N\delta t$. The factor $2/N$ scales the transform when it is obtained from a data set of finite length (we scale with $1/N$ for $k = 0$ only). The $k = 0$ or zero-frequency term $Y(f_0)$ equals the mean value of $\{y(r\delta t)\}$; you can compare it with Equation 2.5.

The DFT as expressed by Equation 2.33 yields $N/2$ discrete values of the Fourier transform of $\{y(r\delta t)\}$. This is the so-called one-sided or half-transform, because it assumes that the data set is one-sided, extending from 0 to $t_f$, and it returns only positive-valued frequencies. More discussion is provided in Bracewell [1] and Champeney [2].

Equation 2.33 performs the numerical integration required by the Fourier integral. Equations 2.30, 2.31, and 2.33 demonstrate that the application of the DFT on the discrete series of data, $y(r\delta t)$, permits the decomposition of the discrete data in terms of frequency and amplitude content. Hence, by using this method, a measured discrete signal of unknown functional form can be reconstructed as a Fourier series through Fourier transform techniques.

Software for computing the Fourier transform of a discrete signal is included in the companion software and in packages such as Matlab® and Excel®. The time required to compute directly the DFT algorithm described in this section increases at a rate that is proportional to $N^2$. This makes it inefficient for use with data sets of large $N$. A fast algorithm for computing the DFT, known as the *fast Fourier transform* (FFT), was developed by Cooley and Tukey [4]. This method is widely available and is the basis for most Fourier analysis software packages, such as those used in Matlab®. The FFT algorithm is discussed in most advanced texts on signal analysis [5,6]. The accuracy of discrete Fourier analysis depends on the frequency content[7] of $y(t)$ and on the Fourier transform frequency resolution. An extensive discussion of these interrelated parameters is given in Chapter 7.

---

[7]The value of $1/\delta t$ must be greater than twice the highest frequency contained in $y(t)$.

**EXAMPLE 2.6**

Convert the continuous signal described by $y(t) = 10 \sin 2\pi t$ [V] into a discrete set of eight numbers using a time increment of 0.125 s.

**KNOWN** The signal has the form $y(t) = C_1 \sin 2\pi f_1 t$

$$f_1 = \omega_1/2\pi = 1 \text{ Hz}$$
$$C(f_1 = 1 \text{ Hz}) = 10 \text{ V}$$
$$\text{where} \quad \phi(f_1) = 0$$
$$\delta t = 0.125 \text{ s}$$
$$N = 8$$

**FIND** $\{y(r\delta t)\}$.

**SOLUTION**
Measuring $y(t)$ every 0.125 s over 1 s produces the discrete data set $\{y(r\delta t)\}$ given in Table 2.2. Note that the measurement produces four complete periods of the signal and that the signal duration is given by $N\delta t = 1$ s. The signal and its discrete representation as a series of impulses in a time domain are plotted in Figure 2.17. See Example 2.7 for more on how to create this discrete series by using common software.

Table 2.2   **Discrete Data Set for $y(t) = 10 \sin 2\pi t$**

| r | y(rδt) | r | y(rδt) |
|---|--------|---|--------|
| 0 | 0.000  | 4 | 0.000  |
| 1 | 7.071  | 5 | −7.071 |
| 2 | 10.000 | 6 | −10.000 |
| 3 | 7.071  | 7 | −7.071 |

**FIGURE 2.17**   Representation of a simple periodic function as a discrete signal.

## EXAMPLE 2.7

Estimate the amplitude spectrum of the discrete data set in Example 2.6.

**KNOWN** Discrete data set $\{y(r\delta t)\}$ of Table 2.2

$$\delta t = 0.125 \text{ s}$$
$$N = 8$$

**FIND** $C(f)$

## SOLUTION

We can use the companion software or a spreadsheet program to find $C(f)$ (see the *Comment* for details). For $N = 8$, a one-sided Fourier transform algorithm will return $N/2$ values for $C(f)$, with each successive amplitude corresponding to a frequency spaced at intervals of $1/N\delta t = 1$ Hz. The amplitude spectrum is shown in Figure 2.18 and has a spike of 10 V centered at a frequency of 1 Hz.

**FIGURE 2.18** Amplitude as a function of frequency for a discrete representation of $10 \sin 2\pi t$ resulting from a discrete Fourier transform algorithm.

**COMMENT** The discrete series and the amplitude spectrum are conveniently reproduced by using the Matlab program files *FunSpect, Signal,* or *DataSpect*, the LabView program *Waveform-Generation*, or spreadsheet software.

The following discussion of Fourier analysis using a spreadsheet program makes specific references to procedures and commands from Microsoft® Excel[8]; similar functions are available in other engineering analysis software. When a spreadsheet is used as shown, the $N = 8$ data point sequence $\{y(r\delta t)\}$ is created as in column 3. Under Data/Data Analysis, select Fourier Analysis. At the prompt, define the $N$ cells containing $\{y(r\delta t)\}$ and define a cell destination (column 4). The analysis executes the DFT of Equation 2.33 by using the FFT algorithm, and it returns $N$ complex numbers, the Fourier coefficients $Y(f)$ of Equation 2.27, to column 4. The analysis returns the two-sided transform. However, we are interested in only the first $N/2$ Fourier coefficients of column 4 (the second $N/2$ coefficients just mirror the first $N/2$). To find the coefficient magnitude $C(f)$ as given in Equation 2.28, use the Excel function IMABS($= \sqrt{A^2 + B^2}$) on each of the first $N/2$ coefficients, and then scale each coefficient magnitude by dividing by $N/2$ for $k = 1$ to $(N/2 - 1)$ and divide by $N$ for $k = 0$ only as noted in Equation 2.33. The resulting $N/2$ scaled coefficients (column 5) now represent the discrete amplitudes corresponding to $N/2$ discrete frequencies (column 6) extending from $f = 0$ to $(N/2 - 1)/N\delta t$ Hz, with each frequency separated by $1/N\delta t$.

---

[8]Microsoft® and Excel are either registered trademarks or trademarks of Microsoft Corporation in the United States and/or other countries.

|  | | Column | | | |
|---|---|---|---|---|---|
| 1 | 2 | 3 | 4 | 5 | 6 |
| $r$ | $t(s)$ | $y(r\delta t)$ | $Y(f)=A-Bi$ | $C(f)$ | $f(Hz)$ |
| 0 | 0 | 0 | 0 | 0 | 0 |
| 1 | 0.125 | 7.07 | $-40i$ | 10 | 1 |
| 2 | 0.25 | 10 | 0 | 0 | 2 |
| 3 | 0.375 | 7.07 | 0 | 0 | 3 |
| 4 | 0.5 | 0 | 0 | | |
| 5 | 0.625 | $-7.07$ | 0 | | |
| 6 | 0.75 | $-10$ | 0 | | |
| 7 | 0.875 | $-7.07$ | $40i$ | | |

These same operations in Matlab code are as follows:

| | |
|---|---|
| $t = 1/8: 1/8: 1$ | defines time from 0.125 s to 1 s in increments of 0.125 s |
| $y = 10*\sin(2*pi*t)$ | creates the discrete time series with $N = 8$ |
| ycoef $= $ fft$(y)$ | performs the Fourier analysis; returns $N$ coefficients |
| $c = $ coef/4 | divides by $N/2$ to scale and determine the magnitudes |

When signal frequency content is not known before conversion to a discrete signal, it may be necessary to experiment with the parameters of frequency resolution, $N$ and $\delta t$, to obtain an unambiguous representation of the signal. Techniques for this are discussed in Chapter 7.

## Analysis of Signals in Frequency Space

Fourier analysis is a tool for extracting details about the frequencies that are present in a signal. Frequency analysis is routinely used in vibration analysis and isolation, determining the condition of bearings, and a wide variety of acoustic applications. An example from acoustics follows.

### EXAMPLE 2.8

Examine the spectra of representative brass and woodwind instruments to illustrate why the characteristic sound of these instruments is so easily distinguished.

**KNOWN**  Figures 2.19 and 2.20 provide a representative amplitude spectrum for a clarinet, a woodwind instrument having a reed, and for brass instruments.

**DISCUSSION**  Figure 2.19 displays the amplitude spectrum for a clarinet playing a B flat. The base tone, $f_o$, is at 231 Hz, with the harmonics having frequencies of $3f_o$, $5f_o$, and so forth. Contrast this with the amplitude spectrum, shown in Figure 2.20, of the first few seconds of Aaron Copland's "Fanfare for the Common Man," which is played by an ensemble of brass instruments, such as French horns, and which has a fundamental frequency of approximately 700 Hz. This spectrum contains harmonics that are $2f_o$, $3f_o$, $4f_o$, and so forth.

**FIGURE 2.19**  Amplitude spectrum for a clarinet playing the note one whole tone below middle C. (Adapted from Campbell, D. M., Nonlinear Dynamics of Musical Reed and Brass Wind Instruments, *Contemporary Physics*, 40(6): 415–431, November 1999.)

**FIGURE 2.20**  Amplitude spectrum for the first three seconds of Aaron Copland's "Fanfare for the Common Man," played by the brass section of the orchestra.

**COMMENT**  Fourier analysis and the examination of a signal's frequency content have myriads of applications. In this example, we demonstrated through frequency analysis a clear reason why clarinets and trumpets sound quite different. The presence of only odd harmonics, or both even and odd harmonics, is a very recognizable difference, easily perceived by the human ear.

**EXAMPLE 2.9**    **Frequency Spectra Applications**

Frequency spectra (amplitude, phase, and power spectra) can be used in a variety of engineering applications. In the past, engineers employed scheduled periodic maintenance to ensure reliable operation of machinery, including components such as bearings. Now the operating condition or "health" of a bearing is monitored through the measurement and frequency analysis of acoustic or vibration signals. Characteristic frequencies are associated with defects in the bearing race or rollers. Thus, bearings are only replaced when necessary.

Figure 2.21 shows a power spectrum (i.e., $C^2(f)/2$) measured on a race car traveling on an oval track. The range of frequencies from 30 to 37 Hz shows increased power compared to the trend in the data for other frequencies. The standard tire circumference for this class of race car is 88.6 in. Therefore, the speeds at 30 and 37 revolutions per second are 151 and 186 mph, respectively. This anomaly corresponds to the tire rotation frequency and indicates that this tire has a defect, most likely a flat spot in the tire created by skidding during braking.

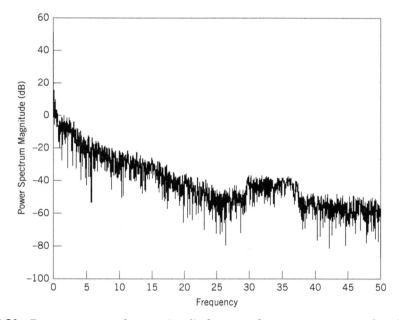

**FIGURE 2.21**    Power spectrum of suspension displacement for a race car on an oval track.

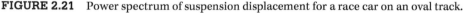

## SUMMARY

This chapter has provided a basis for the description of signals. The capabilities of a measurement system can be properly specified when the nature of the input signal is known. The descriptions of general classes of input signals will be found in Chapter 3 to allow universal descriptions of measurement system dynamic behavior.

Any signal can be represented by a static magnitude and a series of varying frequencies and amplitudes. This being the case, measurement system selection and design must consider the frequency content of the input signals the system is intended to measure. Fourier analysis was introduced to allow a precise definition of the

frequencies and the phase relationships among various frequency components within a particular signal. In Chapter 7, it will be shown as a tool for the accurate interpretation of discrete signals.

Signal characteristics form an important basis for the selection of measurement systems and the interpretation of measurement system output. In Chapter 3, these ideas are combined with the concept of a generalized set of measurement system behaviors. The combination of generalized measurement system behavior and generalized descriptions of input waveforms provides an understanding of a wide range of instruments and measurement systems.

# REFERENCES

1. Bracewell, R. N., *The Fourier Transform and Its Applications*, 3rd ed., rev., McGraw Hill, New York, 1999.

2. Champeney, D. C., *Fourier Transforms and Their Physical Applications*, Academic, London, 1973.

3. Bracewell, R. N., The Fourier Transform, *Scientific American*, 260(6): 86, 1989.

4. Cooley, J. W., and J. W. Tukey, An Algorithm for the Machine Calculation of Complex Fourier Series, *Mathematics of Computation* 19: 207, April 1965 (see also Special Issue on the fast Fourier transform, *IEEE Transactions on Audio and Electroacoustics* AU-2, June 1967).

5. Bendat, J. S., and A. G. Piersol, *Random Data: Analysis and Measurement Procedures*, 3rd ed., Wiley, New York, 2000 (see also Bendat, J. S., and A. G. Piersol, *Engineering Applications of Correlation and Spectral Analysis*, 2nd ed., Wiley, New York, 1993).

6. Cochran, W. T., et al. What Is the Fast Fourier Transform?, *Proceedings of the IEEE* 55(10): 1664, 1967.

## SUGGESTED READING

Halliday, D., R. Resnick, and J. Walker. *Fundamentals of Physics*, 10th ed., Wiley, New York, 2013.

Kreyszig, E., *Advanced Engineering Mathematics*, 10th ed., Wiley, New York, 2011.

## NOMENCLATURE

| | | | | |
|---|---|---|---|---|
| $f$ | frequency, in Hz ($t^{-1}$) | | $N$ | total number of discrete data points; integer |
| $k$ | spring constant ($mt^{-2}$) | | $T$ | period ($t$) |
| $m$ | mass ($m$) | | $U$ | unit step function |
| $t$ | time ($t$) | | $Y$ | Fourier transform of $y$ |
| $y$ | dependent variable | | $\alpha$ | angle (rad) |
| $y_m$ | discrete data points | | $\beta$ | angle (rad) |
| $A$ | amplitude | | $\delta f$ | frequency resolution ($t^{-1}$) |
| $B$ | amplitude | | $\delta t$ | sample time increment ($t$) |
| $C$ | amplitude | | $\phi$ | phase angle (rad) |
| $F$ | force ($mlt^{-2}$) | | $\omega$ | circular frequency in rad/s ($t^{-1}$) |

# 3 Measurement System Behavior

## 3.1 Introduction

This chapter introduces the concept of simulating measurement system behavior through mathematical modeling. Modeling provides details on how a measurement system's design affects its response to an input signal. If the response is insufficient, the output signal will not be the intended copy of the input signal, and a particular system may not be suitable for measuring certain signals. To explore this concept, this chapter discusses system response behavior to various input signals.

Throughout this chapter, we refer either to the response of the measurement system as a whole or to the response of any individual component or instrument that makes up that system. Either is important, and both are interpreted in similar ways. Each individual stage of the measurement system has its own response to a given input. The overall system response is affected by the response of each stage of the complete system.

Upon completion of this chapter, the reader will be able to:

- Relate generalized measurement system models to dynamic response

- Describe and analyze models of zero-, first-, and second-order measurement systems and predict their general behavior

- Calculate static sensitivity, magnitude ratio, and phase shift for a range of systems and input waveforms

- State the importance of phase linearity in reducing signal distortion

- Analyze the response of a measurement system to a complex input waveform

- Determine the response of coupled measurement systems

## 3.2 General Model for a Measurement System

As pointed out in Chapter 2, all input and output signals can be broadly classified as being static, dynamic, or some combination of the two. For a static signal, only the signal magnitude is needed to reconstruct the input signal. Consider measuring the length of a board using a ruler. Once the ruler is positioned, the indication (output) of the magnitude of length is immediately displayed, because the board length (input) does not change over the time required to make the measurement. The board length represents a static input signal that is interpreted through the static output magnitude indicated by the ruler. But consider measuring the vibration displacement of a motor. Vibration signals vary in amplitude and time and at one or more frequencies. These provide a dynamic input signal to the measuring instrument. But a device intended for measuring static information is not very useful in determining dynamic information, so we need different instruments that can follow the input time-dependent signal faithfully.

# Dynamic Measurements

Because dynamic signals vary with time, the measurement system must be able to respond fast enough to keep up with the input signal. The ability of any measurement system to follow dynamic signals is a characteristic of the design of the measuring system components. Consider the time response of a common bulb thermometer for measuring body temperature. The thermometer, initially at approximately room temperature, is placed under the tongue. But even after several seconds, the thermometer does not indicate the expected value of body temperature, and its display continually changes. What has happened? Surely your body temperature is not changing. Experience shows that within a few minutes, the correct body temperature will be indicated, so we wait. Experience also tells us that if we need accurate information faster, we would need a different type of temperature sensor. Electronic medical-grade thermometers report the correct temperature in 10 s. In this example, body temperature itself is constant (static) during the measurement, but the input signal to the thermometer is suddenly changed from room temperature to body temperature—that is, mathematically, the applied waveform is a step change between these two temperatures. This is a dynamic event as the thermometer (the measurement system) sees it! The thermometer must gain energy from its new environment to reach thermal equilibrium, and this takes a finite amount of time.

Now consider the task of assessing the ride quality of an automobile suspension system. A simplified view for one wheel of this system is shown in Figure 3.1. As a tire moves along the road, the road surface provides the time-dependent input signal, $F(t)$, to the suspension at the tire contact point. The motion sensed by the passengers, $y(t)$, is a basis for the ride quality and can be described by a waveform that depends on the input from the road and the behavior of the suspension. An engineer must anticipate the form of the input signals so as to design the suspension to attain a desirable output signal.

Measurement systems play a key role in documenting ride quality. But just as the road and car interact to provide ride quality, the input signal and the measurement system interact in creating the output signal. In many situations, the goal of the measurement is to deduce the input signal based on the output signal. Either way, we see that it is important to understand how a measurement system responds to different forms of input signals.

With the previous discussion in mind, consider that the primary task of a measurement system is to sense an input signal and to translate that information into a readily understandable and quantifiable output form. We can reason that a measurement system performs some mathematical

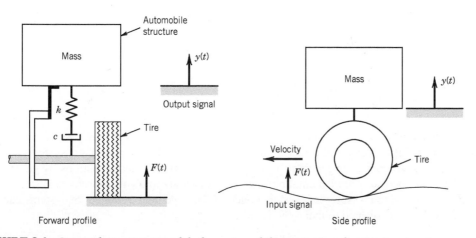

**FIGURE 3.1**  Lumped parameter model of an automobile suspension showing input and output signals.

**FIGURE 3.2**   Measurement system operation on an input signal, $F(t)$, provides the output signal, $y(t)$.

operation on a sensed input. In fact, a general measurement system can be represented by a differential equation that describes the operation that a measurement system performs on the input signal. This concept is illustrated in Figure 3.2. For an input signal, $F(t)$, the system performs some operation that yields the output signal, $y(t)$. Then we use $y(t)$ to infer $F(t)$. We will propose a general mathematical model for a measurement system. Then, by representing a typical input signal as some function that acts as an input to the model, we can study just how the measurement system would behave by solving the model equation. In essence, we perform the analytical equivalent of a system calibration. This information can then be used to determine those input signals for which a particular measurement system is best suited.

## Measurement System Model

In this section, we apply lumped parameter modeling to measurement systems. In lumped parameter modeling, the spatially distributed physical attributes of a system are modeled as discrete elements. The automotive suspension model discussed earlier, and illustrated in Figure 3.1, is a lumped parameter model. In the automotive suspension, the mass, stiffness, and damping of a coil spring are properties spatially distributed along its length, but in a lumped parameter model these are represented by the discrete elements of a mass, spring, and damper. An advantage is that the governing equations of the models reduce from partial (time and space dependent) to ordinary differential equations (time dependent).

Consider the following general model of a measurement system, which consists of an $n$th-order linear ordinary differential equation in terms of a general output signal, represented by variable $y(t)$, and subject to a general input signal, represented by the forcing function, $F(t)$:

$$a_n \frac{d^n y}{dt^n} + a_{n-1} \frac{d^{n-1} y}{dt^{n-1}} + \cdots + a_1 \frac{dy}{dt} + a_0 y = F(t) \qquad (3.1)$$

where

$$F(t) = b_m \frac{d^m x}{dt^m} + b_{m-1} \frac{d^{m-1} x}{dt^{m-1}} + \cdots + b_1 \frac{dx}{dt} + b_0 x \quad m \leq n$$

The coefficients $a_0, a_1, a_2, \ldots, a_n$ and $b_0, b_1, \ldots, b_m$ represent physical system parameters whose properties and values depend on the measurement system itself. Real measurement systems can be modeled this way by considering their governing system equations. These equations are generated by application of pertinent fundamental physical laws of nature to the measurement system. Our discussion is limited to measurement system concepts, but a general treatment of systems can be found in text books dedicated to that topic [1–3].

**EXAMPLE 3.1**

As an illustration, consider the seismic accelerometer depicted in Figure 3.3a. Various configurations of this instrument are used in seismic and vibration engineering to determine the motion of large bodies to which the accelerometer is attached. Basically, as the small accelerometer mass reacts to motion, it places the piezoelectric crystal into compression or tension, causing a surface charge to develop on the crystal. The charge is proportional to the motion. As the large body moves, the mass of the accelerometer will move with an inertial response. The stiffness of the spring, $k$, provides a restoring force to move the accelerometer mass back to equilibrium while internal frictional damping, $c$, opposes any displacement away from equilibrium. A model of this measurement device in terms of ideal lumped elements of stiffness, mass, and damping is given in Figure 3.3b and the corresponding free-body diagram in Figure 3.3c.

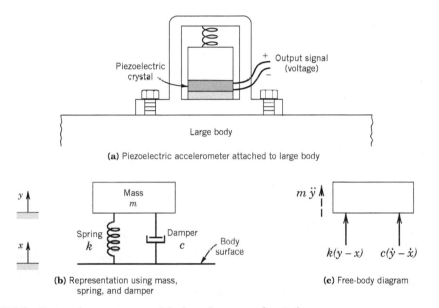

**(a)** Piezoelectric accelerometer attached to large body

**(b)** Representation using mass, spring, and damper

**(c)** Free-body diagram

**FIGURE 3.3**   Lumped parameter model of accelerometer (Ex. 3.1).

Let $y$ denote the position of the small mass within the accelerometer and $x$ denote the displacement of the body. Solving Newton's second law for the free body yields the second-order linear, ordinary differential equation the displacement $y$ is the pertinent output from the accelerometer due to displacement $x$, the equation has been written such that all output terms—that is, all the $y$ terms, are on the left side. All other terms are to be considered as input signals and are placed on the right side. Comparing this to the general form for a second-order equation ($n = 2$; $m = 1$) from Equation 3.1,

$$a_2 \frac{d^2y}{dt^2} + a_1 \frac{dy}{dt} + a_0 y = b_1 \frac{dx}{dt} + b_0 x$$

we can see that $a_2 = m$, $a_1 = b_1 = c$, $a_0 = b_0 = k$, and that the forces developed due to the velocity and displacement of the body become the input signals to the accelerometer. If we could anticipate the waveform of $x$, for example, $x(t) = x_0 \sin \omega t$, we could solve for $y(t)$, which gives the measurement system response.

Fortunately, many measurement systems can be modeled by zero-, first-, or second-order linear, ordinary differential equations. More complex systems can usually be simplified to these lower orders. Our intention here is only to attempt to understand how systems behave and how such response is closely related to the design features of a measurement system. The exact input–output relationship is found from calibration. But modeling guides us in choosing specific instruments and measuring methods by predicting system response to signals and in determining the type, range, and specifics of calibration. Next, we examine several special cases of Equation 3.1 that model the most important concepts of measurement system behavior.

## 3.3 Special Cases of the General System Model

### Zero-Order Systems

The simplest model of a measurement system and one used with static signals is the zero-order system model. This is represented by the zero-order differential equation:

$$a_0 y = F(t)$$

Dividing through by $a_0$ gives

$$y(t) = KF(t) \tag{3.2}$$

where $K = 1/a_0$. $K$ is called the static sensitivity or steady gain of the system. This system property was introduced in Chapter 1 as the relation between the change in output associated with a change in static input. In a zero-order model, the system output is considered to respond to the input signal instantaneously. If an input signal of magnitude $F(t) = A$ were applied, the instrument would indicate $KA$, as modeled by Equation 3.2. The scale of the measuring device would be calibrated to indicate $A$ directly.

For real systems, the zero-order system concept is used to model the non-time-dependent measurement system response to static inputs. In fact, the zero-order concept appropriately models any system during a static calibration. When dynamic input signals are involved, a zero-order model is valid only at static equilibrium. Real measurement systems possess inertial or storage capabilities that require higher-order differential equations to correctly model their time-dependent response to dynamic input signals.

### Determination of $K$

The static sensitivity is found from the static calibration of the measurement system. It is the slope of the calibration curve, $K = dy/dx$, evaluated at a pertinent value of $x$.

---

**EXAMPLE 3.2**

A pencil-type pressure gauge commonly used to measure tire pressure can be modeled at static equilibrium by considering the force balance on the gauge sensor, a piston that slides up and down a cylinder. An exploded view of an actual gauge is shown in Figure 3.4c. In Figure 3.4a, we model the piston motion[1] as being restrained by an internal spring of stiffness, $k$, so that at static equilibrium the absolute pressure force, $F$, bearing on the piston equals the force exerted on the piston by the spring, $F_s$, plus the atmospheric pressure force, $F_{atm}$. In this manner, the transduction of

---

[1]In a common gauge, there may or may not be the mechanical spring shown.

Display scale

$ky = F_s$

Piston of area $A$

$F_{atm}$

$+ y$

$F$

**(b)** Free-body diagram

Spring

Sliding piston

Intake valve

**(a)** Pencil-style pressure gauge

**(c)** Photograph of the components within a pencil-type pressure gauge

Photo courtesy of Richard S. Figliola.

**FIGURE 3.4**    Lumped parameter model of pressure gauge (Ex. 3.2). (Photo courtesy of Richard S. Figliola.)

pressure into mechanical displacement occurs. By keeping the piston of very small mass, we can neglect inertia. Considering the piston free-body at static equilibrium in Figure 3.4b, the static force balance, $\sum F = 0$, gives

$$ky = F - F_{atm}$$

where $y$ is measured relative to some static reference position marked as zero on the output display. Pressure creates a force acting over the piston surface area, $A$. Dividing through by area provides the zero-order response equation between output displacement and input pressure and gives

$$y = (A/k)(p - p_{atm})$$

The term $(p - p_{atm})$ represents the pressure relative to atmospheric pressure. It is the pressure indicated by this gauge. Direct comparison with Equation 3.2 shows that the input pressure magnitude equates to the piston displacement through the static sensitivity, $K = A/k$. Drawing from the concept of Figure 3.2, the system operates on pressure so as to bring about the relative displacement of the piston, the magnitude of which is used to indicate the pressure. The equivalent of spring stiffness and piston area affect the magnitude of this displacement—factors considered in its design. The exact static input–output relationship is found through calibration of the gauge. Because elements such as piston inertia and frictional dissipation were not considered, this model would not be appropriate for studying the dynamic response of the gauge.

## First-Order Systems

Measurement systems that contain storage elements do not respond instantaneously to changes in input. The bulb thermometer discussed in Section 3.2 is a good example. The bulb exchanges energy with its environment until the two are at the same temperature, storing or releasing energy during the exchange. The temperature of the bulb sensor changes with time until this equilibrium is reached, which accounts physically for its lag in response. The rate at which temperature changes with time can be modeled with a first-order derivative and the thermometer behavior modeled as a first-order equation. In general, systems with a storage or dissipative capability but negligible inertial forces may be modeled using a first-order differential equation of the form

$$a_1 \dot{y} + a_0 y = F(t) \tag{3.3}$$

with $\dot{y} = dy/dt$. Dividing through by $a_0$ gives

$$\tau \dot{y} + y = KF(t) \tag{3.4}$$

where $\tau = a_1/a_0$. The parameter $\tau$ is called the *time constant* of the system. Regardless of the physical dimensions of $a_0$ and $a_1$, their ratio will always have the dimensions of time. The time constant provides a measure of the speed of system response and thus is an important specification in measuring dynamic input signals. To explore this concept more fully, consider the response of the general first-order system to the following two forms of an input signal: the step function and the simple periodic function.

## Step Function Input

The step function, $AU(t)$, is defined as

$$AU(t) = 0 \; t \leq 0^-$$
$$AU(t) = A \; t \geq 0^+$$

where $A$ is the amplitude of the step function and $U(t)$ is defined as the unit step function as depicted in Figure 3.5. Physically, this function describes a sudden change in the input signal from a constant value of one magnitude to a constant value of some other magnitude, such as a sudden change in loading, displacement, or any physical variable. When we apply a step function input to a measurement system, we obtain information about how quickly a system will respond to a sudden change in input signal. To illustrate this, let us apply a step function as an input to the general first-order system. Setting $F(t) = AU(t)$ in Equation 3.4 gives

$$\tau \dot{y} + y = KAU(t) = KF(t)$$

with an arbitrary initial condition denoted by, $y(0) = y_0$. Solving for $t \geq 0^+$ yields

$$\underbrace{y(t)}_{\text{Time responnse}} = \underbrace{KA}_{\text{Steady response}} + \underbrace{(y_0 - KA)e^{-t/\tau}}_{\text{Transient response}} \tag{3.5}$$

**FIGURE 3.5**    The unit step function, $U(t)$.

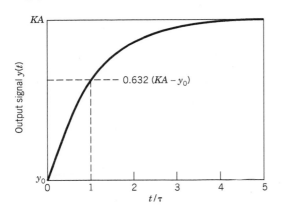

FIGURE 3.6  First-order system time response to a step function input: the time response, $y(t)$.

The solution of the differential equation, $y(t)$ is the time response (or simply the response) of the system. Equation 3.5 describes the behavior of the system to a step change in input value. The value $y(t)$ is in fact the output value indicated by the display stage of the system. It should represent the time variation of the output display of the measurement system if an actual step change were applied to the system. We have simply used mathematics to simulate this response.

Equation 3.5 consists of two parts. The first term is known as the steady response because, as $t \to \infty$, the response of $y(t)$ approaches this steady value. The steady response is that portion of the output signal that remains after the transient response has decayed to zero. The second term on the right side of Equation 3.5 is known as the transient response of $y(t)$ because it is both time-dependent and, as $t \to \infty$, the magnitude of this term eventually reduces to zero.

For illustrative purposes, let $y_0 < A$ so that the time response becomes as shown in Figure 3.6. Over time, the indicated output value rises from its initial value, the value at the instant the change in input is applied, to an eventual constant value, $y_\infty = KA$, the value at steady response. As an example, compare this general time response to the recognized behavior of the bulb thermometer when measuring body temperature as discussed earlier. We see a qualitative similarity. In fact, in using a bulb thermometer to measure body temperature, this is a real step function input to the thermometer, itself a first-order measuring system.

Suppose we rewrite the response Equation 3.5 in the form

$$\Gamma(t) = \frac{y(t) - y_\infty}{y_0 - y_\infty} = e^{-t/\tau} \tag{3.6}$$

The term $\Gamma(t)$ is called the *error fraction* of the output signal. Equation 3.6 is plotted in Figure 3.7, where the time axis is nondimensionalized by the time constant. We see that the error fraction decreases from a value of 1 and approaches a value of 0 with increasing $t/\tau$. At the instant the step change in input is introduced, $\Gamma = 1.0$, so that the indicated output from the measurement system is 100% in error. This means that the system has responded 0% to the input change—that is, it has not changed. From Figure 3.7, it is apparent that as time moves forward, the system does respond, and with decreasing error in its indicated value. The percent response of the system to a step change is $(1 - \Gamma) \times 100$. Then when $t = \tau$, where $\Gamma = 0.368$, the system will have responded to 63.2% of the step change. Furthermore, when $t = 2.3\tau$, the system will have responded ($\Gamma = 0.10$) to 90% of the step change; by $t = 5\tau$, we find the response to be 99.3%. Values for the percent response and the corresponding error as functions of $t/\tau$ are summarized in Table 3.1. The time required for a system to respond to a value that is 90% of the step input, $y_\infty - y_0$, is an instrument specification and is called the *rise time* of the system.

Based on this behavior, we can see that the time constant is in fact a measure of how quickly a first-order measurement system will respond to a change in input value. A smaller time constant

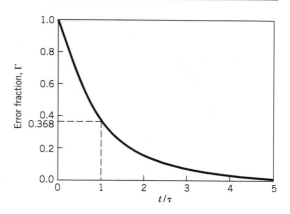

**FIGURE 3.7** First-order system time response to a step function input: the error fraction, $\Gamma$.

**Table 3.1   First-Order System Response and Error Fraction**

| $t/\tau$ | % Response | $\Gamma$ | % Error |
|---|---|---|---|
| 0 | 0.0 | 1.0 | 100.0 |
| 1 | 63.2 | 0.368 | 36.8 |
| 2 | 86.5 | 0.135 | 13.5 |
| 2.3 | 90.0 | 0.100 | 10.0 |
| 3 | 95.0 | 0.050 | 5.0 |
| 5 | 99.3 | 0.007 | 0.7 |
| $\infty$ | 100.0 | 0.0 | 0.0 |

indicates a shorter time between the instant that an input is applied and when the system reaches an essentially steady output. We define the *time constant* as the time required for a first-order system to achieve 63.2% of the step change magnitude, $y_\infty - y_0$. The time constant is a system property.

**Determination of $\tau$** From the development above, the time constant can be experimentally determined by recording the system's response to a step function input of a known magnitude. In practice, it is best to record that response from $t = 0$ until steady response is achieved. The data can then be plotted as error fraction versus time on a semilog plot, shown in Figure 3.8. This type of plot is equivalent to the transformation

$$\ln \Gamma = 2.3 \log \Gamma = -(1/\tau)t \tag{3.7}$$

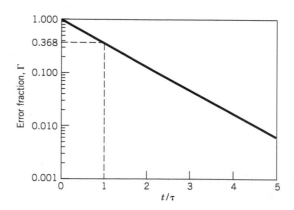

**FIGURE 3.8** The error fraction plotted on semilog coordinates.

which is of the linear form $Y = mX + B$ (where $Y = \ln \Gamma$, $m = -(1/\tau)$, $X = t$, and $B = 0$ here). A linear curve fit through the data will provide a good estimate of the slope, $m$, of the resulting plot. From Equation 3.7, we see that $m = -1/\tau$, which yields the estimate for $\tau$.

This method offers advantages over attempting to compute $\tau$ directly from the time required to achieve 63.2% of the step-change magnitude. First, real systems may deviate somewhat from perfect first-order behavior. On a semilog plot (Figure 3.8), such deviations are readily apparent as trends away from a straight line. Modest deviations do not pose a problem. But strong deviations indicate that the system is not behaving as expected, thus requiring a closer examination of the system operation, the step function experiment used, or the assumed form of the system model. Second, acquiring data during the step function experiment is prone to some random error in each data point. The use of a data curve fit to determine $\tau$ uses all the data over time so as to minimize the influence of an error in any one data point. Third, the method eliminates the need to determine the $\Gamma = 1.0$ and 0.368 points exactly, which are difficult to establish in practice and so are prone to systematic error.

## EXAMPLE 3.3

Suppose a bulb thermometer originally indicating 20°C is suddenly exposed to a fluid temperature of 37°C. Develop a model that simulates the thermometer output response.

**KNOWN**

$$T(0) = 20°C$$
$$T_\infty = 37°C$$
$$F(t) = [T_\infty - T(0)]U(t)$$

**ASSUMPTIONS** To keep things simple, assume the following: no installation effects (neglect conduction and radiation effects); sensor mass is mass of liquid in bulb only; uniform temperature within bulb (lumped mass); thermometer scale is calibrated to indicate temperature

**FIND** $T(t)$

**SOLUTION**
Consider the energy balance developed in Figure 3.9. According to the first law of thermodynamics, the rate at which energy is exchanged between the sensor and its environment through convection, $\dot{Q}$, must be balanced by the storage of energy within the thermometer, $dE/dt$. This conservation of energy is written as

$$\frac{dE}{dt} = \dot{Q}$$

Control volume

$T_\infty$

$\dfrac{dE}{dt}$

$m$, $c_v$, $T(t)$

Bulb sensor

$\dot{Q}_{in}$

**FIGURE 3.9** Lumped parameter model of thermometer and its energy balance (Ex. 3.3).

Energy storage in the bulb is manifested by a change in bulb temperature so that for a constant mass bulb, $dE(t)/dt = mc_v dT(t)/dt$. Energy exchange by convection between the bulb at $T(t)$ and an environment at $T_\infty$ has the form $\dot{Q} = hA_s \Delta T$. The first law can be written as

$$mc_v \frac{dT(t)}{dt} = hA_s \left[ T_\infty - T(t) \right]$$

This equation can be written in the form

$$mc_v \frac{dT(t)}{dt} + hA_s [T(t) - T(0)] = hA_s F(t) = hA_s \left[ T_\infty - T(0) \right] U(t)$$

with initial condition $T(0)$ and where

$m$ = mass of liquid within thermometer

$c_v$ = specific heat of liquid within thermometer

$h$ = convection heat transfer coefficient between bulb and environment

$A_s$ = thermometer surface area

The term $hA_s$ controls the rate at which energy can be transferred between a fluid and a body; it is analogous to electrical conductance. By comparison with Equation 3.3, $a_0 = hA_s$, $a_1 = mc_v$, and $b_0 = hA_s$. Rewriting for $t \geq 0^+$ and simplifying yields

$$\frac{mc_v}{hA_s} \frac{dT(t)}{dt} + T(t) = T_\infty$$

From Equation 3.4, this determines that the time constant and static sensitivity are

$$\tau = \frac{mc_v}{hA_s} \quad K = \frac{hA_s}{hA_s} = 1$$

Direct comparison with Equation 3.5 yields this thermometer response:

$$T(t) = T_\infty + [T(0) - T_\infty]e^{-t/\tau}$$
$$= 37 - 17e^{-t/\tau} [^\circ C]$$

**COMMENT** Two interactive examples of the thermometer problem (from Exs. 3.3–3.5) are available using the online supplements. In the program *FirstOrd*, the user can choose input functions and study the system response. In the LabView program *Temperature_response*, the user can apply interactively a step change in temperature and study the first-order system response of a thermal sensor.

Clearly the time constant, $\tau$, of the thermometer can be reduced by decreasing its mass-to-area ratio or by increasing $h$ (for example, increasing the fluid velocity around the sensor). Without modeling, such information could be ascertained only by trial and error, a time-consuming and costly method with no assurance of success. Also, it is significant that we found that the response of the temperature measurement system in this case depends on the environmental conditions of the measurement that control $h$, because the magnitude of $h$ affects the magnitude of $\tau$. If $h$ is not controlled during response tests (i.e., if it is an extraneous variable), ambiguous results will occur. For example, the curve of Figure 3.8 could become nonlinear, or replications will not yield the same values for $\tau$.

This example also indicates that the results of a well-executed step-response calibration may not be directly indicative of an instrument's performance during a measurement if the measurement conditions differ from those existing during the step calibration. A carefully thought out test plan can minimize such differences.

## EXAMPLE 3.4

For the thermometer in Example 3.3 subjected to a step change in input, calculate the 90% rise time in terms of $t/\tau$.

**KNOWN**  Same as Example 3.3

**ASSUMPTIONS**  Same as Example 3.3

**FIND**  90% response time in terms of $t/\tau$

### SOLUTION

The percent response of the system is given by $(1 - \Gamma) \times 100$ with the error fraction, $\Gamma$, defined by Equation 3.6. From Equation 3.5, we note that at $t = \tau$, the thermometer will indicate $T(t) = 30.75°C$, which represents only 63.2% of the step change from 20°C to 37°C. The 90% rise time represents the time required for $\Gamma$ to drop to a value of 0.10. Then

$$\Gamma = 0.10 = e^{-t/\tau}$$

or $t/\tau = 2.3$.

**COMMENT**  In general, a time equivalent to $2.3\tau$ is required to achieve 90% of the applied step input for a first-order system.

---

## EXAMPLE 3.5

A particular thermometer is subjected to a step change, such as in Example 3.3, in an experimental exercise to determine its time constant. The temperature data are recorded with time and presented in Figure 3.10. Determine the time constant for this thermometer. In the experiment, the heat transfer coefficient, $h$, is estimated to be 6 W/m²- °C from engineering handbook correlations.

**KNOWN**  Data of Figure 3.10

$$h = 6 \text{ W/m}^2\text{- }°C$$

**ASSUMPTIONS**  First-order behavior using the model of Example 3.3; constant properties

**FIND**  $\tau$

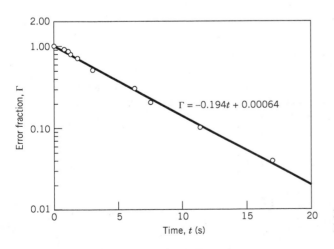

**FIGURE 3.10**  Temperature–time history of Example 3.5.

Plot annotation: $\Gamma = -0.194t + 0.00064$

## SOLUTION

According to Equation 3.7, the time constant should be the negative reciprocal of the slope of a line drawn through the data of Figure 3.10. Aside from the first few data points, the data appear to follow a linear trend, indicating a nearly first-order behavior and validating our model assumption. The data are fit to the first-order equation[2]

$$2.3 \log \Gamma = (-0.194)t + 0.00064$$

With $m = -0.194 = -1/\tau$, the time constant is calculated to be $\tau = 5.15$ seconds.

**COMMENT** If the experimental data were to deviate significantly from first-order behavior, this would be a clue either that our assumptions do not fit the real-problem physics or that the test conduct has control or execution problems.

### Simple Periodic Function Input

Periodic signals are commonly encountered in engineering processes. Examples include vibrating structures, vehicle suspension dynamics, biological circulations, and reciprocating pump flows. When periodic inputs are applied to a first-order system, the input signal frequency has an important influence on measuring system time response and affects the output signal. This behavior can be studied effectively by applying a simple periodic waveform to the system. Consider the first-order measuring system to which an input of the form of a simple periodic function, $F(t) = A \sin \omega t$, is applied for $t \geq 0^+$:

$$\tau \dot{y} + y = KA \sin \omega t$$

with initial conditions $y(0) = y_0$. Note that $\omega$ in [rad/s] $= 2\pi f$ with $f$ in [Hz]. The general solution to this differential equation yields the measurement system's output signal—that is, the time response to the applied input, $y(t)$:

$$y(t) = Ce^{-t/\tau} + \frac{KA}{\sqrt{1 + (\tau\omega)^2}} \sin(\omega t - \tan^{-1}\tau\omega) \tag{3.8}$$

where the value for $C$ depends on the initial conditions.

So what has happened? The output signal, $y(t)$, of Equation 3.8 consists of a transient and a steady periodic response. The first term on the right side is the transient response. As $t$ increases, this term decays to zero and no longer influences the output signal. Transient response is important only during the initial period following the application of the new input. We already have information about the system transient response from the step function study, so we focus our attention on the second term, the steady response. The steady response is time-dependent but periodically repeats itself. This term persists for as long as the periodic input is maintained. From Equation 3.8, we see that the frequency of the steady response term remains the same as the input signal frequency, but note that the amplitude of the steady response depends on the value of the applied frequency, $\omega$. Also, the phase angle of the periodic function has changed.

Equation 3.8 can be rewritten in a general form:

$$y(t) = Ce^{-t/\tau} + B(\omega) \sin(\omega t + \Phi)$$

$$B(\omega) = \frac{KA}{\sqrt{1 + (\tau\omega)^2}} \tag{3.9}$$

$$\Phi(\omega) = -\tan^{-1}(\tau\omega)$$

---

[2]The least-squares approach to curve fitting is discussed in detail in Chapter 4.

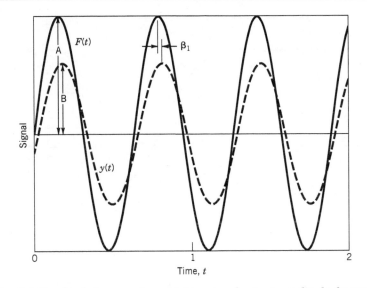

**FIGURE 3.11** Relationship between a sinusoidal input and output: amplitude, frequency, and time delay.

where $B(\omega)$ represents the amplitude of the steady response and the angle $\Phi(\omega)$ represents the *phase shift*. A comparison between the input signal and the system output response is given in Figure 3.11 for an arbitrary frequency and system time constant. From Equation 3.9, both $B$ and $\Phi$ are frequency dependent. Hence the exact form of the output response depends on the value of the frequency of the input signal. The steady response of any system to which a periodic input of frequency, $\omega$, is applied is known as the *frequency response* of the system. The frequency affects the magnitude of amplitude $B$ and also can bring about a time delay. This time delay, $\beta_1$, is seen in the phase shift, $\Phi(\omega)$, of the steady response. For a phase shift given in radians, the time delay in units of time is

$$\beta_1 = \frac{\Phi}{\omega}$$

that is, we can write

$$\sin(\omega t + \Phi) = \sin\left[\omega\left(t + \frac{\Phi}{\omega}\right)\right] = \sin[\omega(t + \beta_1)]$$

The value for $\beta_1$ will be negative, indicating that the time shift is a delay between the output and input signals. Since Equation 3.9 applies to all first-order measuring systems, the magnitude and phase shift by which the output signal differs from the input signal are predictable.

We define a *magnitude ratio*, $M(\omega)$, as the ratio of the output signal amplitude to the input signal amplitude, $M(\omega) = B/KA$. For a first-order system subjected to a simple periodic input, the magnitude ratio is

$$M(\omega) = \frac{B}{KA} = \frac{1}{\sqrt{1 + (\tau\omega)^2}} \tag{3.10}$$

The magnitude ratio for a first-order system is plotted in Figure 3.12, and the corresponding phase shift is plotted in Figure 3.13. The effects of both system time constant and input signal frequency on frequency response are apparent in both figures. This behavior can be interpreted in the following manner. For those values of $\tau\omega$ for which the system responds with values of $M(\omega)$ near unity, the measurement system transfers all or nearly all of the input signal amplitude to the output and with very little time delay; that is, $B$ will be nearly equal to $KA$ in magnitude, and $\Phi(\omega)$ will be near zero degrees. At large values of $\tau\omega$ the measurement system filters out any frequency

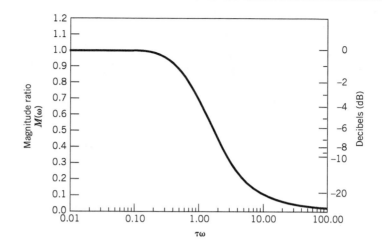

**FIGURE 3.12**    First-order system frequency response: magnitude ratio.

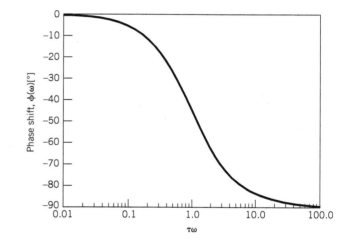

**FIGURE 3.13**    First-order system frequency response: phase shift.

information of the input signal by responding with very small amplitudes, which is seen by the small $M(\omega)$, and by large time delays, as evidenced by increasingly nonzero $\beta_1$.

Any combination of $\omega$ and $\tau$ produces the same results. If we wanted to measure signals with high-frequency content, then we would need a system having a small $\tau$. On the other hand, systems of large $\tau$ may be adequate to measure signals of low-frequency content. Often the trade-offs compete available technology against cost.

The *dynamic error*, $\delta(\omega)$, of a system is defined as $\delta(\omega) = M(\omega) - 1$. It is a measure of the inability of a system to adequately reconstruct the amplitude of the input signal for a particular input frequency. We normally want measurement systems to have a magnitude ratio at or near unity over the anticipated frequency band of the input signal to minimize $\delta(\omega)$. Perfect reproduction of the input signal is not possible, so some dynamic error is inevitable. We need some way to state this. For a first-order system, we define a *frequency bandwidth* as the frequency band over which $M(\omega) \geq 0.707$; in terms of the decibel (plotted in Figure 3.12) defined as

$$dB = 20 \ \log M(\omega) \tag{3.11}$$

this is the band of frequencies within which $M(\omega)$ remains above $-3$ dB, that is, $-3$ dB $\leq M(\omega) \leq 0$ dB.

*The functions $M(\omega)$ and $\Phi(\omega)$ represent the frequency response of a first-order measurement system to periodic inputs.* These equations and universal curves provide guidance in selecting measurement systems and system components.

*Determination of Frequency Response* The frequency response of a measurement system is found physically by a dynamic calibration. In this case, the calibration would entail applying a simple periodic waveform of known amplitude and frequency to the system sensor stage and measuring the corresponding output stage amplitude and phase shift. In practice, developing a method to produce a periodic input signal in the form of a physical variable may demand considerable ingenuity and effort. Hence, in many situations an engineer elects to rely on modeling to infer system frequency response behavior. We can predict the dynamic behavior if the time constant and static sensitivity of the system and the range of input frequencies are all known.

### EXAMPLE 3.6

A temperature sensor is to be selected to measure temperature within a reaction vessel. It is suspected that the temperature will behave as a simple periodic waveform with a frequency somewhere between 1 and 5 Hz. Sensors of several sizes are available, each with a known time constant. Based on time constant, select a suitable sensor, assuming that a dynamic error of $\pm 2\%$ is acceptable.

**KNOWN**  $1 \leq f \leq 5\text{Hz}$

$|\delta(\omega)| \leq 0.02$

**ASSUMPTIONS** First-order system

$F(t) = A \sin \omega t$

**FIND** Time constant, $\tau$

**SOLUTION**
With $|\delta(\omega)| \leq 0.02$, we would set the magnitude ratio between $0.98 \leq M \leq 1$. Then,

$$0.98 \leq M(\omega) = \frac{1}{\sqrt{1 + (\tau\omega)^2}} \leq 1$$

From Figure 3.12, this constraint is maintained over the range $0 \leq \tau\omega \leq 0.2$. We can also see in this figure that for a system of fixed time constant, the smallest value of $M(\omega)$ will occur at the largest frequency. So with $\omega = 2\pi f = 2\pi(5)\text{rad/s}$ and solving for $M(\omega) = 0.98$ yields, $\tau \leq 6.4$ ms. Accordingly, a sensor having a time constant of 6.4 ms or less will work.

## Second-Order Systems

Systems possessing inertia contain a second-derivative term in their model equation (e.g., see Ex. 3.1). A system modeled by a second-order differential equation is called a second-order system. Examples of some second-order instruments include accelerometers and pressure transducers (including microphones and loudspeakers).

In general, a second-order measurement system subjected to an arbitrary input, $F(t)$, can be described by an equation of the form

$$a_2\ddot{y} + a_1\dot{y} + a_0y = F(t) \tag{3.12}$$

where $a_0$, $a_1$, and $a_2$ are physical parameters used to describe the system and $\ddot{y} = d^2y/dt^2$. This equation can be rewritten as

$$\frac{1}{\omega_n^2}\ddot{y} + \frac{2\zeta}{\omega_n}\dot{y} + y = KF(t) \tag{3.13}$$

where

$$\omega_n = \sqrt{\frac{a_0}{a_2}} = \text{natural frequency of the system}$$

$$\zeta = \frac{a_1}{2\sqrt{a_0 a_2}} = \text{damping ratio of the system}$$

Consider the homogeneous solution to Equation 3.13. Its form depends on the roots of the characteristic equation of Equation 3.13:

$$\frac{1}{\omega_n^2}\lambda^2 + \frac{2\zeta}{\omega_n}\lambda + 1 = 0$$

This quadratic equation has two roots,

$$\lambda_{1,2} = -\zeta\omega_n \pm \omega_n\sqrt{\zeta^2 - 1}$$

Depending on the value of $\zeta$ three forms of homogeneous solution are possible:

$0 \le \zeta < 1$ (underdamped system solution)

$$y_h(t) = Ce^{-\zeta\omega_n t}\sin(\omega_n\sqrt{1 - \zeta^2}t + \Theta) \tag{3.14a}$$

$\zeta = 1$ (critically damped system solution)

$$y_h(t) = C_1 e^{\lambda_1 t} + C_2 e^{\lambda_2 t} \tag{3.14b}$$

$\zeta > 1$ (overdamped system solution)

$$y_h(t) = C_1 e^{\lambda_1 t} + C_2 e^{\lambda_2 t} \tag{3.14c}$$

The homogeneous solution determines the transient response of a system. The damping ratio, $\zeta$, is a measure of system damping, a property of a system that enables it to dissipate energy internally. For systems with $0 \le \zeta \le 1$, the transient response will be oscillatory, whereas for $\zeta \ge 1$, the transient response will not oscillate. The critically damped solution, $\zeta = 1$, denotes the demarcation between oscillatory and non-oscillatory behavior in the transient response.

## Step Function Input

Again, the step function input is applied to determine the general behavior and speed at which the system will respond to a change in input. The response of a second-order measurement system to a step function input is found from the solution of Equation 3.13, with $F(t) = AU(t)$, to be

$$y(t) = KA - KAe^{-\zeta\omega_n t}\left[\frac{\zeta}{\sqrt{1 - \zeta^2}}\sin(\omega_n t\sqrt{1 - \zeta^2}) + \cos(\omega_n t\sqrt{1 - \zeta^2})\right] \quad 0 \le \zeta < 1 \tag{3.15a}$$

$$y(t) = KA - KA(1 + \omega_n t)e^{-\omega_n t} \quad\quad \zeta = 1 \tag{3.15b}$$

$$y(t) = KA - KA\left[\frac{\zeta + \sqrt{\zeta^2 - 1}}{2\sqrt{\zeta^2 - 1}}e^{(-\zeta + \sqrt{\zeta^2 - 1})\omega_n t} - \frac{\zeta + \sqrt{\zeta^2 - 1}}{2\sqrt{\zeta^2 - 1}}e^{(-\zeta - \sqrt{\zeta^2 - 1})\omega_n t}\right] \quad \zeta > 1 \tag{3.15c}$$

where we set the initial conditions, $y(0) = \dot{y}(0) = 0$ for convenience.

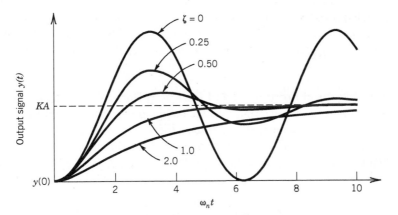

**FIGURE 3.14**  Second-order system time response to a step function input.

Equations 3.15a–c are plotted in Figure 3.14 for several values of $\zeta$. The interesting feature is the transient response. For underdamped systems, the transient response is oscillatory about the steady value and occurs with a period

$$T_d = \frac{2\pi}{\omega_d} = \frac{1}{f_d} \tag{3.16}$$

$$\omega_d = \omega_n \sqrt{1 - \zeta^2} \tag{3.17}$$

where $\omega_d$ is called the *damped natural frequency*, or *ringing frequency*. In instruments, this oscillatory behavior is called "ringing." The ringing phenomenon and the associated ringing frequency are properties of the measurement system and are independent of the input signal. It is the free oscillation frequency of a system displaced from its equilibrium.

The duration of the transient response is controlled by the $\zeta\omega_n$ term. In fact, its influence is similar to that of a time constant in a first-order system, such that we could define a second-order time constant as $\tau = 1/\zeta\omega_n$. The system settles to $KA$ more quickly when it is designed with a larger $\zeta\omega_n$ (i.e., smaller $\tau$). Nevertheless, for all systems with $\zeta > 0$, the response eventually indicates the steady value of $y_\infty = KA$ as $t \to \infty$.

Recall that the *rise time* is defined as the time required to achieve a value within 90% of the step input. For a second-order system, the rise time is the time required to first achieve 90% of $(KA - y_0)$. Rise time is reduced by decreasing the damping ratio, as seen in Figure 3.14. However, the severe ringing associated with very lightly damped systems actually delays the time to achieve a steady value compared to systems of higher damping. This is demonstrated by comparing the response at $\zeta = 0.25$ with the response at $\zeta = 1$ in Figure 3.14. With this in mind, the time required for a measurement system's oscillations to settle to within $\pm 10\%$ of the steady value, $KA$, is defined as its *settling time*. The settling time is an approximate measure of the time to achieve a steady response. A damping ratio of about 0.7 appears to offer a good compromise between ringing and settling time. If an error fraction ($\Gamma$) of a few percent is acceptable, then a system with $\zeta = 0.7$ will settle to steady response in about half the time of a system having $\zeta = 1$. For this reason, most measurement systems intended to measure sudden changes in input signal are typically designed such that parameters $a_0$, $a_1$, and $a_2$ provide a damping ratio of between 0.6 and 0.8.

**Determination of Ringing Frequency and Rise and Settling Times**  The experimental determination of the ringing frequency associated with underdamped systems is performed by applying a step input to the second-order measurement system and measuring the response. This type of calibration also yields information concerning the time to steady response of the system, which includes rise and settling times. Example 3.8 describes such a test. Typically, measurement systems suitable for dynamic signal measurements have specifications that include 90% rise time and settling time. *Adjust Second Order Parameters.vi* explores system values and response.

**Determination of Natural Frequency and Damping Ratio**   From the underdamped system response to a step function test, values for the damping ratio and natural frequency can be extracted. From Figure 3.14, we see that with ringing the amplitude decays logarithmically with time toward a steady-state value. Let $y_{max}$ represent the peak amplitude occurring with each cycle. Then for the first two successive peak amplitudes, let $y_1 = (y_{max})_1 - y_\infty$ and $y_2 = (y_{max})_2 - y_\infty$. The damping ratio is found from

$$\zeta = \frac{1}{\sqrt{1 + \left(2\pi / \ln\left(\frac{y_1}{y_2}\right)\right)^2}} \tag{3.18}$$

From the calculation to find the ringing frequency using Equation 3.16, the natural frequency is found using Equation 3.17. Alternately, we could perform the step function test with $y(0) = KA$ and $y_\infty = 0$ and still use the same approach.

### EXAMPLE 3.7

Determine the physical parameters that affect the natural frequency and damping ratio of the accelerometer of Example 3.1.

**KNOWN**   Accelerometer shown in Figure 3.3

**ASSUMPTIONS**   Second-order system as modeled in Example 3.1

**FIND** $\omega_n, \zeta$

**SOLUTION**
A comparison between the governing equation for the accelerometer in Example 3.1 and Equation 3.13 gives

$$\omega_n = \sqrt{\frac{k}{m}} \quad \zeta = \frac{c}{2\sqrt{km}}$$

The physical parameters of mass, spring stiffness, and frictional damping control the natural frequency and damping ratio of this measurement system.

### EXAMPLE 3.8

The curve shown in Figure 3.15 is the recorded voltage output signal of a diaphragm pressure transducer subjected to a step change in input. From a static calibration, the pressure–voltage relationship was found to be linear over the range 1 atmosphere (atm) to 4 atm with a static sensitivity of 1 V/atm. For the step test, the initial pressure was atmospheric pressure, $p_a$, and the final pressure was $2p_a$. Estimate the rise time, the settling time, and the ringing frequency of this measurement system.

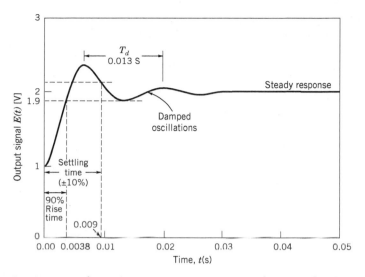

**FIGURE 3.15** Pressure transducer time response to a step input for Example 3.8.

**KNOWN**   $p(0) = 1$ atm

$p_\infty = 2$ atm

$K = 1$V/atm

**ASSUMPTIONS** Second-order system behavior

**FIND** Rise and settling times; $\omega_d$

**SOLUTION**

The ringing behavior of the system noted on the trace supports the assumption that the transducer can be described as having a second-order behavior. From the given information,

$$E(0) = Kp(0) = 1\text{V}$$
$$E_\infty = Kp_\infty = 2\text{V}$$

the step change observed on the trace should appear as a magnitude of 1 V. The 90% rise time occurs when the output first achieves a value of 1.9 V. The 90% settling time will occurs when the output settles between $1.9 < E(t) < 2.1$ V. From Figure 3.15, the rise occurs is about 4 ms, and the settling time is about 9 ms. The period of the ringing behavior, $T_d$, is judged to be about 13 ms for an $\omega_d \approx 485$ rad/s.

## Simple Periodic Function Input

The response of a second-order system to a simple periodic function input of the form $F(t) = A \sin \omega t$ is given by

$$y(t) = y_h + \frac{KA \sin[\omega t + \Phi(\omega)]}{\left\{ \left[ 1 - (\omega/\omega_n)^2 \right]^2 + \left[ 2\zeta\omega/\omega_n \right]^2 \right\}^{1/2}} \tag{3.19}$$

with frequency-dependent phase shift

$$\Phi(\omega) = \tan^{-1} \left( -\frac{2\zeta\omega/\omega_n}{1 - (\omega/\omega_n)^2} \right) \tag{3.20}$$

The exact form for $y_h$ is found from Equations 3.14a–c and depends on the value of $\zeta$. The steady response, the second term on the right side, has the general form

$$y_{\text{steady}}(t) = y(t \to \infty) = B(\omega) \sin[\omega t + \Phi(\omega)] \tag{3.21}$$

with amplitude $B(\omega)$. Comparing Equations 3.19 and 3.21 shows that the amplitude of the steady response of a second-order system subjected to a sinusoidal input is also dependent on $\omega$. So the amplitude of the output signal is frequency-dependent. In general, we can define the magnitude ratio, $M(\omega)$, for a second-order system as

$$M(\omega) = \frac{B(\omega)}{KA} = \frac{1}{\left\{ \left[ 1 - (\omega/\omega_n)^2 \right]^2 + \left[ 2\zeta\omega/\omega_n \right]^2 \right\}^{1/2}} \tag{3.22}$$

The frequency response of a second-order system is given by Equations 3.20 and 3.22 and this is plotted in Figures 3.16 and 3.17. The magnitude ratio-frequency dependence for a second-order system is plotted in Figure 3.16 for several values of damping ratio. A corresponding plot of the phase-shift dependency on input frequency and damping ratio is shown in Figure 3.17. For an ideal measurement system, $M(\omega)$ would equal unity and $\Phi(\omega)$ would equal zero for all values of measured frequency. Instead, $M(\omega)$ approaches zero and $\Phi(\omega)$ approaches $-\pi$ as $\omega/\omega_n$ becomes large. Keep in mind that $\omega_n$ is a property of the measurement system, whereas $\omega$ is a property of the input signal.

### System Characteristics

Several tendencies are apparent in Figures 3.16 and 3.17. For a system of zero damping, $\zeta = 0$, $M(\omega)$ will approach infinity and $\Phi(\omega)$ jumps to $-\pi$ in the vicinity of $\omega = \omega_n$. This behavior is characteristic of system resonance. Real systems possess some amount of damping, which modifies the abruptness and magnitude of resonance, but underdamped systems may still achieve a damped resonance. This region around $\omega = \omega_n$ on Figures 3.16 and 3.17 is called the *resonance*

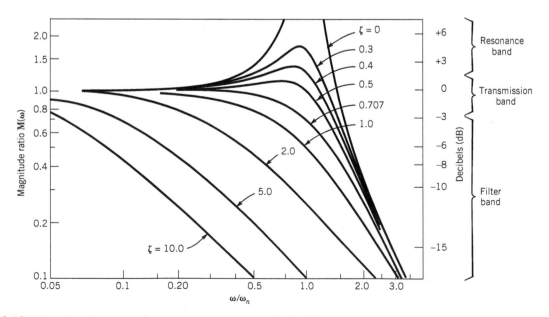

**FIGURE 3.16**    Second-order system frequency response: magnitude ratio.

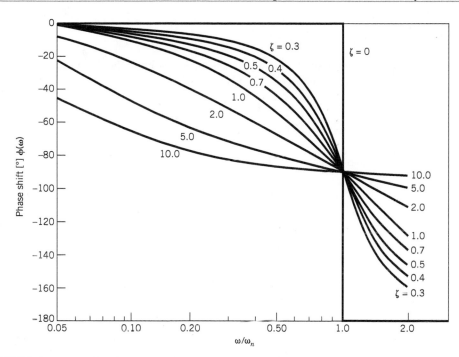

**FIGURE 3.17**   Second-order system frequency response: phase shift.

*band* of the system, referring to the range of frequencies over which the system shows $M(\omega) > 1$. Peak resonance in underdamped systems occurs at the *resonance frequency*,

$$\omega_R = \omega_n \sqrt{1 - 2\zeta^2} \qquad (3.23)$$

The resonance frequency is a property of the measurement system. Resonance behavior is excited by a periodic input signal frequency. The resonance frequency differs from the ringing frequency of free oscillation. Resonance behavior results in considerable phase shift. For most applications, operating at frequencies within the resonance band is confusing and could even be damaging to some delicate sensors. Resonance behavior is very nonlinear and results in distortion of the signal. On the other hand, systems having $\zeta > 0.707$ do not resonate.

At small values of $\omega/\omega_n$, $M(\omega)$ remains near unity and $\Phi(\omega)$ near zero. This means that information concerning the input signal of frequency $\omega$ will be passed through to the output signal with little alteration in the amplitude or phase shift. This region on the frequency response curves is called the *transmission band*. The actual extent of the frequency range for near unity gain depends on the system damping ratio. The transmission band of a system is either specified by its frequency bandwidth, typically defined for a second-order system as $-3 \text{ dB} \leq M(\omega) \leq 3 \text{ dB}$, or otherwise specified explicitly. One needs to operate within the transmission band of a measurement system to measure the dynamic content of the input signal accurately.

At large values of $\omega/\omega_n$, $M(\omega)$ approaches zero. In this region, the measurement system attenuates the amplitude information in the input signal. A large phase shift occurs. This region is known as the *filter band*, typically defined as the frequency range over which $M(\omega) \leq -3 \text{ dB}$. Most readers are familiar with the use of a filter to remove undesirable features from a desirable product. The amplitudes of the input signal corresponding to those frequencies within the filter band are reduced or eliminated completely—they are filtered out. So you need to match carefully the measurement system characteristics with the signal being measured; or else, desirable signal information may be filtered out.

## EXAMPLE 3.9

Determine the frequency response of a pressure transducer that has a damping ratio of 0.5 and a ringing frequency (found by a step test) of 1200 Hz.

**KNOWN** $\zeta = 0.5$

$$\omega_d = 2\pi(1200\,\text{Hz}) = 7540\,\text{rad/s}$$

**ASSUMPTIONS** Second-order system behavior

**FIND** $M(\omega)$ and $\Phi(\omega)$

## SOLUTION

The frequency response of a measurement system is determined by $M(\omega)$ and $\Phi(\omega)$ as defined in Equations 3.20 and 3.18. Since $\omega_d = \omega_n\sqrt{1-\zeta^2}$, the natural frequency of the pressure transducer is found to be $\omega_n = 8706$ rad/s. The frequency response at selected frequencies is computed from Equations 3.20 and 3.22:

| $\omega$ (rad/s) | $M(\omega)$ | $\Phi(\omega)$ [°] |
|---|---|---|
| 500 | 1.00 | −3.3 |
| 2,600 | 1.04 | −18.2 |
| 3,500 | 1.07 | −25.6 |
| 6,155 | 1.15 | −54.7 |
| 7,540 | 1.11 | −73.9 |
| 8,706 | 1.00 | −90.0 |
| 50,000 | 0.05 | −170.2 |

**COMMENT** The resonance behavior in the transducer response peaks at $\omega_R = 6155$ rad/s. As a rule, resonance effects can be minimized by operating at input frequencies of less than ~30% of the system's natural frequency. The response of second-order systems can be studied in more detail using the companion software. Try the Matlab program *SecondOrd.m* and LabView program *Second_order.vi*.

## EXAMPLE 3.10

An accelerometer is to be selected to measure a time-dependent motion. In particular, input signal frequencies below 100 Hz are of prime interest. Select a set of acceptable parameter specifications for the instrument assuming a dynamic error of ±5%.

**KNOWN** $f \leq 100$ Hz (i.e., $\omega \leq 628$ rad/s)

**ASSUMPTIONS** Second-order system
   Dynamic error of ±5% acceptable

**FIND** Select $\omega_n$ and $\zeta$

## SOLUTION

To meet a $\pm 5\%$ dynamic error constraint, we want $0.95 \leq M(\omega) \leq 1.05$ over the frequency range $0 \leq \omega \leq 628$ rad/s. This solution is open ended in that a number of instruments with different $\omega_n$ will do the task. So as one solution, let us set $\zeta = 0.7$ and then solve for the required $\omega_n$ using Equation 3.22:

$$0.95 \leq M(\omega) = \frac{1}{\left\{ \left[ 1 - (\omega/\omega_n)^2 \right]^2 + \left[ 2\zeta\omega/\omega_n \right]^2 \right\}^{1/2}} \leq 1.05$$

With $\omega = 628$ rad/s, these equations give $\omega_n \geq 1047$ rad/s. We could plot Equation 3.22 with $\zeta = 0.7$, as shown in Figure 3.18. In Figure 3.18, we find that $0.95 \leq M(\omega) \leq 1.05$ over the frequency range $0 \leq \omega/\omega_n \leq 0.6$. Again, this makes $\omega_n \geq 1047$ rad/s acceptable. So as one solution, an instrument having $\zeta = 0.7$ and $\omega_n \geq 1047$ rad/s meets the problem constraints.

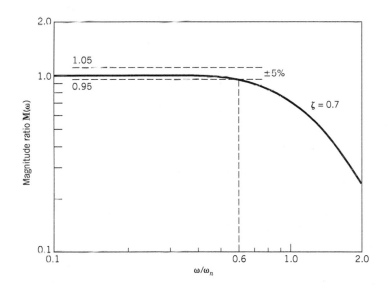

**FIGURE 3.18**    Magnitude ratio for second-order system with $\zeta = 0.7$ for Example 3.10.

---

**EXAMPLE 3.11**    **Measurements Isolation Table**

A common approach to vibration control in state-of-the-art metrology and for performing any measurement sensitive to building vibrations is to mount test equipment on an isolation table. These are practical solutions when using sensitive microscopes, interferometers, delicate instruments, and balance scales. An isolation table is simply a table surface that is isolated from floor vibrations through a pneumatic suspension (mass-spring-damper) system mounted within each of the table legs. The goal is to design a suspension system that has a very low resonance frequency so as to effectively operate in the filter band of system response for the anticipated excitation frequencies. The response of a commercially available table is shown in Figure 3.19. Its resonance frequency is about 1 Hz. As such, excitation floor vibrations above 2 Hz will be highly attenuated at the table surface.

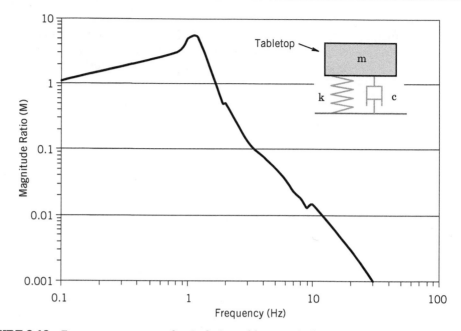

**FIGURE 3.19**  Frequency response of an isolation table to vertical ground-induced vibrations. Inset depicts a single leg isolation system.

**COMMENT**  Similarly in Example 1.2, engineers added more mass, $m$, to the steering column of the golf cart. The steering column natural frequency is described by $\omega_n = \sqrt{k/m}$. Adding mass lowered its natural frequency. In this way, the engineers were able to move the steering column response to the engine's speed-based excitation frequencies well into its filter band, thus attenuating the vibrations felt by the driver.

## 3.4  Transfer Functions

Consider the schematic representation in Figure 3.20. The measurement system operates on the input signal, $F(t)$, by some function, $G(s)$, so as to indicate the output signal, $y(t)$. This operation can be explored by taking the Laplace transform of both sides of Equation 3.4, which describes the general first-order measurement system. One obtains

$$Y(s) = \frac{1}{\tau s + 1} KF(s) + \frac{y_0}{\tau s + 1}$$

where $y_0 = y(0)$. This can be rewritten as

$$Y(s) = G(s)KF(s) + G(s)Q(s) \tag{3.24}$$

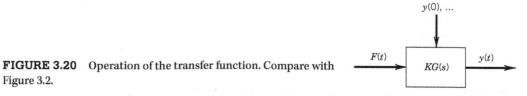

**FIGURE 3.20**  Operation of the transfer function. Compare with Figure 3.2.

where $G(s)$ is the transfer function of the first-order system given by

$$G(s) = \frac{1}{\tau s + 1} \tag{3.25}$$

and $Q(s)$ is the system initial state function. Because it includes $KF(s)$, the first term on the right side of Equation 3.24 contains the information that describes the steady response of the measurement system to the input signal. The second term describes its transient response. Included in both terms, the transfer function $G(s)$ plays a role in the complete time response of the measurement system. As indicated in Figure 3.20, the *transfer function* defines the mathematical operation that the measurement system performs on the input signal $F(t)$ to yield the time response (output signal) of the system. The text website provides a table of common Laplace transforms with their application to the solution of ordinary differential equations.

The system frequency response, which has been shown to be given by $M(\omega)$ and $\Phi(\omega)$, can be found by finding the value of $G(s)$ at $s = i\omega$. This yields the complex number

$$G(s = i\omega) = G(i\omega) = \frac{1}{\tau i \omega + 1} = M(\omega)e^{i\Phi(\omega)} \tag{3.26}$$

where $G(i\omega)$ is a vector on the real-imaginary plane having a magnitude, $M(\omega)$, and inclined at an angle, $\Phi(\omega)$, relative to the real axis as indicated in Figure 3.21. For the first-order system, the magnitude of $G(i\omega)$ is simply that given by $M(\omega)$ from Equation 3.10 and the phase shift angle by $\Phi(\omega)$ from Equation 3.9.

For a second-order or higher system, the approach is the same. The governing equation for a second-order system is defined by Equation 3.13, with initial conditions $y(0) = y_0$ and $\dot{y}(0) = \dot{y}_0$ The Laplace transform yields

$$Y(s) = \frac{1}{(1/\omega_n^2)s^2 + (2\zeta/\omega_n)s + 1} KF(s) + \frac{s\dot{y}_0 + y_0}{(1/\omega_n^2)s^2 + (2\zeta/\omega_n)s + 1} \tag{3.27}$$

which can again be represented by

$$Y(s) = G(s)KF(s) + G(s)Q(s) \tag{3.28}$$

By inspection, the transfer function is given by

$$G(s) = \frac{1}{(1/\omega_n^2)s^2 + (2\zeta/\omega_n)s + 1} \tag{3.29}$$

Solving for $G(s)$ at $s = i\omega$, we obtain for a second-order system

$$G(s = i\omega) = \frac{1}{(i\omega)^2/\omega_n^2 + 2\zeta i \omega/\omega_n + 1} = M(\omega)e^{i\Phi(\omega)} \tag{3.30}$$

which gives exactly the same magnitude ratio and phase shift relations as given by Equations 3.20 and 3.22.

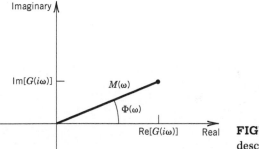

**FIGURE 3.21** Complex plane approach to describing frequency response.

## 3.5 Phase Linearity

We can see from Figures 3.16 and 3.17 that systems having a damping ratio near 0.7 possess the broadest frequency range over which $M(\omega)$ will remain at or near unity and that over this same frequency range the phase shift essentially varies in a linear manner with frequency. Although it is not possible to design a measurement system without accepting some amount of phase shift, it is desirable to design a system such that the phase shift varies linearly with frequency. This is because a nonlinear phase shift is accompanied by a significant distortion in the waveform of the output signal. *Distortion* refers to a notable change in the shape of the waveform from the original, as opposed to simply an amplitude alteration or time delay. To minimize distortion, many measurement systems are designed with $0.6 \leq \zeta \leq 0.8$.

Signal distortion can be illustrated by considering a particular complex waveform represented by a general function, $u(t)$:

$$u(t) = \sum_{n=1}^{\infty} \sin\, n\omega t = \sin\, \omega t + \sin\, 2\omega t + \cdots \tag{3.31}$$

Suppose during a measurement a phase shift of this signal were to occur such that the phase shift remained linearly proportional to the frequency; that is, the measured signal, $v(t)$, could be represented by

$$v(t) = \sin(\omega t - \Phi) + \sin(2\omega t - 2\Phi) + \cdots \tag{3.32}$$

Or, by setting

$$\theta = (\omega t - \Phi) \tag{3.33}$$

we write

$$v(t) = \sin \theta + \sin 2\theta + \cdots \tag{3.34}$$

We see that $v(t)$ in Equation 3.22 is equivalent to the original signal, $u(t)$. If the phase shift were not linearly related to the frequency, this would not be so. This is demonstrated in Example 3.12.

### EXAMPLE 3.12

Consider the effect of the variations in phase shift with frequency on a measured signal by examination of the signal defined by the function

$$u(t) = \sin t + \sin 5t$$

Suppose that this signal is measured in such a way that a phase shift that is linearly proportional to the frequency occurs in the form

$$v(t) = \sin(t - 0.35) + \sin[5t - 5(0.35)]$$

Both $u(t)$ and $v(t)$ are plotted in Figure 3.22. We can see that the two waveforms are identical except that $v(t)$ lags $u(t)$ by some time increment.

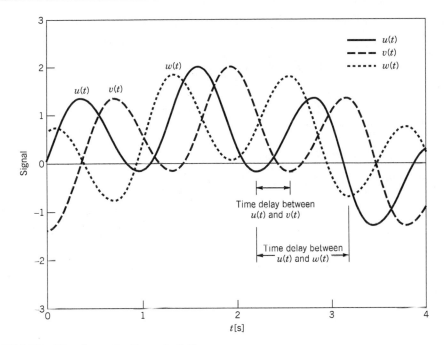

**FIGURE 3.22** Waveforms for Example 3.12.

Now suppose this signal is measured in such a way that the relation between phase shift and frequency was nonlinear, such as in the signal output form

$$w(t) = \sin(t - 0.35) + \sin(5t - 5)$$

The $w(t)$ signal is also plotted in Figure 3.22. It behaves differently from $u(t)$, and this difference is the signal distortion. In comparing $u(t)$, $v(t)$, and $w(t)$, it is apparent that a nonlinear relation between phase shift and frequency brings about signal distortion. This is an undesirable result.

## 3.6 Multiple-Function Inputs

So far, we have discussed measurement system response to a signal containing only a single frequency. What about measurement system response to multiple input frequencies? Or to an input that consists of both a static and a dynamic part, such as a periodic strain signal from an oscillating beam? When using models that are linear, such as ordinary differential equations subjected to inputs that are linear in terms of the dependent variable, the principle of superposition of linear systems applies in the solution of these equations. *The principle of superposition* states that a linear combination of input signals applied to a linear measurement system produces an output signal that is simply the linear addition of the separate output signals that would result if each input term had been applied separately. Because the form of the transient response is not affected by the input function, we can focus on the steady response. In general, we can write that if the forcing function of a form

$$F(t) = A_0 + \sum_{k=1}^{\infty} A_k \sin \omega_k t \tag{3.35}$$

is applied to a system, then the combined steady response will have the form

$$y_{\text{steady}}(t) = KA_0 + \sum_{k=1}^{\infty} B(\omega_k)\sin[\omega_k t + \Phi(\omega_k)] \tag{3.36}$$

where $B(\omega_k) = KA_k M(\omega_k)$. The development of the superposition principle can be found in basic texts on dynamic systems [4].

### EXAMPLE 3.13

Predict the steady output signal from a second-order instrument having $K = 1$ unit/unit, $\zeta = 2$, and $\omega_n = 628$ rad/s, which is used to measure the input signal

$$F(t) = 5 + 10\sin 25t + 20\sin 400t$$

**KNOWN** Second-order system
   $K = 1$ unit/unit; $\zeta = 2.0$; $\omega_n = 628$ rad/s
   $F(t) = 5 + 10\sin 25t + 20\sin 400t$

**ASSUMPTIONS** Linear system (superposition holds)

**FIND** $y(t)$

**SOLUTION**
Because $F(t)$ has a form consisting of a linear addition of multiple input functions, the steady response signal will have the form of Equation 3.33 of $y_{\text{steady}}(t) = KF(t)$ or

$$y(t) = 5K + 10KM\,(25\text{ red/s})\sin[25t + \Phi(25\text{ red/s})]$$
$$+ 20KM\,(400\text{red/s})\sin[400t + \Phi(400\text{ rad/s})]$$

Using Equations 3.20 and 3.22, or, alternatively, using Figures 3.16 and 3.17, with $\omega_n = 628$ rad/s and $\zeta = 2.0$, we calculate

$$M(25\text{ rad/s}) = 0.99 \qquad \Phi(25\text{ rad/s}) = -9.1°$$
$$M(400\text{ rad/s}) = 0.39 \qquad \Phi(400\text{ rad/s}) = -77°$$

so that the steady output signal will have the form

$$y(t) = 5 + 9.9\sin(25t - 9.1°) + 7.8\sin(400t - 77°)$$

The output signal is plotted against the input signal in Figure 3.23. The amplitude spectra for both the input signal and the output signal are shown in Figure 3.24. Spectra can also be generated by using the accompanying software programs *FunSpect.m* and *DataSpect.m*.

**COMMENT** The static magnitude and the amplitude of the 25 rad/s component of the input signal are passed along to the output signal closely. However, the amplitude of the 400 rad/s component is severely reduced (down 61%). This is a filtering effect due to the measurement system frequency response.

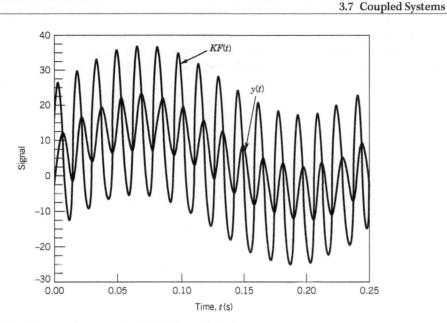

**FIGURE 3.23** Input and output signals for Example 3.13.

**FIGURE 3.24** Amplitude spectrum of Example 3.13.

## 3.7 Coupled Systems

As instruments in each stage of a measurement system are connected (transducer, signal conditioner, output device, etc.), the output from one stage becomes the input to the next stage to which it is connected, and so forth. The overall measurement system will have a coupled output response to the original input signal that is a combination of each individual response to the input. However, the system concepts of zero-, first-, and second-order systems studied previously can still be used for a case-by-case study of the coupled measurement system.

This concept is easily illustrated by considering a first-order sensor—transducer that may be connected to a second-order recording device. Suppose that the input to the sensor is a simple periodic waveform, $F_1(t) = A \sin \omega t$. The transducer responds with an output signal of the form of Equation 3.8:

$$y_t(t) = Ce^{-t/\tau} + \frac{K_t A}{\sqrt{1 + (\tau\omega)^2}} \sin(\omega t + \Phi_t)$$

$$\Phi_t = -\tan^{-1}\tau\omega$$

(3.37)

where the subscript $t$ refers to the transducer. However, the sensor-transducer output signal now becomes the input signal, $F_2(t) = y_t$, to the second-order recording device. The output from the second-order device, $y_s(t)$, will be a second-order response appropriate for input $F_2(t)$,

$$y_s(t) = y_h(t) + \frac{K_t K_s A \sin[\omega t + \Phi_t + \Phi_s]}{\left[1 + (\omega\tau^2)\right]^{1/2} \left\{ \left[1 - (\omega/\omega_n)^2\right]^2 + \left[2\zeta\omega/\omega_n\right]^2 \right\}^{1/2}}$$

$$\Phi_s = -\tan^{-1}\frac{2\zeta\omega/\omega}{1 - (\omega/\omega_n)^2}$$

(3.38)

where subscript $s$ refers to the signal recording device and $y_h(t)$ is the transient response. The output signal displayed on the recorder, $y_s(t)$, is the measurement system response to the original input signal to the transducer, $F_1(t) = A \sin \omega t$. The steady amplitude of the recording device output signal is the product of the static sensitivities and magnitude ratios of the first- and second-order systems. The phase shift is the sum of the phase shifts of the two systems.

Based on the concept behind Equation 3.28, we can make a general observation. Consider the schematic representation in Figure 3.25, which depicts a measurement system consisting of $H$ interconnected devices, $j = 1, 2, \ldots, H$, each device described by a linear system model. The overall transfer function of the combined system, $G(s)$, is the product of the transfer functions of each of the individual devices, $G_j(s)$, such that

$$KG(s) = K_1 G_1(s) K_2 G_2(s) \ldots K_H G_H(s)$$

(3.39)

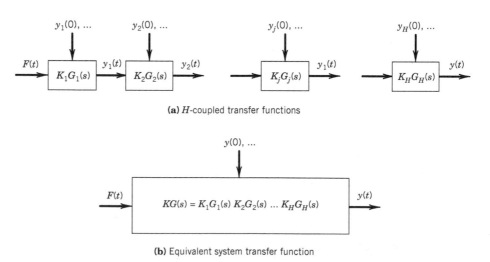

(a) $H$-coupled transfer functions

(b) Equivalent system transfer function

**FIGURE 3.25**   Coupled systems: describing the system transfer function.

At $s = i\omega$, Equation 3.37 becomes

$$KG(i\omega) = (K_1 K_2 \ldots K_H) \times [M_1(\omega) M_2(\omega) \ldots M_H(\omega)] e^{i[\Phi_1(\omega) + \Phi_2(\omega) + \cdots + \Phi_H(\omega)]} \qquad (3.40)$$

According to Equation 3.40, given an input signal to device 1, the system steady output signal at device $H$ will be described by the system frequency response $G(i\omega) = M(\omega) e^{i\Phi(\omega)}$, with an overall system static sensitivity described by

$$K = K_1 K_2 \ldots K_H \qquad (3.41)$$

The overall system magnitude ratio is the product

$$M(\omega) = M_1(\omega) M_2(\omega) \ldots M_H(\omega) \qquad (3.42)$$

and the overall system phase shift is the sum

$$\Phi(\omega) = \Phi_1(\omega) + \Phi_2(\omega) + \cdots + \Phi_H(\omega) \qquad (3.43)$$

This holds true provided that significant loading effects do not exist, a situation discussed in Chapter 6.

## 3.8 Summary

Just how a measurement system responds to a time-dependent input signal depends on the properties of the signal and the frequency response of the particular system. Modeling has enabled us to develop and to illustrate these concepts. Those system design parameters that affect system response are exposed through modeling, which assists in instrument selection. Modeling has also suggested the methods by which measurement system specifications such as time constant, response time, frequency response, damping ratio, and resonance frequency can be determined both analytically and experimentally. Interpretation of these system properties and their effect on system performance was determined.

The response rate of a system to a change in input is estimated by use of the step function input. The system parameters of time constant, for first-order systems, and natural frequency and damping ratio, for second-order systems, are used as indicators of system response rate. The frequency response of a system, through the magnitude ratio and phase shift, and are found by applying an input signal of a periodic waveform. Figures 3.12, 3.13, 3.16, and 3.17 are universal frequency response curves for first- and second-order systems. These curves can be found in most engineering and mathematical handbooks and can be applied to any first- or second-order system, as appropriate.

REFERENCES

1. Close, C. M., D. K. Frederick, J. Newell, *Modeling and Analysis of Dynamic Systems*, 3rd ed., Wiley, Boston, 2001.

2. Doebelin, E. O., *Measurement Systems: Applications and Design*, McGraw-Hill, New York, 2003.

3. Ogata, K., *System Dynamics*, 4th ed., Prentice-Hall, Englewood Cliffs, NJ, 2003.

4. Palm, W. J., III,, *Modeling, Analysis and Control of Dynamic Systems*, 2nd ed., Wiley, New York, 2000.

5. Burgess, J.C., A quick estimation of damping from free damped oscillograms, *Wright Air Development Center, WADC TR* 59-676, March 1961, pp. 457–460.

## N O M E N C L A T U R E

| | |
|---|---|
| $a_0, a_1, \ldots, a_n$ | physical coefficients |
| $b_0, b_1, \ldots, b_n$ | physical coefficients |
| $c$ | damping coefficient $(mt^{-1})$ |
| $f$ | cyclical frequency $(f = \omega/2\pi)$ (Hz) |
| $k$ | spring constant or stiffness $(mt^{-2})$ |
| $m$ | mass $(m)$ |
| $p(t)$ | pressure $(ml^{-1}t^{-2})$ |
| $t$ | time $(t)$ |
| $x(t)$ | independent variable |
| $y(t)$ | dependent variable |
| $y^n$ | $n$th time derivative of $y(t)$ |
| $y^n(0)$ | initial condition of $y^n$ |
| $A$ | input signal amplitude |
| $B(\omega)$ | output signal amplitude |
| $C$ | constant |
| $E(t)$ | voltage (V) or energy |
| $F$ | force $(mlt^{-2})$ |
| $F(t)$ | forcing function |
| $G(s)$ | transfer function |
| $K$ | static sensitivity |

| | |
|---|---|
| $M(\omega)$ | magnitude ratio, $B/KA$ |
| $T(t)$ | temperature $(°)$ |
| $T_d$ | ringing period $(t)$ |
| $U(t)$ | unit step function |
| $\beta_1$ | time lag $(t)$ |
| $\delta(\omega)$ | dynamic error |
| $\tau$ | time constant $(t)$ |
| $\Phi(\omega)$ | phase shift |
| $\omega$ | signal (circular) frequency $(t^{-1})$ |
| $\omega_n$ | natural frequency magnitude $(t^{-1})$ |
| $\omega_d$ | ringing frequency $(t^{-1})$ |
| $\omega_R$ | resonance frequency $(t^{-1})$ |
| $\zeta$ | damping ratio |
| $\Gamma$ | error fraction |

## Subscripts

| | |
|---|---|
| 0 | initial value |
| $\infty$ | final or steady value |
| $h$ | homogeneous solution |

# Probability and Statistics

# 4

## 4.1 Introduction

Suppose we had a large box containing a population of thousands of similar-sized round bearings. To get an idea of the size of the bearings and variations in size, we might measure two dozen from these thousands. The resulting sample of diameter values forms a data set, which we then use to imply something about the entire population of the bearings in the box, such as the average diameter size and size variation. But how close are these values taken from our small data set to estimating the actual average size and variation of all the bearings in the box? If we selected a different two dozen bearings, should we expect the size values to be exactly the same? These are questions that surround most engineering measurements. We make some measurements that we then use to infer things about the variable measured. What is the mean value of the variable based on a finite number of measurements and how well does this value represent the entire population? Do the variations found ensure that we can meet tolerances for the population as a whole? How good are these results? These questions have answers in probability and statistics.

For a given set of measurements, we want to be able to quantify (1) a single representative value that best characterizes the average of the measured data set, (2) a representative value that provides a measure of the variation in the measured data set, and (3) how well the average of the data set represents the true average value of the entire population of the measured variable. The value of (1) can vary with repeated data sets, so the difference between it and the true average value of the population is a type of random error. Additionally, (3) requires establishing an interval within which the true average value of the population is expected to lie. This interval quantifies the probable range of this random error and is called a random uncertainty.

This chapter presents an introduction to the concepts of probability and statistics at a level sufficient to provide information for a large class of engineering judgments. Such material allows for the reduction of raw data into results. We use terminology common to engineering, and it is not meant to replace a sound study of inferential statistics. Upon completion of this chapter, the reader will be able to:

- Quantify the statistical characteristics of a data set as it relates to the population of the measured variable

- Explain and use probability density functions to describe the behavior of variables

- Quantify a confidence interval about the measured mean value at a given probability and to apply hypothesis tests about the behavior of engineering measurements

- Perform regression analysis on a data set and quantify the confidence intervals for the parameters of the resulting curve fit or response surface

- Execute a Monte Carlo simulation that predicts the behavior expected in a result due to variations in the variables involved in computing that result

## 4.2  Statistical Measurement Theory

Sampling refers to obtaining a set of data through repeated measurements of a variable under fixed operating conditions. This variable is known as the *measured variable* or, in statistical language, the *measurand*. We take the sampling from the population of all possible values of the measured variable. In engineering practice, truly fixed operating conditions are often difficult to maintain, so when we use the term we will mean that the conditions under which the measurements are taken are maintained as close to the same as possible between measurements; any deviations in these conditions will show up as variations in the data set.

This chapter considers the effects of random errors and how to quantify them. Recall that random errors are manifested through data scatter and by the statistical limitations of a finite sampling to predict the behavior of a population. In this chapter, we assume that any systematic errors in the measurement are negligible.[1] Recall from Chapter 1 that this is the case when the average error in a data set is zero. We begin by considering the following measurement problem: Estimate the true value $x'$ based on the information derived from the repeated measurement of variable $x$. In the absence of systematic error, the true value of $x$ is the mean value of all possible measured values of $x$. This is the value we want to estimate from the measurement. A sampling of the variable $x$ under controlled, fixed operating conditions renders a sample, a finite number of data points. We use these limited data to infer $x'$ based on the calculated sample mean value, $\bar{x}$. We can imagine that if the number of data points, $N$, is very small, then the estimation of $x'$ from the data set could be heavily influenced by the value of any one data point. If the data set were larger, then the influence of any one data point would be offset by the larger influence of the other data. As $N \to \infty$ or toward the total number in the population, all the possible variations in $x$ become included in the data set. From a practical point of view, finite-sized data sets are common, in which case the measured data can provide only an estimate of the true value.

From a statistical analysis of the data set and an analysis of sources of error that influence these data, we can estimate $x'$ as

$$x' = \bar{x} \pm u_{\bar{x}} \quad (P\%) \tag{4.1}$$

where $\bar{x}$ represents the most probable estimate of $x'$ based on the available data and $\pm u_{\bar{x}}$ represents the *uncertainty interval* in that estimate at some probability level, *P%*. *Uncertainties are numbers that quantify the possible range of the effects of the errors.* The uncertainty interval is the range about $\bar{x}$ within which we would expect $x'$ to lie. It combines the uncertainty estimates of the random error and of systematic error in the measurement of $x$.[2] This chapter discusses ways to estimate the uncertainty arising from the effects of random error, called the random uncertainty. Chapter 5 discusses systematic errors and how to combine the uncertainties of both random and systematic errors into the uncertainty interval of Equation 4.1.

### Probability Density Functions

Random scatter of the measured data occurs regardless of the care taken to obtain the set from independent measurements under identical conditions. Thus the measured variable behaves as a random variable. A *random variable* is one whose value is subject to the effects of random chance. A continuous random variable is one that is continuous in time or space, such as the value of a motor's speed. A discrete random variable is one that is composed of discrete values, such as the values of the diameters for each of the individual bearings mentioned in Section 4.1. During repeated measurements of a variable, each data point may tend to assume one preferred value or lie within some interval about this value more often than not, when all the data are compared.

---

[1] Systematic error does not vary with repeated measurements and thus does not affect the statistics of a measurement. Systematic errors are discussed in Chapter 5.

[2] Texts that ignore systematic (bias) error refer to this uncertainty interval as a "confidence interval."

Table 4.1     **Sample of Random Variable $x$**

| $i$ | $x_i$ | $i$ | $x_i$ |
|---|---|---|---|
| 1 | 0.98 | 11 | 1.02 |
| 2 | 1.07 | 12 | 1.26 |
| 3 | 0.86 | 13 | 1.08 |
| 4 | 1.16 | 14 | 1.02 |
| 5 | 0.96 | 15 | 0.94 |
| 6 | 0.68 | 16 | 1.11 |
| 7 | 1.34 | 17 | 0.99 |
| 8 | 1.04 | 18 | 0.78 |
| 9 | 1.21 | 19 | 1.06 |
| 10 | 0.86 | 20 | 0.96 |

This tendency toward one central value about which all the other values are scattered is known as a *central tendency* of a random variable.[3] Probability deals with the concept that certain values of a variable may be measured more frequently relative to other values.

The central value and those values scattered about it can be determined from the probability density of the measured variable. The frequency with which the measured variable assumes a particular value or interval of values is described by its probability density. Consider a sample of $x$ shown in Table 4.1, which consists of $N$ individual measurements, $x_i$, where $i = 2, \ldots, N$, each measurement taken at random but under identical test-operating conditions. The measured values of this variable are plotted as data points along a single axis as shown in Figure 4.1.

In Figure 4.1, we can see a portion of the axis on which the data points tend to clump; this region contains the central value. Such behavior is typical of most engineering variables. We expect that the true mean value of the population of variable $x$ is contained somewhere in this clump.

This description for variable $x$ can be extended. Suppose we plot the data of Table 4.1 in a different way. The abscissa will be divided between the maximum and minimum measured values of $x$ into $K$ small intervals. Let the number of times, $n_j$, that a measured value assumes a value within an interval defined by $x - \delta x \leq x \leq x + \delta x$ be plotted on the ordinate. The resulting plot of $n_j$ versus $x$ is called a *histogram* of the variable [1]. The histogram is just another way of viewing both the tendency and the probability density of a variable. The ordinate can be nondimensionalized as $f_j = n_j/N$, converting the histogram into a *frequency distribution*.

For small $N$, the interval number $K$ should be conveniently chosen with a good rule that $n_j \geq 5$ for at least one interval. One suggestion for setting the minimum number of intervals $K$ uses a correlation developed from the data of Bendat and Piersol [2] as

$$K = 1.87(N - 1)^{0.40} + 1 \tag{4.2}$$

As $N$ becomes very large, a value of $K \approx N^{1/2}$ works reasonably well [1, 2]. The concept of the histogram is illustrated in Example 4.1.

**FIGURE 4.1**     Concept of density in reference to a measured variable (from Example 4.1).

---

[3]Not all random variables display a central tendency; the value in a fair roll of a die, for example, would show values from 1 through 6, each with an equal frequency of 1/6. But the fair roll of two dice will show a central tendency to the combined value of 7. Try it!

**EXAMPLE 4.1**

Construct a histogram and frequency distribution for the data in Table 4.1.

**KNOWN** Data of Table 4.1

$$N = 20.$$

**ASSUMPTIONS** Fixed operating conditions.

**FIND** Histogram and frequency distribution.

**SOLUTION**
To develop the histogram, compute a reasonable number of intervals for this data set. For $N = 20$, a convenient estimate of $K$ is found from Equation 4.2 to be

$$K = 1.87(N - 1)^{0.40} + 1 = 7$$

Next, determine the maximum and minimum values of the data set and divide this range into $K$ intervals. For a minimum of 0.68 and a maximum of 1.34, a value of $\delta x = 0.05$ is chosen. The intervals are as follows:

| $j$ | Interval | $n_j$ | $f_j = n_j/N$ |
|---|---|---|---|
| 1 | $0.65 \le x_i < 0.75$ | 1 | 0.05 |
| 2 | $0.75 \le x_i < 0.85$ | 1 | 0.05 |
| 3 | $0.85 \le x_i < 0.95$ | 3 | 0.15 |
| 4 | $0.95 \le x_i < 1.05$ | 7 | 0.35 |
| 5 | $1.05 \le x_i < 1.15$ | 4 | 0.20 |
| 6 | $1.15 \le x_i < 1.25$ | 2 | 0.10 |
| 7 | $1.25 \le x_i < 1.35$ | 2 | 0.10 |

The results are plotted in Figure 4.2. The plot displays a definite central tendency seen as the maximum frequency of occurrence falling within the interval 0.95 to 1.05.

**COMMENT** The total number of measurements, $N$, equals the sum of the number of occurrences,

$$N = \sum_{j=1}^{K} n_j$$

The area under the percent frequency distribution curve equals 100%; that is,

$$100 \times \sum_{j=1}^{K} f_j = 100\%$$

*Probability-density.vi* and *Running-histogram.vi* demonstrate the influence of population size and interval numbers on the histogram.

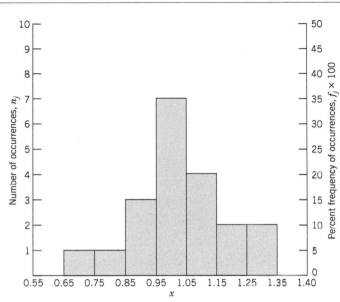

**FIGURE 4.2** Histogram and frequency distribution for data in Table 4.1.

As $N \rightarrow \infty$, the probability density function, $p(x)$, of the population of variable $x$ is developed. In the limit as $\delta x \rightarrow 0$

$$p(x) = \lim_{\delta x \to 0} \frac{n_j}{N(2\delta x)} \tag{4.3}$$

The *probability density function* is a statistical model of the population of a variable. It defines the probability that a variable might assume a particular value upon any individual measurement. It also describes the central tendency of the variable and its variation. This central tendency is the desired representative value that gives the best estimate of the true mean value.

The actual shape that the probability density function takes depends on the nature of the variable it represents and the circumstances surrounding the process in which the variable is involved. There are a number of standard distribution shapes that suggest how a variable could be distributed. Some common engineering examples are shown in Table 4.2. The specific values of the variable and the width of the distribution depend on the actual process, but the overall shape of the plot will most likely fit one of these. Often, experimentally determined histograms are used to infer guidance on the most likely model for $p(x)$. Of course, the list in Table 4.2 is not inclusive, and the reader is referred to more complete treatments of this subject [3–5].

Regardless of its probability density form, a variable that shows a central tendency can be described and quantified through its mean value and variance. In the absence of systematic errors, the population *true mean value* or central tendency of a random variable, $x(t)$, which is continuous in either time or space, and having a probability density function $p(x)$, is given by the parameter

$$x' = \lim_{T \to \infty} \frac{1}{T} \int_0^T x(t)dt \tag{4.4a}$$

which for any continuous random variable $x$ is equivalent to

$$x' = \int_{-\infty}^{\infty} xp(x)dx \tag{4.4b}$$

**Table 4.2    Standard Statistical Distributions and Relations Common to Measurements**

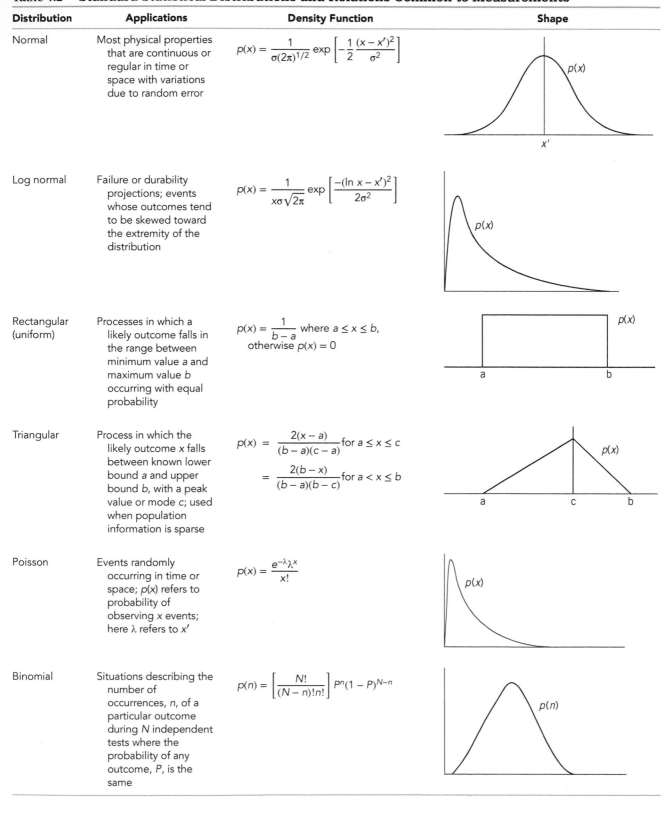

| Distribution | Applications | Density Function | Shape |
|---|---|---|---|
| Normal | Most physical properties that are continuous or regular in time or space with variations due to random error | $p(x) = \dfrac{1}{\sigma(2\pi)^{1/2}} \exp\left[-\dfrac{1}{2}\dfrac{(x - x')^2}{\sigma^2}\right]$ | |
| Log normal | Failure or durability projections; events whose outcomes tend to be skewed toward the extremity of the distribution | $p(x) = \dfrac{1}{x\sigma\sqrt{2\pi}} \exp\left[\dfrac{-(\ln x - x')^2}{2\sigma^2}\right]$ | |
| Rectangular (uniform) | Processes in which a likely outcome falls in the range between minimum value $a$ and maximum value $b$ occurring with equal probability | $p(x) = \dfrac{1}{b - a}$ where $a \le x \le b$, otherwise $p(x) = 0$ | |
| Triangular | Process in which the likely outcome $x$ falls between known lower bound $a$ and upper bound $b$, with a peak value or mode $c$; used when population information is sparse | $p(x) = \dfrac{2(x - a)}{(b - a)(c - a)}$ for $a \le x \le c$<br>$= \dfrac{2(b - x)}{(b - a)(b - c)}$ for $a < x \le b$ | |
| Poisson | Events randomly occurring in time or space; $p(x)$ refers to probability of observing $x$ events; here $\lambda$ refers to $x'$ | $p(x) = \dfrac{e^{-\lambda}\lambda^x}{x!}$ | |
| Binomial | Situations describing the number of occurrences, $n$, of a particular outcome during $N$ independent tests where the probability of any outcome, $P$, is the same | $p(n) = \left[\dfrac{N!}{(N - n)!n!}\right] P^n(1 - P)^{N-n}$ | |

If the measured variable is described by discrete data, the mean value of measured variable, $x_i$, where $i = 1, 2, \ldots, N$, is given by

$$x' = \lim_{N \to \infty} \frac{1}{N} \sum_{i=1}^{N} x_i \tag{4.5}$$

Physically, the width of the density function reflects the data variation. For a continuous random variable, the population *true variance* is given by

$$\sigma^2 = \lim_{T \to \infty} \frac{1}{T} \int_0^T [x(t) - x']^2 dt \tag{4.6a}$$

which is equivalent to

$$\sigma^2 = \int_{-\infty}^{\infty} (x - x')^2 p(x) dx \tag{4.6b}$$

or for discrete data, the population *variance* is given by

$$\sigma^2 = \lim_{N \to \infty} \frac{1}{N} \sum_{i=1}^{N} (x_i - x')^2 \tag{4.7}$$

The population *standard deviation*, $\sigma$, a commonly used statistical parameter, is defined as the square root of the variance—that is, $\sigma = \sqrt{\sigma^2}$.

The fundamental difficulty in using Equations 4.3 to 4.7 is that we assume a knowledge of the values for the entire population of the variable. But what if the data set represents only a finite portion of that population? Do these relations change? Real data set sizes may range from a single value to a large finite number. The next section discusses the interpretation of the behavior of the entire population of a variable with a connection between probability and statistics. Then we focus on trying to infer the behavior of a population based on the statistics of finite data sets.

## 4.3 Describing the Behavior of a Population

This section discusses the relation between probability and statistics. To do this, we assume a particular distribution for $p(x)$ to characterize the behavior of the population of $x$. Table 4.2 lists a few of the many distributions we could use, but to develop our relations, we use the *normal* (or *Gaussian*) *distribution*.[4] Much of the existing theory of statistics was developed using this distribution. It is particularly suited to describing the behavior of the continuous random variables common to engineering measurements. The normal distribution predicts that the scatter seen in a measured data set will be distributed symmetrically about some central tendency. Its shape is the familiar bell curve, as seen in Table 4.2.

The probability density function for a random variable, $x$, having a normal distribution is defined by

$$p(x) = \frac{1}{\sigma \sqrt{2\pi}} \exp\left[-\frac{1}{2} \frac{(x - x')^2}{\sigma^2}\right] \tag{4.8}$$

Where $x'$ is defined as the true mean value of $x$ and $\sigma^2$ as the true variance of $x$. Hence, $p(x)$ depends on the specific values for $x'$ and $\sigma^2$. A maximum in $p(x)$ occurs at $x = x'$, the true mean value.

---

[4]This distribution was independently suggested in the 18th century by Gauss, LaPlace, and DeMoivre. However, Gauss retains the eponymous honor.

This indicates that in the absence of systematic error the central tendency of a random variable having a normal distribution is toward its true mean value. The variance reflects the width or range of variation of $p(x)$.

Given $p(x)$, how can we predict the probability that any future measurement will fall within some stated interval of $x$ values? The probability that $x$ will assume a value within the interval $x' \pm \delta x$ is given by the area under $p(x)$, which is found by integrating over this interval. Thus, this probability is given by

$$P(x' - \delta x \le x \le x' + \delta x) = \int_{x'-\delta_x}^{x'+\delta_x} p(x)dx \qquad (4.9)$$

Using Equation 4.9 is made easier with the following transformations. Begin by defining the terms $\beta = (x - x')/\sigma$ as the *standardized normal variate* for any value $x$, with

$$z_1 = (x_1 - x')/\sigma \qquad (4.10)$$

as the $z$ variable, which refers to a specific interval on $p(x)$ between $x'$ and $x_1$. It follows that $dx = \sigma d\beta$, so that Equation 4.9 can be written as

$$P(-z_1 \le \beta \le z_1) = \frac{1}{\sqrt{2\pi}} \int_{-z_1}^{z_1} e^{-\beta^2/2} d\beta \qquad (4.11)$$

Because for a normal distribution, $p(x)$ is symmetrical about $x'$, we can write

$$\frac{1}{\sqrt{2\pi}} \int_{-z_1}^{z_1} e^{-\beta^2/2} d\beta = 2 \times \left[ \frac{1}{\sqrt{2\pi}} \int_{0}^{z_1} e^{-\beta^2/2} d\beta \right] \qquad (4.12)$$

The value in brackets in Equation 4.12 is the *normal error function*. The value $P(z_1) = \frac{1}{\sqrt{2\pi}} \int_{0}^{z_1} e^{-\beta^2/2} d\beta$ is tabulated in Table 4.3 for the interval defined by $z_1$ shown in Figure 4.3. As written and tabulated, the integral is one-sided (i.e., it is evaluated from 0 to $z_1$), so the normal error function provides half the probability expressed by Equation 4.11.

We note that algorithms that provide probability values without the need for interpolation are widely available in statistics texts and reputable internet sites or easily computed using built-in functions in spreadsheets and symbolic solvers (such as Excel® and Matlab®). The limited tables provided in this chapter are only intended to support the examples that illustrate the theory.

The parameters defined by Equations 4.4 to 4.7 are statements associated with the probability of a population. The area under the portion of the probability density function curve, $p(x)$, defined by the interval $x' - z_1\sigma \le x \le x' + z_1\sigma$, gives the probability that a measurement will assume a value within that interval. Direct integration of $p(x)$ for a normal distribution between the limits $x' \pm z_1\sigma$ with $z_1 = 1$ covers 68.26% of the area under $p(x)$. This equates to a 68.26% chance that a measurement of $x$ will have a value within the interval $x' \pm 1\sigma$. As the interval defined by $z_1$ is increased, the probability of occurrence increases. For

$z_1 = 1$,    68.26% of the area under $p(x)$ lies within $\pm z_1\sigma$ of $x'$.

$z_1 = 2$,    95.45% of the area under $p(x)$ lies within $\pm z_1\sigma$ of $x'$.

$z_1 = 3$,    99.73% of the area under $p(x)$ lies within $\pm z_1\sigma$ of $x'$.

This concept is illustrated in Figure 4.4.

**Table 4.3    Probability Values for Normal Error Function: One-Sided Integral Solutions for**

$$P(z_1) = \frac{1}{\sqrt{2\pi}} \int_0^{z_1} e^{-\beta^2/2} d\beta$$

| $z_1 = \dfrac{x_1 - x'}{\sigma}$ | 0.00 | 0.01 | 0.02 | 0.03 | 0.04 | 0.05 | 0.06 | 0.07 | 0.08 | 0.09 |
|---|---|---|---|---|---|---|---|---|---|---|
| 0.0 | 0.0000 | 0.0040 | 0.0080 | 0.0120 | 0.0160 | 0.0199 | 0.0239 | 0.0279 | 0.0319 | 0.0359 |
| 0.1 | 0.0398 | 0.0438 | 0.0478 | 0.0517 | 0.0557 | 0.0596 | 0.0636 | 0.0675 | 0.0714 | 0.0753 |
| 0.2 | 0.0793 | 0.0832 | 0.0871 | 0.0910 | 0.0948 | 0.0987 | 0.1026 | 0.1064 | 0.1103 | 0.1141 |
| 0.3 | 0.1179 | 0.1217 | 0.1255 | 0.1293 | 0.1331 | 0.1368 | 0.1406 | 0.1443 | 0.1480 | 0.1517 |
| 0.4 | 0.1554 | 0.1591 | 0.1628 | 0.1664 | 0.1700 | 0.1736 | 0.1772 | 0.1809 | 0.1844 | 0.1879 |
| 0.5 | 0.1915 | 0.1950 | 0.1985 | 0.2019 | 0.2054 | 0.2088 | 0.2123 | 0.2157 | 0.2190 | 0.2224 |
| 0.6 | 0.2257 | 0.2291 | 0.2324 | 0.2357 | 0.2389 | 0.2422 | 0.2454 | 0.2486 | 0.2517 | 0.2549 |
| 0.7 | 0.2580 | 0.2611 | 0.2642 | 0.2673 | 0.2794 | 0.2734 | 0.2764 | 0.2794 | 0.2823 | 0.2852 |
| 0.8 | 0.2881 | 0.2910 | 0.2939 | 0.2967 | 0.2995 | 0.3023 | 0.3051 | 0.3078 | 0.3106 | 0.3133 |
| 0.9 | 0.3159 | 0.3186 | 0.3212 | 0.3238 | 0.3264 | 0.3289 | 0.3315 | 0.3340 | 0.3365 | 0.3389 |
| 1.0 | 0.3413 | 0.3438 | 0.3461 | 0.3485 | 0.3508 | 0.3531 | 0.3554 | 0.3577 | 0.3599 | 0.3621 |
| 1.1 | 0.3643 | 0.3665 | 0.3686 | 0.3708 | 0.3729 | 0.3749 | 0.3770 | 0.3790 | 0.3810 | 0.3830 |
| 1.2 | 0.3849 | 0.3869 | 0.3888 | 0.3907 | 0.3925 | 0.3944 | 0.3962 | 0.3980 | 0.3997 | 0.4015 |
| 1.3 | 0.4032 | 0.4049 | 0.4066 | 0.4082 | 0.4099 | 0.4115 | 0.4131 | 0.4147 | 0.4162 | 0.4177 |
| 1.4 | 0.4192 | 0.4207 | 0.4222 | 0.4236 | 0.4251 | 0.4265 | 0.4279 | 0.4292 | 0.4306 | 0.4319 |
| 1.5 | 0.4332 | 0.4345 | 0.4357 | 0.4370 | 0.4382 | 0.4394 | 0.4406 | 0.4418 | 0.4429 | 0.4441 |
| 1.6 | 0.4452 | 0.4463 | 0.4474 | 0.4484 | 0.4495 | 0.4505 | 0.4515 | 0.4525 | 0.4535 | 0.4545 |
| 1.7 | 0.4554 | 0.4564 | 0.4573 | 0.4582 | 0.4591 | 0.4599 | 0.4608 | 0.4616 | 0.4625 | 0.4633 |
| 1.8 | 0.4641 | 0.4649 | 0.4656 | 0.4664 | 0.4671 | 0.4678 | 0.4686 | 0.4693 | 0.4699 | 0.4706 |
| 1.9 | 0.4713 | 0.4719 | 0.4726 | 0.4732 | 0.4738 | 0.4744 | 0.4750 | 0.4756 | 0.4761 | 0.4767 |
| 2.0 | 0.4772 | 0.4778 | 0.4783 | 0.4788 | 0.4793 | 0.4799 | 0.4803 | 0.4808 | 0.4812 | 0.4817 |
| 2.1 | 0.4821 | 0.4826 | 0.4830 | 0.4834 | 0.4838 | 0.4842 | 0.4846 | 0.4850 | 0.4854 | 0.4857 |
| 2.2 | 0.4861 | 0.4864 | 0.4868 | 0.4871 | 0.4875 | 0.4878 | 0.4881 | 0.4884 | 0.4887 | 0.4890 |
| 2.3 | 0.4893 | 0.4896 | 0.4898 | 0.4901 | 0.4904 | 0.4906 | 0.4909 | 0.4911 | 0.4913 | 0.4916 |
| 2.4 | 0.4918 | 0.4920 | 0.4922 | 0.4925 | 0.4927 | 0.4929 | 0.4931 | 0.4932 | 0.4934 | 0.4936 |
| 2.5 | 0.4938 | 0.4940 | 0.4941 | 0.4943 | 0.4945 | 0.4946 | 0.4948 | 0.4949 | 0.4951 | 0.4952 |
| 2.6 | 0.4953 | 0.4955 | 0.4956 | 0.4957 | 0.4959 | 0.4960 | 0.4961 | 0.4962 | 0.4963 | 0.4964 |
| 2.7 | 0.4965 | 0.4966 | 0.4967 | 0.4968 | 0.4969 | 0.4970 | 0.4971 | 0.4972 | 0.4973 | 0.4974 |
| 2.8 | 0.4974 | 0.4975 | 0.4976 | 0.4977 | 0.4977 | 0.4978 | 0.4979 | 0.4979 | 0.4980 | 0.4981 |
| 2.9 | 0.4981 | 0.4982 | 0.4982 | 0.4983 | 0.4984 | 0.4984 | 0.4985 | 0.4985 | 0.4986 | 0.4986 |
| 3.0 | 0.49865 | 0.4987 | 0.4987 | 0.4988 | 0.4988 | 0.4988 | 0.4989 | 0.4989 | 0.4989 | 0.4990 |

It follows directly that the representative value that characterizes a measure of the variation in a measured data set is the standard deviation. The probability that some $i$th measured value of $x$ will have a value between $x' \pm z_1\delta$ is $2 \times P(z_1) \times 100 = P\%$. This is written as

$$x_i = x' \pm z_1\sigma \ (P\%) \tag{4.13}$$

Thus, simple statistical analyses provide useful quantification of a measured variable in terms of probability. This in turn can be useful in engineering situations where the probable

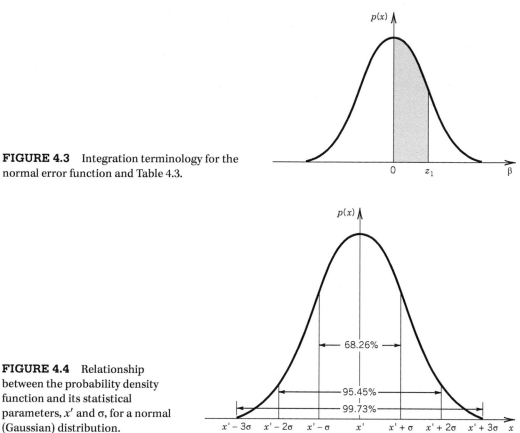

**FIGURE 4.3** Integration terminology for the normal error function and Table 4.3.

**FIGURE 4.4** Relationship between the probability density function and its statistical parameters, $x'$ and $\sigma$, for a normal (Gaussian) distribution.

outcome of a measured variable needs to be predicted or specified. These ideas are explored in Examples 4.2 and 4.3.

### EXAMPLE 4.2

Using the probability values in Table 4.3, show that the probability that a measurement will yield a value within $x' \pm \sigma$ is 0.6826 or 68.26%.

**KNOWN** Table 4.3.

$$z_1 = 1$$

**ASSUMPTIONS** Data follow a normal distribution.

**FIND** $P(x' - \sigma \leq x \leq x' + \sigma)$.

**SOLUTION**
To estimate the probability that a single measurement will have a value within some interval, we need to solve the integral

$$\frac{1}{\sqrt{2\pi}} \int_0^{z_1=1} e^{-\beta^2/2} d\beta$$

over the interval defined by $z_1$. Table 4.3 lists the solutions for this integral. Using Table 4.3 for $z_1 = 1$, we find $P(z_1) = 0.3413$. However, because $z_1 = (x_1 - x')/\sigma$, the probability that any measurement of $x$ will produce a value within the interval $0 \leq x \leq x' + \sigma$ is 34.13%. Since the normal distribution is symmetric about $x'$, the probability that $x$ will fall within the interval defined between $-z_1\delta$ and $+z_1\delta$ for $z_1 = 1$ is $(2)(0.3413) = 0.6826$ or 68.26%. Accordingly, if we made a single measurement of $x$, the probability that the value found would lie within the interval $x' - \sigma \leq x \leq x' + \sigma$ would be 68.26%.

**COMMENT** Similarly, for $z_1 = 1.96$, the probability would be 95.0%.

---

**EXAMPLE 4.3**

The population of a well-defined varying voltage signal are described by $x' = 8.5$ V and $\sigma^2 = 2.25$ $V^2$. If a single measurement of the voltage signal is made, determine the probability that this measured value will have a value of between 10.0 and 11.5 V.

**KNOWN** $x' = 8.5$ V.

$$\sigma^2 = 2.25 \ V^2.$$

**ASSUMPTIONS** Signal has a normal distribution about $x'$.

**FIND** $P(10.0 \leq x \leq 11.5)$.

**SOLUTION**
To find the probability that $x$ will fall into the interval $10.0 \leq x \leq 11.5$ requires finding the area under $p(x)$ bounded by this interval. The standard deviation of the variable is $\sigma = \sqrt{\sigma^2} = 1.5$V, so our interval falls under the portion of the $p(x)$ curve bounded by $z_1 = (10.0 - 8.5)/1.5 = 1$ and $z_1 = (11.5 - 8.5)/1.5 = 2$. From Table 4.3, the probability that a value will fall between $8.5 \leq x < 10.0$ is $P(8.5 \leq x \leq 10.0) = P(z_1 = 1) = 0.3413$. For the interval defined by $8.5 \leq x \leq 11.5$, $P(8.5 \leq x \leq 11.5) = P(z_1 = 2) = 0.4772$. The area we need is just the overlap of these two intervals, so

$$P(10.0 \leq x \leq 11.5) = P(8.5 \leq x \leq 11.5) - P(8.5 \leq x < 10.0)$$
$$= 0.4772 - 0.3413 = 0.1359$$

So there is a 13.59% probability that the measurement will yield a value between 10.0 and 11.5 V.

**COMMENT** In general, the probability that a measured value will lie within an interval defined by any two values of $z_1$, such as $z_a$ and $z_b$, is found by integrating $p(x)$ between $z_a$ and $z_b$. For a normal density function, this probability is identical to the operation, $P(z_b) - P(z_a)$.

---

## 4.4 Statistics of Finite-Sized Data Sets

We now try to predict the behavior of a population of variable $x$ based on knowledge obtained from a finite-sized random sample of $x$. For example, if we recall the box of bearings discussed in Section 4.1, some two dozen bearings were measured, each having been randomly selected from a population numbering in the thousands. Within the constraints imposed by probability, and if we assume a probability density function for the population, it is possible to *estimate* the true mean and true variance of the population of all the bearings from the statistics of the small sample. The science behind using a small sample to characterize the behavior of an entire population is called *inferential statistics*.

Suppose we examine the case where we obtain $N$ measurements of $x$, each measurement represented by $x_i$, where $i = 1, 2, ...,N$, and $N$ is a finite value. In cases where $N$ does not represent the total population, the statistical values calculated from such finite data sets are only estimates of the true statistics of the population of $x$. We refer to such statistical estimates as the *finite statistics*. An important point: Whereas infinite statistics describe the true behavior of the population of a variable, finite statistics describe only the behavior of the sampled data set, which we then use to infer the behavior of the population.

Finite-sized data sets provide the statistical estimates known as the *sample mean value* $(\bar{x})$, the *sample variance* $(s_x^2)$, and its outcome, the *sample standard deviation* $(s_x)$, defined by

$$\bar{x} = \frac{1}{N} \sum_{i=1}^{N} x_i \tag{4.14a}$$

$$s_x^2 = \frac{1}{N-1} \sum_{i=1}^{N} (x_i - \bar{x})^2 \tag{4.14b}$$

$$s_x = \sqrt{s_x^2} = \left( \frac{1}{N-1} \sum_{i=1}^{N} (x_i - \bar{x})^2 \right)^{1/2} \tag{4.14c}$$

where $(x_i - \bar{x})$ is called the *deviation* of $x_i$. For some problems, the sample standard deviation is more easily computed by using an equivalent form given by

$$s_x = \sqrt{\frac{\sum_{i=1}^{N} x_i^2 - N\bar{x}^2}{N-1}} \tag{4.14d}$$

The sample mean value provides a probable estimate of the true mean value, $x'$. The sample variance represents a probable measure of the variation found in a data set. The *degrees of freedom*, $\nu$, in a statistical estimate equate to the number of data points minus the number of previously determined statistical parameters used in estimating that value. For example, the degrees of freedom in the sample variance is $\nu = N - 1$, as seen in the denominator of Equations 4.14b, c and d. These equations are robust and are used regardless of the actual probability density function of the population.

The relations between probability and statistics previously developed can be extended to data sets of finite sample size with only some modification. When the parameters of mean value and variance are based on a sample rather than the entire population, the $z$ variable does not give a reliable weighting of the true probability of the population. Instead we introduce the Student's $t$ variable. For a normal distribution of $x$ about some sample mean value, $\bar{x}$, we can state that statistically

$$x_i = \bar{x} \pm t_{\nu,P} s_x \quad (P\%) \tag{4.15}$$

where the variable $t_{\nu,P}$ provides a coverage factor used for finite data sets and which replaces the $z$ variable. The Student's $t$ variable is defined as

$$t = \frac{\bar{x} - x'}{s_x / \sqrt{N}} \tag{4.16}$$

The interval $\pm t_{\nu,P} s_x$ represents a precision interval, given at probability $P\%$, within which one should expect any measured value to fall.

The value for the $t$ estimator provides a coverage factor, similar to the use of a $z$-variable, that is a function of the probability, $P$, and the degrees of freedom, $\nu = N - 1$. These $t$ values can be obtained from Table 4.4, a tabulation from the *Student's t distribution* as published by

Table 4.4    **Student's *t* Distribution (Two-Tailed)**

| $\nu$ | $t_{50}$ | $t_{90}$ | $t_{95}$ | $t_{99}$ |
|---|---|---|---|---|
| 1 | 1.000 | 6.314 | 12.706 | 63.657 |
| 2 | 0.816 | 2.920 | 4.303 | 9.925 |
| 3 | 0.765 | 2.353 | 3.182 | 5.841 |
| 4 | 0.741 | 2.132 | 2.776 | 4.604 |
| 5 | 0.727 | 2.015 | 2.571 | 4.032 |
| 6 | 0.718 | 1.943 | 2.447 | 3.707 |
| 7 | 0.711 | 1.895 | 2.365 | 3.499 |
| 8 | 0.706 | 1.860 | 2.306 | 3.355 |
| 9 | 0.703 | 1.833 | 2.262 | 3.250 |
| 10 | 0.700 | 1.812 | 2.228 | 3.169 |
| 11 | 0.697 | 1.796 | 2.201 | 3.106 |
| 12 | 0.695 | 1.782 | 2.179 | 3.055 |
| 13 | 0.694 | 1.771 | 2.160 | 3.012 |
| 14 | 0.692 | 1.761 | 2.145 | 2.977 |
| 15 | 0.691 | 1.753 | 2.131 | 2.947 |
| 16 | 0.690 | 1.746 | 2.120 | 2.921 |
| 17 | 0.689 | 1.740 | 2.110 | 2.898 |
| 18 | 0.688 | 1.734 | 2.101 | 2.878 |
| 19 | 0.688 | 1.729 | 2.093 | 2.861 |
| 20 | 0.687 | 1.725 | 2.086 | 2.845 |
| 21 | 0.686 | 1.721 | 2.080 | 2.831 |
| 30 | 0.683 | 1.697 | 2.042 | 2.750 |
| 40 | 0.681 | 1.684 | 2.021 | 2.704 |
| 50 | 0.680 | 1.679 | 2.010 | 2.679 |
| 60 | 0.679 | 1.671 | 2.000 | 2.660 |
| $\infty$ | 0.674 | 1.645 | 1.960 | 2.576 |

William S. Gosset[5] (1876–1937). Gossett recognized that the use of the $z$ variable did not yield accurate estimates of the precision interval at small degrees of freedom. Careful inspection of Table 4.4 shows that the $t$ value inflates the size of the interval required to attain a percent probability, $P\%$, at small degrees of freedom. That is, it has the effect of increasing the magnitude of $t_{\nu,P}s_x$ relative to $z_1\sigma$ at a desired probability. However, as the value of $N$ increases, $t$ approaches those values given by the $z$ variable just as the value of $s_x$ must approach $\sigma$.

The behavior of the Student's $t$ probability density function is illustrated in Figure 4.5(a) and (b). In Figure 4.5(a), we observe that the distribution is bell shaped and symmetric about $x = 0$ (i.e., in Equation 4.16, $\bar{x} - x' = 0$) but changes in width depending on the degrees of freedom, $\nu$. Smaller values of $\nu$ have a wider curve (with wider tails than with the $z$-variable) but the distribution narrows as it approaches the exact shape of the normal distribution as $\nu \to \infty$. The total area under the two-tailed curve is 1 regardless of the degrees of freedom. Figure 4.5(b) illustrates that the relationship between probability and the probability density function remains related to

---

[5] At the time, Gosset was employed as a brewer and statistician by a well-known Irish brewery. You might pause to reflect on his multifarious contributions.

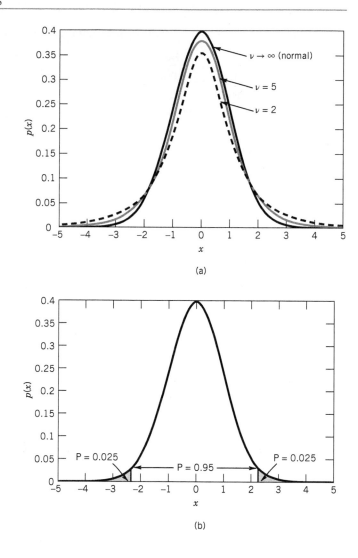

**FIGURE 4.5** Behavior of the probability density function for the Student's *t* estimator. (a) The probability density function for the *t* estimator and degrees of freedom, $\nu$. As $\nu \to \infty$, p(x) approaches that of the normal distribution function.
(b) Depiction of a two-tailed *t* distribution showing the $t_{95}$ limits for $\nu = 10$.

the area under the curve, just as it is with the normal distribution. Here for $\nu = 10$, 95% of the area falls between $-t_{10,95}$ and $+t_{10,95}$ with 2.5% of the area to the right of $+t_{10,95}$ and 2.5% to the left of $-t_{10,95}$.

We point out that for very small sample sizes ($N \leq 10$), the calculated statistics can be quite misleading. In that situation, other information regarding the measurement may be required, including additional measurements.

## Standard Deviation of the Means

If we were to take a second batch of two dozen of the bearings discussed in Section 4.1 and measure each bearing's diameter, we would expect the statistics from this new sample to differ somewhat from the first sample. This is due to the variation in each bearing's diameter arising from manufacturing tolerances (i.e., due to random chance). The difference between the calculated mean diameter of either batch and the mean diameter of the entire box is a random error due to using small data sets to infer the statistics of an entire population. So how can we quantify how good our estimate is of the true mean based on a sample mean?

Suppose we were to take $N$ measurements from a large population of variable $x$, each measurement taken under fixed operating conditions. If we were then to duplicate this procedure $M$ times (i.e., $M$ sets of $N$ measurements), we would calculate slightly different values for the sample mean and sample variance for each data set. Why? The chance occurrence of events in any finite-sized sample affects the estimate of sample statistics; this is easily demonstrated. From these $M$ replications of the $N$ measurements, we can compute a set of mean values. We would find that the population of mean values would themselves each be normally distributed about some central value. In fact, regardless of the shape of $p(x)$ assumed, the distribution of mean values obtained from all possible random samples will follow a normal distribution, $p(\bar{x})$.[6] This process is visualized in Figure 4.6. The amount of variation possible in the sample means would depend on two values: the sample variance, $s_x^2$, and the sample size, $N$. The discrepancy tends to increase with variance and decrease with $N^{1/2}$.

The variation in the sample statistics gathered from each data set is characterized by a normal distribution of the sample mean values about the true mean. This variation is estimated from a single finite data set using the *standard deviation of the means*, $s_{\bar{x}}$:

$$s_{\bar{x}} = \frac{s_x}{\sqrt{N}} \tag{4.17}$$

This statistic is also known as the standard error (of the means).

An illustration of the relation between the standard deviation of a data set and the standard deviation of the means is given in Figure 4.7. The *standard deviation of the means provides a measure of how well a measured mean value from a small sample size represents the true mean of the entire population.* The range over which the possible values of the true mean value might lie at some probability level $P$ based on the information from a sample data set is given as

$$\bar{x} \pm t_{v,P} s_{\bar{x}} \quad (P\%) \tag{4.18}$$

where $\pm t_{v,P} s_{\bar{x}}$ expresses a confidence interval about the mean value, with coverage factor $t$ at the assigned probability, $P\%$, within which one should expect the true value of $x$ to fall. This *confidence interval* is a quantified measure of the random error in the estimate of the true value

**FIGURE 4.6** The normal distribution tendency of the sample means about a true value in the absence of systematic error.

---

[6]This is a consequence of what is proved in the *central limit theorem* [3, 4].

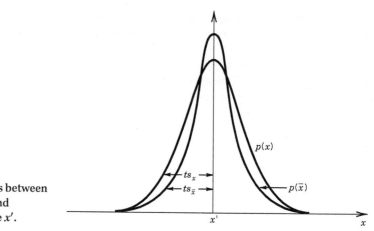

**FIGURE 4.7**   Relationships between $s_x$ and the distribution of $x$ and between $s_{\bar{x}}$ and the true value $x'$.

of variable $x$. The value $s_{\bar{x}}$ represents the random standard uncertainty in the mean value, and the value $t_{v,P}s_{\bar{x}}$ represents the *random uncertainty* in the mean value at $P\%$ confidence due to variation in the measured data set. *In the absence of systematic error in a measurement*, the confidence interval assigns the true value within a likely range about the sample mean value (one due solely on the small sample size). The estimate of the true mean value based on the statistics of a finite data set is then stated as

$$x' = \bar{x} \pm t_{v,P}s_{\bar{x}} \quad (P\%) \tag{4.19}$$

Equation 4.19 is an important and powerful equation in engineering measurements.

## EXAMPLE 4.4

Consider the data of Table 4.1. (a) Compute the sample statistics for this data set. (b) Estimate the interval of values over which 95% of the measurements of $x$ should be expected to lie. (c) Estimate the true mean value of $x$ at 95% probability based on this finite data set.

**KNOWN** Table 4.1.

$\qquad N = 20.$

**ASSUMPTIONS** Data set follows a normal distribution.

$\qquad\qquad$ No systematic errors.

**FIND** $\bar{x}, \bar{x} \pm t_{v,95}s_x$,    and    $\bar{x} \pm t_{v,95}s_{\bar{x}}$.

**SOLUTION**
The sample mean value is computed for the $N = 20$ values by the relation

$$\bar{x} = \frac{1}{20}\sum_{i=1}^{20} x_i = 1.02$$

This, in turn, is used to compute the sample standard deviation

$$s_x = \left[\frac{1}{19}\sum_{i=1}^{20}(x_i - \bar{x})^2\right]^{1/2} = 0.16$$

The degrees of freedom in the data set is $\nu = N - 1 = 19$. From Table 4.4 at 95% probability, $t_{19,95} = 2.093$. Then, the interval of values in which 95% of the measurements of $x$ should lie is predicted by Equation 4.15:

$$x_i = 1.02 \pm (2.093 \times 0.16) = 1.02 \pm 0.33 \quad (95\%)$$

Accordingly, if a twenty-first data point were taken, there is a 95% probability that its value would lie between 0.69 and 1.35.

The true mean value is estimated by the sample mean value. However, the random uncertainty at 95% probability for this estimate is $t_{19,95}s_{\bar{x}}$, where

$$s_{\bar{x}} = \frac{s_x}{\sqrt{N}} = \frac{0.16}{\sqrt{20}} = 0.036 \approx 0.04$$

Then, in the absence of systematic errors, we write, from Equation 4.19,

$$x' = \bar{x} \pm t_{19,95}s_{\bar{x}} = 1.02 \pm 0.08 \quad (95\%)$$

So at 95% confidence, the true mean value lies between 0.94 and 1.10, assuming that the only error is due to random variations of the measured value. Program *Finite-population.vi* demonstrates the effect of sample size on the histogram and the statistics of the data set.

## Repeated Tests and Pooled Data

Repeated tests provide separate data sets comprised of independent measurements of the same measured variable. These data can be combined to provide a better statistical estimate of a measured variable than are obtained from a single sampling. The important restriction we make here is that each test sample is drawn from the same population. If that is not true, a different statistical analysis may be required [4, 6].

Consider $M$ repeated tests that provide $M$ independent data sets of a measurement of variable $x$. Each test, denoted by subscript $j$, consists of $N_j$ repeated measurements. The pooled mean value combines each sample mean value weighted by the number of measurements $N_j$ for each test relative to the total measurements as

$$\langle \bar{x} \rangle = \frac{\displaystyle\sum_{j=1}^{M} N_j \bar{x}_j}{\displaystyle\sum_{j=1}^{M} N_j} = \frac{N_1 \bar{x}_1 + N_2 \bar{x}_2 + \ldots + N_M \bar{x}_M}{N_1 + N_2 + \ldots + N_M} \tag{4.20}$$

Similarly, the pooled variance combines the sample variance for each test weighted by the degrees of freedom for that test as

$$\langle s_x \rangle^2 = \frac{\displaystyle\sum_{j=1}^{M} \nu_j s_{x_j}^2}{\displaystyle\sum_{j=1}^{M} \nu_j} = \frac{\nu_1 s_{x_1}^2 + \nu_2 s_{x_2}^2 + \ldots + \nu_M s_{x_M}^2}{\nu_1 + \nu_2 + \ldots + \nu_M} \tag{4.21}$$

where $\nu_j = N_j - 1$. The pooled standard deviation is the square root of the pooled variance

$$\langle s_x \rangle = \sqrt{\frac{\sum\limits_{j=1}^{M} \nu_j s_{x_j}^2}{\sum\limits_{j=1}^{M} \nu_j}} \tag{4.22}$$

The pooled standard deviation of the means (or pooled standard error) is computed by

$$\langle s_{\bar{x}} \rangle = \langle s_x \rangle \sqrt{\sum\limits_{j=1}^{M} \frac{1}{N_j}} \tag{4.23}$$

A statistic used to test the significance of variations between the sample means is the mean-square value

$$\langle s_x \rangle_B^2 = \frac{1}{(M-1)} \sum\limits_{j=1}^{M} N_j \left( \bar{x}_j - \langle \bar{x} \rangle \right)^2 \tag{4.24}$$

The ratio $F = \langle s_x \rangle_B^2 / \langle s_x \rangle^2$ is used as the test statistic (called the $F$ statistic) for confirming the assumed equality of the population means under the null hypothesis test introduced in the next section. The test statistic for comparing the equivalence between two mean values of repeated tests is

$$t = \frac{(\bar{x}_1 - \bar{x}_2)}{\langle s_p \rangle \sqrt{\frac{1}{N_1} + \frac{1}{N_2}}} \tag{4.25}$$

Further use of the $t$ statistic is also described in the next section. More details regarding the analysis of pooled data obtained from repeated tests and Analysis of Variance (ANOVA), which provides analysis of multiple sample means, are discussed in basic statistics texts (e.g., [4, 6, 7]).

## 4.5 Hypothesis Testing

A hypothesis is a statement that something is believed to be true. A hypothesis test uses the tools of statistics to test it. The hypothesis might be that the difference between some observed (measured) behavior of a population and an assumed behavior is small enough to be considered as due to random variations (i.e., chance) alone. This is the null hypothesis, $H_0$. The alternative hypothesis, $H_a$, is that the difference results from something other than random variation, such as a systematic effect. So we might test whether measured differences in values between sample data sets arise from variations alone or whether something has changed between the data sets. The hypothesis test is used to decide whether the null hypothesis should be rejected in favor of the alternative hypothesis.

Suppose we want to test whether a measured sample mean is a reasonable proxy for the population mean comparing information of the observed scatter in the data set against the assumed distribution of the population. For this test, the null hypothesis is expressed as $H_0: x' = x_0$, where $x_0$ is a number. Now the alternative hypothesis will depend on the exact purpose of the hypothesis test. When we want to determine whether the sample mean, $\bar{x}$, differs from the population mean, the alternative hypothesis is expressed as $H_a: x' \neq x_0$. This is called the two-tailed test, because $H_a$ could be true if either $x' < x_0$ or $x' > x_0$. If the sample data are inconsistent with the null hypothesis, we reject it in favor of the alternative hypothesis. Otherwise, we do not reject the null hypothesis. In other situations, we might be interested in knowing whether $x' < x_0$, in which case we set $H_a: x' < x_0$. If the interest is in knowing whether $x' > x_0$, we set $H_a: x' > x_0$. Each of these is called a single-tailed test. We assume in each of these tests that $p(\bar{x})$ has a normal distribution.

**z-Test** The $z$-test is used when the population $\sigma$ is known. This occurs in situations where a value for $\sigma$ is specified, such as a value we might assign for quality control in manufacturing, or when the population distribution is already established. The test statistic is given by the $z$-variable. With $x' = x_0$, the observed $z$-variable is

$$z_0 = \frac{\bar{x} - x_0}{\sigma/\sqrt{N}} \tag{4.26}$$

This statistic is evaluated against critical values of $z$ at a desired *level of significance*, $\alpha$, where $P(z) \equiv 1 - \alpha$. We do not reject the null hypothesis if the values of the test statistic lie within a defined acceptance region. For the two-tailed test, the acceptance region is defined by the interval $P(-z_{\alpha/2} \le z_0 \le z_{\alpha/2})$ where $\pm z_{\alpha/2}$ are the critical values. Similarly, for the one-tailed tests, the acceptance region is defined as $P(-z_\alpha \le z_0)$ or $P(z_0 \le z_\alpha)$. We reject the null hypothesis if the value lies outside the acceptance region, a region we call the rejection region. The concept is shown in Figure 4.8 with common critical values listed in Table 4.5. When applying a hypothesis test to most engineering situations, testing at a two-tailed level of significance of 0.05 is adequate.

**t-Test** The $t$-test is used when $\sigma$ is not known. The test statistic is given by the $t$-variable. With $x' = x_0$, the observed $t$-variable is

$$t_0 = \frac{x - x_0}{s_x/\sqrt{N}} \tag{4.27}$$

This statistic is evaluated against critical values for $t$ at the desired level of significance, $\alpha$. For the two-tailed test, the acceptance region is given by the interval $P(-t_{\alpha/2} \le t_0 \le t_{\alpha/2})$. Here $+t_{\alpha/2}$ are the critical values. Similarly for the one-tailed tests, the acceptance region is $P(-t_\alpha \le t_0)$ or $P(t_0 \le t_\alpha)$. The concept is also shown in Figure 4.8.

**p-Value** The *p-value* indicates the probability that the numerical difference between two tested values, such as those of the critical statistic and test statistic, is due only to random chance. It is the observed level of significance. A very small $p$-value indicates that effects other than random

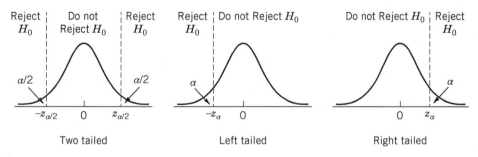

**FIGURE 4.8**    Critical values for a hypothesis test at level of significance, $\alpha$.

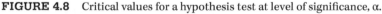

**Table 4.5    Critical Values Used in the z-test**

| Level of significance, $\alpha$ | 0.10 | 0.05 | 0.01 |
|---|---|---|---|
| Critical values of $z$ One-tailed tests | −1.28 or 1.28 | −1.645 or 1.645 | −2.33 or 2.33 |
| Critical values of $z$ Two-tailed tests | −1.645 and 1.645 | −1.96 and 1.96 | −2.58 and 2.58 |

chance are influencing the variations in a sampling. To calculate the $p$-value, assume that the null hypothesis is true and compute the observed level of significance corresponding to the test statistic. If this $p$-value is less than $\alpha$, we reject $H_0$; otherwise we do not reject $H_0$.

**Hypothesis Test** (1) Establish the null hypothesis, such as $H_0$: $x' = x_0$, where $x_0$ is the population or a target value. Then, establish the appropriate alternative hypothesis, $H_a$. (2) Assign a level of significance, $\alpha$, to determine the critical values; this sets the rejection region. (3) Calculate the observed value of the test statistic. (4) Compare the observed test statistic to the critical values. If the observed statistic is in the rejection region, reject the null hypothesis; otherwise, do not reject.

### EXAMPLE 4.5

From experience, a bearing manufacturer knows that it can produce roller bearings to the stated dimensions of 2.00 mm within a standard deviation of 0.061 mm. As part of its quality assurance program, engineering takes samples from boxes of bearings before shipment and measures them. On one occasion, a sample size of 25 bearings is obtained with the result $\bar{x} = 2.03$ mm. Do the measurements support the hypothesis that $x' = 2.00$ mm for the whole box at a 5% ($\alpha = 0.05$) level of significance (i.e., a 95% confidence level)?

**KNOWN** $\bar{x} = 2.03$ mm; $x_0 = 2.00$ mm; $\sigma = 0.061$ mm; $N = 25$.

**FIND** Apply hypothesis test at $\alpha = 0.05$.

### SOLUTION

Use the four steps for the hypothesis test: (1) We set the null hypothesis at $H_0$: $x' = x_0 = 2.00$ mm. The alternative hypothesis will be $H_a$: $x' \neq 2.00$ mm. (2) With $\alpha = 0.05$, the acceptance region is the area under the probability curve, $P(-z_{0.025} \leq z_0 \leq z_{0.025}) = 1 - \alpha = 0.95$. Assuming that the bearing population is normally distributed, we can write $P(-z_{0.025} \leq z \leq z_{0.025}) = 2P(z \leq z_{0.025}) = 0.475$. Then, using Table 4.3, we find $z$ by matching with the probability, $P(z = 1.960) = 0.475$. So the two-tailed critical value is $|\pm z_c| = 1.96$ (as noted in Table 4.5). (3) With $x_0 = 2.00$ mm, $\bar{x} = 2.03$ mm, $N = 25$, and $\sigma = 0.061$ mm; then $z_0 = (\bar{x} - x_0)/\sigma/\sqrt{N} = 2.459$. (4) We find that $z_0 > z_c$. So $z_0$ falls outside the acceptance region, and we reject $H_0$ at the 5% level of significance. The difference between the sample mean and the target population mean is larger than would likely occur by chance alone. Conclusion: This batch of bearings has a manufacturing defect!

**COMMENT** We know that random variations in a small sample can cause the sample mean to differ from the population mean. Here we find that the difference is larger than random variation alone can explain at $\alpha = 0.05$ or, with $P(z) \equiv 1 - \alpha$, at the 95% probability confidence level. There is likely a systematic error affecting the mean bearing diameter. Alternatively: We can also compute a $p$-value. The observed value of $z_0 = 2.459$ corresponds to a two-tailed, observed level of significance of $p = 0.0138$. That is, from Table 4.3, $2P(2.459) = 0.9862$, so $p = 1 - 0.9862 = 0.0138$. Because $p < \alpha$, we can reject $H_0$.

### EXAMPLE 4.6

Golf clubs and balls are tested for performance using a robotic swing device called the Iron Byron. A golf tour professional requests a 6-iron golf club from a manufacturer to be tuned to carry the ball a distance of 180 yards on average when struck with a club head speed of 85 mph. Accuracy is important, so the club should meet the requirement within 5 yards to either side of 180 yards.

The company engineer tests the selected club using a tour-style golf ball. Six tests of the club provide a mean carry of 182.7 yards with a sample standard deviation of 2.7 yards. Should the engineer certify that the club meets the tour professional's specifications at the 5% level of significance (1 yard = 0.914 m)?

**KNOWN** $\bar{x} = 182.7$ yards; $x_0 = 180$ yards; $s_x = 2.7$ yards; $N = 6$.

**FIND** Apply hypothesis test at $\alpha = 0.05$.

**SOLUTION**
(1) We test the null hypothesis, $H_0$: $x' = x_0 = 180$ yards against the alternative hypothesis, $H_a$: $x' \neq 180$ yards. (2) For a two-sided $t$-test with $\alpha = 0.05$, $P = 1 - \alpha = 0.95$. So with $N = 6$, $\nu = N - 1 = 5$, the critical values (Table 4.4) will be $|\pm t_{5,95}| = |\pm t_c| = 2.571$. (3) The test statistic is $t_0 = (\bar{x} - x_0)/s_x/\sqrt{N} = (182.7 - 180.0)/2.7/\sqrt{6} = 2.449$. (4) The observed test statistic, $t_0$, lies within the acceptance region, the area under the probability curve defined between $\pm t_c$. We do not reject the null hypothesis at the 5% level of significance. The data do not provide sufficient evidence to conclude that the carry distance of this golf club will differ from 180 yards when striking a ball at 85 mph. Conclusion: The club meets specifications.
**COMMENT** Alternatively: Using more extensive $t$-tables (e.g., reference 4), we can also find a two-tailed $t = 2.449$ and $\nu = 5$, which corresponds to $p = 0.0580$. As $p > \alpha$, do not reject $H_0$.

# 4.6 Chi-Squared Distribution

If we plotted the sample standard deviation for many data sets, each having $N$ data points, we would generate the probability density function, $p(\chi^2)$. The $p(\chi^2)$ follows the so-called *chi-squared* ($\chi^2$) *distribution* depicted in Figure 4.9.

For the normal distribution, the $\chi^2$ statistic [1, 3, 4] is

$$\chi^2 = \nu s_x^2 / \sigma^2 \qquad (4.28)$$

with degrees of freedom $\nu = N - 1$. We can use this to test how well $s_x^2$ predicts $\sigma^2$.

**FIGURE 4.9**  The $\chi^2$ distribution with its dependency on degrees of freedom.

## Precision Interval in a Sample Variance

A precision interval for the sample variance can be formulated by the probability statement

$$P\left(\chi^2_{1-\alpha/2} \leq \chi^2 \leq \chi^2_{\alpha/2}\right) = 1 - \alpha \tag{4.29}$$

with a probability of $P(\chi^2) = 1 - \alpha$. Combining Equations 4.28 and 4.29 gives

$$P\left(vs_x^2/\chi^2_{\alpha/2} \leq \sigma^2 \leq vs_x^2/\chi^2_{1-\alpha/2}\right) = 1 - \alpha \tag{4.30}$$

For example, the 95% precision interval by which $s_x^2$ estimates $\sigma^2$ is given by

$$vs_x^2/\chi^2_{0.025} \leq \sigma^2 \leq vs_x^2/\chi^2_{0.975} \quad (95\%) \tag{4.31}$$

Note that this interval is bounded by the 2.5% and 97.5% levels of significance (for 95% coverage).

The $\chi^2$ distribution estimates the discrepancy expected due to random chance. Values for $\chi^2_\alpha$ are tabulated in Table 4.6 as a function of the degrees of freedom. The $P(\chi^2)$ value equals the area under $p(\chi^2)$ as measured from the left, and the $\alpha$ value is the area as measured from the right, as noted in Figure 4.10. The total area under $p(\chi^2)$ is equal to unity.

**Table 4.6    Values for $\chi^2_\alpha$**

| $\nu$ | $\chi^2_{0.99}$ | $\chi^2_{0.975}$ | $\chi^2_{0.95}$ | $\chi^2_{0.90}$ | $\chi^2_{0.50}$ | $\chi^2_{0.05}$ | $\chi^2_{0.025}$ | $\chi^2_{0.01}$ |
|---|---|---|---|---|---|---|---|---|
| 1 | 0.000 | 0.001 | 0.004 | 0.016 | 0.455 | 3.84 | 5.02 | 6.63 |
| 2 | 0.020 | 0.051 | 0.103 | 0.211 | 1.39 | 5.99 | 7.38 | 9.21 |
| 3 | 0.115 | 0.216 | 0.352 | 0.584 | 2.37 | 7.81 | 9.35 | 11.3 |
| 4 | 0.297 | 0.484 | 0.711 | 1.06 | 3.36 | 9.49 | 11.1 | 13.3 |
| 5 | 0.554 | 0.831 | 1.15 | 1.61 | 4.35 | 11.1 | 12.8 | 15.1 |
| 6 | 0.872 | 1.24 | 1.64 | 2.20 | 5.35 | 12.6 | 14.4 | 16.8 |
| 7 | 1.24 | 1.69 | 2.17 | 2.83 | 6.35 | 14.1 | 16.0 | 18.5 |
| 8 | 1.65 | 2.18 | 2.73 | 3.49 | 7.34 | 15.5 | 17.5 | 20.1 |
| 9 | 2.09 | 2.70 | 3.33 | 4.17 | 8.34 | 16.9 | 19.0 | 21.7 |
| 10 | 2.56 | 3.25 | 3.94 | 4.78 | 9.34 | 18.3 | 20.5 | 23.2 |
| 11 | 3.05 | 3.82 | 4.57 | 5.58 | 10.3 | 19.7 | 21.9 | 24.7 |
| 12 | 3.57 | 4.40 | 5.23 | 6.30 | 11.3 | 21.0 | 23.3 | 26.2 |
| 13 | 4.11 | 5.01 | 5.89 | 7.04 | 12.3 | 22.4 | 24.7 | 27.7 |
| 14 | 4.66 | 5.63 | 6.57 | 7.79 | 13.3 | 23.7 | 26.1 | 29.1 |
| 15 | 5.23 | 6.26 | 7.26 | 8.55 | 14.3 | 25.0 | 27.5 | 30.6 |
| 16 | 5.81 | 6.91 | 7.96 | 9.31 | 15.3 | 26.3 | 28.8 | 32.0 |
| 17 | 6.41 | 7.56 | 8.67 | 10.1 | 16.3 | 27.6 | 30.2 | 33.4 |
| 18 | 7.01 | 8.23 | 9.39 | 10.9 | 17.3 | 28.9 | 31.5 | 34.8 |
| 19 | 7.63 | 8.91 | 10.1 | 11.7 | 18.3 | 30.1 | 32.9 | 36.2 |
| 20 | 8.26 | 9.59 | 10.9 | 12.4 | 19.3 | 31.4 | 34.2 | 37.6 |
| 30 | 15.0 | 16.8 | 18.5 | 20.6 | 29.3 | 43.8 | 47.0 | 50.9 |
| 60 | 37.5 | 40.5 | 43.2 | 46.5 | 59.3 | 79.1 | 83.3 | 88.4 |

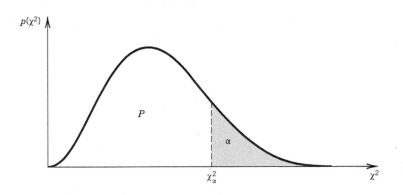

**FIGURE 4.10** The $\chi^2$ distribution as it relates to probability $P$ and to the level of significance, $\alpha\,(=1-P)$.

## EXAMPLE 4.7

A manufacturer knows from experience that the variance in the diameter of the roller bearings used in its bearings is 3.15 $\mu m^2$. Rejecting bearings drives up the unit cost. However, the manufacturer rejects any batch of roller bearings if the sample variance of 20 pieces selected at random exceeds 5.00 $\mu m^2$. Assuming a normal distribution, what is the probability that any given batch might be rejected even though its true variance is actually within the tolerance limits?

**KNOWN** $\sigma^2 = 3.15\ \mu m^2$

$\qquad s_x^2 = 5.00\ \mu m^2$ based on $N = 20$.

**FIND** $\chi_a^2$.

**SOLUTION**
For $\nu = N - 1 = 19$, and using Equation 4.28, the observed $\chi^2$ value is

$$\chi_\alpha^2(\nu) = \nu s_x^2/\sigma^2 = 30.16$$

For this value of $\chi^2$ with $\nu = 19$, we find that the observed value of $\alpha \approx 0.05$ (Table 4.6). Taking $\chi_{0.05}^2$ as a measure of discrepancy due to random chance, we interpret this result as there being only a 5% chance that a new batch within tolerance might be rejected. Conclusion: Rejecting a batch on the basis of $s_x^2 > 5.00\ \mu m^2$ is a good test at a 5% level of significance.

## Goodness-of-Fit Test

Just how well does a set of measurements follow an assumed distribution function? In Example 4.4, we assumed that the data of Table 4.1 followed a normal distribution based only on the rough form of its histogram (Figure 4.2). A more rigorous approach would apply the chi-squared test using the chi-squared distribution. The chi-squared test provides a measure of the discrepancy between the measured variation of a data set and the variation due to chance as predicted by the assumed density function.

To begin, construct a histogram of $K$ intervals from a data set of $N$ measurements. This establishes the number of measured values, $n_j$, that lie within the $j$th interval. Then calculate the degrees of freedom in the variance for the data set, $\nu = N - m$, where $m$ is the number of restrictions imposed. From $\nu$, estimate the predicted number of occurrences, $n'_j$, to be expected from the distribution function. For this test, the $\chi^2$ value is calculated from the entire histogram by

$$\chi^2 = \frac{\displaystyle\sum_{j=1}^{K}(n_j - n'_j)}{n'_j} \qquad j = 1, 2, \ldots, K \tag{4.32}$$

The goodness-of-fit test evaluates the null hypotheses that the data are described by the assumed distribution against the alternative that the data are not sampled from the assumed distribution. For the given degrees of freedom, the better a data set fits the assumed distribution function, the lower its observed $\chi^2$ value (left side of Table 4.6), whereas the higher the observed $\chi^2$ value (right side of Table 4.6), the more dubious the fit.

## EXAMPLE 4.8

Test the hypothesis that the variable $x$ as given by the measured data of Table 4.1 is described by a normal distribution at the 5% level of significance ($\alpha = 0.05$).

**KNOWN** Table 4.1 and histogram of Figure 4.2.

From Example 4.4: $\bar{x} = 1.02$; $s_x = 0.16$; $N = 20$; $K = 7$.

**FIND** Apply the chi-squared test to the data set to test for normal distribution.

**SOLUTION**
Figure 4.2 (Example 4.1) provides a histogram for the data of Table 4.1 giving the values for $n_j$ for $K = 7$ intervals. To evaluate the hypothesis, we must find the predicted number of occurrences, $n'_j$, for each of the seven intervals based on a normal distribution. To do this, substitute $x' = \bar{x}$ and $\sigma = s_x$ and compute the probabilities based on the corresponding $z$ values.

For example, consider the second interval ($j = 2$). Using Table 4.3, the predicted probabilities are

$$
\begin{aligned}
P(0.75 \leq x_i \leq 0.85) &= P(0.75 \leq x_i < x') - P(0.85 \leq x_i < x') \\
&= P(z_a) - P(z_b) \\
&= P(1.6875) - P(1.0625) \\
&= 0.454 - 0.356 = 0.098
\end{aligned}
$$

So for a normal distribution, we should expect the measured value to lie within the second interval for 9.8% of the measurements. With

$$n'_2 = N \times P(0.75 \leq x_i < 0.85) = 20 \times 0.098 = 1.96$$

which is to say that 1.96 occurrences are expected out of 20 measurements in this second interval. The actual measured data set shows $n_2 = 1$.

The results are summarized in Table 4.7 with $\chi^2$ based on Equation 4.32. Because two calculated statistical values ($\bar{x}$ and $s_x$) are used in the computations, the degrees of freedom in $\chi^2$ are restricted by 2. So, $\nu = K - 2 = 7 - 2 = 5$ and, from Table 4.6, for the observed $\chi^2_a(\nu = 5) = 1.91$ so $\alpha = 0.86$ (note: $\alpha = 1 - P$ is found here by interpolation between columns). At the tested level of significance, $\chi^2_{0.05}(5) = 11.1$. The observed level of significance of 0.86 is greater than the tested level of significance of 0.05, so we do not reject the null hypothesis. The difference between the measured distribution and a normal distribution is not statistically significant. Conclusion: The data set is described by a normal distribution at the 5% level of significance.

**Table 4.7** $\chi^2$ **Test for Example 4.8**

| j | nj | $n'_j$ | $(n_j - n'_j)^2/n'_j$ |
|---|-----|--------|----------------------|
| 1 | 1 | 0.92 | 0.008 ~ 0.01 |
| 2 | 1 | 1.96 | 0.47 |
| 3 | 3 | 3.76 | 0.15 |
| 4 | 7 | 4.86 | 0.93 |
| 5 | 4 | 4.36 | 0.03 |
| 6 | 2 | 2.66 | 0.16 |
| 7 | 2 | 1.51 | 0.16 |
| Totals | 20 | 20 | $\chi_a^2 = 1.91$ |

# 4.7 Regression Analysis

A measured variable is often a function of one or more independent variables. We can use regression analysis to establish a functional relationship between the dependent variable and the independent variable. A regression analysis assumes that the variation found in the dependent measured variable follows a normal distribution about each fixed value of the independent variable. This behavior is illustrated in Figure 4.11 by considering the dependent variable $y_{i,j}$ consisting of $N$ measurements, $i = 1, 2, ..., N$, of $y$ at each of $n$ values of independent variable, $x_j, j = 1, 2,...,$ $n$. This type of behavior is common during calibrations and in many types of measurements in which the dependent variable $y$ is measured under controlled values of $x$. Repeated measurements of $y$ yield a normal distribution with variance $s_y^2(x_j)$, about some mean value, $\bar{y}(x_j)$.

Most spreadsheet and engineering software packages can perform a regression analysis on a data set. The following discussion presents the concepts of a particular type of regression analysis, its interpretation, and its limitations. More information on regression analysis, including multiple variable regression, can be found elsewhere [4, 6].

## Least-Squares Regression Analysis

The regression analysis for a single variable of the form $y = f(x)$ provides an $m$th-order polynomial fit of the data in the form

$$y_c = a_0 + a_1 x + a_2 x^2 + \cdots + a_m x^m \tag{4.33}$$

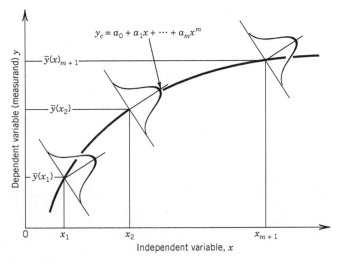

**FIGURE 4.11** Distribution of measured value $y$ about each fixed value of independent variable $x$. The curve $y_c$ represents a possible functional relationship.

where $y_c$ refers to the value of $y$ predicted by the polynomial equation for a given value of $x$. For $n$ different values of the independent variable included in the analysis, the highest order, $m$, of the polynomial that can be determined is restricted to $m \leq n - 1$. The values of the $m$ coefficients $a_0, a_1, \ldots, a_m$ are determined by the analysis. A common regression analysis for engineering applications is the method of least-squares. The *method of least-squares* attempts to minimize the sum of the squares of the deviations between the actual data and the polynomial fit of a stated order by adjusting the values of the polynomial coefficients.

Consider the situation in which there are $N$ values of $x$ and $y$, referred to as $x_i, y_i$, where $i = 1, 2, \ldots, N$. We seek an $m$th-order polynomial based on a set of $N$ data points of the form $(x, y)$ in which $x$ and $y$ are the independent and dependent variables, respectively. The task is to find the $m + 1$ coefficients, $a_0, a_1, \ldots, a_m$, of the polynomial of Equation 4.31. We define the deviation between any dependent variable $y_i$ and the polynomial as $y_i - y_{c_i}$, where $y_{c_i}$ is the value of the polynomial evaluated at the data point $(x_i, y_i)$. The sum of the squares of this deviation for all values of $y_i$, $i = 1, 2, \ldots, N$, is

$$D = \sum_{i=1}^{N} (y_i - y_{c_i})^2 \tag{4.34}$$

The goal of the method of least-squares is to reduce $D$ to a minimum for a given order of polynomial. This is done by inserting Equation 4.33 into 4.34 and equating the total derivative of $D$ to zero (i.e., $dD = 0$). This develops a set of $m + 1$ equations that are solved simultaneously to yield the unknown regression coefficients, $a_0, a_1, \ldots, a_m$. These are reinserted into Equation 4.33 for $y_c$. Intermediate details are provided on the text website or can be found in [1, 3, 4, 6].

In general, the regression polynomial will not pass through every data point $(x_i, y_i)$ exactly. So with any data pair $(x_i, y_i)$ there will be some deviation, $y_i - y_{c_i}$. We can compute a standard deviation based on these differences by

$$s_{yx} = \sqrt{\frac{\sum_{i=1}^{N} (y_i - y_{c_i})^2}{\nu}} \tag{4.35}$$

where $\nu$ is the degrees of freedom of the fit and $\nu = N - (m + 1)$. The statistic $s_{yx}$ is referred to as the *standard error of the fit* and is related to how closely a polynomial fits the data set.

The best order of polynomial fit to apply to a particular data set is that *lowest* order of fit that maintains a logical physical sense between the dependent and independent variables and reduces $s_{yx}$ to an acceptable value. This first point is important. If the underlying physics of a problem implies that a certain order relationship should exist between dependent and independent variables, there is no sense in forcing a curve fit using higher order polynomials. Because the method of least-squares tries to minimize the sum of the squares of the deviations, it forces inflections in the curve fit that may not be real. Consequently, although higher-order curve fits generally reduce $s_{yx}$, they likely do not reflect the physics behind the data set very well. In any event, it is good practice to have at least two independent data points for each order of polynomial attempted—that is, at least two data points for a first-order curve fit, four for a second-order, and so on.

If we consider variability in both the independent and dependent variables, then the random uncertainty arising from random data scatter about the curve fit at any value of $x$ is estimated by (1, 4)

$$t_{\nu,P} s_{yx} \left[ \frac{1}{N} + \frac{(x - \bar{x})^2}{\sum_{i=1}^{N} (x_i - \bar{x})^2} \right]^{1/2} \quad (P\%) \tag{4.36}$$

which is known as the Scheffe band [10], and where

$$\bar{x} = \sum_{i=1}^{N} x_i / N$$

and $x$ is the value used to estimate $y_c$ in Equation 4.33. Hence the curve fit, $y_c$, with confidence interval is given as

$$y_c(x) \pm t_{v,P} s_{yx} \left[ \frac{1}{N} + \frac{(x - \bar{x})^2}{\sum_{i=1}^{N} (x_i - \bar{x})^2} \right]^{1/2} \quad (P\%) \tag{4.37}$$

The effect of the second term in the brackets of either Equations 4.36 or 4.37 increases the confidence interval toward the outer limits of the polynomial.

The error estimated by the random uncertainty in the curve fit is usually one of several elements of error in an engineering test and may not be a dominant source of uncertainty. Further, often in engineering measurements, the independent variable is a well-controlled value and the principal source of variation in the curve fit is due to the random error in the dependent (measured) variable. In such cases, we might want a single representative value for the uncertainty due to regression analysis. A simplification is to approximate the random standard uncertainty due to the curve fit as a single value, given by

$$\frac{s_{yx}}{\sqrt{N}} \tag{4.38}$$

This is the value of the Scheffe band at its waist (where $x = \bar{x}$). This approximation can be weighted to an appropriate confidence level as

$$t_{v,P} \frac{s_{yx}}{\sqrt{N}} \tag{4.39}$$

This is a major simplification but provides a convenient representative value in tedious uncertainty analysis calculations (Chapter 5). The approximation reduces the interval to a single value. This simplification (Equation 4.38) should only be used when the curve fit uncertainty is not the dominant one in a complete uncertainty analysis so that differences between Equations 4.36 and 4.39 are not significant in reporting the value of a test outcome.

Multiple regression analysis involving multiple variables of the form $y = f(x_{1_i}, x_{2_i}, \cdots)$ is also possible, leading to a multidimensional response surface. The details are not discussed here, but the concepts generated for the single-variable analysis are carried through for multiple-variable analysis. The interested reader is referred elsewhere [3, 4, 6].

The companion LabView program *Polynomial_Fit* performs a least-squares regression analysis. It allows the user to enter data points manually or to read data from a file. Other software packages can also do this processing.

**EXAMPLE 4.9**

The following data are suspected to follow a linear relationship. Find an appropriate equation of the first-order form.

| x (cm) | y (V) |
|--------|-------|
| 1.0    | 1.2   |
| 2.0    | 1.9   |
| 3.0    | 3.2   |
| 4.0    | 4.1   |
| 5.0    | 5.3   |

**KNOWN** Independent variable $x$.

Dependent measured variable $y$.

$N = 5$.

**FIND** $y_c = a_0 + a_1 x$.

**SOLUTION**

We seek a polynomial of the form $y_c = a_0 + a_1 x$, that minimizes the term $D = \sum_{i=1}^{N} (y_i - y_{c_i})^2$. With the total derivative expressed as $dD = \dfrac{\partial D}{\partial a_0} da_0 + \dfrac{\partial D}{\partial a_1} da_1$, then setting each partial derivative to zero gives

$$\frac{\partial D}{\partial a_0} = 0 = -2 \left\{ \sum_{i=1}^{N} \left[ y_i - (a_0 + a_1 x_i) \right] \right\}$$

$$\frac{\partial D}{\partial a_1} = 0 = -2 \left\{ \sum_{i=1}^{N} \left[ y_i - (a_0 + a_1 x_i) \right] \right\}$$

so that $dD = 0$. Solving simultaneously for the coefficients $a_0$ and $a_1$ yields

$$a_0 = \frac{\sum x_i \sum x_i y_i - \sum x_i^2 \sum y_i}{\left( \sum x_i \right)^2 - N \sum x_i^2}$$

$$a_1 = \frac{\sum x_i \sum y_i - N \sum x_i y_i}{\left( \sum x_i \right)^2 - N \sum x_i^2}$$

(4.40)

Substituting the data set into Equation 4.40 gives $a_0 = 0.02$ and $a_1 = 1.04$. Hence

$$y_c = 0.02 + 1.04x \text{ V}$$

## Linear Polynomials

For linear polynomials a correlation coefficient $r$ can be found by

$$r = \frac{N \sum_{i=1}^{N} x_i y_i - \sum_{i=1}^{N} x_i \sum_{i=1}^{N} y_i}{\sqrt{N \sum_{i=1}^{N} x_i^2 - \left( \sum_{i=1}^{N} x_i \right)^2} \sqrt{N \sum_{i=1}^{N} y_i^2 - \left( \sum_{i=1}^{N} y_i \right)^2}}$$

(4.41)

The correlation coefficient provides a measure of the association between $x$ and $y$ as predicted by the form of the curve fit equation. It is bounded by $\pm 1$, which represents perfect correlation; the sign indicates that $y$ increases or decreases with $x$. For $\pm 0.9 < r \leq \pm 1$, a linear regression can be considered a reliable relation between $y$ and $x$. Alternatively, the $r^2$ value (called the coefficient of determination) is often reported, which is indicative of how well the variance in $y$ is accounted for by the fit. The correlation coefficient only indicates the hypothesis that $y$ and $x$ are associated; correlation does not imply cause and effect. The $r$ and $r^2$ values are not effective estimators of the random error in $y_c$; instead the $s_{yx}$ value is used.

The precision estimate in the slope of the fit $t_{\nu,P} s_{a1}$ can be calculated from

$$s_{a1} = s_{yx} \sqrt{\frac{1}{\sum\limits_{i=1}^{N} (x_i - x)}}$$

(4.42)

The precision estimate of the zero intercept $t_{\nu,P} s_{a0}$ can be calculated from

$$s_{a0} = s_{yx} \sqrt{\frac{\sum\limits_{i=1}^{N} x_i^2}{\left[ N \sum\limits_{i=1}^{N} (x_i - x) \right]}}$$

(4.43)

An error in $a_0$ would offset a calibration curve from its $y$ intercept. The derivation and further discussion on Equations 4.41 to 4.43 can be found elsewhere [1, 3, 4, 6].

## EXAMPLE 4.10

Compute the correlation coefficient and the standard error of the fit for the data in Example 4.9. Estimate the random uncertainty associated with the fit. Plot the curve fit with its 95% confidence interval.

**KNOWN** $y_c = 0.02 + 1.04x$ V

**ASSUMPTIONS** No systematic errors.

**FIND** $r$ and $s_{yx}$.

## SOLUTION
Direct substitution of the data set into Equation 4.41 yields the correlation coefficient of $r = 0.997$. An equivalent estimator is $r^2$. Here $r^2 = 0.99$, which indicates that 99% of the variance in $y$ is accounted for by the fit. The correlation supports the linear relation between $x$ and $y$.

The random uncertainty between the data and this fit is given by

$$t_{3,95} s_{yx} \left[ \frac{1}{N} + \frac{(x - \bar{x})^2}{\sum\limits_{i=1}^{N} (x_i - \bar{x})^2} \right]^{1/2} .$$

Using Equation 4.35, we compute the standard error of the fit, $s_{yx}$

$$s_{yx} = \sqrt{\frac{\sum\limits_{i=1}^{N} (y_i - y_{ci})^2}{\nu}} = 0.16$$

with degrees of freedom $v = N - (m + 1) = 3$. The $t$ estimator is, $t_{3,\,95} = 3.18$.

The curve fit is plotted in Figure 4.12 both using Equations 4.37 and 4.39. The approximation of equation 4.39 gives $t_{3,95}s_{yx}/\sqrt{N} = 0.23$. The 95% confidence interval is found using equation 4.37, which shows a slightly broader uncertainty band than equation 4.39 with uncertainties of $\pm 0.39$ at $x = 1$ and $x = 5$ and reducing to $t_{3,95}s_{yx}/\sqrt{N} = 0.23$ at the waist (where $x = \bar{x}$ ).

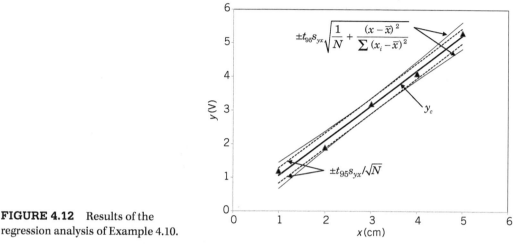

**FIGURE 4.12**   Results of the regression analysis of Example 4.10.

# 4.8 Data Outlier Detection

It is not uncommon to find a spurious data point that does not appear to fit the tendency of the data set. Data that lie outside the probability of normal variation incorrectly offset the sample mean value estimate, inflate the random uncertainty estimates, and influence a least-squares correlation. Statistical techniques can be used to detect such data points, which are known as *outliers*. Outliers may be the result of simple measurement glitches, such as a random spike in operating condition, or may reflect a more fundamental problem with test variable controls. Tools for outlier detection are discussed next. Once detected, the decision to remove the data point from the data set must be made carefully. If outliers are removed from a data set, the statistics are recomputed using the remaining data. The outlier tests are only used on the data set one time.

One approach to outlier detection is Chauvenet's criterion, which identifies outliers having less than a $1/2N$ probability of occurrence. To apply this criterion, let the test statistic be $z_0 = |x_i - \bar{x}|/s_x$ where $x_i$ is a suspected outlier in a data set of $N$ values. In terms of the probability values of Table 4.3, the data point is a potential outlier if

$$(1 - 2 \times P(z_0)) < \frac{1}{2N} \tag{4.44}$$

For large data sets, another approach, the *three-sigma test*,[7] identifies those data points that lie outside the range of 99.73% probability of occurrence, $\bar{x} \pm t_{v,99.7}s_x$, as being potential outliers. Either approach assumes that the sample set follows a normal distribution, which may not be true. Other methods for outlier detection are discussed elsewhere [3, 7].

---

[7]So named for as $v \to \infty$, the value of $t$ approaches 3 for 99.7% probability.

## EXAMPLE 4.11

Consider the data given below for 10 measurements of a tire's pressure made using an inexpensive handheld gauge (note: 14.5 psi = 1 bar). Test for outliers using the Chauvenet's criterion.

| $i$ | 1 | 2 | 3 | 4 | 5 | 6 | 7 | 8 | 9 | 10 |
|---|---|---|---|---|---|---|---|---|---|---|
| $x_i$ (psi) | 28 | 31 | 27 | 28 | 29 | 24 | 29 | 28 | 18 | 27 |

**KNOWN** Data values for $N = 10$.

**ASSUMPTIONS** Each measurement is obtained under fixed conditions.

**FIND** Apply outlier detection tests.

### SOLUTION

Based on the 10 data points, the sample mean and sample standard deviation are found from Equations 4.14a and c to be $\bar{x} = 27$ psi with an $s_x = 3.8$. But the tire pressure should not vary much between two readings beyond the precision capabilities of the measurement system and technique, so data point 9 ($x = 18$) is a potential outlier.

Apply Chauvenet's criterion to this data point. For $N = 10$, Equation 4.44 gives $(1 - 2 \times P(z_0)) < \frac{1}{2N} = \frac{1}{20} = 0.050$. So the criterion identifies the data point as an outlier if it lies outside $1 - 0.050 = 0.950$, which for 10 data points is the 95% probability spread of the data. For $x_i = 18$, $z_0 = 2.368$. Then $P(z_0) = P(2.368) = 0.4910$, so that $(1 - 2 \times 0.4910) = 0.018$. As $0.018 < 0.050$, this test identifies the data point $x = 18$ as a potential outlier.

## 4.9 Number of Measurements Required

We can use the previous discussions to assist in the design and planning of a test program. For example, how many measurements, $N$, are required to reduce the estimated value for random error in the sample mean to an acceptable level? To answer this question, begin with Equation 4.19:

$$x' = \bar{x} \pm t_{v,P} s_{\bar{x}} \quad (P\%) \tag{4.19}$$

Let CI be the confidence interval in Equation 4.18; that is,

$$\text{CI} = \pm t_{v,P} s_{\bar{x}} = \pm t_{v,P} \frac{s_x}{\sqrt{N}} \quad (P\%) \tag{4.45}$$

This interval is two-sided about the mean, defining a range from $-t_{v,P} \frac{s_x}{\sqrt{N}}$ to $+t_{v,P} \frac{s_x}{\sqrt{N}}$. We introduce the precision value $d$ as

$$d = \frac{\text{CI}}{2} = t_{v,P} \frac{s_x}{\sqrt{N}} \tag{4.46}$$

For example, if the confidence interval is $\pm 1$ units, an interval of width 2 units, then $d = 1$ unit. It follows that the required number of measurements is estimated by

$$N \approx \left( \frac{t_{v,P} s_x}{d} \right)^2 \quad (P\%) \tag{4.47}$$

Because the degrees of freedom in $t$ depends on $N$, solving Equation 4.47 requires iteration. Equation 4.47 provides a first estimate for the number of measurements needed. How closely it reduces the range of random error to the constraint depends on how well the assumed value of $s_x$ approximates the population $\sigma$.

A shortcoming of this method is the need to estimate $s_x$. In some cases, it can be based on experience of the population by using a known $\sigma$. Alternatively, the approach is to make a preliminary number of measurements, $N_1$, to obtain a sample estimate of $s_1$. Then use this $s_1$ to estimate the number of measurements required. The total number of measurements, $N_T$, will be estimated by

$$N_T \approx \left( \frac{t_{N-1,95} s_1}{d} \right)^2 \quad (95\%) \tag{4.48}$$

This establishes that $N_T - N_1$ additional measurements are required.

## EXAMPLE 4.12

Determine the number of measurements required to reduce the 95% confidence interval of the mean value of a variable to within $\pm 1$ unit if the variance of the population is estimated to be 64 units.

**KNOWN**    CI = $\pm 1$ units = 2 units        $P$ = 95%
             $d$ = 1                              $\sigma^2$ = 64 units·

**FIND** $N$ required.

**SOLUTION**
Equation 4.47 has two unknowns in $N$ and $t$

$$N \approx \left( \frac{t_{v,95} s_x}{d} \right)^2 \quad (95\%)$$

We iterate by using a trial-and-error approach until convergence. Suppose we begin by guessing that $N = 61$ so that $t_{v,95} = 2.00$. Then we solve Equation 4.47 with $t_{v,95} = 2.00$ and $s_x = 8$ units to find $N = 256$. We now use $N = 256$, so that $v = 255$ and $t_{v,95} = 1.96$. This gives the new estimate of $N = 245$. Repeat again with $N = 245$, so that $v = 244$, $t_{v,95} = 1.96$. We again get $N = 245$. The analysis is converged at 245 measurements. Check the results after 245 measurements to ensure that the variance level used was representative of the actual population.
**COMMENT** Since the confidence interval is reduced as $N^{1/2}$, the procedure of increasing $N$ to decrease this interval becomes one of diminishing returns.

## EXAMPLE 4.13

From 21 preliminary measurements of a variable, the standard deviation of the data set is 160 units. We want to reduce the 95% confidence interval in the mean value to $\pm 30$ units. Estimate the total number of measurements required.

**KNOWN**    $s_1$ = 160 units        $N_1$ = 21
             $d$ = CI/2 = 30          $t_{20,95}$ = 2.093·

**FIND** $N_T$.

## SOLUTION

At 21 measurements, the confidence interval in the mean value is $\pm(2.093)(160)/\sqrt{21} = \pm 73$ units. We need to reduce this to $\pm 30$ units. The total number of measurements required is estimated by

$$N_T \approx \left(\frac{t_{N-1,95}s_1}{d}\right)^2 = \left(\frac{2.093 \times 160}{30}\right)^2 = 125 \quad (95\%)$$

Thus, as a first guess, a total of 125 measurements are estimated to be necessary. Take an additional 104 measurements, then reanalyze to be certain the constraint is met.

## 4.10 Monte Carlo Simulations

When a result is computed from the values of one or more independent random variables, the variability in each of the independent variables directly affects the variability in the result. *Monte Carlo simulations* provide one way to incorporate such variability into predicting the behavior of the result. The simulation outcome is the predicted probability density function of the result, $p(R)$, and its associated statistics. This outcome makes it a very useful sampling method.

To illustrate a Monte Carlo simulation, consider result $R$ that is a known function of two variables through the parametric relationship, $R = f(x, y)$, as in Figure 4.13. Each variable is defined by its own probability density function, $p(x)$ and $p(y)$. Each iteration of a Monte Carlo simulation randomly draws one probable value for $x = x_i$ and for $y = y_i$ from their respective density functions to compute a value for $R = R_i$. This iteration process continues updating the data set for $R$ until the predicted standard deviation for $R$, that is, $s_R$, converges to an asymptotic value [8]. At convergence, the value of $s_R$ is assumed to represent $\sigma_R$ of the population. The quality of convergence must be traded against the cost of the simulation, but values within 1% to 5% suffice for most applications.

A Monte Carlo simulation is based on assumed distributions of variables, and so it is an approximation. Yet if the assumed distributions are reasonable, then the result of the approximation will be very good. The drawback is that the typical number of iterations required can be on the order of $10^4$ to $10^6$. The number of iterations can be reduced using improved sampling techniques [9].

Simulations can be run within spreadsheets, Matlab, or similar programs using built-in routines to facilitate sampling from a density function. For example, in spreadsheets, the RAND function samples from a rectangular distribution to generate a random number between 0 and 1, which can then be scaled to a population range. The NORMINV function samples from a normal distribution. In Matlab, these same operations use the RAND and NORMRND (or RANDN) functions, respectively.

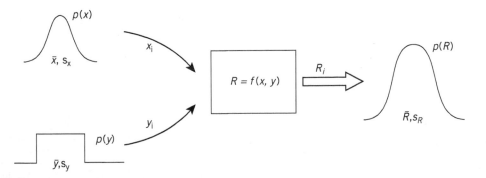

**FIGURE 4.13** Elements of a Monte Carlo simulation of $R = f(x, y, \ldots)$.

A spreadsheet add-in called **MCsim.xla** [11] and a Matlab m-file **MC.m** are included in the accompanying online files to serve as building blocks for coding Monte Carlo analyses.

**EXAMPLE 4.14**

A small current is passed through a resistance circuit board to achieve a prescribed voltage for a certain application. From experience with one particular board design, a manufacturer knows that it can expect a mean resistance of 1000 $\Omega$ with a standard deviation of 100 $\Omega$ with a population best described by a normal distribution. A nominal 100 mA current is passed through the circuit, which can be set to within 5 mA, a process described by a rectangular distribution. Each circuit is tested for tolerance control. Model the expected population for this voltage test using a Monte Carlo simulation.

**KNOWN** $\overline{\mathfrak{R}} = 1000\ \Omega$   $s_{\mathfrak{R}} = 100\ \Omega$ normal distribution.
$\overline{I} = 0.100$ A   $I_{max} = 0.105$ A   $I_{min} = 0.095$ A rectangular distribution.

**SOLUTION**
The parametric model is given by Ohm's law

$$E = f(I, \mathfrak{R}) = I\mathfrak{R}$$

where $E$ is the result and $\mathfrak{R}$ represents the resistance. The simulation starts by generating a random value for $I$ and $\mathfrak{R}$ based on a sampling of their respective density functions. A value for result $E$ is then computed. This process repeats itself throughout the simulation.

For a rectangular function, sampling is done using a random number generator set between $I_{max}$ and $I_{min}$. The standard deviation of a rectangular distribution is $\sqrt{(I_{max} - I_{min})/\sqrt{12}}$. In a spreadsheet, each $i$th random sample of current is given by

$$I_i = I_{min} + RAND(\ ) * (I_{max} - I_{min})$$

Here $(I_{max} - I_{min})$ appropriately scales the RAND value to the population. Similarly, the resistance is determined by sampling from a normal distribution. In a spreadsheet, this can be done with the NORMINV function. Each $i$th random sample for resistance is given by

$$\mathfrak{R}_i = NORMINV(RAND(\ ), \overline{\mathfrak{R}}, s_{\mathfrak{R}})$$

As new values for $I$ and $\mathfrak{R}$ are created, a new voltage is computed as

$$E_i = I_i\mathfrak{R}_i$$

This builds the population for voltage. In Figure 4.14, we show the result from 100,000 iterations showing a normal distribution with the following statistics:

| Variable, $x$ | $\overline{x}$ | $s_x$ |
|---|---|---|
| $E$ [V] | 100.003 | 10.411 |
| $I$ [A] | 0.100 | 0.0029 |
| $\mathfrak{R}$ [$\Omega$] | 999.890 | 99.940 |

So the voltage test population is normally distributed about a mean of 100.003 V with a standard deviation of 10.411 V.

**FIGURE 4.14** Predicted histogram for voltage (V) based on assumed distributions in current and resistance. Results of 100,000 Monte Carlo trials in Example 4.14.

## SUMMARY

The behavior of a random variable is defined by its unique probability density function, which provides exact information about its mean value and variability. The purpose of measurements is to infer the parameters of this density function from the acquired limited data set. The normal scatter of data about some central mean value arises from several contributing factors, including the process variable's own variation under nominally fixed operating conditions and random errors in the measurement system and in the measurement procedure. Furthermore, a finite number of measurements of a variable can only go so far in estimating the behavior of an entire population of values. We discuss how estimates based on a limited data set introduce another random error into predicting the true value of the measurement, something addressed by the sample statistics.

Statistics is a powerful tool used to interpret and present data. In this chapter, we developed the most basic methods used to understand and quantify finite-sized data sets. Methods of estimating the true mean value based on a limited number of data points and the resulting random uncertainty in such estimates were presented along with the treatment of data curve fitting. A summary table of these statistical estimators is given as Table 4.8. Monte Carlo methods were presented as an alternative tool to predict how variations in independent variables will affect the variation in a result. Throughout this chapter we have assumed negligible systematic error in the data set. In the next chapter, we expand our error considerations in our estimate of the true value of a measurement.

**Table 4.8    Summary Table for a Sample of *N* Data Points**

| | |
|---|---|
| Sample mean | $\bar{x} = \dfrac{1}{N} \sum\limits_{i=1}^{N} x_i$ |
| Sample standard deviation | $s_x = \sqrt{\dfrac{1}{N-1} \sum\limits_{i=1}^{N} (x_i - \bar{x})^2}$ |
| Standard deviation of the means[a] | $s_{\bar{x}} = \dfrac{s_x}{\sqrt{N}}$ |
| Precision interval for a single data point, $x_i$ | $\pm t_{v,P} s_x$ (P%) |
| Confidence interval[b, c] for a mean value, $\bar{x}$ | $\pm t_{v,P} s_{\bar{x}}$ (P%) |
| Confidence interval[b, d] for curve fit, $y = f(x)$ | $\pm t_{v,P} s_{yx} \left[ \dfrac{1}{N} + \dfrac{(x - \bar{x})^2}{\sum\limits_{i=1}^{N} (x_i - \bar{x})^2} \right]^{\frac{1}{2}}$ |

[a]Measure of random standard uncertainty in x.
[b]In the absence of systematic errors.
[c]Measure of random uncertainty in $\bar{x}$.
[d]Measure of random uncertainty in curve fit (see conditions of Equations. 4.37–4.39).

## REFERENCES

1. Kendal, M. G., and A. Stuart, *Advanced Theory of Statistics*, Vol. 2, Griffin, London, 1961.

2. Bendat, J., and A. Piersol, *Random Data Analysis*, Wiley, New York, 1971.

3. Lipson, C., and N. J. Sheth, *Statistical Design and Analysis of Engineering Experiments*, McGraw-Hill, New York, 1973.

4. Miller, I., and J. E. Freund, *Probability and Statistics for Engineers*, 3rd ed., Prentice Hall, Englewood Cliffs, NJ, 1985.

5. Bury, K., *Statistical Distributions in Engineering*, Cambridge Press, Cambridge, UK, 1999.

6. Vardeman, S. B., *Statistics for Engineering Problem Solving*, PWS Publishing Company, Boston, 1994.

7. Scheffe, H., *The Analysis of Variance*, Wiley Classics Series, Wiley Interscience, New York, 1999.

8. ASME/ANSI Power Test Codes, *Test Uncertainty PTC 19.1-2013*, American Society of Mechanical Engineers, New York, 2013.

9. Joint Committee for Guides in Metrology (JCGM), *Propagation of Distributions Using the Monte Carlo Method*, JCGM 101: JCGM, Sevres Cedex, France, 2008.

10. Devroye, L., *Non-Uniform Random Variate Generation*, Springer, New York, 1986.

11. Barreto, H., and Howland, F., *Introductory Econometrics: Using Monte Carlo Simulation with Microsoft Excel*, Cambridge Press, New York, 2005.

## NOMENCLATURE

| | | | |
|---|---|---|---|
| $a_0, a_1, \ldots, a_m$ | polynomial regression coefficients | $\langle s_x \rangle$ | pooled sample standard deviation of $x$ |
| $f_j$ | frequency of occurrence of a value | $\langle s_{\bar{x}} \rangle$ | pooled standard deviation of the means of $x$ |
| $x$ | measured variable; measurand | $\langle s_x \rangle^2$ | pooled sample variance of $x$ |
| $x_i$ | $i$th measured value in a data set | $s_{yx}$ | standard error of the (curve) fit between $y$ and $x$; |
| $x'$ | true mean value of the population of $x$ | | standard random uncertainty in a curve fit |
| $\bar{x}$ | sample mean value of $x$ | $t$ | Student's $t$ variable |
| $\langle \bar{x} \rangle$ | pooled sample mean of $x$ | $z$ | $z$ variable |
| $p(x)$ | probability density function of $x$ | $\beta$ | normalized standard variate |
| $s_x$ | sample standard deviation $x$ | $\sigma$ | true standard deviation of the population of $x$ |
| $s_{\bar{x}}$ | sample standard deviation of the means of $x$; standard random uncertainty in $\bar{x}$ | $\sigma^2$ | true variance of the population of $x$ |
| $s_x^2$ | sample variance of $x$ | $\chi^2$ | chi-squared value |
| | | $\nu$ | degrees of freedom |

# Uncertainty Analysis

**5**

## 5.1 Introduction

Whenever we plan a test or later report a test result, we want to know something about the quality of the result. Uncertainty analysis provides a methodical approach to estimating the quality of the result from a test. This chapter focuses on how to estimate the "± what?" in a test result.

Suppose the competent dart thrower of Chapter 1 tossed several practice rounds of darts at a bull's-eye. This would give us a good idea of the thrower's tendencies. Then, let the thrower toss another round. Without looking, can you guess where the darts will hit? Test measurements that include systematic and random error components are much like this. We can calibrate a measurement system to get a good idea of its behavior and accuracy. However, from the calibration we can only estimate how well any measured value might estimate the true value in a subsequent measurement.

*Errors are a property of the measurement*. Measurement is the process of assigning a value to a physical variable based on a sampling from the population of that variable. Error causes a difference between the value assigned by measurement and the true value of the population of the variable measured. Measurement errors are introduced from various elements, for example, the individual instrument calibrations, the data set finite statistics, and the approach used.

If we knew the exact value of an error, we would just correct for it. But most often we do not know the true value of the measured variable but rather only know its measured value instead. So we will not know the exact values of the errors affecting the measurement. Instead, we draw from what we do know to estimate a probable range to define the limits of error in the measurement. This estimate of the limits of the error is an assigned value called the *uncertainty*. The uncertainty is associated with an interval about the measured value within which we suspect that the true value must fall at a stated level of probability. *Uncertainty analysis* is the process of identifying, quantifying, and combining the estimated values of the errors in reporting test results.

*Uncertainty is a property of the result*. The outcome of a measurement is a result, and the uncertainty quantifies the quality of that result. Uncertainty analysis provides a powerful design tool useful for evaluating different measurement systems and methods, designing a test plan, and/or reporting uncertainty. This chapter presents a systematic approach for identifying, quantifying, and combining the estimates of the errors in a measurement. While the chapter stresses the methodology of analyses, we emphasize the concomitant need for an equal application of critical thinking and professional judgment in applying the analyses. The quality of an uncertainty analysis depends on the engineer's knowledge of the test, the measured variables, the equipment, and the measurement procedures [1].

*Errors are effects, and uncertainties are numbers*. In the context of an uncertainty analysis, an error is an effect that causes a measured value to differ from the true value. The uncertainty is an assigned numerical value that quantifies the probable limits of the error.

This chapter approaches uncertainty analysis as an evolution of information from test design through final data analysis. While the structure of the analysis remains the same at each step, the number of errors identified and their uncertainty values may change as more information becomes available. In fact, the uncertainty in the result may increase as the analysis considers more information. There is no exact answer to an analysis, just the result from a reasonable approach using honest numbers. This is the nature of an uncertainty analysis.

There are two accepted professional documents on uncertainty analysis. The American National Standards Institute/American Society of Mechanical Engineers (ANSI/ASME) Power Test Codes (PTC) 19.1 Test Uncertainty [2] is the United States engineering test standard, and our approach most closely follows that method. The International Organization on Standardization's "Guide to the Expression of Uncertainty in Measurement" (ISO GUM) [1] is an international metrology standard. The two differ in some terminology and how errors and their uncertainties are cataloged. For example, PTC 19.1 refers to random and systematic errors, terms that classify errors by how they influence the measurement. Random and systematic uncertainties are the values assigned to these errors. Knowing how an error affects a measurement provides useful insight into how to best interpret the results or improve the test methods. ISO GUM assigns type A and type B uncertainties to errors, terms that reflect only how the values assigned to errors are estimated. This information is often useful in calibration reports but also alerts the engineer to methods used. The differences in these two documents are real and valuable but they are not significant to the numerical outcome. Once past the classifications, the two methods are quite similar and can even be used together [2]. The important point is that the end outcome of an uncertainty analysis by either method will yield a similar numerical result!

Upon completion of this chapter, the reader will be able to

- Explain the relation between an error and an uncertainty

- Execute an appropriate uncertainty analysis corresponding to the level and quantity of information available

- Explain the differences between systematic and random errors and how to treat their assigned uncertainties

- Analyze a test system and approach from test design through data presentation to assign and propagate uncertainties

- Propagate uncertainties throughout a test to understand their impact on the final reported result

## 5.2 Measurement Errors

In the discussion that follows, we discuss how errors affect a result. For this, errors are grouped into two categories: systematic error and random error.

Consider the repeated measurement of a variable under conditions that are expected to produce the same value of the measured variable. The relationship between the true value of the population and the measured data set, containing both systematic and random errors, can be illustrated as in Figure 5.1. The total error in a set of measurements obtained under seemingly fixed conditions can be described by the systematic errors and the random errors in those measurements. The systematic errors shift the sample mean away from the true mean by a fixed amount, and within a sample of many measurements, the random errors bring about a distribution of measured values about the sample mean. Even a so-called accurate measurement contains small amounts of systematic and random errors. We do not consider the effects of measurement blunders that result in obviously fallacious data—such data should be discarded.

Measurement errors enter during all aspects of a test and obscure our ability to ascertain the information that we desire: the true value of the variable measured. If the result depends on more than one measured variable or test parameter, the errors associated with each entity will

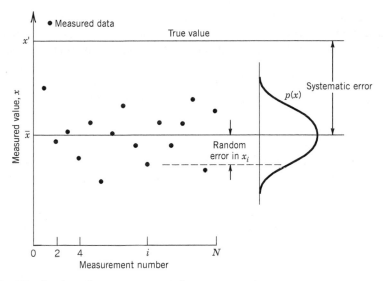

**FIGURE 5.1** Distribution of errors on repeated measurements.

propagate through the measurements and affect the result. In Chapter 4, we stated that the best estimate of the true value sought in a measurement is provided by its sample mean value and the uncertainty in that value,

$$x' = \bar{x} \pm u_x \quad (P\%) \tag{4.1}$$

Equation 4.1 remains correct, but in Chapter 4 we considered only the random uncertainty based on the statistics calculated from a measured data set in the $u_x$ term. In this chapter, we extend this so that the $u_x$ term includes the uncertainties assigned to all known errors, both systematic and random. Certain assumptions are implicit in an uncertainty analysis:

1. The test objectives are known and the measurement itself is a clearly defined process.

2. Any known corrections for systematic error have been applied to the data set, in which case any systematic uncertainty assigned to that error is the uncertainty of the correction.

3. Except where stated otherwise, we assume a normal distribution of errors and reporting of uncertainties.

4. Unless stated otherwise and for simplicity, the errors are assumed to be independent (uncorrelated) of each other. But errors are often correlated and we discuss how to handle these in Section 5.9.

5. The engineer has some "experience" with the system components.

With regard to point 3, we usually assume that the population of errors is best described by a normal distribution. But this is not always so. For example, when manufacturer specifications cite only "limits of error," the error source population might be better described by a rectangular distribution. This is discussed more in Section 5.7.

With regard to point 5, by "experience" we mean that the engineer either is able to assign an uncertainty value based on tests to enable assigning a value, based on past experience of what to expect from a particular measurement, or uses other trusted information, such as manufacturer's specifications or the technical literature.

We might begin the design of an engineering test with an idea and some catalogs, and end the project after data have been obtained and analyzed. As with any part of the design process, the uncertainty analysis evolves as the design of the measurement system and process matures. In the next several sections, we discuss uncertainty analysis for the following measurement situations: (1) design stage, where tests are being planned but information is limited; (2) advanced stage and single measurement analysis, where additional information about process control is used to improve a design-stage uncertainty estimate; and (3) multiple measurement analysis, where all available test information and test data are combined to assess the uncertainty in the measured test result. The methods we use for situation 3 follow current engineering standards.

**General versus Detailed Uncertainty Analysis** In this chapter, we discuss how uncertainty is propagated from individual contributions of error to the uncertainty in the final result. We are particularly interested in how each error affects the result. However, in performing an uncertainty analysis for measurement situations (1) and (2) discussed in the above paragraph, we often begin with a "general" analysis, one in which no distinction is made between random and systematic errors. In a general analysis, we simply assign uncertainty values as indicated by a value $u$. This is particularly useful when we have little information about the basis of each uncertainty value or its effect on the test outcome or are making a quick analysis perhaps so as to guide further test decisions.

For situation (3) above, and often for situation (2), we have sufficient information to distinguish between systematic and random errors and so we can assign uncertainty values separately for each error, thus enabling a "detailed" analysis. In a detailed analysis, systematic and random uncertainties are designated as values $b$ and $s$, respectively, until the end result is computed. This is the preferred approach for analyzing test information for product performance assurance and for improving test methods to achieve better results. The added value of using this detailed approach is that by understanding how individual errors affect a result, we can better understand how to interpret the results and/or improve the test.

## 5.3 Design-Stage Uncertainty Analysis

Design-stage uncertainty analysis refers to an analysis performed early on in the formulation of a test. It is intended to provide a quick estimate of the minimum uncertainty to be expected based on available information, which in the planning stage is usually just information on the instruments for the measurement technique planned or past experience with such measurements. If this uncertainty value is too large, then alternate approaches will need to be found. So, it is useful for selecting instruments and selecting between alternative measurement techniques. Identifying random and systematic errors separately in this preliminary analysis is unnecessary, so we suggest using a general analysis and designate each uncertainty value in this analysis with the symbol $u$.

The *zero-order uncertainty* of the instrument, $u_0$, attempts to estimate the expected variation in the measured values. As a minimum, this variation will be an amount on the level of the instrument resolution (i.e., interpolation error). Assign a numerical value to $u_0$ of either one-half of the analog instrument resolution[1] or its digital least count. This value will reasonably represent the uncertainty interval on either side of the reading with a probability of 95%. Then,

$$u_0 = \frac{1}{2}\text{resolution} = 1 \text{ LSD} \tag{5.1}$$

where LSD refers to the least significant digit of the readout. If information exists to better predict the expected variation in a measured value, such as from past experience with such measurements, use it instead.

---

[1]It is possible to assign a value for $u_0$ that differs from half the scale resolution. Discretion should be used.

The second piece of information that is usually available is the manufacturer's statement concerning instrument error. We can assign this stated value as the *instrument uncertainty*, $u_c$. If no probability level is provided with such information, then assuming a 95% level is reasonable. Of course, if additional information exists to better predict instrument uncertainty, use it. Often, experienced users have a good idea of what to expect based on their prior use of equipment and techniques and these values should be used at the design-stage. We next look at a way to combine individual uncertainty values into a single design-stage analysis estimate.

## Combining Elemental Errors: RSS Method

Each individual measurement error interacts with other errors to affect the overall uncertainty of a measurement. This is called *uncertainty propagation*. Each individual error is called an "elemental error." Errors are often delineated into elemental parts, each part due to an individual contributing factor (recall Table 1.1). For example, the sensitivity error and linearity error of a transducer are two elemental errors, and the numbers associated with these are their uncertainties. These individual uncertainties must be combined in some reasonable manner. An accepted approach of combining uncertainties is termed the *root-sum-squares* (RSS) method.

Consider a measurement of $x$ that is subject to some $K$ elements of error, each of uncertainty $u_k$, where $k = 1, 2, \ldots, K$. If we know that (1) each error is independent of the other elemental errors and (2) assume that the measured value is equally sensitive to each of the elemental errors, then the uncertainty in the measured variable, $u_x$, due to these elemental errors can be computed using the RSS *method*:

$$
\begin{aligned}
u_x &= \sqrt{u_1^2 + u_2^2 + \cdots + u_k^2} \\
&= \sqrt{\sum_{k=1}^{K} u_k^2} \quad (P\%)
\end{aligned}
\tag{5.2}
$$

The RSS method of combining uncertainties is based on the assumption that the square of an uncertainty is a measure of the variance (i.e., $s^2$) assigned to an error, and the propagation of these variances yields a probable estimate of the total uncertainty. Note that it is imperative to maintain consistency in the units of each uncertainty in Equation 5.2 and to assign each uncertainty term at the same probability level.

When a probability level equivalent to a spread of one standard deviation is used, this uncertainty is called the "*standard*" *uncertainty* (1, 2). For a normal distribution, a standard uncertainty equates to a 68% probability level. In test engineering, it is common to report final uncertainties at a 95% probability level ($P\% = 95\%$), and this is equivalent to assuming the probability covered by two standard deviations. Whatever level is used, consistency is important.

## Design-Stage Uncertainty

The *design-stage uncertainty*, $u_d$, for an instrument or measurement method is an interval found by combining the instrument uncertainty with the zero-order uncertainty,

$$
u_d = \sqrt{u_0^2 + u_c^2} \quad (P\%)
\tag{5.3}
$$

This procedure for estimating the design-stage uncertainty is outlined in Figure 5.2. The design-stage uncertainty for a test system is arrived at by combining each of the design-stage uncertainties for each component in the system using the RSS method while maintaining consistency of units and confidence levels.

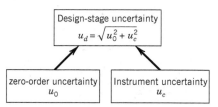

**FIGURE 5.2**  Design-stage uncertainty procedure in combining uncertainties.

Due to the limited information used, a design-stage uncertainty estimate is intended only as a guide for selecting equipment and procedures before a test, and is never used for reporting results. *If additional information about other measurement errors is known at the design stage, then their uncertainties can and should be used instead in Equation 5.3.* So Equation 5.3 provides a minimum value for uncertainty. In later sections of this chapter, we move toward more thorough uncertainty analyses.

### EXAMPLE 5.1

A force-measuring instrument comes with a certificate of calibration that identifies two instrument errors and assigns each an uncertainty at 95% confidence over its range. Provide an estimate of the instrument design-stage uncertainty. What is the instrument standard uncertainty?

| | |
|---|---|
| Resolution: | 0.25 N |
| Range: | 0 to 200 N |
| Linearity error: | within 0.20 N over range |
| Hysteresis error: | within 0.30 N over range |

**KNOWN**  Catalog specifications.

**ASSUMPTIONS**  Instrument uncertainty at 95% level; normal distribution.

**FIND**  $u_c, u_d$.

**SOLUTION**
We follow the procedure outlined in Figure 5.2. An estimate of the instrument uncertainty depends on the uncertainty assigned to each of the contributing elemental errors of linearity, $e_1$, and hysteresis, $e_2$, respectively assigned as

$$u_1 = 0.20 \text{ N} \quad (95\%) \qquad u_2 = 0.30 \text{ N} \quad (95\%)$$

Then using Equation 5.2 with $K = 2$ yields the instrument uncertainty

$$u_c = \sqrt{(0.20)^2 + (0.30)^2}$$
$$= 0.36 \text{ N}$$

The instrument resolution is given as 0.25 N, from which we assume $u_0 = 0.125$ N. From Equation 5.3, the design-stage uncertainty of this instrument when used in a measurement would be

$$u_d = \sqrt{u_0^2 + u_c^2} = \sqrt{(0.125)^2 + (0.36)^2}$$
$$= \pm 0.38 \text{ N} \quad (95\%)$$

We assume that the 95% confidence interval stated in the specifications reflects a coverage factor of two standard deviations of a normal distribution. That is, a 95% coverage equates to 1.96 standard deviations and this value is rounded to 2. A standard uncertainty reflects a coverage factor of one standard deviation. Hence, the instrument standard uncertainty at a 68% confidence level would be

$$u_c = 0.36/2 = 0.18\text{N} \quad (68\%)$$

## EXAMPLE 5.2

A voltmeter is used to measure the electrical output signal from a pressure transducer. The nominal pressure is expected to be about 3 psi (3 lb/in.$^2$ = 0.2 bar). Estimate the design-stage uncertainty in this combination. The following information is available:

**Voltmeter**

| | |
|---|---|
| Resolution: | 10 μV |
| Accuracy: | within 0.001% of reading |

**Transducer**

| | |
|---|---|
| Range: | ±5 psi (~ ±0.35 bar or ~ ±260 mm Hg) |
| Sensitivity: | 1 V/psi |
| Input power: | 10 VDC ± 1% |
| Output: | ±5 V |
| Linearity error: | within 2.5 mV/psi over range |
| Sensitivity error: | within 2 mV/psi over range |
| Resolution: | Negligible |

**KNOWN** Instrument specifications.

**ASSUMPTIONS** Values at 95% probability; normal distribution of errors.

**FIND** $u_c$ for each device and $u_d$ for the measurement system.

**SOLUTION**
The procedure in Figure 5.2 is used for both instruments to estimate the design-stage uncertainty in each. The resulting uncertainties are then combined using the RSS approximation to estimate the system $u_d$.

The uncertainty in the voltmeter at the design stage is given by Equation 5.3 as

$$(u_d)_E = \sqrt{(u_o)_E^2 + (u_c)_E^2}$$

From the information available,

$$(u_0)_E = 5 \text{ μV}$$

For a nominal pressure of 3 psi, we expect to measure an output of 3 V. Then,

$$(u_c)_E = (3 \text{ V} \times 1.0 \times 10^{-5}) = 30 \text{ μV}$$

so that the design-stage uncertainty in the voltmeter is

$$(u_d)_E = 30.4 \ \mu V$$

The uncertainty in the pressure transducer output at the design stage is also found using Equation 5.2. Assuming that we operate within the input power range specified, the instrument output uncertainty can be estimated by considering the uncertainty in each of the instrument elemental errors of linearity, $e_1$, and sensitivity, $e_2$:

$$(u_c)_p = \sqrt{u_1^2 + u_2^2}$$
$$= \sqrt{(2.5 \ mV/psi \times 3 \ psi)^2 + (2 \ mV/psi \times 3 \ psi)^2}$$
$$= 9.61 \ mV$$

Since $(u_0) \approx 0$ V/psi, the design-stage uncertainty in the transducer in terms of indicated voltage is $(u_d)_p = 9.61$ mV.

Finally, $u_d$ for the combined system is found by using the RSS method for the design-stage uncertainties of the two devices. The design-stage uncertainty in pressure as indicated by this measurement system is estimated to be

$$u_d = \sqrt{(u_d)_E^2 + (u_d)_p^2}$$
$$= \sqrt{(0.030 \ mV)^2 + (9.61 \ mV)^2}$$
$$= 9.6 \ mV$$

But since the sensitivity is 1 V/psi, the uncertainty in terms of pressure can be stated as

$$u_d = \pm 0.0096 \ psi \quad (95\%)$$

This is equivalent to $\pm 0.5$ mm Hg (95%).

**COMMENT** Design-stage uncertainty analysis shows us that the transducer, not the voltmeter, dominates the uncertainty in this setup!

## 5.4 Identifying Error Sources

Design-stage uncertainty is intended to provide a first estimate into the uncertainty of a test. It does not explicitly address all of the possible known errors that influence a measured result. Before proceeding to more detailed analyses, we provide a checklist to assist in identifying common errors. It is not necessary to classify error sources as we do here, but it is helpful.

Suppose we break down the measurement process into three distinct stages: calibration, data acquisition (measurement), and data reduction. Errors that enter during each of these steps can be grouped under their respective error source heading: (1) calibration errors, (2) data-acquisition errors, and (3) data-reduction errors. Within each of these three *error source groups*, we can list the types of errors encountered. Such errors are the elemental errors of the measurement. Later, we will assign uncertainty values to each error.

Do not become preoccupied with these groupings. Use them only as a guide. If you place an error in different group, it is okay. The important thing is that you recognize each elemental error and consider its potential uncertainty The group you place it in does not affect the final uncertainty!

Table 5.1  **Calibration Error Source Group: Common errors**

|  | Element of error source[a] |
|---|---|
| 1 | Standard or reference value errors |
| 2 | Instrument or system errors |
| 3 | Calibration process errors |
| 4 | Calibration curve fit (or see Table 5.3) |

[a]Systematic error or random error in each element.

## Calibration Errors

Calibration does not eliminate system errors but it provides a way to quantify the uncertainty in the particular pieces of equipment used. We assign the calibration errors to those elemental errors that enter the measuring system during its calibration. *Calibration errors* tend to enter through three sources: (1) the standard or reference value used in the calibration, (2) the instrument or system under calibration, and (3) the calibration process. For example, the laboratory standard used for calibration contains some inherent uncertainty, and this is passed along to the calibrated instrument. Measuring system errors, such as linearity, repeatability, hysteresis, contribute to overall instrument uncertainty. A curve fit of the calibration data contains uncertainty (Section 4.6). All of these effects are built into the calibration data. In Table 5.1, we list the common elemental errors that commonly contribute to this error source group. As a note, the number shown associated with each element of error is just an arbitrary counter—it has no other significance or meaning.

## Data-Acquisition Errors

Errors arising during the act of acquiring data are listed as a data-acquisition error. These errors include other sensor and instrument errors unaccounted for by calibration; uncontrolled variables, such as changes or unknowns in measurement system operating conditions; and sensor installation effects on the measured variable. In addition, the inferential statistics of sampled variables and any other temporal and spatial variations that contribute to data scatter, all contribute to uncertainty. We list some common elemental errors from this source in Table 5.2.

Table 5.2  **Data-Acquisition Error Source Group: Common errors**

|  | Element of error source[a] |
|---|---|
| 1 | Measurement system operating conditions |
| 2 | Sensor–transducer stage (instrument error) |
| 3 | Signal–conditioning stage (instrument error) |
| 4 | Output stage (instrument error) |
| 5 | Process operating conditions |
| 6 | Sensor installation effects |
| 7 | Environmental effects |
| 8 | Spatial variation error |
| 9 | Temporal variation error |
| etc. |  |

[a]Systematic error or random error in each element.
*Note*: A total input-to-output measurement system calibration combines elements 2, 3, 4, and possibly 1 within this error source group.

**Table 5.3    Data-Reduction Error Source Group: Common errors**

|  | Element of error source[a] |
|---|---|
| 1 | Curve fit error |
| 2 | Truncation error |
| 3 | Modeling error |
| etc. | |

[a]Systematic error or random error in each element.

## Data-Reduction Errors

Curve fits and correlations of data (Section 4.6) introduce data-reduction errors into test results. Also, truncation errors, interpolation errors, and errors from assumed models or functional relationships affect the quality of a result. We list some elemental errors typical of this error source group in Table 5.3.

## 5.5  Systematic and Random Errors and Standard Uncertainties

### Systematic Error

A systematic error[2] remains constant in repeated measurements under fixed operating conditions. A systematic error may cause either a high or a low offset in the estimate of the true value of a measured variable. The estimate of the probable range of a systematic error is assigned a value called the systematic uncertainty. The *systematic standard uncertainty*, is represented by the symbol, $b$. It can be assigned over an interval, $\pm b$, which has a confidence level of one standard deviation, equivalent to a probability level of 68% for a normal distribution (see Section 5.10 for assigning nonsymmetrical intervals). The *systematic uncertainty* assigned at some desired confidence level is given by $t_{v,\ p}b$, or simply $tb$. When stated at a 95% probability level, it is defined by the interval, $\pm B$, where

$$B = 2b \tag{5.4}$$

which assigns a value of $t = 2$ or two standard deviations. This $t$ value assumes large degrees of freedom in the assigned systematic uncertainty for which $t_{95} = 1.96$, which is rounded to 2 for convenience [2].

*Correction for a known systematic error* If the value of a particular systematic error is known then the data can be corrected to account for it. The value of the systematic standard uncertainty assigned to the corrected measurement would be the uncertainty in the correction applied.

The reader has probably experienced systematic errors in common measurements. For example, improperly using the floating tang[3] on the end of a metal tape measure will offset the measurement, a systematic error. The inability to exactly zero a gauge or device before using it offsets its measured value. In fact, the action of zeroing a gauge is an example of correcting for a known systematic error and the systematic uncertainty assigned is associated with how closely one can set zero.

---

[2]This error was called "bias" error in engineering documents prior to the 1990s.
[3]The tang is the hook at the end of the tape usually attached by loose rivets.

Consider a home bathroom scale indicating your body weight; does the reading contain a systematic error? How might we assign a systematic uncertainty to its indicated weight? Perhaps we notice that it measures differently from the measurement taken in a physician's office or at the gym. By comparing the readings (i.e., a sort of interlaboratory comparison), we can estimate the limits of possible systematic error. Or maybe we decide to calibrate the scale using calibrated standard masses, account for local gravitational acceleration, and correct the output, thereby estimating the systematic error of the measurement and assigning an uncertainty to this value (i.e., direct calibration against a local standard). Or we could carefully measure body volume displacement in water and compare the results to estimate differences (i.e., concomitant methodology). Or, we can use the specification provided by the manufacturer (i.e., experience). Without any of the above, what value would we assign? Would we even suspect a systematic error?

But let us think about this. The insidious aspect of a systematic error has been revealed: it is difficult to recognize because it is an offset, it does not change on repeated measurements. Why suspect a systematic error? Experience teaches us to think through each measurement carefully because systematic error is always present at some magnitude. We see that it is difficult to recognize systematic errors without comparison, so a good test design includes some means to estimate them. Various methodologies can be utilized: (1) calibration, (2) concomitant methodology, (3) interlaboratory comparisons, or (4) judgment/experience. Calibration can reveal and reduce instrument systematic errors and assist in assigning values to its associated uncertainty. Concomitant methodology uses different methods for estimating the same thing allowing the means to compare the results [e.g., see reference 3]. Lastly, interlaboratory comparisons, an excellent replication method, introduce different instruments, facilities, and/or personnel into an otherwise identical measurement procedure. The variations in the results between several facilities can provide a statistical estimate of the systematic uncertainty [2].

In lieu of the above, a judgment value based on past experience may have to be assigned; judgment values are usually understood to be made at the 95% confidence level. For example, the value that first came to mind to you in the bathroom scale example above likely covered a 95% interval.

Note that calibration cannot eliminate systematic error, but it may reduce its uncertainty. Consider the calibration of a temperature transducer against a National Institute of Standards and Technology (NIST) standard certified to be correct to within 0.01°C. If the calibration data show that the transducer output has a systematic offset of 0.2°C relative to the standard, then we would just correct all the data obtained with this transducer by 0.2°C. Simple enough, we correct it! But the standard itself still has an intrinsic systematic uncertainty of 0.01°C, and this uncertainty plus any uncertainty in the correction value applied remains in the calibrated transducer.

## Random Error

When repeated measurements are made under fixed operating conditions, random errors show themselves as scatter of the measured data. Hence, the effects of these errors are easily observed. Random error[4] is introduced through the repeatability of the measurement system components, calibration, and measurement procedure and technique; by the measured variable's own natural variability as estimated by sample statistics; and by the variations in the process operating and environmental conditions from which the measurements are taken.

The estimate of the probable range of a random error is a value called the random uncertainty. The *random standard uncertainty* is defined by the symbol $s_{\bar{x}}$. It is defined over an interval given by $\pm s_{\bar{x}}$, where

$$s_{\bar{x}} = s_x / \sqrt{N} \tag{5.5}$$

---

[4]This error was called a "precision" error in engineering documents prior to the 1990s.

with degrees of freedom $\nu = N - 1$ and assumes that the errors are normally distributed.[5] The interval has a confidence level of one standard deviation, equivalent to a probability of 68% for a population of $x$ having a normal distribution. The *random uncertainty* at a desired confidence level is given by $t_{\nu, p} s_{\bar{x}}$, or simply $ts$, where $t$ is found from Table 4.4.

## Other Ways Used to Classify Error and Uncertainty

In the ISO GUM (1), an uncertainty assigned based on a statistical analysis of a data set is called a "Type A" uncertainty. An uncertainty assigned based on any nonstatistical approach is called a "Type B" uncertainty. The intent is to inform the ultimate user as to the method used to assign the value. The ISO GUM recommends tagging assigned uncertainties as "A" or "B" to keep track. The designations are particularly targeted to manufacturer calibration and instrumentation specifications so that calibration details passed on to a user are clear.

The terms systematic error and random error describe how identified errors affect a test result. In ASME PTC 19.1 (2), the terms systematic uncertainty and random uncertainty are the values assigned to these errors. The ISO GUM does not use these uncertainty terms. An error assigned an uncertainty value classified as either a Type A or Type B could have either a systematic or random effect on a test result [1, 2]. There is no direct correlation between the different terms and this makes it possible to use both methods simultaneously [2]. For example, a statistical analysis of available data used to estimate an average offset within a set of data, that is, a value used to assign a systematic standard uncertainty, $b$, would also be tagged as a Type A uncertainty.

The computational community has introduced other terms in describing error and uncertainty in their models. *Aleatory uncertainty* refers to that caused by the natural variability within a system or a parameter. *Epistemic uncertainty* refers to that arising due to a lack of knowledge, such as due to deficiencies in the model or simplifications used. Epistemic errors are quantified through parametric variation and sensitivity analyses, probability modeling, or by other means. There are no direct correlations with the other terms described.

Each approach has its place [2]. Tagging uncertainties as Type A and Type B in instrument calibration certification eliminates ambiguity in interpreting specifications. Separating random and systematic effects in reporting test results is a valuable aid for interpreting the results, as well as providing a best practice to improve on the test methods used.

## 5.6 Uncertainty Analysis: Multi-Variable Error Propagation

Suppose we want to determine how long it would take to fill a small swimming pool from a garden hose. One way is to measure the time required to fill a bucket of known volume to estimate the flow rate from the garden hose. Armed with a measurement of the volume of the pool, we can calculate the time to fill the pool. Clearly, small errors in estimating the flow rate from the garden hose would translate into large differences in the time required to fill the pool! Here we are using measured values, the flow rate, and volume, to estimate a result, the time required to fill the pool.

Very often in engineering, a result is determined through a functional relationship between measured values. For example, we just calculated a flow rate above by measuring time, $t$, and bucket volume, $\forall$, since $Q = f(t, \forall) = \forall/t$. But how do uncertainties in either measured quantity contribute to the uncertainty in flow rate? Is the uncertainty in $Q$ more sensitive to uncertainty in volume or in time? More generally, how are uncertainties in variables propagated to a calculated result? We now explore these questions.

---

[5]The standard uncertainty when estimated from a rectangular distribution (1) is discussed in Section 5.7.

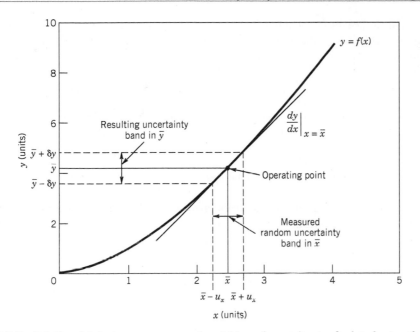

**FIGURE 5.3**   Relationship between a measured variable and a resultant calculated using the value of that variable.

## Propagation of Error

A general relationship between some dependent variable $y$ and a measured variable $x$, that is, $y = f(x)$, is illustrated in Figure 5.3. Now suppose we measure $x$ a number of times at some operating condition so as to establish its sample mean value and the uncertainty $u_x$. This implies that the true value for $x$ lies somewhere within the interval $\bar{x} \pm u_x$. It is reasonable to assume that the true value of $y$, which is determined from the measured values of $x$, falls within the interval defined by

$$\bar{y} \pm \delta y = f(\bar{x} \pm u_x) \tag{5.6}$$

Expanding this as a Taylor series yields

$$\bar{y} \pm \delta y = f(\bar{x}) \pm \left[ \left( \frac{dy}{dx} \right)_{x=\bar{x}} u_x + \frac{1}{2} \left( \frac{d^2 y}{dx^2} \right)_{x=\bar{x}} (u_x)^2 + \cdots \right] \tag{5.7}$$

By inspection, the mean value for $y$ must be $f(\bar{x})$ so that the term in brackets estimates $\pm \delta y$. A linear approximation for $\delta y$ can be made, which is valid when $u_x$ is small and neglects the higher order terms in Equation 5.7, as

$$\delta y \approx \left( \frac{dy}{dx} \right)_{x=\bar{x}} u_{\bar{x}} \tag{5.8}$$

The derivative term, $(dy/dx)_{x=\bar{x}}$, defines the slope of a line that passes through the point specified by $\bar{x}$. For small deviations from the value of $\bar{x}$, this slope predicts an acceptable, approximate relationship between $u_x$ and $\delta y$. The derivative term is a measure of the sensitivity of $y$ to changes in $x$. The width of the interval defined by $\pm u_x$ corresponds to $\pm \delta y$, within which we should expect the true value of $y$ to lie. Then, the uncertainty in $y$ due to $x$ is $u_y = \delta y$.

Figure 5.3 illustrates the concept that errors in an independent variable are propagated through to a resultant variable in a predictable way. In general, the uncertainty in $x$ affects the uncertainty in the resultant $y$ as

$$u_y = \left(\frac{dy}{dx}\right)_{x=\bar{x}} u_x \qquad (5.9)$$

Compare the similarities between Equations 5.8 and 5.9 and in Figure 5.3.

This idea can be extended to multivariable relationships. Consider a result $R$, which is determined through some functional relationship between independent variables $x_1, x_2, \ldots, x_L$ defined by

$$R = f_1\{x_1, x_2, \ldots, x_L\} \qquad (5.10)$$

where $L$ is the number of independent variables involved. Each variable contains some measure of uncertainty that affects the result. The best estimate of the true mean value $R'$ would be stated as

$$R' = \bar{R} \pm u_R \quad (P\%) \qquad (5.11)$$

where the sample mean of $R$ is found from

$$\bar{R} = f_1\{\bar{x}_1, \bar{x}_2, \ldots, \bar{x}_L\} \qquad (5.12)$$

and the uncertainty in $\bar{R}$ is found from

$$u_R = f_1\{u_{\bar{x}_1}, u_{\bar{x}_2}, \ldots, u_{\bar{x}_L}\} \qquad (5.13)$$

In Equation 5.13, each $u_{\bar{x}_i}$, $i = 1, 2, \ldots, L$ represents the uncertainty associated with the best estimate of $x_1$ and so forth through $x_L$. The value of $u_R$ reflects the contributions of the individual uncertainties as they are propagated through to the result.

A general sensitivity index, $\theta_i$, results from the Taylor series expansion, Equation 5.9, and the functional relation of Equation 5.10 and is given by

$$\theta_i = \frac{\partial R}{\partial x_i}\bigg|_{x=\bar{x}} \quad i = 1, 2, \ldots, L \qquad (5.14)$$

Each sensitivity index relates how $R$ is affected by changes in the corresponding variable $x_i$. Each is evaluated at either the mean values of the measurement or, lacking these estimates, the expected values of the corresponding variables. Equation 5.14 can also be estimated numerically using finite differencing methods [5], as discussed next.

The contribution of the uncertainty in $x$ to the result $R$ is estimated by the term $\theta_i u_{\bar{x}_i}$. The most probable estimate of $u_R$ is generally accepted as that value given by the second power relation [4], which is the square root of the sum of the squares (RSS). The propagation of uncertainty in the variables to the result is by

$$u_R = \left[\sum_{i=1}^{L} (\theta_i u_{\bar{x}_i})^2\right]^{1/2} \qquad (5.15)$$

## Approximating a Sensitivity Index

There are ways to estimate a sensitivity index other than by direct differentiation. One useful numerical approach is to use *dithering*. Here a small perturbation is applied to an input variable of a functional model and the effect on the output variable value noted. The change in output value relative to the change in input value provides the estimate of the sensitivity. The method uses a finite difference method to approximate the derivatives. Dithering is particularly useful when the functional information is built into a numerical or experimental model or built within a spreadsheet or whenever analytical differentiation is cumbersome.

Given that $y = f(x)$, then for $x = x_0$, let $y^+ = f(x + \delta x)$, $y^- = f(x - \delta x)$, such that

$$\frac{\partial y}{\partial x}\Big|_{x=x_0} = \frac{y^+ - y^-}{2\delta x} \qquad (5.16)$$

To ensure an accurate estimate of the sensitivity index, the value of $\delta x$ is successively reduced until the value of $\partial y/\partial x$ converges on a single answer, that is, it no longer changes in value within an acceptable tolerance.

### EXAMPLE 5.3

Suppose we have a calibration model given by $y = x^3 + 2x^2 + 2x + 1$. Then we know that $\partial y/\partial x = 3x^2 + 4x + 2$. Estimate the sensitivity index $\theta = \partial y/\partial x$ about the value $x = 4$ by dithering.

### SOLUTION
Analytically we can solve for $y = f(x)$ for any $x$ as

| $x$ | 0.000 | 2.000 | 4.000 | 6.000 | 8.000 | 10.00 |
|---|---|---|---|---|---|---|
| $y$ | 1.00 | 21.00 | 105.0 | 301.0 | 657.0 | 1221.0 |
| $\theta = \partial y/\partial x\big|_{x_0}$ | 2.00 | 22.00 | 66.00 | 134.00 | 226.00 | 342.00 |

But we can also use dithering to approximate the sensitivity index about any point $x$. We do this by successively changing the value of $x$ by an amount $\delta x$ and calculating the finite difference derivative using Equation 5.16. Using the point $x = 4.000$ for illustration:

| $\delta x$ | 0.4000 | 0.2000 | 0.1000 | 0.0500 | 0.0100 |
|---|---|---|---|---|---|
| $y+$ | 133.704 | 118.768 | 111.741 | 108.3351 | 105.6614 |
| $y-$ | 80.776 | 92.352 | 98.539 | 101.7349 | 104.3414 |
| $\theta = \partial y/\partial x\big|_{x_0}$ | 66.16 | 66.04 | 66.01 | 66.00 | 66.00 |

The solution converges for $\delta x \leq 0.05$. The approximate value found by dithering $\theta = \partial y/\partial x\big|_{x=4} = 66.00$ agrees with the value found by exact differentiation.

**COMMENT** The procedure of testing against a known result illustrates a verification test that the numerical technique applied works properly.

In multivariable relationships, such as $R = f_1\{x_1, x_2, \ldots, x_L\}$, the result will be sensitive to changes in the measured variables and test parameters. Each individual sensitivity can be found by dithering the independent variables and test parameters one at a time while holding all other variables constant.

### EXAMPLE 5.4

A measurement model follows $R = Kx^3$, where $x$ is the measured variable and $K$ is a coefficient. Estimate the sensitivities and propagation to uncertainty in the result if: $K = 2.201$ kg/mV, $u_k = 0.0220$ kg/mV; $\bar{x} = 1.234$ mV, $u_k = 0.0100$ mV.

**KNOWN**

$$R = Kx^3$$

$$\bar{x} = 1.234 \text{ mV} \qquad u_{\bar{x}} = 0.0100 \text{ mV} \qquad (95\%)$$

$$K = 2.201 \text{ kg/mV} \qquad u_K = 0.0220 \text{ kg/mV} \qquad (95\%).$$

**SOLUTION**
We evaluate the mean result and its uncertainty as

$$\bar{R} = f(\bar{x}, K) = K\bar{x}^3 = 4.136 \text{ kg} \quad \text{and} \quad u_R = f(u_x, u_K) = \left[\left(\frac{\partial R}{\partial x} u_x\right)^2 + \left(\frac{\partial R}{\partial K} u_K\right)^2\right]^{\frac{1}{2}}$$

We use dithering to estimate the sensitivities, $\theta_x = \partial R/\partial x$ and $\theta_K = \partial R/\partial K$ about the values $\bar{x} = 1.234$ mV and $K = 2.201$ kg/mV, as

| $\delta x$ | 0.100 | 0.010 | 0.001 | $\delta K$ | | 0.100 | 0.022 | 0.002 |
|---|---|---|---|---|---|---|---|---|
| $\delta x^+$ | 5.2250 | 4.2372 | 4.1459 | $\delta K^+$ | | 4.3238 | 4.1772 | 4.1396 |
| $\delta x^-$ | 3.2097 | 4.0361 | 4.1258 | $\delta K^-$ | | 3.9479 | 4.0945 | 4.1321 |
| $\theta_x = \partial R/\partial x\|_{\bar{x},K}$ | 10.077 | 10.055 | 10.055 | $\theta_K = \partial R/\partial K\|_{\bar{x},K}$ | | 1.8791 | 1.8791 | 1.8791 |

So, $\partial R/\partial x = 10.055$ kg/mV and $\partial R/\partial K = 1.8791$ mV. Then,

$$u_R = 0.1087 \text{ kg}$$

$$R' = 4.136 \pm 0.1087 \text{ kg} \quad (95\%)$$

**COMMENT** We used values about one order above and below the stated uncertainty values for dithering. Alternatively, a value of 1% of the variable value is a good start. In highly nonlinear problems, smaller perturbation values may be needed to get convergence.

---

The sensitivity index can also be estimated from any information that conveys functional information between two variables, including tabulated data and graphs (such as Figure 5.3), by using finite differencing methods. Suppose we have functional information in tabular or graphical form, such as $y = f(x)$,

| | $x_1$ | $x_2$ | $x_3$ |
|---|---|---|---|
| $y$ | $y(x_1)$ | $y(x_2)$ | $y(x_3)$ |

where $x_1 < x_2 < x_3$. Then to find the sensitivity about a point $y(x_2)$, we could use a central finite differencing approximation for the derivative

$$\theta = \frac{\partial y}{\partial x}\Big|_{x=x_2} = \frac{y(x_3) - y(x_1)}{x_3 - x_1} \tag{5.17}$$

Similarly, one could use forward- or backward-differencing approximations.

In an experimental setup, sensitivities can be estimated physically by allowing a small change in an input variable or test parameter both above and below a setpoint while noting the change in output value.

# Sequential Perturbation

A similar numerical approach can be used to estimate the propagation of uncertainty through to a result directly and is called *sequential perturbation* [6]. The method uses a finite difference method to approximate the derivatives:

1. Based on measurements for the independent variables under some fixed operating condition, calculate a result $R_o$ where $R_o = f(x_1, x_2, \ldots, x_L)$. This value fixes the operating point for the numerical approximation (e.g., see Figure 5.3).

2. Increase the independent variables by their respective uncertainties and recalculate the result based on each of these new values. Call these values $R_i^+$. That is,

$$
\begin{aligned}
R_1^+ &= f(x_1 + u_{x1}, \ x_2, \ \ldots, \ x_L), \\
R_2^+ &= f(x_1, \ x_2 + u_{x2}, \ \ldots, x_L), \ \ldots \\
R_L^+ &= f(x_1, \ x_2, \ \ldots, \ x_L + u_{xL}),
\end{aligned}
\tag{5.18}
$$

3. In a similar manner, decrease the independent variables by their respective uncertainties and recalculate the result based on each of these new values. Call these values $R_i^-$.

4. Calculate the differences $\delta R_i^+$ and $\delta R_i^-$ for $i = 1, 2, \ldots, L$

$$
\begin{aligned}
\delta R_i^+ &= R_i^+ - R_o \\
\delta R_i^- &= R_i^- - R_o
\end{aligned}
\tag{5.19}
$$

5. Evaluate the approximation of the uncertainty contribution from each variable,

$$
\delta R_i = \frac{\delta R_i^+ - \delta R_i^-}{2} \approx 0_i u_i
\tag{5.20}
$$

Then, the uncertainty in the result is

$$
u_R = \left[ \sum_{i=1}^{L} (\delta R_i)^2 \right]^{1/2}
\tag{5.21}
$$

Equations 5.15 and 5.21 provide two methods for estimating the propagation of uncertainty to a result. In most cases, each equation yields nearly the identical result and the choice of method is left to the user. The method can also be used to estimate just the sensitivity index of Equation 5.14 [2].

We point out that sometimes these methods may calculate unreasonable estimates of $u_R$. When this happens the cause can be traced to a sensitivity index that changes rapidly with small changes in the independent variable $x_i$ coupled with a large value of the uncertainty $u_{x_i}$. This occurs when the operating point is close to an inflection point in the functional relationship. In these situations, reevaluate the cause and extent of the variation in sensitivity index or use a more accurate approximation for the sensitivity, including backward or forward differencing or using the higher order terms in the Taylor series of Equation 5.7. In subsequent sections, we develop methods to estimate the uncertainty values from available information.

**EXAMPLE 5.5**

For a displacement transducer having the calibration curve, $y = KE$, estimate the uncertainty in displacement $y$ for $E = 5.00$ V, if $K = 10.10$ mm/V with $u_K = \pm 0.10$ mm/V and $u_E = \pm 0.01$ V at 95% confidence.

**KNOWN**    $y = KE$

$$E = 5.00 \text{ V} \qquad u_E = 0.01 \text{ V}$$

$$K = 10.10 \text{ mm/V} \quad u_K = 0.10 \text{ mm/V}$$

**FIND** $u_y$.

**SOLUTION**
Based on Equations 5.12 and 5.13, respectively,

$$y = f(E, K) \quad \text{and} \quad u_y = f(u_E, u_K)$$

From Equation 5.15, the uncertainty in the displacement at $y = KE$ is

$$u_y = \left[ (\theta_E u_E)^2 + (\theta_K u_K)^2 \right]^{1/2}$$

where the sensitivity indices are evaluated from Equation 5.14 as

$$\theta_E = \frac{\partial y}{\partial E} = K \quad \text{and} \quad \theta_K = \frac{\partial y}{\partial K} = E$$

or we can write Equation 5.15 as

$$u_y = \left[ (K u_E)^2 + (E u_K)^2 \right]^{1/2}$$

The operating point occurs at the nominal or the mean values of $E = 5.00$ V and $y = 50.50$ mm. With $E = 5.00$ V and $K = 10.10$ mm/V and substituting for $u_E$ and $u_k$, evaluate $u_y$ at its operating point:

$$u_y|_{y=50.5} = \left[ (0.10)^2 + (0.50)^2 \right]^{1/2} = 0.51 \text{ mm}$$

Alternatively, we can use sequential perturbation. The operating point for the perturbation is again $y = R_o = 50.5$ mm. Using Equations 5.16 through 5.18 gives

| $i$ | $x_i$ | $R_i^+$ | $R_i^-$ | $\delta R_i^+$ | $\delta R_i^-$ | $\delta R_i$ |
|-----|-------|---------|---------|----------------|----------------|--------------|
| 1   | $E$   | 50.60   | 50.40   | 0.10           | −0.10          | 0.10         |
| 2   | $K$   | 51.00   | 50.00   | 0.50           | −0.50          | 0.50         |

Then, using Equation 5.21,

$$u_y|_{y=50.5} = \left[ (0.10)^2 + (0.50)^2 \right]^{1/2} = 0.51 \text{ mm}$$

The two methods give the identical result. We state the calculated displacement in the form of Equation 5.11 as

$$y' = 50.50 \pm 0.51 \text{ mm} \quad (95\%)$$

# Monte Carlo Method

A Monte Carlo simulation provides another effective way to estimate either (1) the propagation of the uncertainties in the independent variables to the uncertainty in a result and or (2) the sensitivity of a result to another variable.

*Uncertainty Estimation* As presented in Chapter 4, the statistical outcomes of a converged Monte Carlo simulation are assumed to be the parameters of the predicted population in result $R$ (i.e., $\overline{R} \to R'$ and $s_R \to \sigma_R$). Accordingly, the standard uncertainty in the result is assigned the value equal to $s_R$, the population standard deviation from the simulation. Generally, convergence in a Monte Carlo simulation is claimed when the computed standard deviation no longer changes by 1% to 5%, but there are more stringent tests described in the literature [7].

*Sensitivity Estimation* The sensitivity of a result to a variable about some operating point, $\partial R/\partial x|_{x_0}$ can be found from the Monte Carlo simulation as the ratio of the standard deviation in the population of the result divided by the standard deviation of population of the variable, such as $\partial R/\partial x|_{x_0} \approx s_R/s_x$.

Monte Carlo simulations are easily coded in spreadsheets, Matlab, or other mathematical packages that have statistical analysis capabilities. A spreadsheet add-in called *MCsim.xla* [14] and a Matlab *m*-file *MC.m* are included in the accompanying online files and can be used as building blocks for coding Monte Carlo analyses.

# 5.7 Advanced-Stage Uncertainty Analysis

In designing a measurement system, a pertinent question is, how would it affect the result if this particular aspect of the technique or equipment were changed? In design-stage uncertainty analysis, we only considered the errors due to a measurement system's resolution and estimated instrument calibration errors. But if additional information is available, we can get a better idea of the uncertainty in a measurement. So, an advanced-stage uncertainty analysis permits taking design-stage analysis further by considering procedural and test control errors that affect the measurement. We consider it as a method for a thorough uncertainty analysis when a large data set is not available. This is often the case in the early stages of a test program or for certain tests where repeating measurements may not be possible. Such an advanced-stage analysis, also known as *single-measurement uncertainty analysis* [3, 6], can be used: (1) in the advanced design stage of a test to estimate the expected uncertainty, beyond the initial design-stage estimate; and (2) to report the results of a test program that involved measurements over a range of one or more parameters but with no or relatively few repeated measurements of the pertinent variables at each test condition. Essentially, the method assesses different aspects of the main test by quantifying potential errors though various well-focused verification tests.

In this section, the goals are either to estimate the uncertainty in some measured value $x$ or in some general result $R$ through an estimation of the uncertainty in each of the factors that may affect $x$ or $R$. We present a technique that uses a step-by-step approach for identifying and estimating the uncertainty associated with errors. We seek the combined value of the estimates at each step.

In our discussions, we assume that the errors follow a normal distribution, but some errors might be better described by other distributions and these can be used. For example, calibration errors that specify only a limit of error are well modeled by a rectangular distribution. The standard uncertainty for a rectangular distribution defined over the interval $(m - n)$, where $m$ is the maximum and $n$ is the minimum of the interval (e.g., refer to Table 4.2), is given by

$$u_x = (m - n)/\sqrt{12} \tag{5.22}$$

which covers one standard deviation (or a 58% confidence level). Multiplying by 2 provides for two standard deviations coverage. If we assume that the errors propagate to the result with a normal distribution, then a coverage factor of 2 approximates the 95% confidence level for consistency.

## Zero-Order Uncertainty

At zero-order uncertainty, all variables and parameters that affect the outcome of the measurement, including time, are assumed to be fixed except for the physical act of observation itself. For example, suppose we took a photograph of a gauge reading under some condition and showed it to 20 technically capable individuals to record their interpretation of the reading. The resulting scatter in those 20 readings would be the result of interpolation error due to instrument resolution alone, a value that we assign as $u_0$. The value $u_0$ estimates the extent of variation expected in the measured value when all influencing effects are controlled except for interpolation error and is found using Equation 5.1. Earlier we placed the value for $u_0$ as one-half the instrument resolution or its least significant digit, or a value assigned by discretion. Note that by itself, a zero-order uncertainty is inadequate for reporting the uncertainty in test results.

## Higher-Order Uncertainty

Higher-order uncertainty estimates consider the controllability of the test operating conditions and the variability of all measured variables. For example, at the first-order level, the effect of time as an extraneous variable in the measurement might be considered. That is, what would happen if we started the test, set the operating conditions, and sat back and watched? If a variation in the measured value is observed, then time is a factor in the test, presumably due to some extraneous influence affecting process control or simply inherent in the behavior of the variable being measured.

In practice, the uncertainty at this first level would be evaluated for each particular measured variable by operating the test facility at some single operating condition that would be within the range of conditions to be used during the actual tests. A set of data (say, $N \geq 30$) would be obtained under some set operating condition. The first-order uncertainty of our ability to estimate the true value of a measured value could be estimated as

$$u_1 = t_{\nu,P}s_{\bar{x}} \tag{5.23}$$

The uncertainty at $u_1$ includes the effects of resolution, $u_0$. So only when $u_1 = u_0$ time is not a factor in the test. In itself, the first-order uncertainty is inadequate for reporting of test results.

With each successive order, another factor identified as affecting the measured value is introduced into the analysis, thus giving a higher but more realistic estimate of the uncertainty. For example, at the second level it might be appropriate to assess spatial variations that affect the outcome, such as when a value from a point measurement is assigned to quantify a larger volume. These are a series of verification tests conducted to assess causality.

## Nth-Order Uncertainty

At the $N$th-order estimate, instrument calibration characteristics are entered into the scheme through the instrument uncertainty $u_c$. A practical estimate of the $N$th-order uncertainty $u_N$ is given by

$$u_N = \left[ u_c^2 + \left( \sum_{i=1}^{N-1} u_i^2 \right) \right]^{1/2} \quad (P\%) \tag{5.24}$$

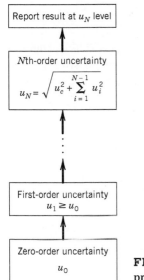

FIGURE 5.4  Advanced-stage and single-measurement uncertainty procedure in combining uncertainties.

Uncertainty estimates at the $N$th order allow for the direct comparison between results of similar tests obtained either using different instruments or at different test facilities. The procedure for a single-measurement analysis is outlined in Figure 5.4.

Note that as a minimum, design-stage analysis includes only the effects found as a result of $u_0$ and $u_c$. It is those in-between levels that allow measurement procedure and control effects to be considered in the uncertainty analysis scheme. When done carefully, the $N$th-order uncertainty estimate provides a reasonable estimate of the uncertainty value sought in advanced design stage or in single-measurement (one-off or unique tests) analyses.

### EXAMPLE 5.6

As an exercise, obtain and examine an inexpensive dial in-oven thermometer. How would you assess the zero- and first-order uncertainty used to determine the temperature of a kitchen oven using the device?

**KNOWN**  Dial thermometer.

**FIND**  Estimate $u_0$ and $u_1$ in oven temperature.

**SOLUTION**
The zero-order uncertainty would be that contributed by the resolution error of the measurement system only. For example, most gauges of this type have a resolution of 10°C. So from Equation 5.1 we estimate

$$u_0 = 5°C$$

At the first order, the uncertainty would be affected by any time variation in the measured value (temperature) and variations in the operating conditions. If we placed the thermometer in the center of an oven, set the oven to some relevant temperature, allowed the oven to preheat to a steady condition, and then proceeded to record the temperature indicated by the thermometer at random intervals, we can estimate the ability of the oven to control its temperature. For $J$ measurements, the first-order uncertainty of the measurement in this oven's mean temperature using this technique and this instrument would be (Equation 5.23),

$$u_1 = t_{J-1,95} s_{\bar{T}}$$

**EXAMPLE 5.7**

A stopwatch is to be used to estimate the time between the start and end of an event. Event duration might range from several seconds to 10 min. Estimate the probable uncertainty in a time estimate using a hand-operated stopwatch that claims an accuracy of 1 min/month (95%) and a resolution of 0.01 s.

**KNOWN** $u_0 = 0.005$ s    (95%)

$$u_c = 60 \text{ s/month} \quad (95\% \text{ assumed})$$

**FIND** $u_d, u_N$.

**SOLUTION**

The design-stage uncertainty gives an estimate of the suitability of an instrument for a measurement. At 60 s/month, the instrument accuracy works out to about 0.01 s/10 min of operation. This gives a design-stage uncertainty of

$$u_d = (u_o^2 + u_c^2)^{1/2} = 0.01 \text{ s} \quad (95\%)$$

for an event lasting 10 min versus $\pm 0.005$ s (95%) for an event lasting 10 s. Note that instrument calibration error controls the longer duration measurement, whereas instrument resolution controls the short duration measurement.

But do instrument resolution and calibration error actually control the uncertainty in this measurement? The design-stage analysis does not include the data-acquisition error involved in the act of physically turning the watch on and off to synchronize with event occurrences. A first-order analysis might be run to estimate the uncertainty that enters through the procedure of using the watch to time an event. Suppose a typical trial run of 30 tries of simply turning a watch on and off suggests that the uncertainty in determining the duration of an occurrence is

$$u_1 = t_{\nu,P} s_{\bar{x}} = 0.05 \text{ s}$$

Then, the uncertainty in measuring the duration of an event would then be better estimated by the $N$th-order uncertainty of Equation 5.24,

$$u_N = (u_1^2 + u_c^2)^{1/2} = 0.05 \text{ s} \quad (95\%)$$

This estimate holds for periods of up to about two hours. Clearly, procedure controls the uncertainty, not the watch. This uncertainty estimate could be further improved by considering how well the operator can synchronize the watch action with the particular event start and finish action.

**COMMENT** In the 2014 Olympics, there was much discussion about events that timed outcomes to 0.01 or 0.001 s, the exact precision used being determined by each individual sport's governing body. In all cases, the timing instruments used were accurate to within $10^{-7}$ s (design stage). But the overall uncertainty in any measurement remains an outcome of the measurement process, of which the instrument accuracy is only one part, and importantly in this case, the peculiarities associated with the timing methods used in each sport.

## EXAMPLE 5.8

A flow meter can be calibrated by providing a known flow rate through the meter and measuring the meter output. One method of calibration with liquid systems is the use of a catch and time technique whereby a volume of liquid, after passing through the meter, is diverted to a tank for a measured period of time from which the flow rate volume/time, is computed. There are two procedures that can be used to determine the known flow rate $Q$ in, say, ft$^3$/min:

1. The volume of liquid, $\forall$, caught in known time $t$ can be measured. Suppose we arbitrarily set $t = 6$ s and assume that our available facilities can determine volume (at $N$th order) to 0.001 ft$^3$. Note: The chosen time value depends on how much liquid we can accommodate in the tank.

2. The time $t$ required to collect 1 ft$^3$ of liquid can be measured.

Suppose we determine an $N$th-order uncertainty in time of 0.15 s. In either case the same instruments are used. Determine which method is better to minimize uncertainty over a range of flow rates based on these preliminary estimates.

**KNOWN**    $u_\forall = 0.001$ ft$^3$

$u_t = 0.15$ s

$Q = f(\forall, \ t) = \forall/t$.

**ASSUMPTIONS** Flow diversion is instantaneous; 95% confidence levels.

**FIND** Preferred method.

**SOLUTION**
From the available information, the propagation of probable uncertainty to the result $Q$ is estimated from Equation 5.15:

$$u_Q = \left[ \left( \frac{\partial Q}{\partial \forall} u_\forall \right)^2 + \left( \frac{\partial Q}{\partial t} u_t \right)^2 \right]^{1/2}$$

By dividing through by $Q$, we obtain the relative (fractional percent) uncertainty in flow rate $u_Q/Q$

$$\frac{u_Q}{Q} = \left[ \left( \frac{u_\forall}{\forall} \right)^2 + \left( \frac{u_t}{t} \right)^2 \right]^{1/2}$$

Representative values of $Q$ are needed to solve this relation. Consider a range of flow rates, say, 1, 10, and 100 ft$^3$/min; the results are listed in the following table for both methods.

| $Q$ (ft$^3$/min) | $t$ (s) | $\forall$ (ft$^3$) | $u_\forall/\forall$ | $u_t/t$ | $\pm u_Q/Q$ |
|---|---|---|---|---|---|
| Method 1 | | | | | |
| 1 | 6 | 0.1 | 0.01 | 0.025 | 0.027 |
| 10 | 6 | 1.0 | 0.001 | 0.025 | 0.025 |
| 100 | 6 | 10.0 | 0.0001 | 0.025 | 0.025 |
| Method 2 | | | | | |
| 1 | 60.0 | 1.0 | 0.001 | 0.003 | 0.003 |
| 10 | 6.0 | 1.0 | 0.001 | 0.025 | 0.025 |
| 100 | 0.6 | 1.0 | 0.001 | 0.250 | 0.250 |

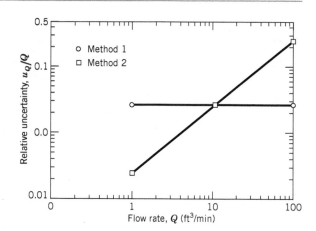

**FIGURE 5.5**   Uncertainty plot for the design analysis of Example 5.8.

In method 1, it is clear the uncertainty in time contributes the most to the relative uncertainty in $Q$, provided that the flow diversion is instantaneous. But in method 2, uncertainty in measuring either time or volume can contribute more to the uncertainty in $Q$, depending on the time sample length. The results for both methods are compared in Figure 5.5. For the conditions selected and preliminary uncertainty values used, method 2 would be a better procedure for flow rates up to 10 ft³/min. At higher flow rates, method 1 would be better. However, the minimum uncertainty in method 1 is limited to 2.5%. The engineer may be able to reduce this uncertainty by improvements in the time measurement procedure.

**COMMENT** These results are without consideration of some other elemental errors that are present in the experimental procedure. For example, the diversion of the flow may not to occur instantaneously. A first-order uncertainty estimate could be used to estimate the added uncertainty to flow rate intrinsic to the time required to divert the liquid to and from the catch tank. If operator influence is a factor, it should be randomized in actual tests. Accordingly, the above calculations may be used as a guide in procedure selection.

## EXAMPLE 5.9

Repeat Example 5.8, using sequential perturbation for the operating conditions $\forall = 1$ ft³ and $t = 6$ s.

**SOLUTION**
The operating point is $R_o = Q = \forall/t = 0.1667$ ft³/s. Then, solving Equations 5.16 to 5.18 gives:

| $i$ | $x_i$ | $R_i^+$ | $R_i^-$ | $\delta R_i^+$ | $\delta R_i^+$ | $\delta R_i$ |
|-----|-------|---------|---------|----------------|----------------|--------------|
| 1 | $\forall$ | 0.1668 | 0.1665 | 0.000167 | −0.000167 | 0.000167 |
| 2 | $t$ | 0.1626 | 0.1709 | −0.00407 | 0.00423 | 0.00415 |

Applying Equation 5.21 gives the uncertainty interval about this operating point (1 ft³, 6 s):

$$u_Q = \left[0.00415^2 + 0.000167^2\right]^{1/2} = \pm 4.15 \times 10^{-3} \text{ ft}^3/\text{s} \quad (95\%)$$

Or on a relative uncertainty basis, an interval of

$$u_Q/Q = \pm 0.025 \quad (95\%)$$

**COMMENT** These last two examples give similar results for the uncertainty aside from insignificant differences due to round-off in the computations.

---

**EXAMPLE 5.10**

A temperature sensor comes only with a manufacturer specification that states its limit of error within its operating range is $\pm 1.5°C$. Establish a value for the instrument standard uncertainty.

**SOLUTION**
When faced with vague manufacturer specifications, the engineer needs to makes some assumptions. Specifications of limits of instrument error refer to the extreme values possible. Accordingly, all values of error within the limit band are equally probable (rectangular distribution). The standard uncertainty of a rectangular distribution is $(m - n)/\sqrt{12}$ (see Equation 5.22). Let $m = 1.5°C$ and $n = -1.5°C$ be the bounds of the distribution. The instrument standard uncertainty is

$$u_c = \frac{[1.5 - (-1.5)]}{\sqrt{12}} = \frac{3.0}{2\sqrt{3}} = 0.87°C \qquad \text{(one standard deviation)}$$

---

# 5.8  Multiple-Measurement Uncertainty Analysis

This section develops a method for estimating the uncertainty in the value assigned to a variable based on a set of measurements obtained under fixed operating conditions. The method parallels the uncertainty standards approved by professional societies in the United States [2, 7] and is in harmony with international guidelines [1]. The procedures assume that the errors follow a normal probability distribution unless otherwise noted, although the procedures are actually quite insensitive to deviations away from such behavior [5].

## Propagation of Elemental Errors

The procedures for multiple-measurement uncertainty analysis consist of the following steps:

- Identify the elemental errors in the measurement. As an aid, consider the errors in each of the three source groups (calibration, data acquisition, and data reduction).

- Estimate the magnitude of systematic and random error in each of the elemental errors.

- Calculate the uncertainty estimate for the result.

Considerable guidance can be obtained from Tables 5.1 to 5.3 for identifying the elemental errors. In multiple-measurement analysis, it is possible to divide the estimates for elemental errors into random and systematic uncertainties.

Consider the measurement of variable $x$, which is subject to elemental random errors each estimated by their random standard uncertainty, $(s_{\bar{x}})_k$, and systematic errors each estimated by their systematic standard uncertainty, $(b_{\bar{x}})_k$. Let subscript $k$, where $k = 1, 2, \ldots, K$, refer to each of up to $K$ elements of error $e_k$. A method to estimate the uncertainty in $x$ based on the uncertainties in each of the elemental random and systematic errors is given below and outlined in Figure 5.6.

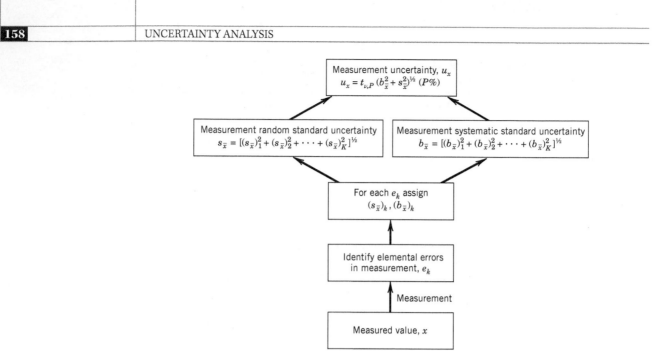

**FIGURE 5.6**    Multiple-measurement uncertainty procedure for combining uncertainties.

The *propagation of elemental random uncertainties* due to the $K$ random errors in measuring $x$ is given by the measurement random standard uncertainty, $s_{\bar{x}}$, as estimated by the RSS method of Equation 5.2:

$$s_{\bar{x}} = \left[ (s_{\bar{x}})_1^2 + (s_{\bar{x}})_2^2 + \cdots + (s_{\bar{x}})_K^2 \right]^{1/2} \tag{5.25}$$

where each

$$(s_{\bar{x}})_k = s_{x_k} / \sqrt{N_k} \tag{5.26}$$

The *measurement random standard uncertainty* represents a basic measure of the uncertainty due to the known elemental errors affecting the variability of variable $x$ at a one standard deviation confidence level. The degrees of freedom, $\nu$, in the standard random uncertainty, is estimated using the Welch–Satterthwaite formula [2]:

$$\nu = \frac{\left( \sum_{k=1}^{K} (s_{\bar{x}}^2)_k \right)^2}{\sum_{k=1}^{K} \left( s_{\bar{x}}^4 \right)_k / \nu_k} \tag{5.27}$$

where $k$ refers to each elemental error with its own $\nu_k = N_k - 1$.

The *propagation of elemental systematic uncertainties* due to $K$ systematic errors in measuring $x$ is treated in a similar manner. The measurement systematic standard uncertainty, $b_{\bar{x}}$, is given by

$$b_{\bar{x}} = \left[ (b_{\bar{x}})_1^2 + (b_{\bar{x}})_2^2 + \cdots + (b_{\bar{x}})_K^2 \right]^{1/2} \tag{5.28}$$

The *measurement systematic standard uncertainty*, $b_{\bar{x}}$, represents a basic measure of the uncertainty due to the elemental systematic errors that affect the measurement of variable $x$ at a one standard deviation confidence level.

The *combined standard uncertainty in x* is reported as a combination of the systematic standard uncertainty and random standard uncertainty in $x$ at a one standard deviation confidence level,

$$u_x = (b_{\bar{x}}^2 + s_{\bar{x}}^2)^{1/2} \tag{5.29}$$

For a normal distribution, this confidence interval is equivalent to a 68% probability level. A more general form of this equation extends it to include the total uncertainty at other confidence levels through the use of an appropriate $t$ value. This is called the *expanded uncertainty* [1, 2] and found at the corresponding level of confidence ($P\%$) by

$$u_x = t_{\nu,P}(b_{\bar{x}}^2 + s_{\bar{x}}^2)^{1/2} \quad (P\%) \tag{5.30}$$

The corresponding degrees of freedom in Equation 5.30 is [2]

$$\nu = \frac{\left( \sum\limits_{k=1}^{K} (s_{\bar{x}}^2)_k + (b_{\bar{x}}^2)_k \right)^2}{\sum\limits_{k=1}^{K} (s_{\bar{x}}^4)_k/\nu_k + \sum\limits_{k=1}^{K} (b_{\bar{x}}^4)_k/\nu_k} \tag{5.31}$$

The degrees of freedom in an uncertainty are calculated from measurements or prior information; otherwise the degrees of freedom may be assumed to be large ($\nu > 30$ in Table 4.4). Reference 1 provides guidance on assigning a value for $\nu$ associated with judgment estimates of uncertainty. When the degrees of freedoms in the systematic uncertainties are large, such as when assigned by judgment or past experience, the second term in the denominator of Equation 5.31 is negligible or essentially zero.

For a 95% confidence level with large degrees of freedom, a value of $t$ of 1.96 is assigned in Equation 5.30. This value can be conveniently rounded off to $t = 2$ (i.e., two standard deviations) [2].

Reference 1 prefers not to assign a probability level to an uncertainty statement but rather to associate a value of the one, two, or three standard deviations, such as we do above for the standard uncertainties. This accommodates different distribution functions and their respective probability coverage in the expanded uncertainty interval. This reflects a difference between the two standards (1, 2) but either approach is acceptable.

## EXAMPLE 5.11

Ten repeated measurements of force $F$ are made over time under fixed operating conditions using the force instrument of Example 5.1. The data are listed below. Estimate the random standard uncertainty due to the elemental error in estimating the true mean value of the force based on the limited data set.

| $N$ | $F$ [N] | $N$ | $F$ [N] |
|-----|---------|-----|---------|
| 1 | 123.2 | 6 | 119.8 |
| 2 | 115.6 | 7 | 117.5 |
| 3 | 117.1 | 8 | 120.6 |
| 4 | 125.7 | 9 | 118.8 |
| 5 | 121.1 | 10 | 121.9 |

**KNOWN** Measured data set $N = 10$.

**FIND** Estimate $s_{\bar{F}}$.

**SOLUTION**
The mean value of the force based on this finite data set is computed to be $\bar{F} = 120.13$ N. A random error is associated with assigning the sample mean value as the true value of force because of the small data set. This error enters the measurement during data acquisition (Table 5.2). The random standard uncertainty in this elemental error is computed through the standard deviation of the means, Equation 5.5:

$$s_{\bar{F}} = \frac{s_F}{\sqrt{N}} = \frac{3.04}{\sqrt{10}} = 0.96 \text{ N} \quad \text{with} \quad \nu = 9$$

**COMMENT** The uncertainty due to data scatter here is Type A and could be classified as being due to a temporal variation error (i.e., variation with time under fixed conditions) as listed in Table 5.2. The resolution error estimated by the zero-order uncertainty $u_0$ assigned in Example 5.1 was just an initial estimate of expected variation in a measurement. We now have information about temporal variations in an actual measurement and we would replace $u_0$ with this new information at the proper level of confidence. For example, in Example 5.1, $u_0 = 0.125$ N (95%). The estimate has now increased to $t_{9,95} s_{\bar{F}} = 2.17$ N (95%).

---

### EXAMPLE 5.12

The force measuring device of Example 5.1 was used in acquiring the data set of Example 5.11. Estimate the systematic uncertainty in the measurement instrument.

**KNOWN** $\begin{aligned} B_1 &= 0.20 \text{ N} \\ B_2 &= 0.30 \text{ N} \end{aligned}$ (95%).

**ASSUMPTIONS** Manufacturer specifications reliable at 95% probability.

**FIND** $B$ (95%).

**SOLUTION**
Lacking the information used to arrive at the uncertainty values associated with the two element errors, $e_1$ and $e_2$, we consider them as Type B uncertainties. We also classify these as systematic uncertainties stated at 95% confidence level. The systematic standard uncertainties are found from Equation 5.4 as $b = B/2$,

$$b_1 = B_1/2 = 0.10 \text{ N} \quad b_2 = B_2/2 = 0.15 \text{ N}$$

The systematic standard uncertainty in the transducer is

$$b = (b_1^2 + b_2^2)^{1/2} = 0.18 \text{ N}$$

at one standard deviation confidence. The expanded systematic uncertainty with $t_{\nu,95} = 2$ at the 95% confidence level is

$$B = 2(b_1^2 + b_2^2)^{1/2} = 0.36 \text{ N}$$

Elemental errors due to instruments can be considered data-acquisition source errors (Table 5.2). If they were found through calibration by the end-user, we would list them as calibration source errors. In the end, as long as they are accounted for, the error source grouping assigned is not important.

## EXAMPLE 5.13

A laboratory temperature standard is used to calibrate an automated temperature measuring system from sensor through indicated output value. The measuring system sensor is placed in the identical environment as the standard sensor. The indicated output values from the system are compared over a range of temperatures based on many readings at each of six setpoint temperatures. Estimate a value for the systematic standard uncertainty of the temperature measuring system.

| Setpoint (°C) | 0.00 | 20.0 | 40.0 | 60.0 | 80.0 | 100.0 |
|---|---|---|---|---|---|---|
| Indicated mean (°C) | 0.046 | 20.067 | 40.072 | 60.133 | 80.057 | 100.116 |
| Offset (°C) | 0.046 | 0.067 | 0.072 | 0.133 | 0.057 | 0.116 |

**KNOWN** Systems-level calibration at different temperatures.

**FIND** Systematic standard uncertainty of the temperature system.

**SOLUTION**
The difference between the applied "known" setpoint temperature and the indicated temperature is listed as the offset, which is a systematic error due to the conversion errors within the measuring unit. The mean of the offset found over the full range of temperatures is $< \bar{b} > = \sum_{i=1}^{6} \bar{b}_i = 0.082°C$

with standard deviation, $SD = \sqrt{\sum_{i=1}^{6} (\bar{b}_i - <\bar{b}>)^2 / 5} = 0.034°C$. We correct the readings at each setpoint temperature using this mean offset. This reduces the average systematic error but there is still uncertainty in the applied correction. The systematic standard uncertainty in the correction is taken as $b = SD/\sqrt{6} = 0.014$ or

$$b = 0.014°C \quad (68\%) \quad (\text{Type A})$$

**COMMENT** There is also a random component that affects each mean setpoint temperature. This random component is not affected by the offset correction.

## EXAMPLE 5.14

In Examples 4.9 and 4.10, a set of measured data were used to generate a polynomial curve fit equation between variables $x$ and $y$. If during a test, variable $x$ were to be measured and $y$ computed from $x$ through the polynomial, estimate the random standard uncertainty due to the data-reduction error in the curve fit at a value of $x = 3$ cm.

**KNOWN** Data set of Example 4.8 ($N = 5$).
First-order polynomial fit of Example 4.9.

**ASSUMPTIONS** $y = 1.04x + 0.02$.

**FIND** Random standard uncertainty in the curve fit at $x = 3$ cm.

## SOLUTION

The curve fit was found to be given by $y = 1.04x + 0.02$ with a standard error of the fit, $s_{yx} = 0.16$. At $x = 3$ cm, $x = \bar{x}$. The random standard uncertainty due to the curve fit is then

$$s_{\bar{x}} = s_{yx} \sqrt{\frac{1}{N} + \frac{(x - \bar{x})^2}{\sum\limits_{i=1}^{5}(x_i - \bar{x})^2}} = s_{yx}/\sqrt{N} = 0.072 \text{ V} \quad \text{with} \quad \nu = 3$$

This is a Type A uncertainty.

---

## EXAMPLE 5.15

The measurement of $x$ contains three elemental random errors from data acquisition. Each elemental error is determined by statistical analysis (therefore, Type A) and assigned a random standard uncertainty as follows:

$$(s_{\bar{x}})_1 = 0.60 \text{ units}, \ \nu_1 = 29$$
$$(s_{\bar{x}})_2 = 0.80 \text{ units}, \ \nu_2 = 9$$
$$(s_{\bar{x}})_3 = 1.10 \text{ units}, \ \nu_3 = 19$$

Assign the random standard uncertainty due to these data-acquisition errors.

**KNOWN** $(s_{\bar{x}})_k$ with $k = 1, 2, 3$.

**FIND** $s_{\bar{x}}$.

## SOLUTION

The random standard uncertainty due to data-acquisition errors is

$$s_{\bar{x}} = \left[(s_{\bar{x}})_1^2 + (s_{\bar{x}})_2^2 + (s_{\bar{x}})_3^2\right]^{1/2} = (0.60^2 + 0.80^2 + 1.10^2)^{1/2} = 1.49 \text{ units}$$

at one standard deviation confidence. This is a Type A uncertainty with degrees of freedom

$$\nu = \frac{\left(\sum\limits_{k=1}^{3}(s_{\bar{x}})_k^2\right)^2}{\sum\limits_{k=1}^{3}(s_{\bar{x}})_k^4/\nu_k} = \frac{(0.6^2 + 0.8^2 + 1.1^2)^2}{(0.6^4/29) + (0.8^4/9) + (1.1^4/19)} = 38.8 \approx 39$$

---

## EXAMPLE 5.16

In reporting the results of an experiment to measure stress in a loaded beam, an uncertainty analysis identifies three elemental errors in the stress measurement having the following values of uncertainty:

| | | |
|---|---|---|
| $(b_{\bar{\sigma}})_1 = 0.5$ N/cm$^2$ | $(b_{\bar{\sigma}})_2 = 1.05$ N/cm$^2$ | $(b_{\bar{\sigma}})_3 = 0$ N/cm$^2$ |
| $(s_{\bar{\sigma}})_1 = 4.6$ N/cm$^2$, $\nu_1 = 14$ | $(s_{\bar{\sigma}})_2 = 10.3$ N/cm$^2$, $\nu_2 = 37$ | $(s_{\bar{\sigma}})_3 = 1.2$ N/cm$^2$, $\nu_3 = 8$ |

where the degrees of freedom in the systematic uncertainties are very large. If the mean value of the stress in the measurement is $\bar{\sigma} = 223.4$ N/cm$^2$, determine the best estimate of the stress at a 95% confidence level, assuming all errors are accounted for.

**KNOWN** Experimental errors with assigned uncertainties.

**ASSUMPTIONS** All elemental errors ($K = 3$) have been included.

**FIND** $\bar{\sigma} \pm u_\sigma$ (95%).

**SOLUTION**
We seek values for the statement, $\sigma' = \bar{\sigma} \pm u_\sigma(95\%)$, given that $\bar{\sigma} = 223.4$ N/cm$^2$. The measurement random standard uncertainty is

$$s_{\bar{\sigma}} = \left[ (s_{\bar{\sigma}})_1^2 + (s_{\bar{\sigma}})_2^2 + (s_{\bar{\sigma}})_3^2 \right]^{1/2} = 11.3 \, \text{N/cm}^2$$

The measurement systematic standard uncertainty is

$$b_{\bar{\sigma}} = \left[ (b_{\bar{\sigma}})_1^2 + (b_{\bar{\sigma}})_2^2 + (b_{\bar{\sigma}})_3^2 \right]^{1/2} = 1.16 \, \text{N/cm}^2$$

The combined standard uncertainty is

$$u_\sigma = (b_{\bar{\sigma}}^2 + s_{\bar{\sigma}}^2)^{1/2} = 11.4 \, \text{N/cm}^2$$

The expanded uncertainty requires the combined degrees of freedom. The degrees of freedom is calculated to be

$$v = \frac{\left( \displaystyle\sum_{k=1}^{K} (s_{\bar{x}}^2)_k + (b_{\bar{x}}^2)_k \right)^2}{\displaystyle\sum_{k=1}^{K} (s_{\bar{x}}^4)_k / v_k) + \sum_{k=1}^{K} (b_{\bar{x}}^4)_k / v_k)} = 49$$

where the degrees of freedoms in the systematic uncertainties are assumed to be very large so that the second term in the denominator is essentially zero. Therefore, $t_{49, 95} \sim 2$ and the expanded uncertainty is

$$u_\sigma = 2[b_{\bar{\sigma}}^2 + s_{\bar{\sigma}}^2]^{1/2} = 2\left[ (1.2 \, \text{N/cm}^2)^2 + (11.3 \, \text{N/cm}^2)^2 \right]^{1/2}$$
$$= 22.7 \, \text{N/cm}^2 \quad (95\%)$$

The best estimate of the stress measurement is

$$\sigma' = 223.4 \pm 22.7 \, \text{N/cm}^2 \quad (95\%)$$

## Propagation of Uncertainty to a Result

Now consider the result $R$, which is determined through the functional relationship between the measured independent variables $x_i$, $i = 1, 2, \ldots, L$ as defined by Equation 5.10. Again, $L$ is the number of independent variables involved and each $x_i$ has an associated measurement systematic standard uncertainty $b_{\bar{x}}$, determined for that variable by Equation 5.28, and a measurement random standard uncertainty $s_{\bar{x}}$, determined using Equation 5.25. The best estimate of the true value $R'$ is given as

$$R' = \bar{R} \pm u_R \quad (P\%) \tag{5.32}$$

where the mean value of the result is determined by

$$\bar{R} = f(\bar{x}_1, \bar{x}_2, \ldots, \bar{x}_L) \tag{5.33}$$

and where the uncertainty in the result $u_R$ is given by

$$u_R = f\left(b_{\bar{x}_1}, b_{\bar{x}_2}, \ldots, b_{\bar{x}_L}; s_{\bar{x}_1}, s_{\bar{x}_2}, \ldots, s_{\bar{x}_L}\right) \tag{5.34}$$

where subscripts $x_1$ through $x_L$ refer to the measurement systematic uncertainties and measurement random uncertainties in each of these $L$ variables.

The propagation of random uncertainty through the variables to the result gives the random standard uncertainty in the result

$$s_R = \left(\sum_{i=1}^{L} [\theta_i s_{\bar{x}_i}]^2\right)^{1/2} \tag{5.35}$$

where $\theta_i$ is the sensitivity index as defined by Equation 5.14.

The degrees of freedom in the random uncertainty is estimated by the Welch–Satterthwaite formula:

$$\nu_s = \frac{\left\{\sum_{i=1}^{L} (\theta_i s_{\bar{x}_i})^2\right\}^2}{\sum_{i=1}^{L} \{(\theta_i s_{\bar{x}_i})^4 / \nu_{\bar{x}_i}\}} \tag{5.36}$$

By propagation of the systematic standard uncertainties of the variables, the systematic standard uncertainty in the result is

$$b_R = \left(\sum_{i=1}^{L} \left[\theta_i b_{x_i}\right]^2\right)^{1/2} \tag{5.37}$$

The terms $\theta_i s_{\bar{x}_i}$ and $\theta_i b_{x_i}$ represent the individual contributions of the $i$th variable to the uncertainty in $R$. Comparing the magnitudes of each individual contribution identifies the relative importance of the uncertainty terms on the result.

The *combined standard uncertainty in the result*, written as $u_R$, is

$$u_R = \left[b_R^2 + s_R^2\right]^{1/2} \tag{5.38}$$

with a confidence level of one standard deviation.

The *expanded uncertainty* in the result is given as

$$u_R = t_{\nu,P}\left[b_R^2 + s_R^2\right]^{1/2} \qquad (P\%) \tag{5.39}$$

The $t$ value is used to provide a reasonable weight to the interval defined by the random uncertainty to achieve the desired confidence. If the degrees of freedom in each variable is large ($N \geq 30$), then a reasonable approximation is to take $t_{\nu,95} = 2$. When the degrees of freedom in each of the variables, $x_i$, is not the same, the Welch–Satterthwaite formula is used to estimate the degrees of freedom in the result expressed as

$$\nu_R = \frac{\left(\sum_{i=1}^{L} (\theta_i s_{\bar{x}_i})^2 + (\theta_i b_{\bar{x}_i})^2\right)^2}{\sum_{i=1}^{L}(\theta_i s_{\bar{x}_i}^4 / \nu_{s_i}) + \sum_{i=1}^{L}(\theta_i b_{\bar{x}}^4 / \nu_{b_i})} \tag{5.40}$$

When the individual degrees of freedom in each of the systematic uncertainties are very large, as is often the case, the second term in the denominator is essentially zero.

## EXAMPLE 5.17

The density of a gas, $\rho$, which follows the ideal gas equation of state, $\rho = p/RT$, is estimated through separate measurements of pressure $p$ and temperature $T$. The gas is housed within a rigid impermeable vessel. The literature accompanying the pressure measurement system states an instrument uncertainty to within $\pm 1\%$ of the reading (95%), and that accompanying the temperature measuring system indicates $\pm 0.6°R$ (95%) uncertainty. Twenty measurements of pressure, $N_p = 20$, and 10 measurements of temperature, $N_T = 10$, are made with the following statistical outcome:

$$\bar{p} = 2253.91 \text{ psfa} \qquad s_p = 167.21 \text{ psfa}$$
$$\bar{T} = 560.4°R \qquad s_T = 3.0°R$$

where psfa refers to $lb/ft^2$ absolute. Determine a best estimate of the density. The gas constant is $R = 54.7$ ft-lb/lb$_m$-°R.

**KNOWN** $\bar{p}, s_p, \bar{T}, s_T$
$\qquad\quad \rho = p/RT; R = 54.7 \text{ ft-lb/lb}_m\text{-}°R$

**ASSUMPTIONS** Gas behaves as an ideal gas.

**FIND** $\rho' = \bar{\rho} \pm u_\rho$    (95%).

## SOLUTION
The measurement objective is to determine the density of an ideal gas through temperature and pressure measurements. The independent and dependent variables are related through the ideal gas law, $\rho = p/RT$. The mean value of density is

$$\bar{\rho} = \frac{\bar{p}}{R\bar{T}} = 0.0735 \text{ lb}_m/ft^3$$

The next step must be to identify and estimate the errors and determine how they contribute to the uncertainty in the mean value of density. Since no calibrations are performed and the gas is considered to behave as an ideal gas in an exact manner, the measured values of pressure and temperature are subject only to elemental errors within the data-acquisition error source group (Table 5.2): instrument errors and temporal variation errors.

The tabulated value of the gas constant is not without error. However, estimating the possible error in a tabulated value is sometimes difficult. According to Kestin [8], the systematic uncertainty in the gas constant is on the order of $\pm(0.33$ J/kg-K)/(gas molecular weight) or $\pm 0.06$ (ft-lb/lbm-°R)/(gas molecular weight). Since this yields a small value for a reasonable gas molecular weight, here we assume a zero (negligible) systematic error in the gas constant.

Consider the pressure measurement. The uncertainty assigned to the temporal variation (data scatter) error is based on the variation in the measured data obtained during presumably fixed operating conditions. The instrument error is assigned a systematic uncertainty based on the manufacturer's statement, which is assumed to be stated at 95% confidence:

$$(b_{\bar{p}})_1 = (B_{\bar{p}}/2)_1 = (0.01 \times 2253.51/2) = 11.28 \text{ psfa} \qquad (s_{\bar{p}})_1 = 0$$

where the subscript keeps track of the error identity. The temporal variation causes a random uncertainty in establishing the mean value of pressure and is calculated as

$$(s_{\bar{p}})_2 = s_{\bar{p}} = \frac{s_p}{\sqrt{N}} = \frac{167.21 \text{ psfa}}{\sqrt{20}} = 37.39 \text{ psfa} \qquad v_{s_p} = 19$$

and assigning no systematic uncertainty to this error gives

$$(b_{\bar{p}})_2 = 0$$

In a similar manner, we assign the uncertainty in the data-acquisition source error in temperature. The instrument errors are considered as systematic only and assigned a systematic uncertainty based on the manufacturer's statement as

$$(b_{\bar{T}})_1 = (b_{\bar{T}})_1/2 = 0.3°\text{R} \qquad (s_{\bar{T}})_1 = 0$$

The temporal variation causes a random uncertainty in establishing the mean value of temperature and this is calculated as

$$(s_{\bar{T}})_2 = s_{\bar{T}} = \frac{s_T}{\sqrt{N}} = \frac{3.0°\text{R}}{\sqrt{10}} = 0.9°\text{R} \qquad v_{s_T} = 9$$
$$(b_{\bar{T}})_2 = 0$$

The uncertainties in pressure and temperature due to the data-acquisition source errors are combined using the RSS method,

$$b_{\bar{p}} = \left[(11.28)^2 + (0)^2\right]^{1/2} = 11.28 \ \text{psfa}$$
$$s_{\bar{p}} = \left[(0)^2 + (37.39)^2\right]^{1/2} = 37.39 \ \text{psfa}$$

Similarly,

$$b_{\bar{T}} = 0.3°\text{R}$$
$$s_{\bar{T}} = 0.9°\text{R}$$

with degrees of freedom in the random standard uncertainties to be

$$(v)_{s_p} = N_p - 1 = 19$$
$$(v)_{s_T} = 9$$

The degrees of freedom in the systematic standard uncertainties, $v_{b_T}$ and $v_{b_p}$, are assumed large.

The systematic and random standard uncertainties propagate through to the result, calculated about the operating point as established by the mean values for temperature and pressure. That is,

$$s_{\bar{\rho}} = \left[\left(\frac{\partial \rho}{\partial T}s_{\bar{T}}\right)^2 + \left(\frac{\partial \rho}{\partial p}s_{\bar{p}}\right)^2\right]^{1/2}$$
$$= \left[(1.3 \times 10^{-4} \times 0.9)^2 + (3.26 \times 10^{-5} \times 37.39)^2\right]^{1/2}$$
$$= 0.0012 \ \text{lb}_\text{m}/\text{ft}^3$$

and

$$b_{\bar{\rho}} = \left[\left(\frac{\partial \rho}{\partial T}b_{\bar{T}}\right)^2 + \left(\frac{\partial \rho}{\partial p}b_{\bar{p}}\right)^2\right]^{1/2}$$
$$= \left[(1.3 \times 10^{-4} \times 0.3)^2 + (3.26 \times 10^{-5} \times 11.28)^2\right]^{1/2}$$
$$= 0.0004 \ \text{lb}_\text{m}/\text{ft}^3$$

The degrees of freedom in the density calculation is determined to be

$$
\nu = \frac{\left[\left(\frac{\partial \rho}{\partial T}s_{\overline{T}}\right)^2 + \left(\frac{\partial \rho}{\partial p}s_{\overline{p}}\right)^2 + \left(\frac{\partial \rho}{\partial T}b_{\overline{T}}\right)^2 + \left(\frac{\partial \rho}{\partial T}b_{\overline{p}}\right)^2\right]^2}{\left[\left(\frac{\partial \rho}{\partial p}s_{\overline{p}}\right)^4/\nu_{s_p} + \left(\frac{\partial \rho}{\partial T}s_{\overline{T}}\right)^4/\nu_{s_T}\right] + \left[\left(\frac{\partial \rho}{\partial T}b_{\overline{T}}\right)^4/\nu_{b_T} + \left(\frac{\partial \rho}{\partial p}b_{\overline{p}}\right)^4/\nu_{b_p}\right]} = 23
$$

The expanded uncertainty in the mean value of density, using $t_{23,\,95} = 2.06$, is

$$
u_\rho = t_{23,95}[b_{\overline{\rho}}^2 + s_{\overline{\rho}}^2]^{1/2} = 2.06 \times [0.0004^2 + 0.0012^2]^{\frac{1}{2}}
$$
$$
= 0.0026 \; \text{lb}_{\text{m}}/\text{ft}^3 \quad (95\%)
$$

The best estimate of the density is reported as

$$
\rho' = 0.0735 \pm 0.0026 \; \text{lb}_{\text{m}}/\text{ft}^3 \quad (95\%)
$$

This measurement of density has an uncertainty of about 3.4%.

**COMMENT** (1) We did not consider the uncertainty associated with our assumption of exact ideal gas behavior, a potential modeling error (see Table 5.3). (2) Note how pressure contributes more to standard uncertainty than does temperature and that the systematic uncertainty is small compared to the random uncertainty. The uncertainty in density is best reduced by actions to reduce the effects of the random errors on the pressure measurements.

---

| Example 5.18 | Changing Values: Uncertainty in Published Properties |
|---|---|

A common feature found in handbooks and textbook appendices is the tabulation of material properties. We use this data in engineering calculations as if these were a matter of fact. However, these data are the result of measurements and so are prone to measurement limitations. Consider the following.

Property reference standards serve to confirm the accuracy of new instrumentation and reproducibility of existing equipment. For example, toluene can be obtained with a very high purity and so serves as a reference standard of choice for calibrating instruments used in measuring the material properties of electrically insulating liquids.

Figure 5.7 shows the thermal conductivity of liquid toluene, $k$, as officially reported over the last century. Has toluene been changing? Of course not, but measured property values will change over time due to improvements in the methodology used to measure them. Up until the mid-1960s, steady-state thermal methods were used for this measurement. The methods were prone to various systematic errors, particularly those due to convection and radiation heat transport, which interfere and offset the calculated conductivity to a larger value. Transient techniques are now used and better able to reduce and to correct for these residual errors. As such, the uncertainty in the reported property data has been reducing with time. Since the 1970s, the reported mean value for thermal conductivity has remained stable to within 0.07%. In 2000, the accepted value was $k$ (298.15 K, 0.10 MPa) = $0.13088 \pm 0.00085$ W/m-K (95%). The exact property value varies with operating temperature and its uncertainty varies from 0.6 to 2% over the full usable temperature range. Whenever accuracy is important, be sure to research the source of the information as part of your test plan.

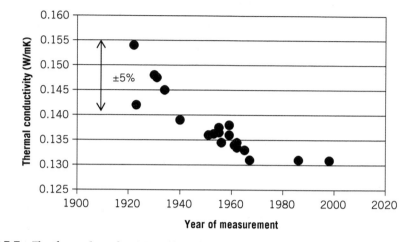

**FIGURE 5.7**    The thermal conductivity of liquid toluene at 298.15 K and 0.10 MPa abs (compiled from 11, 12).

## 5.9 Correction for Correlated Errors

So far, we have assumed that all of the different elements of error in a test are independent from the others. If two errors are not independent, they are "correlated."

For example, when the same instrument is used to measure different variables, the instrument errors in those variables are correlated. So if the same voltmeter was used to measure the bridge deflection voltage from several different strain gauge setups, the instrument errors assigned to the voltmeter are correlated for the various strain measurements. Or if multiple instruments are calibrated against the same standard, then the uncertainty in the standard used is passed on to each instrument and these errors are correlated. The numerical effect of correlated errors on the uncertainty depends on the functional relationship between the variables and the magnitudes of the elemental errors that are correlated. We now introduce a correction for treating the correlated errors.

Consider the result $R$, which is determined through the functional relationship between the measured independent variables $x_i$, $i = 1, 2, \ldots, L$ where $L$ is the number of independent variables and parameters involved. Each $x_i$ is subject to elemental systematic errors with standard uncertainties, $b_k$, where $k = 1, 2, \ldots, K$, refer to each of up to any of $K$ elements of error. Now allow that as many as $H$ of these $K$ elemental errors are correlated between variables. When correlated errors are included, the systematic standard uncertainty in a result is estimated by

$$b_R = \left[ \sum_{i=1}^{L} (\theta_i b_{\bar{x}_i})^2 + 2 \sum_{i=1}^{L-1} \sum_{j=i+1}^{L} \theta_i \theta_j \delta_{ij} b_{\bar{x}_i \bar{x}_j} \right]^{1/2} \tag{5.41}$$

where index $j$ is a counter equal to $i + 1$ and with

$$\theta_i = \frac{\partial R}{\partial x_i}\bigg|_{x=\bar{x}} \tag{5.14}$$

Equation 5.42 introduces the covariance, $b_{\bar{x}_i \bar{x}_j}$, to account for correlated errors and this is found from

$$b_{\bar{x}_i \bar{x}_j} = \sum_{h=1}^{H} (b_{\bar{x}_i})_h (b_{\bar{x}_j})_h \tag{5.42}$$

where $H$ is the number of elemental errors that are correlated between variables $x_i$ and $x_j$ and $h$ is a counter for each correlated error. The value $\delta_{ij}$ is a covariance coefficient with a value between 0 (no correlation) and 1 (correlated). Normally its value is either 0 or 1 but could be assigned a value in between. Note that Equation 5.41 reduces to Equation 5.37 when no errors are correlated (i.e., when $H = 0$, or $\delta_{ij} = 0$). References 2 and 9 discuss treatment of correlated systematic errors in extended detail.

There are situations where random errors can be correlated [10]. In those cases, the random standard uncertainty in a result is estimated by

$$s_R = \left[ \sum_{i=1}^{L} (\theta_i s_{\bar{x}_i})^2 + 2 \sum_{i=1}^{L-1} \sum_{j=i+1}^{L} \theta_i \theta_j r_{\bar{x}_i \bar{x}_j} s_{\bar{x}_i \bar{x}_j} \right]^{1/2} \tag{5.43}$$

where $r_{\bar{x}_i \bar{x}_j}$ is the correlation coefficient between $x_i$ and $x_j$ as given by equation 4.41 and

$$s_{\bar{x}_i \bar{x}_j} = \sum_{h=1}^{H} (s_{\bar{x}_i})_h (s_{\bar{x}_j})_h \tag{5.44}$$

In a general analysis, the correlated error is handled by the general standard uncertainty equation:

$$u_R = \left[ \sum_{i=1}^{L} (\theta_i u_{\bar{x}_i})^2 + 2 \sum_{i=1}^{L-1} \sum_{j=i+1}^{L} \theta_i \theta_j \delta_{ij} u_{\bar{x}_i} u_{\bar{x}_j} \right]^{1/2} \tag{5.45}$$

where $\delta_{ij}$ is assigned a value between 0 and 1.

### EXAMPLE 5.19

Suppose a result is a function of three variables, $x_1, x_2, x_3$. There are four systematic errors associated with both $x_1$ and $x_2$ and five associated with $x_3$. Only the first and second systematic elemental errors associated with the second and third variable ($x_2$ and $x_3$) are determined to be correlated because these errors arise from common sources. Express the systematic standard uncertainty in the result and find the covariance term.

### SOLUTION
Equation 5.41 is expanded to the form

$$b_R = \left[ (\theta_1 b_{\bar{x}_1})^2 + (\theta_2 b_{\bar{x}_2})^2 + (\theta_3 b_{\bar{x}_3})^2 + 2\theta_1 \theta_2 \delta_{12} b_{\bar{x}_1 \bar{x}_2} + 2\theta_1 \theta_3 \delta_{13} b_{\bar{x}_1 \bar{x}_3} + 2\theta_2 \theta_3 \delta_{23} b_{\bar{x}_2 \bar{x}_3} \right]^{1/2}$$

For the two correlated errors ($H = 2$, $\delta_{12} = \delta_{13} = 0$, $\delta_{23} = 1$) associated with variables 2 and 3 ($i = 2, j = 3$), the systematic uncertainty in the result reduces to

$$b_R = \left[ (\theta_1 b_{\bar{x}_1})^2 + (\theta_2 b_{\bar{x}_2})^2 + (\theta_3 b_{\bar{x}_3})^2 + 2\theta_2 \theta_3 \delta_{23} b_{\bar{x}_2 \bar{x}_3} \right]^{1/2}$$

where the covariance term is

$$b_{\overline{x}_2\overline{x}_3} = \sum_{h=1}^{2} (b_{\overline{x}_2})_h (b_{\overline{x}_3})_h = (b_{\overline{x}_2})_1 (b_{\overline{x}_3})_1 + (b_{\overline{x}_2})_2 (b_{\overline{x}_3})_2$$

and subscripts $h = 1$ and $h = 2$ refer to the relevant elemental errors.

---

### EXAMPLE 5.20

Result $R$ is a function of two variables, $X$ and $Y$, such that $R = X + Y$, and each variable having one elemental error. If $\overline{X} = 10.1\,\text{V}$ with $b_{\overline{x}} = 1.1$ V and $\overline{Y} = 12.2$ V with $b_{\overline{y}} = 0.8\,V$, estimate the systematic standard uncertainty in the result if the systematic errors are (1) uncorrelated and (2) correlated.

### SOLUTION

From the stated information, $\overline{R} = 10.1 + 12.2 = 22.3$ V. For the uncorrelated case ($\delta_{XY} = 0$), the systematic standard uncertainty is estimated as

$$b_{R_{\text{unc}}} = \left[(\theta_X b_{\overline{X}})^2 + (\theta_Y b_{\overline{Y}})^2\right]^{1/2} = \left[(1 \times 1.1)^2 + ((1 \times 0.8)^2)^2\right]^{1/2} = 1.36\text{ V}$$

For the correlated case ($\delta_{XY} = 1$), with $H = 1$ and $L = 2$, the uncertainty is estimated as

$$b_{R_{\text{cor}}} = \left[(\theta_X b_{\overline{X}})^2 + (\theta_{\overline{Y}} b_{\overline{Y}})^2 + 2\theta_X \theta_Y \delta_{XY} b_{\overline{X}\overline{Y}}\right]^{1/2}$$
$$= \left[(1 \times 1.1)^2 + (1 \times 0.8)^2 + 2(1)(1)(1)(1.1)(0.8)\right]^{1/2} = 3.12\text{ V}$$

which is stated at the one-standard deviation confidence level and where

$$b_{\overline{X}\overline{Y}} = \sum_{h=1}^{1} (b_{\overline{X}})_1 (b_{\overline{Y}})_1 = b_{\overline{X}} b_{\overline{Y}}$$

**COMMENT** In this case, the correlated systematic errors have a large impact on the systematic uncertainty. This is not always the case as it depends on the functional relationship itself. We leave it to the reader to show that if $R = X/Y$, then the covariance term in this problem would have little impact on the systematic standard uncertainty in the result.

---

## 5.10 Nonsymmetrical Systematic Uncertainty Interval

There are situations where the error must be bounded on one side or skewed about the measured mean value such that the systematic uncertainty is better described by a nonsymmetrical interval. To develop this case, we assume that the limits of possible systematic error are known, as is the mean of the measured data set $\overline{x}$. Let $\overline{x} + B_{\overline{x}}^{+}$ and $\overline{x} - B_{\overline{x}}^{-}$ be the upper and lower limits of the systematic uncertainty relative to the measured mean value (Figure 5.8), with $B_{\overline{x}}^{-} = 2b_{\overline{x}}^{-}$ and $B_{\overline{x}}^{+} = 2b_{\overline{x}}^{+}$ where $b_{\overline{x}}^{-}$ and $b_{\overline{x}}^{+}$ are the lower and upper systematic standard uncertainty values. The modeling approach assumes that the errors are symmetric about some mean value of the error distribution but asymmetric about the measured mean value.

**FIGURE 5.8** Relation between the measured mean value, the mean value of the distribution of errors, and the systematic uncertainties for treating nonsymmetrical uncertainties.

If we model the error distribution as a normal distribution, then we assume that the limits defined using $B_{\bar{x}}^-$ to $B_{\bar{x}}^+$ cover 95% of the error distribution. Define the offset between the mean of the error distribution and the mean value of the measurement as

$$q = \frac{(\bar{x} + B_{\bar{x}}^+) + (\bar{x} + B_{\bar{x}}^-)}{2} - \bar{x} = \frac{B_{\bar{x}}^+ - B_{\bar{x}}^-}{2} = b_{\bar{x}}^+ - b_{\bar{x}}^- \tag{5.46}$$

The systematic standard uncertainty has an average width,

$$b_{\bar{x}} = \frac{(\bar{x} + B_{\bar{x}}^+) - (\bar{x} - B_{\bar{x}}^-)}{4} = \frac{(\bar{x} + b_{\bar{x}}^+) - (\bar{x} + b_{\bar{x}}^-)}{2} \tag{5.47}$$

For stating the true value, we use the combined standard uncertainty

$$u_x = \sqrt{b_{\bar{x}}^2 + s_{\bar{x}}^2} \tag{5.48}$$

to achieve the approximate confidence interval, $q \pm t_{\nu,P} u_x$. For large degrees of freedom, we can state

$$x' = \bar{x} + q - t_{\nu,P} u_x, \ \bar{x} + q + t_{\nu,P} u_x \tag{5.49}$$

for which we choose an appropriate value for $t$, such as $t = 2$ for 95% confidence.

If we model the error distribution as a rectangular distribution, then we assume that the limits defined using $B_{\bar{x}}^-$ to $B_{\bar{x}}^+$ specify the limits of that distribution. The systematic standard uncertainty is then (from Equation 5.22),

$$b_{\bar{x}} = \frac{(\bar{x} + B_{\bar{x}}^+) - (\bar{x} - B_{\bar{x}}^-)}{\sqrt{12}} \tag{5.50}$$

for use in Equations 5.47 and 5.48. Regardless of the assumed systematic error distribution, we assume that the uncertainties propagate normally.

A normal or rectangular distribution is not always the appropriate model for treating asymmetric systematic uncertainty. References 1 and 2 provide a number of scenarios to treat nonsymmetrical uncertainties including use of other error distributions. As an example, Monsch et al. [13] apply the above approach to estimating the nonsymmetrical systematic uncertainty in the induced drag on an aircraft wing in which the induced drag cannot be less than zero.

# SUMMARY

Uncertainty analysis provides the "± what" to a test result or the anticipated result from a proposed test plan. This chapter discussed the manner in which various errors can enter into a measurement with emphasis on their influences on the test result. Both random errors and systematic errors are considered. Random errors differ on repeated measurements, lead to data scatter, and can be quantified by statistical methods. Systematic errors remain constant in repeated measurements. Errors are quantified by uncertainties. Procedures for uncertainty analysis were introduced both as a means of estimating uncertainty propagation within a measurement and for the propagation of uncertainty among the independent measured variables that are used to determine a result through some functional relationship. We have discussed the role and use of uncertainty analysis in both the design of measurement systems, through selection of equipment and procedures, as well as in the interpretation of measured data. We have presented procedures for estimating uncertainty at various stages of a test. The procedures provide reasonable estimates of the uncertainty to be expected in measurements.

We advise the reader that test uncertainty by its very nature evades exact values. But the methodology presented gives reasonable estimates provided that the engineer has used sound, impartial judgment in its implementation. The important thing is that the uncertainty analysis be performed during the design and development stages of a test, as well as at its completion to assess the stated results. Only then can the engineer use the results with appropriate confidence and professional and ethical responsibility.

# REFERENCES

1. International Organization for Standardization (ISO), *Guide to the Expression of Uncertainty in Measurement*, Geneva, Switzerland, 1993.

2. ANSI/ASME Power Test Codes-PTC 19.1, *Test Uncertainty*, American Society of Mechanical Engineers, New York, 2005.

3. Smith, R. E., and S. Wenhofer, From Measurement Uncertainty to Measurement Communications, Credibility and Cost Control in Propulsion Ground Test Facilities, *Journal of Fluids Engineering* 107: 165–172, 1985.

4. Kline, S. J., and F. A. McClintock, Describing Uncertainties in Single Sample Experiments, *Mechanical Engineering* 75: 3–8, 1953.

5. Epperson, J. F., *An Introduction to Numerical Methods and Analysis*, Wiley-Interscience, New York, 2007.

6. Moffat, R. J., Uncertainty Analysis in the Planning of an Experiment, *Journal of Fluids Engineering* 107: 173–181, 1985.

7. Taylor, B. N., and C. Kuyatt, *Guidelines for Evaluating and Expressing the Uncertainty of NIST Measurement Results*, NIST Technical Note 1297, Gaithersburg, MD, 1994.

8. Kestin, J., *A Course in Thermodynamics*, revised printing, Hemisphere, New York, 1979.

9. Coleman, H., and W. G. Steele, *Experimentation and Uncertainty Analysis for Engineers*, 2nd ed., Wiley-Interscience, New York, 1999.

10. Dieck, R., *Measurement Uncertainty: Methods and Applications*, 3rd ed., Instrumentation, Systems and Automation Society (ISA), Research Triangle Park, NC, 2002.

11. Ramires, M., C. Nieto de Castro, R. Perkins, Y. Nagasaka, A. Nagashima, M. Assael, and W. Wakeham, Standard Reference Data for Thermal Conductivity of Liquids, *J. Phys. Chem. Ref. Data*, 29(2): 133–139, 2000.

12. Nieto de Castro, C., S. Li, A. Nagashima, R. Trengrove, and W. Wakeham , Standard Reference Data for Thermal Conductivity of Liquids, *J. Phys. Chem. Ref. Data*, 15(3): 1073–1086, 1986.

13. Monsch, S., R. S. Figliola, E. Thompson, and J. Camberos, Induced Drag for 3D Wing with Volume Integral (Trefftz Plane) Technique, AIAA Paper 2007-1079, AIAA, New York, 2007.

14. Barreto, H., and F. Howland, *Introductory Econometrics: Using Monte Carlo Simulation with Microsoft Excel*, Cambridge Press, New York, 2005.

# NOMENCLATURE

| | | | |
|---|---|---|---|
| $a$ | lower bound | $s_{\bar{x}}$ | random standard uncertainty in $x$; sample standard deviation of the means |
| $b$ | systematic standard uncertainty | | |
| $b_{\bar{x}}$ | systematic standard uncertainty in $x$ | $s_{yx}$ | standard deviation of a fit |
| $e_k$ | $k$th elemental error | $t_{v, 95}$ | $t$ variable (at 95% probability) |
| $f\,(\,)$ | general functional relation | $t$ | time ($t$); Student's $t$-variable |
| $p$ | pressure ($ml^{-1}\,t^{-2}$) | $u_x$ | uncertainty of a measured variable |
| $q$ | offset factor | $u_d$ | design-stage uncertainty |
| $s_x$ | sample standard deviation of $x$ | $u_c$ | instrument calibration uncertainty |

| | | | |
|---|---|---|---|
| $u_0$ | zero-order uncertainty | $Q$ | flow rate $(l^3 t^{-1})$ |
| $u_i$ | $i$th-order uncertainty | $T$ | temperature $(°)$ |
| $u_N$ | $N$th-order uncertainty | $\forall$ | volume $(l^{-3})$ |
| $x$ | measured variable | $\delta_{ij}$ | covariance coefficient |
| $x'$ | true (mean) value of the population of $x$ | $\theta_i$ | sensitivity index |
| $\bar{x}$ | sample mean value of $x$ | $\rho$ | gas density $(ml^{-3})$ |
| $R$ | result or resultant value | $\sigma$ | stress $(ml^{-1}t^{-2})$; population standard deviation |
| $P$ | probability | $\nu$ | degrees of freedom |
| $B$ | systematic uncertainty (at 95% or $2\sigma$ probability) | $\langle\ \rangle$ | pooled statistic |
| $(P\%)$ | confidence level | | |

# 6 Analog Electrical Devices and Measurements

## 6.1 Introduction

This chapter provides an introduction to basic electrical analog devices used with analog signals or to display signals in an analog form. Information is often transferred between stages of a measurement system as an analog electrical signal. This signal typically originates from the measurement of a physical variable using a fundamental electromagnetic or electrical phenomenon and then propagates from stage to stage of the measurement system (see Fig. 1.1). Accordingly, we discuss some signal conditioning and output methods.

Although analog devices have been supplanted by their digital equivalents in many applications, they are still widely used and remain integral to a variety of engineered devices. Analog technology forms the basis for power management in cell phones and for innovative cardiac monitoring and is abundant in modern flat panel displays. Also, an analog output format is often ergonomically superior in monitoring, as evidenced by modern car speedometer dials and dial wristwatches. Within a signal chain, it is common to find digital and analog electrical devices being used together; this requires special signal conditioning. An understanding of analog device function provides insight, as well as a historical reference point, into the advantages and disadvantages of digital counterparts; such issues are discussed in this chapter.

Upon completion of this chapter, the reader will be able to:

- Understand the principles behind common analog voltage and current measuring devices

- Understand the operation of balanced and unbalanced resistance bridge circuits

- Define, identify, and minimize loading errors

- Understand the basic principles involved in signal conditioning, especially filtering and amplification

- Apply proper grounding and shielding techniques in measuring system hookups

## 6.2 Analog Devices: Current Measurements

### Direct Current

One way to measure a DC electric current is to use an analog device that responds to the force exerted on a current-carrying conductor in a magnetic field. Because an electric current is a series of moving charges, a magnetic field exerts a force on a current-carrying conductor. This force can be used as a measure of the flow of current in a conductor by moving a pointer on a display.

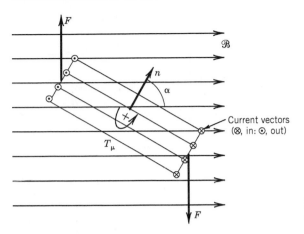

**FIGURE 6.1** Forces and resulting torque on a current loop in a magnetic field.

A current loop in a magnetic field experiences a torque if the loop is not aligned with the magnetic field, as illustrated in Figure 6.1. The torque[1] on a loop composed of $N$ turns is given by

$$T_\mu = NIAB \sin \alpha \qquad (6.1)$$

where

$A$ = cross-sectional area defined by the perimeter of the current loop

$B$ = magnetic field strength (magnitude)

$I$ = current

$\alpha$ = angle between the normal to the cross-sectional area of the current loop and the magnetic field

One approach to measure current is the very common D'Arsonval movement shown in Figure 6.2. In this arrangement, the uniform radial magnetic field and torsional spring result in a steady angular deflection of the coil that corresponds to the existing current through the coil. The coil and fixed permanent magnet are arranged in the normal direction to the current loop—that is, $\alpha$ = 90 degrees. If an analog dial and pointer indicates current, voltage, or resistance, then it is likely to employ this mechanism.

Most devices that use the D'Arsonval movement employ a pointer whose deflection increases with the magnitude of current applied. This mode of operation is called the *deflection mode*.

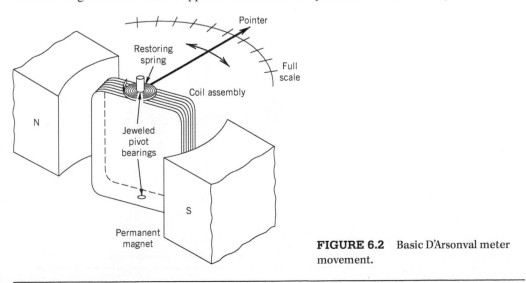

**FIGURE 6.2** Basic D'Arsonval meter movement.

---

[1]In vector form, the torque on a current loop can be written $\mathbf{T}_\mu = \mu \times \mathbf{B}$, where $\mu$ is the magnetic dipole moment.

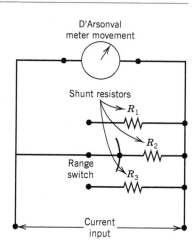

D'Arsonval
meter movement

Shunt resistors

$R_1$

$R_2$

Range
switch    $R_3$

Current
input

**FIGURE 6.3**    Simple multirange ammeter (with make-before-break selector switch). Shunt resistors determine meter range.

A typical circuit for an analog current measuring device, an ammeter, is shown in Figure 6.3. Here the deflection of the pointer indicates the magnitude of the current flow.

**EXAMPLE 6.1    Case Study: Galvanometer Rotary Actuator**

The operating principle of the D'Arsonval movement can be employed to create a high-precision, high performance rotary actuator, termed a galvanometer, or "galvo." The primary area of application is the fast and precise positioning of mirrors for the deflection of laser beams. The system consists of a "motor" section based on moving magnet technology coupled with a high-precision position sensor. These are employed to rotate mirrors in scanning systems, and have step times in the 100 μsec range. Applications include laser scanning for engraving, medical applications including ophthalmology, and laser light shows! Figure 6.4 shows galvanometer actuators with appropriate mirrors for laser scanning.

**COMMENT** In this example the term galvanometer is used to describe an actuator. The term galvanometer is also used to describe a sensor, as will become clear in the following.

**FIGURE 6.4**    Scanning galvanometer heads with mirrors.

Courtesy of SCANLAB AG-Puchheim Germany

In the previous example, we discussed an actuator that is termed a galvanometer, but the term is also used to describe a sensor. In that case, a *galvanometer* is a measuring device used to detect a current flow in a circuit. It is a highly sensitive D'Arsonval movement calibrated about zero current; the indicator normally indicates zero but can deflect to the plus or minus direction. The pointer is used to adjust a circuit to a zero current state. This mode of operation is called the *null mode*. The highest sensitivity of commercial galvanometers is approximately 0.1 μA/division for a galvanometer with a mechanical pointer.

Errors inherent to the D'Arsonval movement include hysteresis and repeatability errors due to mechanical friction in the pointer-bearing movement, and linearity errors in the spring that provides the restoring force for equilibrium. Also, in developing a torque, the D'Arsonval movement must extract energy from the current flowing through it. This draining of energy from the signal being measured changes the measured signal, and such an effect is called a *loading error*. Loading errors occur in all instruments that operate in deflection mode. A quantitative analysis of loading errors is provided later.

## EXAMPLE 6.2

A galvanometer consists of $N$ turns of a conductor wound about a core of length $l$ and radius $r$ that is situated perpendicular to a magnetic field of uniform flux density **B**. A DC current passes through the conductor due to an applied potential, $E_i(t)$. The output of the device is the rotation of the core and pointer, $\theta$, as shown in Figure 6.5. Develop a lumped parameter model relating pointer rotation and current.

**FIND** A dynamic model relating $\theta$ and $I$.

## SOLUTION
The galvanometer is a rotational system consisting of a torsional spring, a pointer, and core that are free to deflect; the rotation is subject to frictional damping. The mechanical portion of the device consists of a coil having moment of inertia $J$, bearings that provide damping from friction, with a damping coefficient $c$, and a torsional spring of stiffness $k$. The electrical portion of the device consists of a coil with a total inductance $L_g$ and a resistance $R_g$.

(a)                                              (b)

**FIGURE 6.5** Circuit and free-body diagram for Example 6.2.

We apply Newton's second law with the mechanical free-body diagram shown in Figure 6.4a.

$$J\frac{d^2\theta}{dt^2} + c\frac{d\theta}{dt} + k\theta = T_\mu \tag{6.2}$$

As a current passes through the coil, a force is developed that produces a torque on the coil:

$$T_\mu = 2NBlrI \tag{6.3}$$

This torque, which tends to rotate the coil, is opposed by an electromotive force $E_m$, due to the Hall effect,

$$E_m = \left(r\frac{d\theta}{dt} \times \mathbf{B}\right) i\hat{\mathbf{k}} = 2NBr\frac{d\theta}{dt} \tag{6.4}$$

produces a current in the direction opposite to that produced by $E_i$. Applying Kirchhoff's law to the electrical free body of Figure 6.5b gives

$$L_g\frac{dI}{dt} + R_gI = E_i - E_m \tag{6.5}$$

Equations 6.3 and 6.4 define the coupling relations between the mechanical Equation 6.2 and the electrical Equation 6.5. From these, we see that the current due to potential $E$ brings about developed torque $T_\mu$ that moves the galvanometer pointer, and that this motion is opposed by the mechanical restoring force of the spring and by the development of an opposing electrical potential $E_m$. The system damping allows the pointer to settle to an equilibrium position.

**COMMENT** The system output, which is the pointer movement here, is governed by the second-order system response described in Chapter 3. However, the torque input to the mechanical system is a function of the electrical response of the system. In this measurement system, the input signal, which is the applied voltage, is transformed into a mechanical torque to provide the measurable output deflection.

## Alternating Current

An AC current can be measured in any number of ways. One technique, found in common deflection meters, uses diodes to form a rectifier that converts the time-dependent AC current into a DC current. This current then can be measured with a calibrated D'Arsonval movement meter as previously described. This is the same technique used in those ubiquitous small transformers used to convert AC wall current into a DC current at a set voltage to power or to charge electronic devices. An *electrodynamometer* is essentially a D'Arsonval movement modified for use with AC current by replacing the permanent magnet with an electromagnet in series with the current coil. These AC meters have upper limits on the frequency of the alternating current that they can effectively measure; most common instruments are calibrated for use with standard line frequency (60 Hz).

An accurate measuring solution for large AC current is the Hall effect probe. This is a probe clamped over the current-carrying wire (conductor) to measure its unknown current flow. To understand its use, let us mention two phenomena. The first is the *Hall effect*, which is a voltage that is developed from any current-carrying conductor placed perpendicular to a magnetic field. For a known current, the magnitude of this voltage directly depends on the magnitude of the magnetic field. The second is that a current passing through a wire generates a magnetic field. So in practice, a Hall-effect probe is realized by coupling these two processes concurrently: Use the unknown current within a wire to generate a magnetic field that develops a measurable voltage across a Hall-effect sensor.

The Hall-effect sensor is a thin conducting semiconductor wafer driven by a known current; this current is unrelated to the unknown current being measured and is provided by a separate source, such as a battery. The Hall-effect probe is an iron-core ring that is placed around the wire carrying the current to be measured. This ring is used to concentrate the magnetic field, and a Hall-effect sensor is attached to the iron core so as to be aligned parallel to the wire carrying the current being measured.

## 6.3 Analog Devices: Voltage Measurements

Often we are interested in measuring either static or dynamic voltage signals. Depending on the source, the magnitude of these signals may range over several orders of magnitude throughout the measured signal chain. The frequency content of dynamic voltage signals is often of interest as well. As such, a wide variety of measurement systems have been developed for voltage measurement of static and dynamic signals. This section discusses several convenient and common methods to indicate voltage in measurement systems.

### Analog Voltage Meters

A DC voltage can be measured in through the analog circuit shown in Figure 6.6, where a D'Arsonval movement is used in series with a resistor. Although fundamentally sensitive to current flow, the D'Arsonval movement can be calibrated in terms of voltage by using an appropriate known fixed resistor and through Ohm's law relating it to the measured current. This basic circuit is employed in the construction of analog dial voltmeters and volt-ohmmeters (VOMs), which for many years served as the common measurement device for current, voltage, and resistance.

An AC voltage can be measured through rectification or through the use of an electromagnet, either in an electrodynamometer or with a movable iron vane. These instruments are sensitive to the *rms* (root-mean-square) value of a simple periodic AC current, and can be calibrated in terms of voltage; shunt resistors can be used to establish the appropriate scale. The circuit shown in Figure 6.6 can also be used to measure an AC voltage if the input voltage is rectified prior to input to this circuit. The waveform output of the rectifier must be considered if a steady meter reading is to be obtained. An AC meter indicates a true rms value for a simple periodic signal only, but a *true* rms AC voltmeter performs the signal integration (e.g., Eq. 2.4) required to accurately determine the rms value in a signal-conditioning stage and indicates true signal rms regardless of waveform.

### Oscilloscope

The *oscilloscope* is a practical graphical display device providing an analog representation of a measured signal. It is used to measure and to visually display voltage magnitude versus time for

**FIGURE 6.6**   A DC voltmeter circuit.

**FIGURE 6.7**  Digital oscilloscope
illustrating oscilloscope output.

dynamic signals over a wide range of frequencies with a signal bandwidth extending commonly into the megahertz range and, with some units, into the 30 GHz range and beyond [1, 2]. A useful diagnostic tool, the oscilloscope provides a visual output of signal magnitude, frequency, distortion, and a delineation of the DC and AC components. The visual image provides a direct means to detect the superposition of noise and interference on a measured signal, something nonvisual metering devices cannot do. In addition to signal versus time, a typical unit can also display two or more signals [$X(t)$ and $Y(t)$], perform addition and subtraction of signals ($X + Y, X - Y$), and display amplitude versus amplitude ($XY$) plots and other features. Some digital oscilloscopes have significant internal storage so as to mimic a data-logging signal recorder. Others have internal fast Fourier transform (FFT) circuitry to provide for direct spectral analysis of a signal (see Chapter 2).

A digital oscilloscope, such as shown in Figure 6.7, provides an analog representation of the measured signal. But it does so by first sampling the signal to convert it into a digital form and then reconstructing the signal on a display as an analog output. In this way, the digital oscilloscope measures and stores voltage in a digital manner but then displays it in an analog format as a sequence of measured points, as shown in Figure 6.7. These common lab devices can be found as small and very portable packages and are often a virtual software component of a 12- to 18-bit analog-to-digital converter data acquisition system. State of the art units include the capability to do signal analysis using application specific software, and some units can run third-party software as well. Analog-to-digital sampling techniques used to measure voltages are discussed in Chapter 7.

The LabView interactive program (virtual instrument) called *Oscilloscope* is available with the companion software. It features a basic two-channel oscilloscope. The user can vary the channel displayed and the time sweep and gain using two built-in signals (sine wave and square wave) and an active signal trigger.

**EXAMPLE 6.3**

We can specify the requirements for an oscilloscope based on the intended signal to be measured. For example, in the USB 1.1 protocol for data transmission, a single frame of data lasts for 1 ms with data transmitted serially at 12 Mbps (million bits per second). We can simplify this as trying to capture a 12-MHz square wave for 1 ms on the oscilloscope screen to base our requirements.

From the discussions of Chapter 2, we know that we need at least five harmonics to reconstruct a square wave with any reasonable fidelity, so the sampling rate required to reconstruct a 12-MHz square wave is at least five times the fundamental frequency or at least 60 MHz (i.e., $60 \times 10^6$ samples/s). So that sets the lowest value on oscilloscope response. The storage capacity of the digital oscilloscope required to capture one frame of the USB data signal is $60 \times 10^6$ samples/s $\times$ 0.001 s or 60,000 samples at 60 MHz. For comparison of potential oscilloscope needs, the more common USB 2.0 transmits at a nominal 480 Mbps. In 2013 a successor standard was issued, USB 3.1, which sets transfer rates up to 10 Gbitps and is called SuperSpeed+. Color coding and the letters SS on connectors identify hardware compatible with this standard.

# Potentiometer

The *potentiometer*[2] is a device used to measure DC voltages that are in the microvolt to millivolt range. Equivalent to a balance scale, a potentiometer balances an unknown input voltage against a known internal voltage until both sides are equal. A potentiometer is a null balance instrument in that it drives the loading error to essentially zero. Potentiometers have been supplanted by digital voltmeters, which are deflection mode devices but have very high input impedances so as to keep loading errors small, even at low-level voltages.

## Voltage Divider Circuit

A general purpose component found in the potentiometer circuit is the *voltage divider circuit* shown in Figure 6.8. The point labeled $A$ in Figure 6.8 represents a sliding contact, which makes an electrical connection with the resistor $R$. The resistance between points $A$ and $B$ in the circuit is a linear function of the distance from $A$ to $B$. So the output voltage sensed by the meter given by

$$E_o = \frac{L_x}{L_T}E_i = \frac{R_x}{R_T}E_i \tag{6.6}$$

holds so long as the internal resistance of the meter, $R_m$, is very large relative to $R_T$.

## Potentiometer Instruments

A simple potentiometer measuring instrument can be derived from the voltage divider circuit as shown in Figure 6.9. In this arrangement a current-sensing device, such as a galvanometer, is used to detect current flow; any current flow through the galvanometer, $G$, would be a result of an imbalance between the measured voltage, $E_m$, and the voltage imposed across points $A$ to $B$, $E_{AB}$. The voltage $E_{AB}$ is adjusted by moving the sliding contact $A$; if this sliding contact is calibrated in terms of a known supply voltage, $E_i$, the circuit may be used to measure voltage by creating a balanced condition, indicated by a zero current flow through the galvanometer. A null balance,

**FIGURE 6.8**  Voltage divider circuit.

---

[2]The term "potentiometer" is used in several different ways: a sliding contact precision variable resistor, the divider circuit in Figure 6.9, and a high-precision circuit used as a voltage-measuring instrument.

**FIGURE 6.9**    Basic potentiometer circuit.

corresponding to zero current flow through $G$, occurs only when $E_m = E_{AB}$. With a known and constant supply voltage $E_i$, the position of $A$ can be calibrated to indicate $E_m$ directly, as suggested by Equation 6.6.

A practical potentiometer includes a means for setting the supply voltage $E_i$. One way uses a separate divider circuit to adjust the output from a cheap, replaceable battery to equal the fixed, known output from a standard voltage cell. This way, the replaceable battery is used for subsequent measurements. Potentiometer instruments can have systematic uncertainties of <1 $\mu$V and random uncertainties <0.2 $\mu$V.

## 6.4 Analog Devices: Resistance Measurements

Resistance measurements may range from simple continuity checks, to determining changes in resistance on the order of $10^{-6}$ $\Omega$, to determining absolute resistance ranging from $10^{-5}$ to $10^{15}$ $\Omega$. As a result of the tremendous range of resistance values that have practical application, numerous measurement techniques that are appropriate for specific ranges of resistance or resistance changes have been developed. Some of these measurement systems and circuits provide necessary protection for delicate instruments, allow compensation for changes in ambient conditions, or accommodate changes in transducer reference points. In addition, the working principle of many transducers is a change in resistance relative to a change in the measured variable. We will discuss the measurement of resistance using basic techniques of voltage and current measurement in conjunction with Ohm's law.

### Ohmmeter Circuits

A simple way to measure resistance is by imposing a voltage across the unknown resistance and measuring the resulting current flow where $R = E/I$. Clearly, from Ohm's law, the value of resistance can be determined in a circuit such as in Figure 6.10, which is the basis of a common analog ohmmeter. Practical analog ohmmeters use circuits similar to those shown in Figure 6.11, which use shunt resistors and a D'Arsonval mechanism for measuring a wide range of resistance while limiting the flow of current through the meter movement. In this technique, the lower limit on the measured resistance $R_1$ is determined by the upper limit of current flow through it, that is, $I_1$. A practical limit to the maximum current flow through a resistance is imposed by the ability of the resistance element to dissipate the power generated by the flow of current ($I^2R$ heating). For example, the limiting ability of a thin metallic conductor to dissipate heat is the principle on which fuses are based. At too large a value of $I_1$, they melt. The same is true for delicate, small resistance sensors!

### Bridge Circuits

A variety of bridge circuits have been devised for measuring capacitance, inductance, and, most often, resistance. A purely resistive bridge, called a *Wheatstone bridge*, provides a means for

**FIGURE 6.10** Basic analog ohmmeter (voltage is adjusted to yield full-scale deflection with terminals shorted).

**FIGURE 6.11** Multirange ohmmeter circuits.

accurately measuring resistance, and for detecting very small changes in resistance. Figure 6.12 shows the basic arrangement for a bridge circuit, where $R_1$ is a sensor that experiences a change in resistance associated with a change in some physical variable. A DC voltage is applied as an input across nodes $A$ to $D$, and the bridge forms a parallel circuit arrangement between these two nodes. The currents flowing through the resistors $R_1$ to $R_4$ are $I_1$ to $I_4$, respectively. Under the condition that the current flow through the galvanometer, $I_g$, is zero, the bridge is in a balanced condition

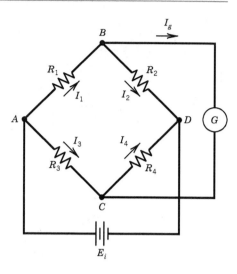

**FIGURE 6.12**   Basic current-sensitive Wheatstone bridge circuit ($G$, galvanometer).

(alternatively, a current-sensing digital voltmeter might be used in place of the galvanometer). A specific relationship exists between the resistances that form the bridge at balanced conditions. To find this relationship, let $I_g = 0$. Under this balanced condition, there is no voltage drop from $B$ to $C$ and

$$I_1 R_1 - I_3 R_3 = 0$$
$$I_2 R_2 - I_4 R_4 = 0 \tag{6.7}$$

Under balanced conditions, the current through the galvanometer is zero, and the currents through the arms of the bridge are equal:

$$I_1 = I_2 \quad \text{and} \quad I_3 = I_4 \tag{6.8}$$

Solving Equation 6.7 simultaneously with the condition stated in Equation 6.8 yields the necessary relationship among the resistances for a balanced bridge:

$$\frac{R_2}{R_1} = \frac{R_4}{R_3} \tag{6.9}$$

Resistance and resistance change can be measured in one of two ways with this bridge circuit. If the resistor $R_1$ varies with changes in the measured physical variable, the resistance of one of the other arms of the bridge can be adjusted to null the circuit and determine resistance. Another method uses a voltage measuring device to measure the voltage unbalance in the bridge as an indication of the change in resistance. We will analyze both of these methods further.

## Null Method

Consider the circuit shown in Figure 6.12, where $R_2$ is an adjustable variable resistance. If the resistance $R_1$ changes due to a change in the measured variable, the resistance $R_2$ can be adjusted to compensate so that the bridge is once again balanced. In this null method of operation, the resistance $R_2$ must be a calibrated variable resistor, such that adjustments to $R_2$ directly indicate the value of $R_1$. The balancing operation may be accomplished either manually or automatically through a closed-loop controller circuit. An advantage of the null method is that the applied input voltage need not be known, and changes in the input voltage do not affect the accuracy of the measurement. In addition, the current detector or controller need only detect if there is a flow of current, not measure its value.

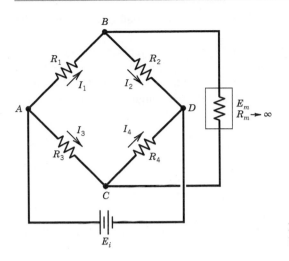

**FIGURE 6.13** Voltage-sensitive Wheatstone bridge.

## Deflection Method

In an unbalanced condition, the magnitude of the current or voltage drop for the meter portion of the bridge circuit is a direct indication of the change in resistance of one or more of the arms of the bridge. Consider first the case in which the voltage drop from node $B$ to node $C$ in the basic bridge is measured by a meter with infinite internal impedance so that there is no current flow through the meter, as shown in Figure 6.13. Knowing the conditions for a balanced bridge given in Equation 6.8, the voltage drop from $B$ to $C$, $E_m$ or $E_o$ can be determined, since the current $I_1$ must equal the current $I_2$, as

$$E_o = I_1 R_1 - I_3 R_3 \tag{6.10}$$

Under these conditions, substituting Equations 6.7 to 6.9 into Equation 6.10 yields

$$E_o = E_i \left( \frac{R_1}{R_1 + R_2} - \frac{R_3}{R_3 + R_4} \right) \tag{6.11}$$

The bridge is usually initially balanced at some reference condition. Any transducer resistance change resulting from a change in the measured variable would then cause a deflection in the bridge voltage away from the balanced condition. Assume that from an initially balanced condition where $E_o = 0$, a change in $R_1$ occurs to some new value, $R_1' = R_1 + \delta R$. The output from the bridge becomes

$$E_o + \delta E_o = E_i \left( \frac{R_1'}{R_1' + R_2} - \frac{R_3}{R_3 + R_4} \right) = E_i \frac{R_1' R_4 - R_3 R_2}{(R_1' + R_2)(R_3 + R_4)} \tag{6.12}$$

In many designs, the bridge resistances are initially equal. Setting $R_1 = R_2 = R_3 = R_4 = R$ allows Equation 6.12 to be reduced to

$$\frac{\delta E_o}{E_i} = \frac{\delta R/R}{4 + 2(\delta R/R)} \tag{6.13}$$

In contrast to the null method of operation of a Wheatstone bridge, the deflection bridge requires a meter capable of accurately indicating the output voltage, as well as a stable and known input voltage. But the bridge output should follow any resistance changes over any frequency input, up to the frequency limit of the detection device. Therefore, a deflection mode is often used to measure time-varying signals.

If the high-impedance voltage measuring device in Figure 6.13 is replaced with a relatively low-impedance current measuring device and the bridge is operated in an unbalanced condition, a current-sensitive bridge circuit results. Consider Kirchhoff's laws applied to the Wheatstone bridge circuit for a current-sensing device with resistance $R_g$. The input voltage is equal to the voltage drop in each arm of the bridge,

$$E_i = I_1 R_1 + I_2 R_2 \tag{6.14}$$

but $I_2 = I_1 - I_g$, which gives

$$E_i = I_1(R_1 + R_2) - I_g R_2 \tag{6.15}$$

If we consider the voltage drops in the path through $R_1$, $R_g$, and $R_3$, the total voltage drop must be zero:

$$I_1 R_1 + I_g R_g - I_3 R_3 = 0 \tag{6.16}$$

For the circuit formed by $R_g$, $R_4$, and $R_2$,

$$I_g R_g + I_4 R_4 - I_2 R_2 = 0 \tag{6.17}$$

or with $I_2 = I_1 - I_g$ and $I_4 = I_3 + I_g$,

$$I_g R_g + (I_3 + I_g)R_4 - (I_1 - I_g)R_2 = 0 \tag{6.18}$$

Equations 6.16 to 6.18 form a set of three simultaneous equations in the three unknowns $I_1$, $I_g$, and $I_3$. Solving these three equations for $I_g$ yields

$$I_g = \frac{E_i(R_1 R_4 - R_2 R_3)}{R_1 R_4(R_2 + R_3) + R_2 R_3(R_1 + R_4) + R_g(R_1 + R_2)(R_3 + R_4)} \tag{6.19}$$

and $E_o = I_g R_g$. Then, the change in resistance of $R_1$ can be found in terms of the bridge deflection voltage, $E_o$, by

$$\frac{\delta R}{R_1} = \frac{(R_3/R_1)\left[E_o/E_i + R_2/(R_2 + R_4)\right]}{1 - E_o/E_i - R_2/(R_2 + R_4)} - 1 \tag{6.20}$$

Consider the case when all of the resistances in the bridge are initially equal to $R$, and subsequently $R_1$ changes by an amount $\delta R$. The current through the meter is given by

$$I_g = E_i \frac{\delta R/R}{4(R + R_g)} \tag{6.21}$$

and the output voltage is given by $E_o = I_g R_g$,

$$E_o = E_i \frac{\delta R/R}{4(1 + R/R_g)} \tag{6.22}$$

The bridge impedance can affect the output from a constant voltage source having an internal resistance $R_s$. The effective bridge resistance, based on a Thévenin equivalent circuit analysis, is given by

$$R_B = \frac{R_1 R_3}{R_1 + R_3} + \frac{R_2 R_4}{R_2 + R_4} \tag{6.23}$$

such that for a power supply of voltage $E_s$

$$E_i = \frac{E_s R_B}{R_s + R_B} \tag{6.24}$$

In a similar manner, the bridge impedance can affect the voltage indicated by the voltage-measuring device. For a voltage-measuring device of internal impedance $R_g$, the actual bridge deflection voltage, relative to the indicated voltage, $E_m$, is

$$E_o = \frac{E_m}{R_g} \left( \frac{R_1 R_2}{R_1 + R_2} + \frac{R_3 R_4}{R_3 + R_4} + R_g \right) \tag{6.25}$$

The difference between the measured voltage $E_m$ and the actual voltage $E_o$ is a loading error, in this case due to the bridge impedance load. Loading errors are discussed next.

### EXAMPLE 6.4

A certain temperature sensor experiences a change in electrical resistance with temperature according to the equation

$$R = R_o \left[ 1 + \alpha(T - T_o) \right] \tag{6.26}$$

where

$R$ = sensor resistance ($\Omega$)
$R_o$ = sensor resistance at the reference temperature, $T_o$ ($\Omega$)
$T$ = temperature (°C)
$T_o$ = reference temperature (0°C)
$\alpha$ = temperature coefficient of resistance 0.00395°C$^{-1}$

This temperature sensor is connected in a Wheatstone bridge such as the one shown in Figure 6.12, where the sensor occupies the $R_1$ location, and $R_2$ is a calibrated variable resistance. The bridge is operated using the null method. The fixed resistances $R_3$ and $R_4$ are each equal to 500 $\Omega$. If the temperature sensor has a resistance of 100 $\Omega$ at 0°C, determine the value of $R_2$ that would balance the bridge at 0°C.

**KNOWN** $R_1 = 100$ $\Omega$.

$$R_3 = R_4 = 500 \ \Omega$$

**FIND** $R_2$ for null balance condition.

### SOLUTION
From Equation 6.9, a balanced condition for this bridge would be achieved when $R_2 = R_1 R_4 / R_3$ or $R_2 = 100$ $\Omega$. Notice that to be a useful circuit, $R_2$ must be adjustable and provide an indication of its resistance value at any setting.

---

### EXAMPLE 6.5

Consider a deflection bridge, which initially has all four arms of the bridge equal to 100 $\Omega$, with the temperature sensor described in Example 6.4 again as $R_1$. The input or supply voltage to the bridge is 10 V. If the temperature of $R_1$ is changed such that the bridge output is 0.569 V, what is the temperature, $T_1$, of the sensor? How much current, $I_1$, flows through the sensor, and how much power, $P_1$, must it dissipate?

**KNOWN** $E_i = 10$ V.

Initial (reference) state: $R_1 = R_2 = R_3 = R_4 = 100\ \Omega$

$$E_o = 0\ \text{V}$$

Deflection state: $E_o = 0.569$ V.

**ASSUMPTION** Voltmeter has infinite input impedance, but source has negligible impedance $(E_s = E_i)$.

**FIND** $T_1, I_1, P_1$.

**SOLUTION**
The change in the sensor resistance can be found from Equation 6.13 as

$$\frac{\delta E_o}{E_i} = \frac{\delta R/R}{4 + 2(\delta R/R)} \Rightarrow \frac{0.569}{10} = \frac{\delta R}{400 + 2\delta R}$$
$$\delta R = 25.67\ \Omega.$$

This gives a total sensor resistance of $R_1 + \delta R = 125.7\ \Omega$, which equates to a sensor temperature of $T_1 = 65°C$.

To determine the current flow through the sensor, consider first the balanced case where all the resistances are equal to 100 $\Omega$. The equivalent bridge resistance $R_B$ is simply 100 $\Omega$ and the total current flow from the supply, $E_i/R_B = 100$ mA. Thus, at the initially balanced condition, the current flow through each arm of the bridge and through the sensor is 50 mA. If the sensor resistance changes to 125.67 $\Omega$, the current will be reduced. If the output voltage is measured by a high-impedance device such that the current flow to the meter is negligible, the current flow through the sensor is given by

$$I_1 = E_i \frac{1}{(R_1 + \delta R) + R_2} \tag{6.27}$$

This current flow is then 44.3 mA.

The power, $P_1 = I_1^2(R_1 + \delta R)$, that must be dissipated from the sensor is 0.25 W, which, depending on the surface area and local heat transfer conditions, may cause a change in the temperature of the sensor.

**COMMENT** The current flow through the sensor results in a sensor temperature higher than would occur with zero current flow due to $I^2R$ heating. This is a loading error that offsets the indicated temperature, a systematic error. There is a trade-off between the increased sensitivity, $dE_o/dR_1$, and the correspondingly increased current associated with $E_1$. The input voltage must be chosen appropriately for a given application.

## 6.5 Loading Errors and Impedance Matching

In an ideal sense, an instrument or measurement system should not in itself affect the variable being measured. Any such effect will alter the variable and be considered as a "loading" that the measurement system exerts on the measured variable. A *loading error* is the difference between the value of the measurand and the indicated value brought on by the act of measurement. Loading effects can be of any form: mechanical, electrical, or optical. When the insertion of a sensor into a process somehow changes the physical variable being measured that is a loading error. A loading error can occur anywhere along the signal path of a measurement system. If the output from one system stage is in any way affected by the subsequent stage, then the signal is affected by *interstage loading error*. Good measurement system design minimizes all loading errors.

To illustrate this idea, consider measuring the temperature of a volume of a high-temperature liquid using a liquid-in-glass thermometer. Some finite quantity of energy must flow from the liquid to the thermometer to achieve thermal equilibrium between the thermometer and the liquid (i.e., the thermometer may cool down or heat up the liquid). As a result of this energy flow, the liquid temperature is changed, and the measured value does not correspond to the initial liquid temperature sought. This measurement has introduced a loading error.

## Loading Errors for Voltage-Dividing Circuit

Consider the voltage-divider circuit shown in Figure 6.8 for the case where $R_m$ is finite. Under these conditions, the circuit can be represented by the equivalent circuit shown in Figure 6.14. As the sliding contact at point $A$ moves, it divides the full-scale deflection resistance $R$ into $R_1$ and $R_2$, such that $R_1 + R_2 = R_T$. The resistances $R_m$ and $R_1$ form a parallel loop, yielding an equivalent resistance, $R_L$, given by

$$R_L = \frac{R_1 R_m}{R_1 + R_m} \tag{6.28}$$

The equivalent resistance for the entire circuit, as seen from the voltage source, is $R_{eq}$:

$$R_{eq} = R_2 + R_L = R_2 + \frac{R_1 R_m}{R_1 + R_m} \tag{6.29}$$

The current flow from the voltage source is then

$$I = \frac{E_i}{R_{eq}} = \frac{E_i}{R_2 + R_1 R_m/(R_1 + R_m)} \tag{6.30}$$

and the output voltage is given by

$$E_o = E_i - IR_2 \tag{6.31}$$

This result can be expressed as

$$\frac{E_o}{E_i} = \frac{1}{1 + (R_2/R_1)(R_1/R_m + 1)} \tag{6.32}$$

The limit of this expression as $R_m$ tends to infinity is

$$\frac{E_o}{E_i} = \frac{R_1}{R_1 + R_2} \tag{6.33}$$

**FIGURE 6.14**  Instruments in parallel to signal path form an equivalent voltage-dividing circuit.

which is just Equation 6.6. Also, as $R_2$ approaches zero, the output voltage approaches the supply voltage, as expected. Expressing the value found from Equation 6.6 as the true value $(E_o/E_i)'$, the loading error $e_I$ may be given by

$$e_I = E_i \left[ \left( \frac{E_o}{E_i} \right) - \left( \frac{E_o}{E_i} \right)' \right] \tag{6.34}$$

Here the loading error goes to zero as $R_m \to \infty$. Hence a large meter resistance reduces loading error in voltage measurements.

## Interstage Loading Errors

Consider a common situation in the measurement chain in which the output voltage signal from one system device provides the input signal to the following device. This following device could be either a measuring device or an interstage device, such as a signal conditioner. An open circuit voltage $E_1$ will be present at the output terminal of device 1 with its equivalent impedance $Z_1$. Device 2 is then attached across the terminals. As shown in Figure 6.15, the Thévenin equivalent circuit consists of a voltage, $E_1$, with internal series impedance $Z_1$. Connecting device 2 across the terminals of device 1 is equivalent to placing an impedance $Z_m$ across the terminals. This causes a current, $I$, to flow in the loop formed by the two terminals, and $Z_m$ acts as a load on the first device. The voltage sensed by device 2 is

$$E_m = IZ_m = E_1 \frac{1}{1 + Z_1/Z_m} \tag{6.35}$$

Consequently, the voltage sensed by device 2 differs from the original voltage owing to the interstage connection. This is a loading error of magnitude, $e_I = E_m - E_1$, or

$$e_I = E_1 \left( \frac{1}{1 + Z_1/Z_m} - 1 \right) \tag{6.36}$$

For a maximum in the voltage potential between stages, it is required that $Z_m \gg Z_1$ such that $e_I \to 0$. The interstage or measuring device should have a high input impedance to minimize loading error.

When the signal is current-driven, such as a 4–20 mA current loop, the maximum current transfer between devices 1 and 2 is desirable. Consider the circuit shown in Figure 6.16. The current through the loop indicated by the device 2 is given by $I_m$, whereas when device 2 is not present in the circuit, the short circuit value is $I'$. That is,

$$I_m = \frac{E_1}{Z_1 + Z_m} \quad \text{and} \quad I' = \frac{E_1}{Z_1} \tag{6.37}$$

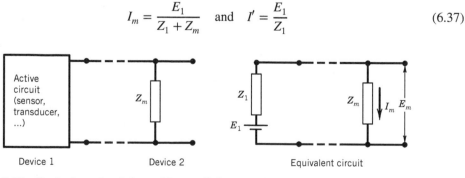

FIGURE 6.15    Equivalent circuit formed by parallel connections.

**FIGURE 6.16** Instruments in series with signal path.

where $Z_m$ represents the total load imposed by device 2 on device 1. The loading error, $e_I = I_m - I'$, is

$$e_I = E_1 \frac{-Z_m}{Z_1^2 + Z_1 Z_m} \qquad (6.38)$$

From Equation 6.38 or 6.37, it is clear that to reduce current-driven loading error, then $Z_m \gg Z_1$, such that $e_I \to 0$.

When a 4–20 mA current loop is used to drive the signal, often a resistor of impedance $Z_m$ is placed across the output terminals and voltage measured. The measured voltage, $E_m$, is

$$E_m = I' Z_m \left( \frac{Z_1}{Z_1 + Z_m} \right) \qquad (6.39)$$

The term in parentheses in Equation 6.39 shows the loading influence, which is minimized as $Z_m \ll Z_1$.

### EXAMPLE 6.6

For the Wheatstone bridge shown in Figure 6.17, find the open circuit output voltage (when $R_m \to \infty$) if the four resistances change by

$$\delta R_1 = +40\,\Omega \quad \delta R_2 = -40\,\Omega \quad \delta R_3 = +40\,\Omega \quad \delta R_4 = -40\,\Omega$$

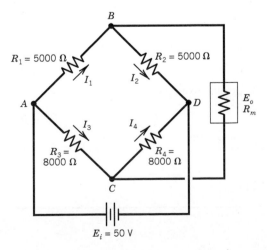

**FIGURE 6.17** Bridge circuit for Example 6.6.

**KNOWN** Measurement device resistance given by $R_m$.

**ASSUMPTIONS** $R_m \to \infty$ Negligible source impedance.

**FIND** $E_o$.

**SOLUTION**
From Equation 6.11,

$$E_o = E_i \left( \frac{R_1}{R_1 + R_2} - \frac{R_3}{R_3 + R_4} \right)$$
$$= 50 \text{ V} \left( \frac{5040 \text{ }\Omega}{5040 \text{ }\Omega + 4960 \text{ }\Omega} - \frac{8040 \text{ }\Omega}{8040 \text{ }\Omega + 7960 \text{ }\Omega} \right) = +0.0750 \text{ V}$$

**EXAMPLE 6.7**

Now consider the case where a meter with internal impedance $R_m = 200$ k$\Omega$ is used in Example 6.6. What is the output voltage with this particular meter in the circuit? Estimate the loading error.

**KNOWN** $R_m = 200,000 \text{ }\Omega$.

**FIND** $E_o, e_I$.

**SOLUTION**
Because the output voltage $E_o$ is equal to $I_m R_m$, where $I_m$ is found from Equation 6.19,

$$E_o = R_g I_m = \frac{E_i R_m (R_1 R_4 - R_2 R_3)}{R_1 R_4 (R_2 + R_3) + R_2 R_3 (R_1 + R_4) + R_m (R_1 + R_2)(R_3 + R_4)}$$

which for $R_m = 200,000 \text{ }\Omega$, gives $E_o = 0.0726$ V.

The loading error $e_I$ is the difference in the two output voltages between infinite meter impedance (Ex. 6.6) and finite meter impedance. This gives $e_I = 0.0726 - 0.0750 = -2.4$ mV with the negative sign reflecting a lower measured value under load.

The bridge output may also be expressed as a ratio of the output voltage with $R_m \to \infty$, $E_o'$, and the output voltage with a finite meter resistance $E_o$:

$$\frac{E_o}{E_o'} = \frac{1}{1 + R_e/R_g}$$

where $R_e = \frac{R_1 R_2}{R_1 + R_2} + \frac{R_3 R_4}{R_3 + R_4}$, which yields for the present example

$$\frac{E_o}{E_o'} = \frac{1}{1 + 6500/200,000} = 0.969$$

The percent loading error, $100 \times \left( \frac{E_o}{E_o'} - 1 \right)$, is $-3.1\%$.

# 6.6 Analog Signal Conditioning: Amplifiers

An amplifier is a device that scales the magnitude of an analog input signal according to the relation

$$E_o(t) = h\{E_i(t)\}$$

where $h\{E_i(t)\}$ defines a mathematical function. The simplest amplifier is the linear scaling amplifier in which

$$h\{E_i(t)\} = GE_i(t) \tag{6.40}$$

where the gain $G$ is a constant that may be any positive or negative value. Many other types of operation are possible, including the "base $x$" logarithmic amplifier in which

$$h\{E_i(t)\} = G\log_x(E_i(t)) \tag{6.41}$$

Amplifiers have a finite frequency response and limited input voltage range.

The most widely used type of amplifier is the solid-state operational amplifier. This device is characterized by a high-input impedance ($Z_i > 10^7 \, \Omega$), a low-output impedance ($Z_o < 100 \, \Omega$), and a high internal gain ($A_o \approx 10^5$ to $10^6$). As shown in the general diagram of Figure 6.18a, an operational amplifier has two input ports, a noninverting and an inverting input, and one output port. The signal at the output port is in phase with a signal passed through the noninverting input port but is 180 degrees out of phase with a signal passed through the inverting input port. The amplifier requires dual-polarity DC excitation power ranging from $\pm 5$ V to $\pm 15$ V. In addition, two DC voltage offset null (bias) input ports provide a means to zero out any output offset signal at zero input; usually a variable 10 k$\Omega$ resistor is placed across these inputs to adjust offset null.

As an example, the pin connection layout of a common operational amplifier circuit, the type 741, is shown in Figure 6.18b in its eight-pin, dual-in-line ceramic package form. This is the familiar rectangular black integrated circuit package seen on circuit boards. Each pin port is numbered and labeled as to function. An internal schematic diagram is shown in Figure 6.18c with the corresponding pin connections labeled. As shown, each input port (i.e., 2 and 3) is attached to the base of an npn transistor.

The high internal open loop gain, $A_o$, of an operational amplifier is given as

$$E_o = A_o(E_{i_2}(t) - E_{i_1}(t)) \tag{6.42}$$

The magnitude of $A_o$, flat at low frequencies, falls off rapidly at high frequencies, but this intrinsic gain curve is overcome by using external input and feedback resistors that set the circuit gain $G$ and circuit response. Some possible amplifier configurations using an operational amplifier are shown in Figure 6.19.

Because the amplifier has a very high internal gain and negligible current draw, resistors $R_1$ and $R_2$ are used to form a feedback loop and control the overall amplifier circuit gain, called the *closed loop gain*, $G$. The noninverting linear scaling amplifier circuit of Figure 6.19a has a closed loop gain of

$$G = \frac{E_o(t)}{E_i(t)} = \frac{R_1 + R_2}{R_2} \tag{6.43}$$

Resistor $R_s$ does not affect the gain but is used to balance out potential variation problems at small currents. Its value is selected such that $R_s \approx R_1 R_2/(R_1 + R_2)$. The inverting linear scaling amplifier circuit of Figure 6.19b provides a gain of

$$G = \frac{E_o(t)}{E_i(t)} = \frac{R_2}{R_1} \tag{6.44}$$

**FIGURE 6.18** Operational amplifier.

By using both inputs, the arrangement forms a differential amplifier, Figure 6.19c, in which

$$E_o(t) = (E_{i_2}(t) - E_{i_1}(t))(R_2/R_1) \qquad (6.45)$$

The differential amplifier circuit is effective as a voltage comparator for many instrument applications (see Section 6.7).

A voltage follower circuit is commonly used to isolate an impedance load from other stages of a measurement system, such as might be needed with a high-output impedance transducer. A schematic diagram of a voltage follower circuit is shown in Figure 6.19d. Note that the feedback signal is sent directly to the inverting port. The output for such a circuit, using Equation 6.42, is

$$E_o(t) = A_o(E_i(t) - E_o(t))$$

**(a)** Noninverting amplifier

**(b)** Inverting amplifier

**(c)** Differential amplifier

**(d)** Voltage follower

**(e)** Integrator

**(f)** Differentiator

**FIGURE 6.19**
Some common amplifier circuits
using operational amplifiers.

or writing in terms of the circuit gain, $G = E_o(t)/E_i(t)$,

$$G = \frac{A_o}{1 + A_o} \approx 1 \tag{6.46}$$

Using Kirchhoff's law about the noninverting input loop yields

$$I_i(t)R_1 + E_o(t) = E_i(t)$$

Then the circuit input resistance $R_i$ is

$$R_i = \frac{E_i(t)}{I_i(t)} = \frac{E_i(t)R_1}{E_i(t) - E_o(t)} \tag{6.47}$$

Likewise the output resistance is found to be

$$R_o = \frac{R_2}{1 + A_o} \qquad (6.48)$$

Because $A_o$ is large, Equations 6.46 to 6.48 show that the input impedance, developed as a resistance, of the voltage follower can be large, its output impedance can be small, and the circuit will have near unity gain. Acceptable values for $R_1$ and $R_2$ range from 10 to 100 k$\Omega$ to maintain stable operation.

Input signal integration and differentiation can be performed with the operational amplifier circuits. For the integration circuit in Figure 6.19e, the currents through $R_2$ and $C$ are given by

$$I_{R_2}(t) = \frac{E_i(t) - E_A(t)}{R_2}$$

$$I_C(t) = C\frac{d}{dt}(E_o(t) - E_A(t))$$

Summing currents at node $A$ yields the integrator circuit operation:

$$E_o(t) = -\frac{1}{R_2 C} \int E_i(t)dt \qquad (6.49)$$

As a special note, if the input voltage is a constant positive DC voltage, then the output signal will be a negative linear ramp, $E_0 = -E_i t / RC$, with $t$ in seconds, a feature used in ramp converters and integrating digital voltmeters (see Section 7.5 in Chapter 7). Integrator circuits are relatively unaffected by high-frequency noise, because integration averages out noise.

Following a similar analysis, the differentiator circuit shown in Figure 6.19f performs the operation

$$E_o(t) = -R_2 C\frac{dE_i(t)}{dt} \qquad (6.50)$$

Differentiator circuits amplify high-frequency noise in signals to the point of masking the true signal. Adding resistor $R_i$ limits high-frequency gain to a $-3$ dB cutoff frequency of $f_c = 1/2\pi R_i C$. In noisy environments, a passive $RC$ differentiator may simply perform better.

## 6.7 Analog Signal Conditioning: Special-Purpose Circuits

### Analog Voltage Comparator

A voltage comparator provides an output that is proportional to the difference between two input voltages. As shown in Figure 6.20a, a basic comparator consists of an operational amplifier operating in a high-gain differential mode. In this case, any difference between inputs $E_{i_1}$ and $E_{i_2}$ is amplified at the large open-loop gain $A_o$. Accordingly, the output from the comparator saturates when $E_{i_1} - E_{i_2}$ is either slightly positive or negative, the actual value set by the threshold voltage $E_T$. The saturation output value $E_{sat}$ is nearly the supply voltage $E_s$. For example, a 741 op-amp driven at $\pm15$ V might saturate at $\pm13.5$ V. The value for $E_T$ can be adjusted by the amplifier bias (offset null) voltage $E_{bias}$. The comparator output is given by

$$\begin{aligned}
E_o &= A_o(E_{i_1} - E_{i_2}) &&\text{for} && |E_{i_1} - E_{i_2}| \le E_T \\
&= +E_{sat} &&\text{for} && E_{i_1} - E_{i_2} > E_T \\
&= -E_{sat} &&\text{for} && E_{i_1} - E_{i_2} < -E_T
\end{aligned} \qquad (6.51)$$

**(a)** Differential amplifier   **(b)** Input–output scheme

**FIGURE 6.20**
Analog voltage comparator.

This input–output relation is shown in Figure 6.20b. For a $\pm 15$ V supply and a gain of about 200,000, the comparator might saturate with a voltage difference of only $\sim 68$ $\mu$V.

Often $E_{i_2}$ is a known reference voltage. This allows the comparator output to be used for control circuits to decide whether $E_{i_1}$ is less than or greater than $E_{i_2}$; a positive difference gives a positive output. One frequent use of the comparator is in an analog-to-digital converter (see Chapter 7). A zener-diode connected between the + input and the output provides a TTL (transistor-transistor logic) compatible output signal for digital system use.

## Sample-and-Hold Circuit

The sample-and-hold circuit (SHC) is used to take a narrow-band measurement of a time-changing signal and to hold that measured value until reset. It is widely used in data acquisition systems using analog-to-digital converters. The circuit tracks the signal until it is triggered to hold it at a fixed value while measuring it. This is illustrated in Figure 6.21a, in which the track-and-hold logic provides the appropriate trigger.

The basic circuit for sample and hold is shown in Figure 6.21b. The switch $S_1$ is a fast analog device. When the switch is closed, the "hold" capacitor $C$ is charged through the source resistor $R_s$. When the capacitor is charged, the switch is opened. The amplifier presents a very high input impedance and very low current, which, together with a very low leakage capacitor, allows for a sufficiently long hold time. The typical SHC is noninverting with a unit gain ($G = 1$).

## Charge Amplifier

A *charge amplifier* is used to convert a high-impedance charge $q$ into an output voltage $E_o$. The circuit consists of a high-gain, inverting voltage operational amplifier such as shown in Figure 6.22. These circuits are commonly used with transducers that utilize piezoelectric crystals or capacitance sensors. A piezoelectric crystal develops a time-dependent surface charge under varying mechanical load.

The circuit output voltage is determined by

$$E_o = -q[C_r + (C_T/A_o)] \tag{6.52}$$

with $C_T = C_t + C_c + C_r$ where $C_t$, $C_c$, and $C_r$ represent the transducer, cable, and feedback capacitances, respectively, $R_c$ is the cable and transducer resistances, and $A_o$ is the amplifier open-loop

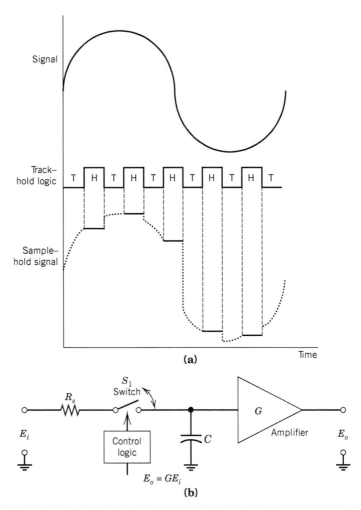

**FIGURE 6.21** Sample-and-hold technique. (a) Original signal and the sample-and-hold signal (time intervals exaggerated for illustrative purposes), (b) circuit.

**FIGURE 6.22** Charge amplifier circuit shown connected to a transducer.

gain. Because the open-loop gain of operational amplifiers is very large, Equation 6.52 can often be simplified to

$$E_o \approx -q/C_r \qquad (6.53)$$

A variable resistance $R_\tau$ and fixed feedback capacitor are used to average out low-frequency fluctuations in the signal.

## 4–20 mA Current Loop

Low-level voltage signals below about 100 mV are quite vulnerable to noise. Examples of transducers having low-level signals include strain gauges, pressure gauges, and thermocouples. One means of transmitting such signals over long distances is by signal boosting through amplification. But amplification also boosts noise. A practical alternative method that is well suited to an industrial environment is a *4-20 mA current loop* (read as "4 to 20"). In this approach, the low-level voltage is converted into a standard current loop signal of between 4 and 20 mA, the lower value for the minimum voltage and the higher value for the maximum voltage in the range. The 4-20 mA current loop can be transmitted over several hundred meters without degradation. The 4-20 mA output is a common option for transducers.

At the output, a receiver converts the current back to a low-level voltage. This can be as simple as a single resistor in series with the loop. For example, a 4–20 mA signal can be converted back to a 1–5 V signal by terminating the loop with a 250 $\Omega$ resistor through which the current flows.

## Multivibrator and Flip-Flop Circuits

*Multivibrator* and *flip-flop* circuits are analog circuits that also form the basis of digital signals. They are useful for system control and triggering of events. The *astable multivibrator* is a switching circuit that toggles on and off continuously between a high- and low-voltage state in response to an applied time-dependent input voltage. It is used to generate a square waveform on a prompt, where typically amplitudes vary between 0 V (low) and 5 V (high) in a manner known as a *TTL signal*. Because the circuit continuously switches between high and low state, it is astable. The heart of this circuit, as shown in Figure 6.23, is the two transistors $T_1$ and $T_2$, which conduct alternately; the circuit is symmetric around the transistors. This generates a square wave signal

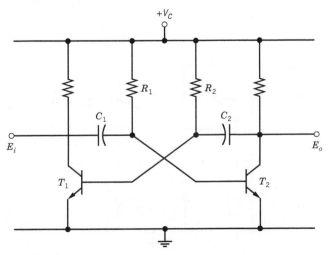

FIGURE 6.23   Basic multivibrator circuit.

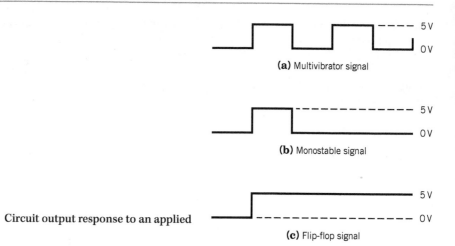

(a) Multivibrator signal

(b) Monostable signal

**FIGURE 6.24**    Circuit output response to an applied input signal.

(c) Flip-flop signal

of fixed period and amplitude at the output as shown in Figure 6.24a. The working mechanism is straightforward. The two transistors change state as the currents through capacitors $C_1$ and $C_2$ increase and decrease due to the applied input. For example, when $T_2$ turns on and its collector switches from $V_c$ toward 0, the base of $T_1$ is driven negative, which turns $T_1$ off. While $C_2$ discharges, $C_1$ charges. As the voltage across $C_1$ increases, the current across it decreases. But so long as the current through $C_1$ is large enough, $T_2$ remains on. It eventually falls to a value that turns $T_2$ off. This causes $C_2$ to charge, which turns $T_1$ on. The period of the resulting square wave is proportional to $R_1 C_1$.

A useful variation of the multivibrator circuit is the *monostable*. In this arrangement, $T_2$ stays on until a positive external pulse or change in voltage is applied to the input. At that moment, $T_2$ turns off for a brief period of time, providing a jump in the output voltage. The monostable circuit cycles just once, awaiting another pulse as shown in Figure 6.24b. Hence, it is often referred to as a *one-shot*. Because it fires just once and resets itself, the monostable is an effective trigger.

Another variation of this circuit is the *flip-flop* or *bistable multivibrator* in Figure 6.25. This circuit is an effective electronic switch. Its operation is analogous to that of a light switch; it is either in an "on" (high) or an "off" (low) state. In practice, transistors $T_1$ and $T_2$ change state every time a pulse is applied. If $T_1$ is on with $T_2$ off, a pulse turns $T_2$ off, turning $T_1$ on, producing the output shown in Figure 6.24c. So the flip-flop output level changes from low to high voltage or

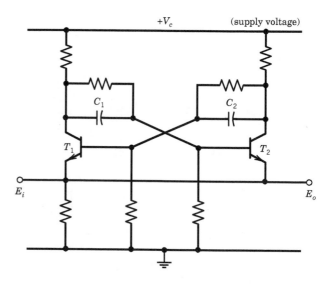

**FIGURE 6.25**    Basic flip-flop circuit.

high to low voltage on command. The flip-flop is also the basic circuit of computer memory chips as it is capable of holding a single bit of information (represented by either a high or low state) at any instant.

## 6.8 Analog Signal Conditioning: Filters

A *filter* is used to remove undesirable frequency information from a dynamic signal. A filter permits signal information associated with a defined band (range) of frequencies to pass, known as the *passband*, while blocking the signal information associated with a band of frequencies, known as the *stopband*. The filter is designed around its cutoff frequency $f_c$, which fixes the boundary between the *passband* and the *stopband*. Filters can be broadly classified as being low-pass, high-pass, bandpass, and notch. The ideal gain characteristics of such filters can be described by the magnitude ratio plots shown in Figure 6.26, which are described as follows. A *low-pass filter* permits frequencies below the prescribed cutoff frequency $f_c$ to pass while blocking the passage of frequency information above the cutoff frequency. Similarly, a *high-pass filter* permits only frequency information above the cutoff frequency to pass. A *bandpass filter* combines features of both the low- and high-pass filters. It is described by a low cutoff frequency $f_{c_1}$ and a high cutoff frequency $f_{c_2}$ to define a band of frequency information that is permitted to pass through the filter. A *notch filter* permits the passage of all frequency information except that within a narrow frequency band. An intensive treatment of filters for analog and digital signals can be found in many specialized texts [3–8].

Filters work by performing a well-defined mathematical operation on the input signal as specified by their transfer function. The transfer function is defined by the position and values of elements in the filter design. *Passive analog filter circuits* consist of combinations of resistors, capacitors, and inductors. *Active filters* incorporate operational amplifiers into the circuit, as well.

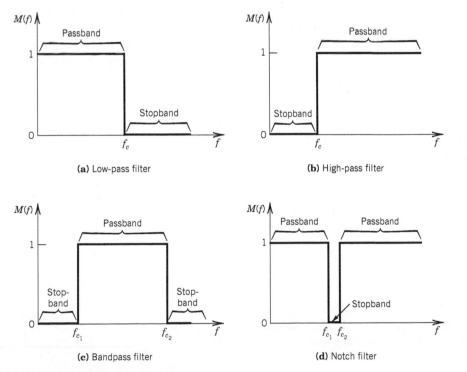

(a) Low-pass filter

(b) High-pass filter

(c) Bandpass filter

(d) Notch filter

**FIGURE 6.26**   Ideal filter characteristics.

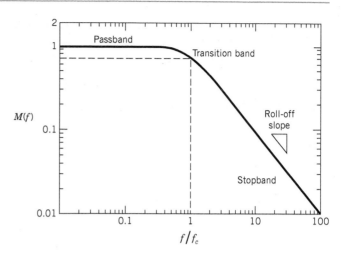

**FIGURE 6.27**   Magnitude ratio for a low-pass Butterworth filter.

The sharp cutoff of the ideal filter cannot be realized in a practical filter. As an example, a plot of the magnitude ratio for a real low-pass filter is shown in Figure 6.27. All filter response curves contain a transition band over which the magnitude ratio decreases with frequency from the passband to the stopband. This rate of transition is known as the filter *roll-off*, usually specified in units of dB/decade. In addition, the filter introduces a phase shift between its input and output signal.

The design of real filters is focused on certain desirable response features. The magnitude and phase characteristics of a real filter can be optimized to meet one of the following: (1) maximum magnitude flatness over the passband, (2) a linear phase response over the passband, or (3) a sharp transition from the passband to stopband with steep roll-off. No one filter can meet all three characteristics, but we can design for any one of them.

For example, a relatively flat magnitude ratio over its passband with a moderately steep initial roll-off and acceptable phase response is a characteristic of a *Butterworth* filter response. On the other hand, a very linear phase shift over its passband but with a relatively gradual initial roll-off is a characteristic of a *Bessel* filter response. The frequency-dependent behavior of low-pass, bandpass, and high-pass filters can be explored in the LabView programs *Butterworth filters* and *Bessel filters*.

## Butterworth Filter Design

A Butterworth filter is optimized to achieve maximum flatness in magnitude ratio over the passband. A simple passive low-pass Butterworth filter can be constructed using the resistor-and-capacitor (RC) circuit of Figure 6.28. By applying Kirchhoff's law about the input loop, we derive the model relating the input voltage $E_i$ to the output voltage $E_o$:

$$RC\dot{E}_o(t) + E_o(t) = E_i(t) \tag{6.54}$$

**FIGURE 6.28**   Simple first-order low-pass resistor-and-capacitor (RC) Butterworth filter circuit.

This real filter model is a first-order system with one reactive (capacitor) component. Its magnitude and phase response has already been given by Equations 3.9 and 3.10 with $\tau = RC$ and $\omega = 2\pi f$, and its magnitude response is shown in Figure 6.28. The roll-off slope is 20 dB/decade.

A filter is designed around its cutoff frequency $f_c$, defined as the frequency at which the signal power is reduced by half. This is equivalent to the magnitude ratio being reduced to 0.707. In terms of the decibel (dB),

$$dB = 20 \log M(f) \tag{3.11}$$

$f_c$ occurs at $-3$ dB—that is, the frequency at which the signal is attenuated by 3 dB. For the filter of Figure 6.28, this requires that $\tau = RC = 1/(2\pi f_c)$.

## Improved Butterworth Filter Designs

With its simplicity, the RC filter begins to roll off well before the cutoff frequency; the roll-off is not very sharp, and the phase shift response (recall Fig. 3.13) is not linear, leading to potential signal distortion. But the flatness in the passband can be extended and the roll-off slope of a filter improved by staging multiple filters in series, called *cascading filters*. This is done by adding reactive elements, such as by alternating inductors (L) and capacitors (C), to the circuit as shown in the resulting LC ladder circuit of Figure 6.29. The resistances $R_s$ and $R_L$ refer to the circuit source and load impedances, respectively. A $k$th-order or $k$-stage filter contains $k$ reactive elements in stages. The magnitude ratio and phase shift for a $k$-stage low-pass Butterworth filter are given by

$$M(f) = \frac{1}{[1 + (f/f_c)^{2k}]^{1/2}} \tag{6.55a}$$

$$\Phi(f) = \sum_{i=1}^{k} \Phi_i(k) \tag{6.55b}$$

where $f / f_c$ is the normalized frequency. The general normalized magnitude response curves for $k$-stage Butterworth filters are shown in Figure 6.30. The roll-off slope is $20 \times k$ (dB/decade) or $6 \times k$ (dB/octave). Frequency-dependent signal attenuation is found from the dynamic error, $\delta (f)$, or directly in decibels by

$$A(f/f_c) = \text{Attenuation(dB)} = 10 \log \left[1 + (f/f_c)^{2k}\right] \tag{6.56}$$

Equation 6.56 is useful for specifying the required order of a filter based on desired response needs. This is emphasized in Figure 6.31, showing how the stages affect attenuation versus normalized frequency.

**FIGURE 6.29**  Ladder circuit for multistage (cascading) low-pass LC filter to achieve a higher-order response.

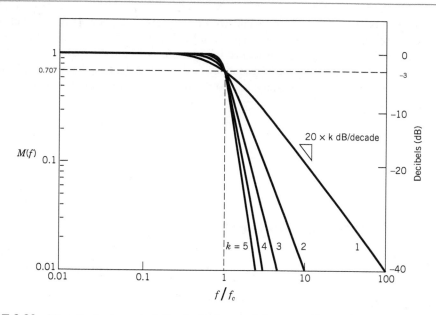

**FIGURE 6.30**    Magnitude characteristics for Butterworth low-pass filters of various stages.

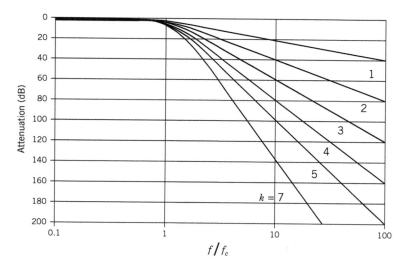

**FIGURE 6.31**    Attenuation with various stages of Butterworth filters.

The specific elements values used in filter design are normalized to accommodate any specific application. To achieve a Butterworth response in the case in which $R_L = R_S$, the ladder circuit of Figure 6.29 uses circuit element values based on the scheme

$$C_i = 2 \sin \left[ \frac{(2i - 1)\pi}{2k} \right] \qquad \text{where } i \text{ is odd} \qquad (6.57a)$$

$$L_i = 2 \sin \left[ \frac{(2i - 1)\pi}{2k} \right] \qquad \text{where } i \text{ is even} \qquad (6.57b)$$

**Table 6.1    Normalized Element Values for Low-Pass LC Butterworth Filter[a] [8]**

| k | $C_1$ | $L_2$ | $C_3$ | $L_4$ | $C_5$ |
|---|-------|-------|-------|-------|-------|
| 2 | 1.414 | 1.414 | | | |
| 3 | 1.000 | 2.000 | 1.000 | | |
| 4 | 0.765 | 1.848 | 1.848 | 0.765 | |
| 5 | 0.618 | 1.618 | 2.000 | 1.618 | 0.618 |

[a]Values for $C_i$ in farads and $L_i$ in henrys are referenced to $R_s = R_L = 1\ \Omega$ and $\omega_c = 1$ rad/s. See discussion for proper scaling. For $k = 1$, use $C_1 = 1$ F or $L_1 = 1$ H.

where $i = 1, 2, \ldots, k$. Specific values are provided in Table 6.1 for elements $C_i$ and $L_i$ for a two-through five-stage Butterworth filter associated with Figure 6.29 [3, 8]. The lead reactive element is assumed to be a capacitor; if the lead reactive element is switched to an inductor (such that $L_i$ is odd), then the table values assigned to $C$ and $L$ switch. The values in Table 6.1 are normalized with $R_s = R_L = 1\ \Omega$ and $\omega_c = 2\pi f_c = 1$ rad/s. For other values, $L_i$ and $C_i$ are scaled by

$$L = L_i R / 2\pi f_c \tag{6.58a}$$

$$C = C_i / (R2\pi f_c) \tag{6.58b}$$

When $R_L \neq R_s$ the impedances can be properly scaled but the reader is referred to published scaling tables, such as given in reference 8. Butterworth filters are quite common and are found in consumer products such as audio equipment. With their flat passband, they are commonly chosen as anti-aliasing filters for data acquisition systems.

A Butterworth high-pass filter is shown in Figure 6.32. In comparing with Figure 6.29, we see that the capacitors and inductors have been swapped. To estimate the high-pass filter values for Figure 6.32, let $(L_i)_{HP} = (1/C_i)_{LP}$ and $(C_i)_{HP} = (1/L_i)_{LP}$ where subscript LP refers to the low-pass values from Table 6.1 and HP refers to the high-pass filter values to be used with Figure 6.32. Reflecting this, the scaling values for a high-pass filter are

$$L = R / 2\pi f_c C_i \tag{6.59a}$$

$$C = 1 / 2\pi f_c R L_i \tag{6.59b}$$

The magnitude ratio for a high-pass filter is given by

$$M(f) = \frac{1}{\left[1 + (f_c/f)^{2k}\right]^{1/2}} \tag{6.60}$$

The phase shift properties described by Equation 6.55b apply.

**FIGURE 6.32**    Ladder circuit for multistage high-pass LC filter.

## EXAMPLE 6.8

Design a one-stage Butterworth RC low-pass filter with a cutoff frequency of 100 Hz at −3 dB if the source and load impedances are 50 Ω. Calculate the expected dynamic error and attenuation at 192 Hz in the realized filter.

**KNOWN** $f_c = 100$ Hz $k = 1$.

**FIND** $C$ and $\delta$.

### SOLUTION

A single-stage low-pass Butterworth RC filter circuit would be similar to that of Figure 6.28, which is just a first-order system with time constant $\tau = RC$. With the relation $\omega = 2\pi f$, the magnitude ratio for this circuit is given by

$$M(f) = \frac{1}{\left[1 + (\omega\tau)^2\right]^{1/2}} = \frac{1}{\left[1 + (2\pi f\tau)^2\right]^{1/2}} = \frac{1}{\left[1 + (f/f_c)^2\right]^{1/2}}.$$

Setting $M(f) = 0.707 = -3$ dB with $f = f_c = 100$ Hz gives

$$\tau = 1/2\pi f_c = RC = 0.0016\,\text{s}$$

With $R = 50$ Ω, we need a capacitor of $C = 32$ µF.

Alternately, we could use Figure 6.29 and Table 6.1 with $k = 1$ for which the normalized value is $C = 1$ F. This value is scaled to $R = 50$ Ω and $f_c = 100$ Hz by

$$C = C_1/(R2\pi f_c) = (1\,F)/(50\Omega)(2\pi)(100\text{Hz}) = 32\,\mu\text{F}$$

A commercially available capacitor size is 33 µF. Using this size in our realized circuit, the cutoff frequency shifts to 96 Hz. We use $f_c = 96$ Hz below.

At $f = 192$Hz, the dynamic error, $\delta(f) = M(f) - 1$ is

$$\delta(192\,\text{Hz}) = M(192\,\text{Hz}) - 1 = -0.55$$

meaning that the input signal frequency content at 192 Hz is reduced by 55%. The attenuation at the normalized frequency of $f/f_c = 2$ is given by equation 6.56 as

$$A(2) = 10\log\left[1 + (2)^2\right] = 46\,\text{dB}$$

## EXAMPLE 6.9

How many filter stages are required to design a low-pass Butterworth filter with a cutoff frequency of 3 kHz if the filter must provide attenuation of at least 60 dB at 30 kHz? If $R_s = R_L = 50$ Ω, find the filter element values.

**KNOWN** $f_c = 3$ kHz; $f = 30$ kHz.

**FIND** $k$ to achieve $A \geq 60$ dB at 30 kHz.

## SOLUTION

The normalized frequency is $f/f_c = 30{,}000 / 3000 = 10$.

$$A(f/f_c = 10) = A(10) = 10\log\left[1 + (f/f_c)^{2k}\right]$$

We can solve directly for $k$ with $A(10) = 60$ dB to get $k = 3$. Or, using Figure 6.31, we find that $A(10) \geq 60$ dB for $k \geq 3$.

Using the ladder circuit of Figure 6.29 with $R_s = R_L = 1\ \Omega$ and $\omega_c = 2\pi f_c = 1$ rad/s, Table 6.1 gives $C_1 = C_3 = 1$ F and $L_2 = 2$ H. Scaling these to $f_c = 3$ kHz and $R_s = R_L = 50\ \Omega$, the element values are

$$L = L_2 R/2\pi f_c = 5.3\ \text{mH}$$
$$C = C_1/(R2\pi f_c) = C_3/(R2\pi f_c) = 106\ \mu\text{F}$$

---

### EXAMPLE 6.10

In a two-speaker (two-way) system, the low-pass circuit passes only low-frequency signal content to the woofer, while the high-pass circuit passes high-frequency content to the wide frequency speaker.

The Linkwitz–Riley high-pass or low-pass filter demonstrates a practical use of Butterworth filters for loudspeaker designs. The 4th order ($k = 4$) design is a cult standard achieving a 24 dB/octave slope. As shown in Figure 6.33 with component design values listed below, the fourth-order design uses two each of inductors and capacitors in conjunction with the natural impedance of the loudspeaker driver to achieve a high-pass or low-pass design, or the two can be coupled as a band-pass filter.

| High-Pass Circuit[a] | Low-Pass Circuit[a] |
|---|---|
| $C_1$ (farad) $= 0.0844/(R_{L_H} \times f_c)$ | $C_2$ (farad) $= 0.2533/(R_{L_L} \times f_c)$ |
| $C_3$ (farad) $= 0.1688/(R_{L_H} \times f_c)$ | $C_4$ (farad) $= 0.0563/(R_{L_L} \times f_c)$ |
| $L_2$ (henry) $= 0.1000 \times R_{L_H}/f_c$ | $L_1$ (henry) $= 0.3000 \times R_{L_L}/f_c$ |
| $L_4$ (henry) $= 0.04501 \times R_{L_H}/f_c$ | $L_3$ (henry) $= 0.1500 \times R_{L_L}/f_c$ |

[a] $f_c$ refers to the cutoff frequency for that particular circuit; $R_{L_H}$ and $R_{L_L}$ refer to the impedances of the high and low frequency speakers, respectively.

Fourth-order high-pass filter

Fourth-order low-pass filter

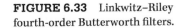

**FIGURE 6.33**  Linkwitz–Riley fourth-order Butterworth filters.

Design a filter for a 120-Hz cut off frequency to be used with a speaker system composed of a woofer and a wide frequency speaker. Assume 8 $\Omega$ drivers.

**KNOWN** $f_c = 120$ Hz; $R_{L_H} = 8\ \Omega$; $R_{L_L} = 8\ \Omega$.

**FIND** $C$ and $L$ for the circuit.

## SOLUTION

Because the filter circuits have identical qualities when using the same $f_c$, the output power to each driver is halved at $f_c$, so the total output sum to each of the two drivers remains at near 0 dB at the crossover frequency. This provides for a desirable nearly flat filter response across the listening frequency range. Applying the information with $f_c = 120$ Hz, we get

| High-Pass Circuit | Low-Pass Circuit |
|---|---|
| $C_1$ (µF) = 87.9 | $C_2$ (µF) = 263.9 |
| $C_3$ (µF) = 175.8 | $C_4$ (µF) = 58.7 |
| $L_2$ (mH) = 6.7 | $L_1$ (mH) = 20.0 |
| $L_4$ (mH) = 30.0 | $L_3$ (mH) = 10.0 |

Commercial values for each element that are close to these calculated would be used in the circuit of Figure 6.33 to achieve a cutoff as close to 120 Hz as possible.

## Bessel Filter Design

A Bessel filter will sacrifice a flat gain over its passband with a gradual initial roll-off in order to maximize a linear phase response over portions of the passband. A linear phase response closely resembles a time delay, reducing distortion. Thus, the attractiveness of a Bessel filter is the ability to pass wideband signals with a minimum of distortion. The filter is widely used for different applications, including audio design where crossover phase shift can have an effect on sound quality.

A $k$-stage low-pass Bessel filter has the transfer function

$$KG(s) = \frac{a_o}{a_o + a_1 s + \cdots + a_k s^k} \tag{6.61}$$

For design purposes, this can be rewritten as

$$KG(s) = \frac{a_o}{D_k(s)} \tag{6.62}$$

where $D_k(s) = (2k - 1) D_{k-1}(s) + s^2 D_{k-2}(s)$, $D_o(s) = 1$ and $D_1(s) = s + 1$.

A $k$-stage LC low-pass Bessel filter can be designed based on the filter circuit shown in Figure 6.29. Table 6.2 lists the normalized corresponding values for elements $L_i$ and $C_i$ to achieve a two- through five-stage Bessel filter corresponding to $\omega_c = 1$ rad with $R_s = R_L = 1\Omega$ [6, 7].

**Table 6.2**    **Normalized Element Values for Low-Pass LC Bessel Filters**[a] [8]

| k | $C_1$ | $L_2$ | $C_3$ | $L_4$ | $C_5$ |
|---|---|---|---|---|---|
| 2 | 0.576 | 2.148 | | | |
| 3 | 0.337 | 0.971 | 2.203 | | |
| 4 | 0.233 | 0.673 | 1.082 | 2.240 | |
| 5 | 0.174 | 0.507 | 0.804 | 1.111 | 2.258 |

[a]Values for $C_i$ in farads and $L_i$ in henrys are referenced to $R_s = R_L = 1\ \Omega$ and $\omega_c = 1$ rad/s. See discussion for proper scaling. For $k = 1$, use $C_1 = 2$ F or $L_1 = 2$ H.

For other values, $L_i$ and $C_i$ are found from Equations 6.61a and 6.61b. Similarly, a high-pass filter with Bessel characteristics and topology similar to Figure 6.32 could be scaled using Table 6.2 and values found from Equations 6.59a and 6.59b.

## Active Filters

An active filter capitalizes on the high-frequency gain characteristics of the operational amplifier to form an effective analog filter. A low-pass active filter is shown in Figure 6.34a using the type 741 operational amplifier. This is a first-order, inverting single-stage, low-pass Butterworth filter. It has a low-pass cutoff frequency given by

$$f_c = \frac{1}{2\pi R_2 C_2} \tag{6.63}$$

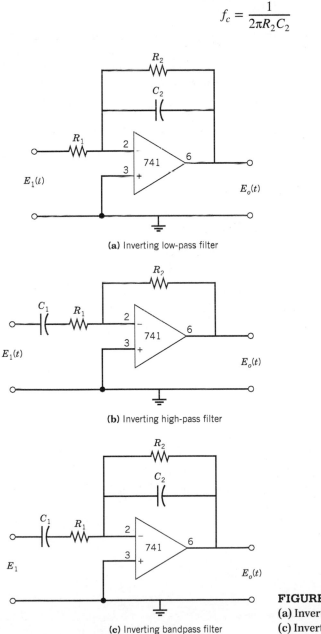

(a) Inverting low-pass filter

(b) Inverting high-pass filter

(c) Inverting bandpass filter

**FIGURE 6.34** Common basic active filters. (a) Inverting low-pass. (b) Inverting high-pass. (c) Inverting bandpass.

The static sensitivity (or gain) of the filter is given by $K = R_2/R_1$. The filter retains the Butterworth characteristics described by Equation 6.58a.

A first-order, inverting single-stage, high-pass Butterworth active filter is shown in Figure 6.34b, using the 741 operational amplifier. The filter has a high-pass cutoff frequency of

$$f_c = \frac{1}{2\pi R_1 C_1} \tag{6.64}$$

with static sensitivity (or gain) of $K = R_2/R_1$. The magnitude ratio is given by

$$M(f) = \frac{1}{\left[1 + (f_c/f)^2\right]^{1/2}} \tag{6.65}$$

An active inverting bandpass filter can be formed by combining the high- and low-pass filters above and is shown in Figure 6.34c. The low cutoff frequency $f_{c_1}$ is given by Equation 6.63 and the high cutoff frequency $f_{c_2}$ by Equation 6.64. This simple circuit is limited in the width of its bandpass. A narrow bandpass requires a higher order filter than the one shown. The design of active filters can be studied with the accompanying LabView program *Lowpass Butterworth Active Filter*.

A commonly used second-order low-pass filter is the unit-gain Sallen-Key topology shown in Figure 6.35a. The generic transfer function is

$$G(s) = \frac{1}{1 + as + bs^2} \tag{6.66}$$

with $a = \omega_c C_1 (R_1 - R_2)$ and $b = \omega_c^2 C_1 C_2 R_1 R_2$ where $\omega_c = 2\pi f_c$. The values for $a$ and $b$ enabling Butterworth or Bessel filter characteristics are listed in Table 6.3. As capacitors come in fewer stock sizes than resistors, usually the capacitor sizes are first selected with $C_2 = nC_1$, where $n$ is a multiple number, such that $C_2 \geq 4bC_1/a^2$, and the resistors are then found by

$$R_{1,2} = \frac{aC_2 \mp \sqrt{(aC_2)^2 - 4bC_1C_2}}{4\pi f_c C_1 C_2} \tag{6.67}$$

**(a)** Low-Pass Filter

**FIGURE 6.35**  Sallen–Key unit-gain filter. (a) Low-pass.
(b) High-pass.                                              **(b)** High-Pass Filter

**Table 6.3    Parameters for Second-Order Sallen–Key Filter Design**

| Parameter | Bessel | Butterworth |
|-----------|--------|-------------|
| $a$ | 1.3617 | 1.4142 |
| $b$ | 0.618 | 1 |
| $Q$ | $1/\sqrt{3}$ | $1/\sqrt{2}$ |

The filter cutoff frequency is set by

$$f_c = \frac{1}{2\pi\sqrt{R_1 R_2 C_1 C_2}} \tag{6.68}$$

The unit-gain Sallen-Key high-pass filter is shown in Figure 6.35b. The design switches the resistors and capacitors from the low-pass filter. For high-pass filters, we can simplify the design and set $C_1 = C_2 = C$. The high-pass filter transfer function is

$$G(s) = \frac{1}{1 + a/s + b/s^2} \tag{6.69}$$

where $a = 2/\omega_c C R_1$ and $b = 1/\omega_c^2 C^2 R_1 R_2$. Values for $a$ and $b$ to meet Butterworth or Bessel filter characteristics are given in Table 6.3 with the resistor values set by $R_1 = 1/\pi f_c a C$ and $R_2 = a/4\pi f_c b C$.

At some frequency, the reactive elements in a filter circuit resonate, leading to a peak in the response behavior. This is behavior predicted by its $Q$-factor, an inverse damping ratio-like parameter. For $Q = 1/2$ response is critically damped, for $Q < 1/2$ response is overdamped, and for $Q > 1/2$ response is underdamped. A higher Q-factor generally allows for sharper roll-off and is used in place of using a higher number of stages in a design. Like its mechanical system analog ($Q = \sqrt{mk}/\zeta$), an underdamped filter shows ringing and has a resonance behavior. The $Q$-factors for Butterworth and Bessel filters are shown in Table 6.3, where

$$Q = \sqrt{b}/a \tag{6.70}$$

# 6.9  Grounds, Shielding, and Connecting Wires

The type of connecting wires used between electrical devices can have a significant impact on the noise level of the signal. Low-level signals of <100 mV are particularly susceptible to errors induced by noise. Some simple rules help to keep noise levels low: (1) Keep the connecting wires as short as possible, (2) keep signal wires away from noise sources, (3) use a wire shield and proper ground, and (4) twist wire pairs along their lengths.

## Ground and Ground Loops

The voltage at the end of a wire that is connected to a rod driven far into the soil would likely be at the same voltage level as the earth—a reference datum called zero or *earth ground*. A *ground* is simply a return path to earth. Now suppose that wire is connected from the rod through an electrical box and then through various building routings to the ground plug of an outlet. Would the ground potential at the outlet still be at zero? Probably not. The network of wires that form the return path to earth would likely act as antennae and pick up some voltage potential relative to earth ground. Any instrument grounded at the outlet would be referenced back to this voltage potential, not to earth ground. The point is that an electrical ground does not represent an absolute value. Ground values vary between ground points because the ground returns pass through different equipment or building wiring on their return path to earth. Thus, if a signal is grounded at two points, say at a power source ground and then at an earth ground, the two grounds could be at different voltage levels. The difference between the two ground point voltages is called the *common-mode voltage* (cmv). This can lead to problems, as discussed next.

*Ground loops* are caused by connecting a signal circuit to two or more grounds that are at different potentials. A ground wire of finite resistance usually carries some current, and so it develops a potential. Thus, two separate and different grounds, even two in close proximity, can be at different potential levels. So when these ground points are connected into a circuit the potential difference itself drives a small unwanted current. The result is a ground loop—an electrical interference superimposed onto the signal. A ground loop can manifest itself in various forms, such as a sinusoidal signal or simply a voltage offset.

Ensure that a system, including its sensors and all electrical components, has only one ground point. Figure 6.36 shows one proper connection between a grounded source, such as a transducer, and a measuring device. In this case, the ground is at the transducer. Note that the common (−) line at the measuring device is not grounded because it is grounded at the source. Thus, it is referred to as being isolated or as a floating ground device. Alternately, the ground can be at the measuring device rather than at the source. The key point is to ground the circuit and all equipment back to only one point, the common ground point. For stationary devices, an excellent ground is created by driving a conducting rod well into the earth for use as the common ground point. Incidentally, many devices are grounded through their AC power lines by means of a third prong. Unless the device power supply is isolated from the circuit, this can set up a ground loop relative to other circuit grounds. To create a floating ground, it is necessary to break this ground connection at this point.

## Shields

Long wires act as antennas and pick up stray signals from nearby electrical fields. The most common problem is AC line noise. Electrical shields are effective against such noise. A *shield* is a piece of metal foil or wire braid wrapped around the signal wires and connected to ground. The shield intercepts external electrical fields, returning them to ground. A shield ground loop is prevented by grounding the shield to only one point, usually the signal ground at the transducer. Figure 6.36 shows such a shield-to-ground arrangement for a grounded voltage source connected to a floating ground measuring device.

A common source of electrical fields is an AC power supply transformer. A capacitance coupling between the 60- or 50-Hz power supply wires and the signal wires is set up. For example, a 1-pF capacitance superimposes a 40-mV interference on a 1-mV signal.

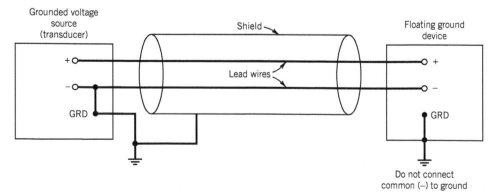

**FIGURE 6.36**
Signal grounding arrangements: grounded source with shield and floating signal at the measuring device. It is important to ground the shield at one end only.

A different source of noise is from a magnetic field. When a lead wire moves within a magnetic field, a voltage is induced in the signal path. The same effect occurs with a lead wire in proximity to an operating motor. The best prevention is to separate the signal lead wires from such sources. Also, twisting the lead wires together tends to cancel any induced voltage, as the currents through the two wires are in opposite directions; the more twists per meter, the better. A final recourse is a magnetic shield made from a material having a high ferromagnetic permeability.

## Connecting Wires

There are several types of wire available. *Single cable* refers to a single length of common wire or wire strand. The conductor is coated for electrical insulation. The wire is readily available and cheap, but should not be used for low-level signal (millivolt level) connections or for connections of more than a few wires. Flat cable is similar but consists of multiple conductors arranged in parallel strips, usually in groups of 4, 9, or 25. Flat cable is commonly used for short connections between adjacent electrical boards, but in such applications the signals are on the order of 1 V or more or are TTL signals. Neither of these two types of wire offers any shielding.

*Twisted pairs* of wires consist of two electrically insulated conductors twisted about each other along their lengths. Twisted pairs are widely used to interconnect transducers and equipment. The intertwining of the wires offers some immunity to noise with that effectiveness increasing with the number of twists per meter. Cables containing several twisted pairs are available; for example, CAT5 network cable (such as Internet cable) consists of four twisted pairs of wires. Shielded twisted pairs wrap the twisted pairs within a metallic foil shield. CAT6 network cable consists of four twisted pairs with a shield separating its wire pairs. Shielded twisted pairs are one of the best choices for applications requiring low-level signals or high data-transfer rates.

*Coaxial cable* consists of an electrically insulated inner single conductor surrounded within an outer conductor made of stranded wire and a metal foil shield. In general, current flows in one direction along the inner wire and the opposite direction along the outer wire. Any electromagnetic fields generated will cancel. Coaxial cable is often the wire of choice for low-level, high-frequency signals. Signals can be sent over very long distances with little loss. A variation of coaxial cable is *triaxial cable*, which contains two inner conductors. It is used in applications as with twisted pairs but offers superior noise immunity.

*Optical cable* is widely used to transmit low-level signals over long distances. This cable may contain one or more fiber-optic ribbons within a polystyrene shell. A transmitter converts the low-level voltage signal into infrared light. The light is transmitted through the cable to a receiver, which converts it back to a low-level voltage signal. The cable is virtually noise-free from magnetic fields and harsh environments and is incredibly light and compact compared to its metal wire counterpart.

## SUMMARY

This chapter has focused on classic but basic analog electrical measurement and signal-conditioning devices. Devices that are sensitive to current, resistance, and voltage were presented. Signal-conditioning devices that can modify or condition a signal include amplifiers, current loops, and filters, and these were also presented. Because all of these devices are widely used, often in combination and often masked within more complicated circuits, the reader should strive to become familiar with the workings of each.

The important issue of loading error was presented. Loading errors, which are due to the act of measurement or to the presence of a sensor, can only be minimized by proper choice of sensor and technique. Loading errors that occur between the connecting stages of a measurement system can be minimized by proper impedance matching. Such impedance matching was discussed, including using the versatile voltage follower circuit for this purpose.

## REFERENCES

1. Hickman, I., *Oscilloscopes*, 5th ed., Newnes, Oxford, UK, 2000.

2. Bleuler, E., and R. O. Haxby, *Methods of Experimental Physics*, 2nd ed., Vol. 2, Academic, New York, NY, 1975.

3. Lacanette, K., *A Basic Introduction to Filters: Active, Passive, and Switched-Capacitor Application Note 779*, National Semiconductor, Santa Clara, CA, 1991.

4. Stanley, W. D., G. R. Dougherty, and R. Dougherty, *Digital Signal Processing*, 2nd ed., Reston (a Prentice Hall company), Reston, VA, 1984.

5. DeFatta, D. J., J. Lucas, and W. Hodgkiss, *Digital Signal Processing*, Wiley, New York, 1988.

6. Lam, H. Y., *Analog and Digital Filters: Design and Realization*, Prentice-Hall, Englewood Cliffs, NJ, 1979.

7. Niewizdomski, S., *Filter Handbook—A Practical Design Guide*, CRC Press, Boca Raton, FL, 1989.

8. Bowick, C., C. Ajluni, and J. Blyler, *RF Circuit Design*, 2nd ed., Newnes, Oxford, UK, 2007.

## NOMENCLATURE

| | | | | |
|---|---|---|---|---|
| $c$ | damping coefficient | | $E_o$ | output voltage (V) |
| $e$ | error | | $F$ | force ($mlt^{-2}$) |
| $f$ | frequency ($t^{-1}$) | | $G$ | amplifier closed-loop gain |
| $f_c$ | filter cutoff frequency ($t^{-1}$) | | $G(s)$ | transfer function |
| $f_m$ | maximum analog signal frequency ($t^{-1}$) | | $I$ | electric current ($C$–$t^{-1}$) |
| $h\{E_i(t)\}$ | function | | $I_e$ | effective current ($C$–$t^{-1}$) |
| $k$ | cascaded filter stage order | | $I_g$ | galvanometer or current device current ($C$–$t^{-1}$) |
| $k$ | spring constant ($mt^{-2}$) | | $K$ | static sensitivity |
| $\hat{k}$ | unit vector aligned with current flow | | $L$ | inductance ($H$) |
| $l$ | length ($l$) | | $M(f)$ | magnitude ratio at frequency $f$ |
| $m$ | mass ($m$) | | $N$ | number of turns in a current-carrying loop |
| $\hat{n}$ | unit vector normal to current loop | | $R$ | resistance ($\Omega$) |
| $q$ | charge ($C$) | | $R_B$ | effective bridge resistance ($\Omega$) |
| $r$ | radius, vector ($l$) | | $R_g$ | galvanometer or detector resistance ($\Omega$) |
| $t$ | time ($t$) | | $R_{eq}$ | equivalent resistance ($\Omega$) |
| $u_R$ | uncertainty in resistance $R$ ($\Omega$) | | $R_m$ | meter resistance ($\Omega$) |
| $A$ | cross-sectional area of a current-carrying loop; attenuation | | $\delta R$ | change in resistance ($\Omega$) |
| | | | $T_\mu$ | torque on a current carrying loop in a magnetic field ($ml^2t^{-2}$) |
| $A_o$ | operational amplifier open-loop gain | | | |
| $E$ | voltage (V) | | $\alpha$ | angle |
| $E_1$ | open circuit potential (V) | | $\tau$ | time constant ($t$) |
| $E_i$ | input voltage (V) | | $\Phi(f)$ | phase shift at frequency $f$ |
| $E_m$ | indicated output voltage (V) | | $\theta$ | angle |

# Sampling, Digital Devices, and Data Acquisition

# 7

## 7.1 Introduction

Integrating analog electrical transducers with digital devices is cost-effective and commonplace. Digital microprocessors are central to most controllers and data acquisition systems (DAS) (the acronym 'DAS' will be used whether singular or plural). There are many advantages to digital data acquisition and control, including the efficient handling and rapid processing of large amounts of data and varying degrees of artificial intelligence. But there are fundamental differences between analog and digital systems that impose some limitations and liabilities upon the engineer. As pointed out in Chapter 2, the most important difference is that analog signals are continuous in both amplitude and time, whereas digital signals are discrete (noncontinuous) in both amplitude and time. It is not immediately obvious how such a discrete signal, which is a series of numbers, can be used to represent the continuous behavior of an analog process variable.

This chapter begins with an introduction to the fundamentals of *sampling*, the process by which continuous signals are made discrete. The major pitfalls are explored. Criteria are presented that circumvent the loss or the misinterpretation of signal information while undergoing the sampling process. We show how a discrete series of data can actually contain all the information available in a continuous signal or at least provide a very good approximation.

The discussion moves on to the devices most often involved in analog and digital systems. Analog devices interface with digital devices through an analog-to-digital (A/D) converter. The reverse process of a digital device interfacing with an analog device occurs through a digital-to-analog (D/A) converter. A digital device interfaces with another digital device through a digital input–output (I/O) port. These interfaces are the major components of computer-based DAS. Necessary components and the basic layout of these systems are introduced, and standard methods for communication between digital devices are presented. Digital image acquisition and processing are introduced because of their increasing importance in a wide variety of applications ranging from quality assurance inspection to high-speed imaging.

Upon completion of this chapter, the reader will be able to:

- Describe analog, discrete time, and digital signals
- Properly choose a sample rate for data acquisition to eliminate aliasing
- Clearly describe the functioning of A/D and D/A converters
- Define and calculate quantization errors
- Understand how instruments interface with DAS
- Perform basic image processing on digital images

## 7.2 Sampling Concepts

Consider an analog signal and its discrete time series representation in Figure 7.1. The information contained in the analog and discrete representations may appear to be quite different. However, the important analog signal information concerning amplitude and frequency can be well represented by such a discrete series—just how well represented depends on

1. The frequency content of the analog signal

2. The size of the time increment between each discrete number

3. The total sample period of the measurement

Chapter 2 discussed how a continuous dynamic signal could be represented by a Fourier series. The discrete Fourier transform (DFT) was also introduced as a method for reconstructing a dynamic signal from a discrete set of data. The DFT conveys all the information needed to reconstruct the Fourier series of a continuous dynamic signal from a representative discrete time series. Hence, Fourier analysis provides certain guidelines for sampling continuous data. This section discusses the concept of converting a continuous analog signal into an equivalent discrete time series. Extended discussions on this subject of signal analysis can be found in many texts [1–3].

### Sample Rate

Just how frequently must a time-dependent signal be measured to determine its frequency content? Consider Figure 7.2a, in which the magnitude variation of a 10-Hz sine wave is plotted versus time over time period $t_f$. Suppose this sine wave is measured repeatedly at successive sample time increments $\delta t$. This corresponds to measuring the signal with a sample frequency (in samples/second) or sample rate (in Hz) of

$$f_s = 1/\delta t \tag{7.1}$$

For this discussion, we assume that the signal measurement occurs at a constant sample rate. For each measurement, the amplitude of the sine wave is converted into a number. For comparison, in Figure 7.2b–d the resulting series versus time plots are given when the signal is measured using

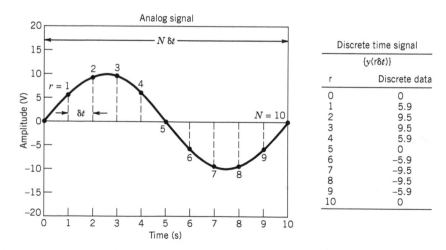

**FIGURE 7.1**   Analog and discrete representations of a time-varying signal.

**(a)** Original 10-Hz sine wave analog signal

**(b)** $f_s$ = 100 Hz

**(c)** $f_s$ = 27 Hz

**(d)** $f_s$ = 12 Hz

**FIGURE 7.2**
The effect of sample rate on signal frequency and amplitude interpretation.

sample time increments (or the equivalent sample rates) of (b) 0.010 s ($f_s$ = 100 Hz), (c) 0.037 s ($f_s$ = 27 Hz), and (d) 0.083 s ($f_s$ = 12 Hz). We can see that the sample rate has a significant effect on our perception and reconstruction of a continuous analog signal in the time domain. As sample rate decreases, the amount of information per unit time describing the signal decreases. In Figure 7.2b,c we can still discern the 10-Hz frequency content of the original signal. But we see in Figure 7.2d that an interesting phenomenon occurs if the sample rate becomes too slow: The sine wave appears to be of a lower frequency.

We can conclude that the sample time increment or the corresponding sample rate plays a significant role in signal frequency representation. The *sampling theorem* states that to reconstruct the frequency content of a measured signal accurately, the sample rate must be more than twice the highest frequency contained in the measured signal.

Denoting the maximum frequency in the analog signal as $f_m$, the sampling theorem requires

$$f_s > 2f_m \tag{7.2}$$

or, equivalently, in terms of sample time increment,

$$\delta t < \frac{1}{2f_m} \tag{7.3}$$

When signal frequency content is important, Equations 7.2 and 7.3 provide a criterion for the minimum sample rate or maximum sample time increment, respectively, to be used in converting data from a continuous to a discrete form. The frequencies that are extracted from the DFT of the resulting discrete series provide an accurate representation of the original signal frequencies regardless of the sample rate used, provided that the requirements of the sampling theorem are satisfied.

## Alias Frequencies

When a signal is sampled at a sample rate that is less than $2f_m$, the higher frequency content of the analog signal takes on the false identity of a lower frequency in the resulting discrete series. This is seen to occur in Figure 7.2d, where, because $f_s < 2f_m$, the 10-Hz analog signal is observed to take on the false identity of a 2-Hz signal. As a result, we misinterpret the frequency content of the original signal! Such a false frequency is called an *alias frequency*.

The alias phenomenon is an inherent consequence of a discrete sampling process. To illustrate this, consider a simple periodic signal:

$$y(t) = C\sin\left[2\pi ft + \phi(f)\right] \tag{7.4}$$

Suppose $y(t)$ is measured with a sample time increment of $\delta t$, so that its discrete time signal is given by

$$\{y(r\delta t)\} = C\sin\left[2\pi fr\delta t + \phi(f)\right] \quad r = 0, 1, \dots, N-1 \tag{7.5}$$

Now using the identity $\sin x = \sin(x + 2\pi q)$, where $q$ is any integer, we rewrite $\{y(r\delta t)\}$ as

$$C\sin\left[2\pi fr\delta t + \phi(f)\right] = C\sin\left[2\pi fr\delta t + 2\pi q + \phi(f)\right]$$
$$= C\sin\left[2\pi \left(f + \frac{m}{\delta t}\right) r\delta t + \phi(f)\right] \tag{7.6}$$

where $m = 0, 1, 2, \dots$ (and hence, $mr = q$ is an integer). This shows that for any value of $\delta t$, the frequencies of $f$ and $f + m/\delta t$ are indistinguishable in a sampled discrete series. Hence, all frequencies given by $f + m/\delta t$ are the alias frequencies of $f$. However, by adherence to the sampling theorem criterion of either Equation 7.2 or 7.3, all $m \geq 1$ are eliminated from the sampled signal, and thus this ambiguity between frequencies is avoided.

This same discussion applies equally to all time-dependent waveforms. This is shown by examining the general Fourier series that can be used to represent such signals. A discrete series such as

$$\{y(r\delta t)\} = \sum_{n=1}^{\infty} C_n\sin\left[2\pi nfr\delta t + \phi_n(f)\right] \tag{7.7}$$

for $r = 0, 1, \dots, N-1$ can be rewritten as

$$\{y(r\delta t)\} = \sum_{n=1}^{\infty} C_n\sin\left[2\pi \left(nf + \frac{nm}{\delta t}\right) r\delta t + \phi_n(f)\right] \tag{7.8}$$

and so it displays the same aliasing phenomenon shown in Equation 7.6.

In general, the Nyquist frequency defined by

$$f_N = \frac{f_s}{2} = \frac{1}{2\delta t} \tag{7.9}$$

represents a folding point for the aliasing phenomenon. All frequency content in the analog signal that is at frequencies above $f_N$ appears as alias frequencies of less than $f_N$ in the sampled signal. That is, such frequencies are folded back and superimposed on the signal as lower frequencies. The aliasing phenomenon occurs in spatial sampling (i.e., $y(x)$ at sampling intervals $\delta x$) as well, and the above discussion applies equally.

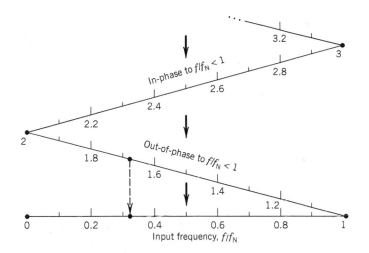

**FIGURE 7.3**  The folding diagram for alias frequencies.

The alias frequency $f_a$, can be predicted from the folding diagram of Figure 7.3. Here the original input frequency axis is folded back over itself at the folding point of $f_N$ and again for each of its harmonics $mf_N$, where $m = 1, 2, \ldots$ For example, and as noted by the solid arrows in Figure 7.3, the frequencies $f = 0.5\,f_N, 1.5\,f_N, 2.5\,f_N, \ldots$ that may be present in the original input signal all appear as the frequency $0.5\,f_N$ in the discrete series $y(r\delta t)$. Use of the folding diagram is illustrated further in Example 7.1.

How does one avoid this alias phenomenon when sampling a signal of unknown frequency content? The preferred option is to choose a sample rate based on the maximum frequency of interest while adhering to the criterion of Equation 7.2 and to pass the signal through a low-pass filter prior to sampling. Based on Equation 7.9, the filter should be set to remove signal content at and above $f_N$. This type of filter is called an *anti-aliasing filter*. Another option is to assign a sample rate so large that the measured signal does not have significant amplitude content above $f_N$. In this way, you choose $f_s$ high enough such that the amplitudes of any frequency content above $f_N$ are small compared to those below $f_N$. With this option, the resulting effects of aliasing are minimized but not eliminated.

## EXAMPLE 7.1

A 10-Hz sine wave is sampled at 12 Hz. Compute the maximum frequency that can be represented in the resulting discrete signal. Compute the alias frequency.

**KNOWN** $f = 10$ Hz
$\quad\quad\quad f_s = 12$ Hz.

**ASSUMPTION** Constant sample rate.

**FIND** $f_N$ and $f_a$.

**SOLUTION**
The Nyquist frequency $f_N$ sets the maximum frequency that can be represented in a resulting data set. Using Equation 7.9 with $f_s = 12$ Hz gives the Nyquist frequency $f_N = 6$ Hz. This is the maximum frequency that can be sampled correctly. So the measured frequency is in error.

All frequency content in the analog signal above $f_N$ appears as an alias frequency $f_a$ between 0 and $f_N$. Because $f/f_N \approx 1.67$, $f$ is folded back and appears in the sampled signal as a frequency between 0 and $f_N$. Reading Figure 7.3, $f = 1.67 f_N$ is located directly above $0.33 f_N$. So $f$ is folded back to appear as $f_a = 0.33 f_N$ or 2 Hz. This folding is indicated by the dashed arrow in Figure 7.3. As a consequence, the 10-Hz sine wave sampled at 12 Hz is completely indistinguishable from a 2-Hz sine wave sampled at 12 Hz in the discrete time series. The 10-Hz signal takes on the identity of a 2-Hz signal—that's aliasing.

**COMMENT** The programs *Sampling.m* and *Aliasing.vi* can be used to explore the effects of sample rate on the resulting signal. *Aliasing.vi* illustrates the folding diagram concept.

---

### EXAMPLE 7.2

A complex periodic signal has the form

$$y(t) = A_1 \sin 2\pi(25)t + A_2 \sin 2\pi(75)t + A_3 \sin 2\pi(125)t$$

If the signal is sampled at 100 Hz, determine the frequency content of the resulting discrete series.

**KNOWN** $f_s = 100$ Hz.
$\qquad\quad f_1 = 25$ Hz $\;f_2 = 75$ Hz $\;f_3 = 125$ Hz.

**ASSUMPTION** Constant sample rate.

**FIND** The alias frequencies $f_{a1}, f_{a2}$, and $f_{a3}$ and discrete set $\{y(r\delta t)\}$.

**SOLUTION**
From Equation 7.9 for a sample rate of 100 Hz, the Nyquist frequency is 50 Hz. All frequency content in the analog signal above $f_N$ takes on an alias frequency between 0 and $f_N$ in the sampled data series. With $f_1 = 0.5 f_N, f_2 = 1.5 f_N$, and $f_3 = 2.5 f_N$, we can use Figure 7.3 to determine the respective alias frequencies.

| $i$ | $f_i$ | $f_{a_i}$ |
|---|---|---|
| 1 | 25 Hz | 25 Hz |
| 2 | 75 Hz | 25 Hz |
| 3 | 125 Hz | 25 Hz |

In the resulting series, $f_{a_1} = f_{a_2} = f_{a_3} = 25$ Hz. Because of the aliasing phenomenon, the 75- and 125-Hz components would be completely indistinguishable from the 25-Hz component. The discrete series would be described by

$$y(r\delta t) = (B_1 + B_2 + B_3) \sin 2\pi(25) r\delta f \quad r = 0, 1, 2, \ldots$$

where $B_1 = A_1, B_2 = -A_2$, and $B_3 = A_3$. Note from Figure 7.3 that when the original frequency component is out of phase with the alias frequency, the corresponding amplitude is negative.

**COMMENT** Example 7.2 illustrates the potential for misrepresenting a signal with improper sampling. With signals for which the frequency content is not known before their measurement, sample frequencies should be increased according to the following scheme:

- Sample the input signal at increasing sample rates using a fixed total sample time, and examine the time plots for each signal. Look for changes in the shape of the waveform (Figure 7.2).

- Compute the amplitude spectrum for each signal at increasing sample rates and compare the resulting frequency content.

- Always use (anti-aliasing) analog filters set at the Nyquist frequency.

## Amplitude Ambiguity

For simple and complex periodic waveforms, the DFT of the sampled discrete time signal remains unaltered by a change in the total sample period $N\delta t$ provided that (1) the total sample period remains an integer multiple of the fundamental period $T_1$ of the measured continuous waveform (i.e., $mT_1 = N\delta t$ where $m$ is an integer) and (2) the sample frequency meets the sampling theorem criterion. If both criteria are met, the amplitudes associated with each frequency in the periodic signal, the spectral amplitudes, are accurately represented by the DFT. This means that an original periodic waveform can be completely reconstructed from a discrete time series regardless of the sample time increment used. The total sample period defines the frequency resolution of the DFT:

$$\delta f = \frac{1}{N\delta t} = \frac{f_s}{N} \tag{7.10}$$

The frequency resolution plays a crucial role in the reconstruction of the signal amplitudes, as will be noted.

An important difficulty arises when $N\delta t$ is not coincident with an integer multiple of the fundamental period of $y(t)$: The resulting DFT cannot *exactly* represent the spectral amplitudes of the sampled continuous waveform. This is exaggerated when $N\delta t$ represents only a relatively few fundamental periods of the signal. The problem is brought on by (1) the truncation of one complete cycle of the signal (Figure 7.4) and (2) from spectral resolution, because the associated fundamental frequency and its harmonics are not coincident with a discrete frequency ($k\delta f$) of the DFT. However, this error decreases either as the value of $N\delta t$ more closely approximates an exact integer multiple of $T_1$ or as $f_s$ becomes very large relative to $f_m$.

This situation is illustrated in Figure 7.4, which compares the amplitude spectrum resulting from sampling the signal $y(t) = 10 \cos 628t$ over different sample periods. Two sample periods of 0.0256 and 0.1024 second were used with $\delta t = 0.1$ ms, and a third sample period of 0.08 s with $\delta t = 0.3125$ ms.[1] These sample periods provide for DFT frequency resolution $\delta f$ of about 39, 9.8, and 12.5 Hz, respectively.

## Leakage

The two spectra shown in Figure 7.4a,b display a spike near the correct 100 Hz with surrounding noise spikes, known as *leakage*, at adjacent frequencies. Note that the original signal $y(t)$ cannot be exactly reconstructed from either of these spectra. The spectrum in Figure 7.4c has been constructed using a longer sample time increment (i.e., lower sample rate) than that of the spectra in Figure 7.4a,b and with fewer data points than the spectrum of Figure 7.4b. Yet the spectrum of Figure 7.4c provides an exact representation of $y(t)$. The $N$ and $\delta t$ combination used has reduced leakage to zero, which maximizes the amplitude at 100 Hz. Its sample period corresponds to exactly eight periods of $y(t)$, and the frequency of 100 Hz corresponds to the center frequency of the eighth frequency interval of the DFT. As seen in Figure 7.4, the loss of accuracy in a DFT

---

[1]Many DFT (including fast Fourier transform [FFT]) algorithms require that $N = 2^M$, where $M$ is an integer. This affects the selection of $\delta t$.

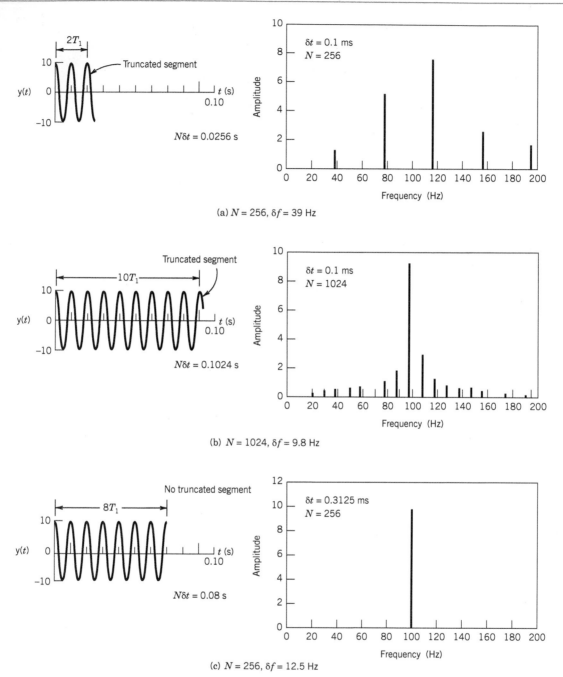

**FIGURE 7.4**    Amplitude spectra for $y(t)$. (a) $\delta f = 39$ Hz. (b) $\delta f = 9.8$ Hz. (c) $\delta f = 12.5$ Hz.

representation occurs in the form of amplitude "leakage" to adjacent frequencies. To the DFT, the truncated segment of the sampled signal appears as an aperiodic signal. The DFT returns the correct spectral amplitudes for both the periodic and aperiodic signal portions. But as a result, the spectral amplitudes for any truncated segment are superimposed onto portions of the spectrum adjacent to the correct frequency. Recall how the amplitude varied in Figure 7.2c. This is how leakage affects the time domain reconstruction. By varying the sample period or its equivalent, the DFT resolution, one can minimize leakage and control the accuracy of the spectral amplitudes.

For an exact discrete representation in both frequency and amplitude of any periodic analog waveform, both the number of data points and the sample rate should be chosen based on the preceding discussion using the criteria of Equations 7.2 and 7.10. Equation 7.2 sets the maximum value for $\delta t$ or, equivalently, the minimum sample rate $f_s$, and Equation 7.10 sets the total sampling time $N\delta t$ from which the data number $N$ is estimated.

We want to also stress that even when a sampled signal is known to be periodic, the engineer may not be able to sample exact integer periods of the signal; it may just be too impractical! But this discussion illustrates why leakage does occur and how it affects the sampled information.

If $y(t)$ is an aperiodic or nondeterministic waveform, there may not be a fundamental period. In such situations, one controls the accuracy of the spectral amplitudes by varying the DFT resolution $\delta f$ to minimize leakage. As $\delta f$ tends toward zero, leakage decreases.

Whether a signal is periodic or aperiodic, leakage can also be controlled by applying a window function, a technique that multiplies the sampled signal point-by-point by a function designed to reduce the signal amplitudes at the beginning and ends of the sampled signal. As truncated segments can occur at the start or end of the sampled signal, the idea is to minimize the influences of any truncated segments while retaining any periodicity in the middle of the signal. Window functions are discussed in references [1–3].

In summary, the reconstruction of a measured waveform from a discrete signal is controlled by the sampling rate and the DFT resolution. By adherence to the sampling theorem, one controls the frequency content of both the measured signal and the resulting spectrum. By variation of $\delta f$, one can control the accuracy of the spectral amplitude representation.

Programs *Sampling.m*, *Aliasing.vi*, and *Aliasing frequency domain.vi* explore the concept of sample rate and amplitude ambiguity. Program *Leakage.2.vi* allows the user to vary sample rate and number of points; it follows the discussion of Figure 7.4. Program *DataSpec.m* explores sampling concepts on user-generated data series. *Signal generation.vi* explores sample rate effects on different wave forms.

## Waveform Fidelity

While choosing the sampling frequency in accordance with the sampling theorem ensures that there will be no aliasing associated with frequency content, it does not address the "shape" of the waveform. If sampling cannot be exact, then what is good enough? As an example, consider the original compact disk for musical recording. The sampling rate was limited by existing technology at the time (ca. 1980), and the sampling rate was chosen so that frequencies up to the $\sim$20 kHz limits of human hearing would not be aliased; thus a sampling rate of 44.1 kHz was chosen.

What does $f_s = 44.1$ kHz imply for the musical tones at 22 kHz? Essentially it means that at the highest frequencies we are representing a sine wave with two discrete points. Thus, there was much criticism and speculation that the higher frequencies present in the music were distorted. Technology has advanced since 1980. High frequency information in digital recordings can now be represented by many data points so that distortion in the recording becomes inaudible.

As a general rule, waveform fidelity is acceptable for $f_s \geq 5f_m$ to $10f_m$ depending on the application.

## 7.3 Digital Devices: Bits and Words

The history of computing hardware began around World War II with mechanical and vacuum-tube–based hardware [4]. Some types of recording devices used holes punched in paper cards or tape; one example is the historically significant player piano rolls. In the 1970s punch cards were used to communicate computer programs to mainframe computers; each letter or number was represented by appropriately punched holes in a single column of the card. The most important concept in digital computing is that all of the required information can be represented as an appropriately structured series of binary representations, either a 0 or a 1. For holes in a paper tape, there are only two possibilities—a hole is there or it is not! All of these representations correspond to a binary numbering system.

Digital systems use some variation of a binary numbering system both to represent and transmit signal information. Binary systems use the binary digit or *bit* as the smallest unit of information. A bit is a single digit, either a 1 or a 0. Bits are like electrical switches and are used to convey both logical and numerical information. From a logic standpoint, the 1 and 0 are represented by the on and off switch settings. By appropriate action a bit can be reset to either on or off, thereby permitting control and logic actions. By combining bits it is possible to define integer numbers greater than 1 or 0. A numerical *word* is an ordered sequence of bits, with a *byte* being a specific sequence usually consisting of 8 bits. Computer memory is usually byte-addressed. The memory location where numerical information is stored is known as a *register*, with each register assigned its own *address*.

A combination of $M$ bits can be arranged to represent $2^M$ different words. For example, a combination of 2 bits can represent $2^2$, or four, possible combinations of bit arrangements: 00, 01, 10, or 11. We can alternatively reset this 2-bit word to produce these four different arrangements that represent the decimal: base 10, integer numbers 0, 1, 2, or 3, respectively. So an 8-bit word can represent the numbers 0 through 255; a 16-bit word can represent 0 through 65,535.

The numerical value for a word is computed by moving by bit from right to left. From the right side, each successive bit increases the value of the word by a unit, a 2, a 4, an 8, and so forth through the progression of $2^M$, provided that the bit is in its on (value of 1) position; otherwise, the particular bit increases the value of the word by zero. A weighting scheme of an $M$-bit word is given as follows:

| Bit $M-1$ | ... | Bit 3 | Bit 2 | Bit1 | Bit 0 |
|---|---|---|---|---|---|
| $2^{M-1}$ | ... | $2^3$ | $2^2$ | $2^1$ | $2^0$ |
| Decimal value | ... | 8 | 4 | 2 | 1 |

Using this scheme, bit $M-1$ is known as the most significant bit (MSB), because its contribution to the numerical level of the word is the largest relative to the other bits. Bit 0 is known as the least significant bit (LSB). In hexadecimal (base 16), a sequence of four bits is used to represent a single digit so that the $2^4$ different digits (0 through 9 plus A through F) form the alphabet for creating a word.

Several binary codes are in common use. In 4-bit straight binary, 0000 is equivalent to analog 0, and 1111 is equivalent to decimal 15 or hexadecimal F. Straight binary is considered to be a unipolar code, because all numbers are of like sign. Bipolar codes allow for sign changes. *Offset binary* is a bipolar code with 0000 equal to $-7$ and 1111 equal to $+7$. The comparison between these two codes and several others is shown in Table 7.1. In offset binary, there are two zero levels, with true zero being centered between them. Note how the MSB is used as a sign bit in a bipolar code.

The *ones-* and *twos-complement binary codes* shown in Table 7.1 are both bipolar. They differ in the positioning of the zero, with the twos-complement code containing one more negative level ($-8$) than positive ($+7$). Twos-complement code is widely used in digital computers. The bipolar AVPS code uses the MSB as the sign bit but uses the lower bits in a straight binary representation; for example, 1100 and 0100 represent $-4$ and $+4$, respectively.

**Table 7.1  Binary Codes (Example: 4-Bit Words)**

| Bits | (Hex) Straight | Offset | Twos complement | Ones complement | AVPS[a] |
|------|----------------|--------|-----------------|-----------------|---------|
| 0000 | 0 | −7 | +0 | +0 | −0 |
| 0001 | 1 | −6 | +1 | +1 | −1 |
| 0010 | 2 | −5 | +2 | +2 | −2 |
| 0011 | 3 | −4 | +3 | +3 | −3 |
| 0100 | 4 | −3 | +4 | +4 | −4 |
| 0101 | 5 | −2 | +5 | +5 | −5 |
| 0110 | 6 | −1 | +6 | +6 | −6 |
| 0111 | 7 | −0 | +7 | +7 | −7 |
| 1000 | 8 | +0 | −8 | −7 | +0 |
| 1001 | 9 | +1 | −7 | −6 | +1 |
| 1010 | A | +2 | −6 | −5 | +2 |
| 1011 | B | +3 | −5 | −4 | +3 |
| 1100 | C | +4 | −4 | −3 | +4 |
| 1101 | D | +5 | −3 | −2 | +5 |
| 1110 | E | +6 | −2 | −1 | +6 |
| 1111 | F | +7 | −1 | −0 | +7 |

[a]AVPS, absolute-value plus-sign code.

A code often used for digital decimal readouts and for communication between digital instruments is binary coded decimal (BCD). In this code, each individual digit of a decimal number is represented separately by its equivalent value coded in straight binary. For example, the three digits of the base 10 number $532_{10}$ are represented by the BCD number 0101 0011 0010 (i.e., binary 5, binary 3, binary 2). In this code each binary number need span but a single decade—that is, have values from 0 to 9. Hence, in digital readouts, when the limits of the lowest decade are exceeded—that is, 9 goes to 10—the code just carries into the next higher decade.

**EXAMPLE 7.3**

A 4-bit register contains the binary word 0101. Convert this to its decimal equivalent assuming a straight binary code.

**KNOWN**  4-bit register.

**FIND**  Convert to base 10.

**ASSUMPTION**  Straight binary code.

**SOLUTION**
The content of the 4-bit register is the binary word 0101. This represents

$$0 \times 2^3 + 1 \times 2^2 + 0 \times 2^1 + 1 \times 2^0 = 5$$

The equivalent of 0101 in decimal is 5.

## 7.4 Transmitting Digital Numbers: High and Low Signals

Electrical devices transmit binary code by using differing voltage levels. Because a bit can have a value of 1 or 0, the presence of a particular voltage (call it HIGH) at a junction could be used to represent a 1-bit value, whereas a different voltage (LOW) could represent a 0. Most simply, this voltage can be affected by use of an open or closed switch using a flip-flop (see Chapter 6) such as depicted in Figure 7.5a.

For example, a signal method common with many digital devices is a +5 V form of TTL (true transistor logic), which uses a nominal +5 V HIGH/0 V LOW scheme for representing the value of a bit. To avoid any ambiguity from voltage fluctuations and provide for more precision, a voltage level between +2 and +5.5 V is taken as HIGH and a bit value of 1, whereas a voltage between −0.6 and +0.8 V is taken as LOW or a 0-bit value. But this scheme is not unique.

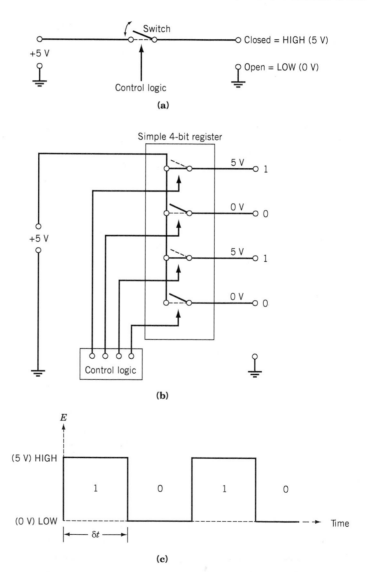

**FIGURE 7.5** Methods for transmitting digital information. (a) Simple on/off (1 or 0) switch. (b) Four-bit register transmitting 1010 in parallel. (c) Serial transmission of 1010.

Binary numbers can be formed through a combination of HIGH and LOW voltages. Several switches can be grouped in parallel to form a register. Such an $M$-bit register forms a number based on the value of the voltages at its $M$ output lines. This is illustrated in Figure 7.5b for a 4-bit register forming the number 1010. So the bit values can be changed simply by changing the opened or closed state of the switches that connect to the output lines. This defines a parallel form in which all of the bits needed to form a word are available simultaneously. But another form is serial, whereby the bits are separated in a time sequence of HIGH/LOW pulses, each pulse lasting for only one predefined time duration, $\delta t$. This is illustrated in the pulse sequence of Figure 7.5c, which also forms the number 1010.

# 7.5 Voltage Measurements

Digital measurement devices consist of several components [5] that interface the digital device with the analog world. In particular, the D/A converter and the A/D converter are discussed next. They form the major components of a digital voltmeter and an analog-to-digital/digital-to-analog DAS.

## Digital-to-Analog Converter

A D/A converter is an $M$-bit digital device that converts a digital binary number into an analog voltage [5]. One possible scheme uses an $M$-bit register with a weighted resistor ladder circuit and operational amplifier, as depicted in Figure 7.6. The circuit consists of $M$ binary weighted resistors having a common summing point. The resistor associated with the register MSB has a

**FIGURE 7.6** Digital-to-analog converter.

value of $R$. At each successive bit the resistor value is doubled, so that the resistor associated with the LSB has a value of $(2^{M-1}R)$. The circuit output is a current, $I$, given by

$$I = E_{\text{ref}} \sum_{m=1}^{M} \frac{c_m}{2^{m-1}R} \qquad (7.11)$$

The values for $c_m$ are either 0 or 1 depending on the associated $m$th bit value of the register that controls the switch setting. The output voltage from the amplifier is

$$E_o = IR_r \qquad (7.12)$$

Note that this is equivalent to an operation in which an $M$-bit D/A converter would compare the magnitude of the actual input binary number, $X$, contained within its register to the largest possible number $2^M$. This ratio determines $E_o$ from

$$E_o = \frac{X}{2^M} \qquad (7.13)$$

The D/A converter will have both digital and analog specifications, the latter expressed in terms of its full-scale analog voltage range output ($E_{\text{FSR}}$). Typical values for $E_{\text{FSR}}$ are 0 to 10 V (unipolar) and $\pm 5$ V (bipolar) and for $M$ are 8, 12, 16, and 18 bits.

## Analog-to-Digital Converter

An A/D converter converts an analog voltage value into a binary number through a process called *quantization*. The conversion is discrete, taking place one number at a time.

The A/D converter is a hybrid device having both an analog side and a digital side. The analog side is specified in terms of a full-scale voltage range, $E_{\text{FSR}}$. The $E_{\text{FSR}}$ defines the voltage span over which the device operates. The digital side is specified in terms of the number of bits of its register. An $M$-bit A/D converter outputs an $M$-bit binary number. It can represent $2^M$ different binary numbers. For example, a typical 8-bit A/D converter with an $E_{\text{FSR}} = 10$ V would be able to represent analog voltages in the range between 0 and 10 V (unipolar) or the range between $\pm 5$ V (bipolar) with $2^8 = 256$ different binary values. Principal considerations in selecting a type of A/D converter include resolution, voltage range, and conversion speed [5].

### Resolution

The A/D converter resolution is defined in terms of the smallest voltage increment that causes a bit change. Resolution is specified in volts per bit and is determined by

$$Q = E_{\text{FSR}}/2^M \qquad (7.14)$$

Primary sources of error intrinsic to any A/D converter are the resolution and associated quantization error, saturation error, and conversion error.

### Quantization Error

An A/D converter has a finite resolution, and any input voltage that falls between two adjacent output codes results in an error. This error is referred to as the *quantization error $e_Q$*. It behaves as noise imposed on the digital signal.

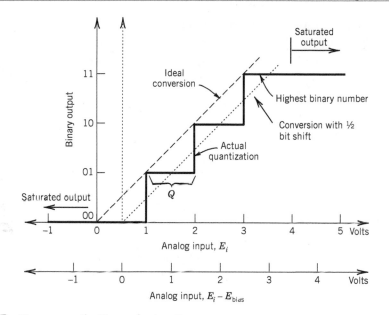

**FIGURE 7.7**   Binary quantization and saturation.

Consider the basic analog input to digital output relationship for a 0- to 4-V, 2-bit A/D converter as represented in Figure 7.7. This scheme is common in digital readout devices. From Equation 7.14, the resolution of this device is $Q = 1$ V/bit. For such a converter, an input voltage of 0.0 V would result in the same binary output of 00 as would an input of 0.9 V. Yet an input of 1.1 V results in an output of 01. Clearly, the output error is directly related to the resolution. In this encoding scheme, $e_Q$ is bounded between 0 and 1 LSB above $E_i$, so we estimate the limit of error as $e_Q = Q$.

Now, consider the scheme most commonly used in the A/D converters within commercial DAS. While the resolution is not changed, this scheme makes the quantization error symmetric about the input voltage. To do this, the analog voltage is shifted internally within the A/D converter by a bias voltage, $E_{bias}$, of an amount equivalent to ½ LSB. This shift is transparent to the user. The effect of such a shift is shown on the second lower axis in Figure 7.7. This now makes $e_Q$ bounded by ±1/2 LSB about $E_i$. So, here the limit of quantization error is $e_Q = \pm 1/2Q$. Regardless of the scheme used, the span of the quantization error remains 1 LSB, and its effect is most significant at smaller measured voltages. In an uncertainty analysis, we would set the resolution uncertainty as the limit of error, such that $u_Q = e_Q$.

The resolution of an A/D converter is sometimes specified in terms of signal-to-noise ratio (SNR). The SNR relates the power of the signal, given by Ohm's law as $E^2/R$, to the power that can be resolved by quantization, given by $E^2/(R \times 2^M)$. The SNR is just the ratio of these values. Defined in terms of the decibel (dB), this gives

$$SNR(\text{dB}) = 20 \log 2^M \tag{7.15}$$

The effect of bit number on resolution and SNR is detailed in Table 7.2. The program *Bits of Resolution.vi* explores the influence of a combination of $M$, $E_{FSR}$, and $Q$ on a measured signal.

## Saturation Error

The analog range of an A/D converter limits the minimum and maximum analog voltage. If either limit is exceeded, the A/D converter output saturates and does not change with a subsequent

Table 7.2    **Conversion Resolution**

| Bits M | $Q^a$ (V/bit) | SNR (dB) |
|--------|---------------|----------|
| 2      | 2.50          | 12       |
| 4      | 0.625         | 24       |
| 8      | 0.0390        | 48       |
| 12     | 0.00244       | 72       |
| 16     | $0.153 \, (10^{-3})$ | 96 |
| 18     | $0.0381 \, (10^{-3})$ | 108 |

$^a$Assumes $E_{FSR} = 10$ V.

increase in input level. As noted in Figure 7.7, an input to the 0- to 4-V, 2-bit A/D converter above 4 V results in an output of binary 11 and below 0 V of binary 00. A saturation error is defined by the difference between the input analog signal level and the equivalent digital value assigned by the A/D converter. Saturation error can be avoided by conditioning signals to remain within the limits of the A/D converter.

## Conversion Error

An A/D converter is not immune to elemental errors arising during the conversion process that lead to misrepresenting the input value. As with any device, the A/D errors can be delineated into hysteresis, linearity, sensitivity, zero, and repeatability errors. The extent of such errors depends on the particular method of A/D conversion. Factors that contribute to conversion error include A/D converter settling time, signal noise during the analog sampling, temperature effects, and excitation power fluctuations [5].

Linearity errors result from the ideal assumption that an $M$-bit A/D converter resolves the analog input range into $2^{M-1}$ equal steps of width $Q$. In practice, the steps may not be exactly equal, which causes a nonlinearity in the ideal conversion line drawn in Figure 7.7. This error is specified in terms of bits.

### EXAMPLE 7.4

Compute the relative effect of quantization error $(e_Q/E_i)$ in the quantization of a 100-mV and a 1-V analog signal using an 8-bit and a 12-bit A/D converter, both having a full-scale range of 0 to 10 V.

**KNOWN** $E_i = 100$ mV and 1V.
    $M = 8$ and 12.
    $E_{FSR} = 0$–10 V.

**FIND** $e_Q/E_i$, where $e_Q$ is the quantization error.

**SOLUTION**
The resolutions for the 8-bit and 12-bit converters can be estimated from Equation 7.14 as

$$Q_8 = \frac{E_{FSR}}{2^8} = \frac{10}{256} = 39 \, \text{mV}$$

$$Q_{12} = \frac{E_{FSR}}{2^{12}} = \frac{10}{4096} = 2.4 \, \text{mV}$$

Assume that the A/D converter is designed so that the limit of quantization error is given by $\pm\frac{1}{2}Q$. The relative effect of the quantization error can be computed by $e_Q/E_i$. The results are tabulated as follows:

| $E_i$ | $M$ | $e_Q = \pm\frac{1}{2}Q$ | $100 \times e_Q/E_i$ |
|-------|-----|-------------------------|----------------------|
| 100 mV | 8 | $\pm19.5$ mV | 19.5% |
| 100 mV | 12 | $\pm1.22$ mV | 1.22% |
| 1 V | 8 | $\pm19.5$ mV | 1.95% |
| 1 V | 12 | $\pm1.22$ mV | 0.122% |

From a relative (i.e., percent) point of view, the quantization error is much greater at lower voltage levels.

## EXAMPLE 7.5

The A/D converter with the specifications listed below is planned to be used in an environment in which the A/D converter temperature may change by $\pm10°C$. Estimate the contributions of conversion and quantization errors to the uncertainty in the digital representation of an analog voltage by the converter.

|  | Analog-to-digital converter |
|--|-----------------------------|
| $E_{FSR}$ | 0–10 V |
| M | 12 bits |
| Linearity error | $\pm3$ bits |
| Temperature drift error | 1 bit/5 °C |

**KNOWN** 12-bit resolution (see Example 7.4).

**FIND** $(u_c)_E$ measured.

**SOLUTION**
We can estimate a design-stage uncertainty as a combination of uncertainty arising from quantization errors $u_Q$ and from conversion errors $u_c$:

$$(u_d)_E = \sqrt{u_o^2 + u_c^2}$$

The resolution of a 12-bit A/D converter with full-scale range of 0 to 10 V is $Q = 2.4$ mV (see Example 7.4), so that the quantization error provides an estimate for the zero-order uncertainty as

$$u_o = \frac{1}{2}Q = 1.2 \text{ mV}$$

Now the conversion error is affected by two elements:

$$\text{Linearity uncertainty} = u_2 = 3 \text{ bits} \times 2.4 \text{ mV}$$
$$= 7.2 \text{ mV}$$
$$\text{Temperature uncertainty} = u_3 = \frac{1 \text{ bit}}{5°C} \times 10°C \times 2.4 \text{ mV}$$
$$= 4.8 \text{ mV}$$

An estimate of the uncertainty resulting from conversion errors is found using the RSS method:

$$u_c = \sqrt{u_2^2 + u_3^2} = \sqrt{(7.2 \text{ mV})^2 + (4.8 \text{ mV})^2} = 8.6 \text{ mV}$$

The combined uncertainty in the digital representation of an analog value due to these uncertainties is an interval described as

$$(u_d)_E = \sqrt{(1.2 \text{ mV})^2 + (8.6 \text{ mV})^2}$$
$$= \pm 8.7 \text{ mV (95\% assumed)}$$

The effects of the conversion errors dominate the uncertainty. A go-no go decision regarding this planned system can be made based on this result.

## Successive Approximation Converters

We next discuss several common methods for converting voltage signals into binary words. More detailed discussions and additional methods for A/D conversion are presented in specialized texts (e.g., 5).

The most common type of A/D converter uses the *successive approximation* technique. This technique uses a trial-and-error approach for converting the input voltage. Basically, the successive approximation A/D converter guesses successive binary values as it narrows in on the appropriate binary representation for the input voltage. As depicted in Figure 7.8, this A/D converter uses an $M$-bit register to generate a trial binary number, a D/A converter to convert the register contents into an analog voltage, and a voltage comparator (see Section 6.7 in Chapter 6) to compare the input voltage to the internally generated voltage. The conversion sequence is as follows:

1. The MSB is set to 1. All other bits are set to 0. This produces a value of $E^*$ at the D/A converter output equivalent to the register setting. If $E^* > E_i$, the comparator goes LOW, causing the MSB to be reset to 0. If $E^* < E_i$, the MSB is kept HIGH at 1.

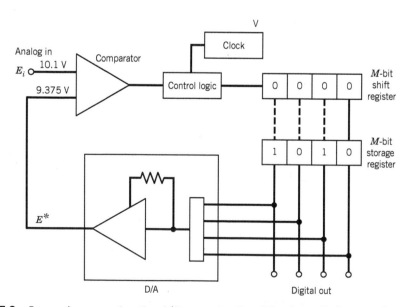

**FIGURE 7.8**    Successive approximation A/D converter. Four-bit scheme is shown with register = 1010 with $E_i = 10.1$ V.

**Table 7.3    Example of Successive Approximation Conversion Sequence for a 4-Bit Converter**

| Sequence | Register | $E^*$ | $E_i$ | Comparator |
|---|---|---|---|---|
| Initial status | 0000 | 0 | 10.1 | |
| MSB set to 1 | 1000 | 7.5 | 10.1 | High |
| Leave at 1 | 1000 | 7.5 | | |
| Next highest bit set to 1 | 1100 | 11.25 | | Low |
| Reset to 0 | 1000 | 7.5 | 10.1 | |
| Next highest bit set to 1 | 1010 | 9.375 | | High |
| Leave at 1 | 1010 | 9.375 | | |
| LSB set to 1 | 1011 | 10.3125 | 10.1 | Low |
| Reset to 0 | 1010 | 9.375 | | |

**2.** The next highest bit (MSB-1) is set to 1. Again, if $E^* > E_i$, it is reset to 0; otherwise its value remains 1.

**3.** The process continues through to the LSB. The final register value gives the quantization of $E_i$.

The process requires a time of one clock tick per bit.

An example of this sequence is shown in Table 7.3 for an input voltage of $E_i = 10.1$ V and using the 0 to 15 V, 4-bit successive approximation A/D converter of Figure 7.8. For this case, the converter has a resolution of 0.9375 V. The final register count of 1010 or its equivalent of 9.375 V is the output from the A/D converter. This measured value differs from the actual input voltage of 10.1 V as a result of the quantization error. In this extreme example, this error can be reduced by increasing the bit size of the register.

The successive approximation converter is typically used when conversion speed at a reasonable cost is important. The number of steps required to perform a conversion equals the number of bits in the A/D converter register. With one clock tick per step, a 12-bit A/D converter operating with a 1-MHz clock would require a maximum time of 12 μs per conversion. But this also reveals the trade-off between increasing the number of bits to lower quantization error and the resulting increase in conversion time. Faster clocks enable higher sample rates. Common maximum sample rates are on the order of 100 kHz to 1 MHz using 12, 16, or 24 bits.

Sources of conversion error originate in the accuracies of the D/A converter and the comparator. Noise is the principal weakness of this type of converter, particularly at the decision points for the higher-order bits. The successive approximation process requires that the voltage remain constant during the conversion process. Because of this, a sample-and-hold circuit (SHC), as introduced in Chapter 6, is used ahead of the converter input to measure and to hold the input voltage value constant throughout the duration of the conversion. The SHC also minimizes noise during the conversion.

## Ramp (Integrating) Converters

Low-level (<1 mV) measurements often rely on ramp converters for their low-noise features. These integrating A/D converters use the voltage level of a linear reference ramp signal to discern the voltage level of the analog input signal and convert it to its binary equivalent. Principal components, as shown in Figure 7.9, consist of an analog comparator, ramp function generator, and counter and $M$-bit register. The reference signal, initially at zero, is increased at set time steps,

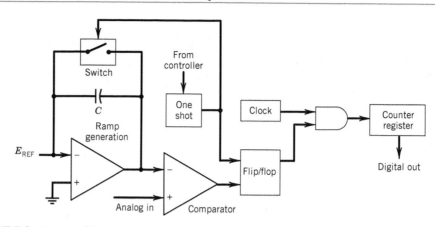

**FIGURE 7.9**    Ramp A/D converter.

within which the ramp level is compared to the input voltage level. This comparison is continued until the two are equal.

The usual method for generating the ramp signal is to use a capacitor of known fixed capacitance $C$, which is charged from a constant current source of amperage, $I$. Because the charge is related to current and time by

$$q = \int_0^t I\,dt \tag{7.16}$$

the reference ramp voltage is linearly related to the elapsed time by

$$E_{\text{ref}} = \frac{q}{C} = \text{constant} \times t \tag{7.17}$$

Time is integrated by a counter that increases the register value by 1 bit at each time step. Time step size depends on the value of $2^M$. When the input voltage and ramp voltage magnitudes cross during a time step, the comparator output goes to zero, which flips a flip-flop halting the process. The register count value then indicates the digital binary equivalent of the input voltage.

The accuracy of this single ramp operation is limited by the accuracy of the timing clock and the known values and constancy of the capacitor and the charging current. Increased accuracy can be achieved using a dual ramp converter, sometimes referred to as a dual-slope integrator, in which the measurement is accomplished in two steps, as illustrated in Figure 7.10. In the first step, the input voltage $E_i$ is applied to an integrator for a fixed time period $t_1$. The integrator output voltage increases in time with a slope proportional to the input voltage. At the end of the time interval, the output of the integrator has reached the level $E^*$. The second step in the process involves the application of a fixed reference voltage $E_{\text{ref}}$ to the integrator. This step occurs immediately at the end of the fixed time interval in the first step, with the integrator voltage at exactly the same level established by the input. The reference voltage has the opposite polarity to the input voltage and thus reduces the output of the integrator at a known rate. The time required for the output to return to zero is a direct measure of the input voltage. The time intervals are measured using a digital counter, which accumulates pulses from a stable oscillator. The input voltage is given by the relationship

$$E_i/E_{\text{ref}} = t_m/t_1 \tag{7.18}$$

where $t_m$ is the time required for step two. Dual-ramp converter accuracy depends on the stability of the counter during conversion and requires a very accurate reference voltage.

Ramp converters are inexpensive but relatively slow devices. However, their integration process tends to average out any noise in the input signal. This is the feature that makes them so

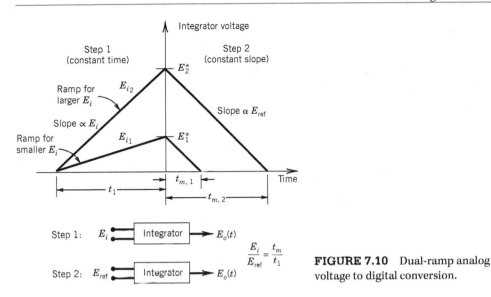

FIGURE 7.10  Dual-ramp analog voltage to digital conversion.

attractive for measuring low-level signals. The maximum conversion time using a ramp converter is estimated as

$$\text{Maximum conversion time} = 2^M/\text{Clock speed} \tag{7.19}$$

For a dual-ramp converter, this time is doubled. If one assumes that a normal distribution of input values are applied over the full range, the average conversion time can be taken to be one-half of the maximum value. For a 12-bit register with a 1-MHz clock and required to count the full $2^{12}$ pulses, a dual-ramp converter would require $\sim 8$ ms for a single conversion, corresponding to a sample rate of 125 Hz.

## Parallel Converters

A very fast A/D converter is the parallel or flash converter depicted in Figure 7.11. These converters are common to high-end stand-alone digital oscilloscopes and spectral analyzers. An $M$-bit parallel converter uses $2^M - 1$ separate voltage comparators to compare a reference voltage to the applied input voltage. As indicated in Figure 7.11, the reference voltage applied to each successive comparator is increased by the equivalent of 1 LSB by using a voltage-dividing resistor ladder. If the input voltage is less than its reference voltage, a comparator will go LOW; otherwise it will go HIGH.

Consider the 2-bit converter shown in Figure 7.11. If $E_{in} \geq \frac{1}{2}E_{ref}$ but $E_{in} < \frac{3}{4}E_{ref}$, then comparators 1 and 2 will be HIGH but comparator 3 will be LOW. In this manner, there can only be $2^M = 2^2$ different HIGH/LOW combinations from $2^2 - 1$ comparators, as noted in Table 7.4 that shows how these combinations correspond to the $2^2$ possible values of a 2-bit binary output. Logic circuits transfer this information to a register.

Because all the comparators act simultaneously, the conversion occurs in a single clock count. Thus, an attraction of this converter is its speed, with typical sample rates on the order of 150 MHz at 8 bits. Its disadvantage lies in the cost associated with its $2^{M-1}$ comparators and the associated logic circuitry. For example, an 8-bit converter requires 255 comparators.

## Digital Voltmeters

The digital voltmeter must convert an analog input signal to a digital output. This conversion can be done through several basic techniques. The most common method used in digital meters uses dual-ramp converters. The limits of performance for digital voltmeters can be significantly higher

**FIGURE 7.11**    Parallel or flash A/D converter. Two-bit scheme is shown.

Table 7.4    **Logic Scheme of a 2-Bit Parallel A/D Converter**

| | Comparator | | |
| --- | --- | --- | --- |
| **1** | **2** | **3** | **Binary output** |
| HIGH | HIGH | HIGH | 11 |
| LOW | HIGH | HIGH | 10 |
| LOW | LOW | HIGH | 01 |
| LOW | LOW | LOW | 00 |

than for analog meters. Because of their very high input impedance, loading errors are very small at low-voltage levels, nearly as low as null balance instruments such as the potentiometer. The resolution (1 LSB) of these meters may be significantly better than their accuracy, so care must be taken in estimating uncertainty in their output. These devices are able to perform integration in both DC and true root-mean-square (rms) AC architecture.

**EXAMPLE 7.6**

A 0- to 10-V, three-digit digital voltmeter (DVM) is built around a 10-bit, single-ramp A/D converter. A 100-kHz clock is used. For an applied input voltage of 6.372 V, what is the DVM output value? How long will the conversion process take?

**KNOWN** $E_{FSR} = 10$ V; $M = 10$.

Clock speed $= 100$ kHz.

Three significant digit readout (x.xx).

Input voltage, $E_i = 6.372$ V.

**FIND**    Digital output display value.

Conversion time.

**SOLUTION**

The resolution of the A/D converter is

$$Q = E_{FSR}/2^M = 10 \text{ V}/1024 = 9.77 \text{ mV/bit}$$

Because each ramp step increases the counter value by 1 bit, the ramp converter has a slope

$$\text{slope} = Q \times 1 \text{ bit/step} = 9.77 \text{ mV/step}$$

The integer number of steps required for conversion is

$$\text{steps required} = \frac{E_i}{\text{slope}} = \frac{6.372 \text{ V}}{0.00977 \text{ V/step}} = 652.49 \text{ or } 653 \text{ steps}$$

so the indicated output voltage is

$$E_0 = \text{slope} \times \text{number of steps}$$
$$= 9.77 \text{ mV/step} \times 653 \text{ steps} = 6.3769 \text{ V} \approx 6.38 \text{ V}$$

A three-digit display rounds $E_0$ to 6.38 V. The difference between the input and output values is attributed to quantization error. With a clock speed of 100 kHz, each ramp step requires 10 μs/step, and conversion time $= 653$ steps $\times$ 10 μs/step $= 6530$ μs.

**COMMENT** A similar dual-ramp converter would require twice as many steps and twice the conversion time.

# 7.6 Data Acquisition Systems

A *data acquisition system* is the portion of a measurement system that quantifies and stores data. There are many ways to do this. The engineer who reads a transducer dial, associates a number with the dial position, and records the number in a log book performs all the tasks germane to a DAS.

This section focuses on microprocessor-based DAS, which are used to perform automated data quantification and storage in engineering testing. The needed microprocessor technology can be found in many devices from specialized control boards to personal mobile devices to laptop, desktop, and dedicated computers. We use the generic term "computer" to refer to any of these devices.

Figure 7.12 shows how a DAS might fit into the general measurement scheme between an analog measurement and its subsequent data reduction. A typical analog signal flow scheme is shown in Figure 7.13 for multiple input signals to a single microprocessor-based/controller DAS.

Dedicated microprocessor systems can continuously perform their programming instructions to measure, store, interpret, and provide process control without any intervention. Such micro-processors use I/O ports to interface with other devices to measure and to output instructions. Programming allows for such operations as which sensors to measure and when and how often,

**FIGURE 7.12**    Typical signal and measurement scheme.

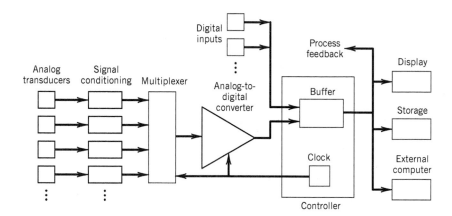

**FIGURE 7.13**    Analog signal flow scheme using an automated data acquisition system.

and data reduction. Programming can allow for decision making and feedback to control process variables. A cost-effective microprocessor system may be derived from a programmable logic controller (PLC).

Computer-based laboratory DAS combine a data acquisition module with both the microprocessor and interface capability of a computer. The interface between external instruments and the computer is done through the dedicated module, which connects to the computer via an external bus port, an I/O board, or wireless port. The computer also provides access to the Internet, an increasingly useful method for transmitting data between locations.

## 7.7 Data Acquisition System Components

### Analog Signal Conditioning: Filters and Amplification

Analog signals usually require some type of signal conditioning to properly interface with a digital system. Filters and amplifiers are the most common components used.

### Filters

*Analog filters* are used to control the frequency content of the signal being sampled. These were discussed in Chapter 6. Anti-alias analog filters reduce signal information above the Nyquist frequency before sampling. Not all data acquisition boards contain analog filters, so these necessary components may need to be added.

*Digital filters* are effective for digital signal analysis. They are not used to prevent aliasing. One digital filtering scheme involves taking the Fourier transform of the sampled signal, multiplying the signal amplitude in the targeted frequency domain to attain the desired frequency response (i.e., type of filter and desired filter settings), and transforming the signal back into time domain using the inverse Fourier transform [1, 2].

A simpler digital filter is the moving average or smoothing filter, which is used for removing noise or showing trends of discrete information. Essentially, this filter replaces a current data point value with an average based on a series of successive data point values. A center-weighted moving averaging scheme takes the form

$$y_i^* = (y_{i-n} + \cdots + y_{i-1} + y_i + y_{i+1} + \cdots + y_{i+n})/(2n + 1) \tag{7.20}$$

where $y_i^*$ is the averaged value that is calculated and used in place of $y_i$, and $(2n + 1)$ equals the number of successive values used for the averaging. For example, if $y_4 = 3$, $y_5 = 4$, $y_6 = 2$, then a three-term average of $y_5$ produces $y_5^* = 3$.

In a similar manner, a forward-moving averaging smoothing scheme takes the form

$$y_i^* = (y_i + y_{i+1} + \cdots + y_{i+n})/(n + 1) \tag{7.21}$$

and a backward-moving averaging smoothing scheme takes the form

$$y_i^* = (y_{i-n} + \cdots + y_{i-1} + y_i)/(n + 1) \tag{7.22}$$

Light filtering might use three terms, whereas heavy filtering might use 10 or more terms. This filtering scheme is easily accomplished within a spreadsheet or programming package.

## Amplifiers

All DAS are input range limited; that is, there is a minimum value of a signal that they can resolve and a maximum value that initiates the onset of saturation. Thus, some transducer signals need amplification or attenuation before conversion. Most DAS contain onboard instrumentation amplifiers as part of the signal-conditioning stage, as depicted in Figure 7.13, with selectable gains ranging from less than to greater than unity. Gain is varied either by logic switches set by software or by resistor jumpers, which effectively reset resistor ratios across op-amplifiers. Section 6.6 in Chapter 6 discusses amplifiers.

Although instrument amplifiers offer excellent output impedance characteristics, voltages can also be attenuated using a voltage divider. The output voltage from the divider circuit of Figure 7.14 is determined by

$$E_o = E_i \frac{R_2}{R_1 + R_2} \tag{7.23}$$

For example, a 0- to 50-V signal can be measured by a 0- to 10-V A/D converter using $R_1 = 40 \text{ k}\Omega$ and $R_2 = 10 \text{ k}\Omega$.

When only the dynamic content of time-dependent signals is important, amplification may require a strategy. For example, suppose the mean value of a voltage signal is large but the dynamic content is small, as with the 5-Hz signal

$$y(t) = 2 + 0.1 \sin 10 \, \pi t (V) \tag{7.24}$$

**FIGURE 7.14** Voltage divider circuit for signal amplitude attenuation.

Setting the amplifier gain at or more than $G = 2.5$ to improve the resolution of the dynamic content would saturate a $\pm 5$-V A/D converter. In such situations, you can remove the mean component from the signal before amplification either by (1) using an offset nulling circuit to zero the mean value as discussed next or, similarly, by (2) adding a mean voltage of equal but opposite sign to the signal, such as by adding $-2$ V in this example or (3) passing the signal through a very low-frequency high-pass filter (also known as *AC coupling*).

### Shunt Resistor Circuits

An A/D converter requires a voltage signal at its input. It is straightforward to convert current signals into voltage signals using a shunt resistor. The circuit in Figure 7.15 provides a voltage

$$E_o = IR_{\text{shunt}} \tag{7.25}$$

for signal current $I$. For a transducer using a standard 4- to 20-mA current loop, a 250-$\Omega$ shunt would convert this current output to a 1- to 5-V signal.

### Offset Nulling Circuit

*Offset nulling* is used to subtract a small constant voltage from a signal, such as to zero out of a transducer output signal. Hence it is an important tool for reducing uncertainty arising from zero error while keeping loading error to a minimum. The technique uses a trim potentiometer ($R_{\text{null}}$) in conjunction with a bridge circuit, such as shown in Figure 7.16, to produce and to apply a null voltage, $E_{\text{null}}$, to the circuit. The available null voltage is

$$E_{\text{null}} = \pm \left[ \frac{E_i}{2} - \frac{E_i R_3 (R_{\text{null}} + R_{\text{sensor}})}{R_{\text{null}} R_{\text{sensor}} + R_3 (R_{\text{null}} + R_{\text{sensor}})} \right] \tag{7.26}$$

The corrected signal voltage is $E_0 = E_i - E_{\text{null}}$. The technique is analogous to nulling (zeroing) the output indicator on a mechanical scale.

**FIGURE 7.15**   Simple shunt resistor circuit.

**FIGURE 7.16**   Circuit for applying a null offset voltage.
(Courtesy of National Instruments.)

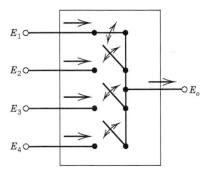

**FIGURE 7.17**   Multiplexer (four-channel shown).

## Components for Acquiring Data

### Multiplexer

When multiple input signal lines are connected by a common throughput line to a single A/D converter, a multiplexer is used to switch between the connections and the A/D converter one at a time. The technique is quite common in commercial data acquisition modules. The approach is illustrated by the multiplexer depicted in Figure 7.17, which uses parallel flip-flops (see Chapter 6) to open and close the connection paths sequentially. The switching rate is determined by the conversion timing control logic.

### A/D Converters

High-speed data acquisition modules employ successive approximation A/D converters with common conversion rates up to the 100-kHz to 10-MHz range or parallel A/D converters for common conversion rates up to and over 10 GHz. Even much higher speeds are achievable. Low-level voltage measurements require the high noise rejection of the dual-ramp A/D converter. Here the trade-off is in speed, with conversion rates of only up to 100 Hz more common.

### D/A Converters

A D/A converter permits a DAS to convert digital numbers into analog voltages, which might be used for process control, such as to change a process variable, activate a device, or to drive a sensor-positioning motor. The D/A signal is initiated by software or a controller.

### Digital Input/Output

Digital signals are composed of discrete states, either high or low. Digital input/output (I/O) connection lines may be used to communicate between instruments (see Section 7.9), control relays (to power equipment on or off), or indicate the state or status of a device. A common way to transmit digital information is the use of a 5-V TTL signal (e.g., see Figure 7.5).

Digital I/O signals may be as a single state (HIGH or LOW; 5 V or 0 V) or as a series of pulses of HIGH/LOW states. The single-state signal might be used to operate a switch or relay, or to signal an alarm. A series of pulses transmits a data series. Gate-pulse counting entails counting pulses that occur over a specified period of time. This enables frequency determination (number of pulses/unit time) and counting/timing applications. Pulse stepping, sending a predetermined number of pulses in a series, is used to drive stepper motors and servos. Several I/O ports can be grouped to send parallel information.

### Closed-Loop Controller

In closed-loop control, the controller is used to compare the state of a process, as determined through the value of a measured variable, with the value of a set condition and to take appropriate

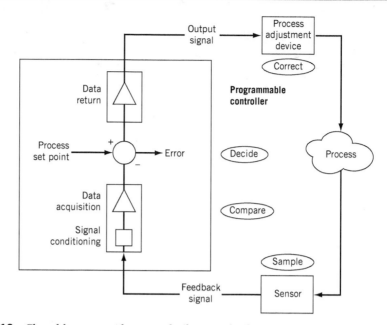

**FIGURE 7.18**    Closed-loop control concept built around a data acquisition–based programmable controller.

action to reduce the difference in value of the two. This difference in value is called the error signal. This error can change as part of the dynamics of the process, which causes the measured variable to change value, or because the set condition itself is changed. The controller action is to adjust the process so as to keep the error within a desired range.

When integrated within a DAS, the control process consists of this sequence: sample, compare, decide, correct. An example of a digital control loop is depicted in Figure 7.18. Here the controller receives input about the process through a sensor that monitors the measured variable. The controller measures the sensor signal by sampling through an A/D converter or some digital input. The controller compares the measured value to the set value, computes the error signal, executes calculations through its control algorithm to decide on the correction needed, and acts to correct the process. In this example, we anticipate that the corrective elements are analog based, and so the controller sends the necessary signal for appropriate corrective action at the output of a D/A converter. The controller repeats this procedure at each cycle. A more extensive treatment of feedback control is provided in Chapter 12.

# 7.8  Analog Input–Output Communication

## Data Acquisition Modules

Data Acquisition Modules consist of a module that supports interfacing with analog and digital external transducers and devices and a microprocessor/computer that controls the process and stores/retrieves information from memory. The most common arrangement uses an external module connected by a wired bus to a tablet, mobile device, laptop, or stationary computer. A plug-in computer board is also common, particularly in industrial environments. Fully self-contained standalone modules exist with an onboard microprocessor and memory. Some modules can interface with other devices using wireless communication, such as Bluetooth or the wireless network. With a wireless or hardwired Internet connection, it is possible to acquire or send data and transmit information to different locations on command.

**FIGURE 7.19**  Generic layout of a multipurpose data acquisition module interfaced with a computer.

Consider the generic layout of a DAS shown in Figure 7.19. Field wiring from/to transducers or other external devices is usually made through a mechanical screw terminal that attaches directly or indirectly to a DAS module. The module allows for multichannel analog and digital input and output. The layout shows the signal path through each of the various module components, which passes onto the module's internal bus. The module interfaces with the controlling computer bus through some communication interface (e.g., a USB port or a Bluetooth connection), labeled here as a "connector." Standalone systems simply port the DAS bus to the onboard microprocessor bus.

A commercially available multichannel high-speed data acquisition module is shown in Figure 7.20. It uses a 16-channel multiplexer and instrument amplifier enabling 8 differential or 16-single-ended analog input channels. Input impedance exceeds 10 G$\Omega$. The 16-bit successive approximation A/D converter is capable of sample rates of up to 250 kHz. Input signals may be unipolar (e.g., 0–10 V) or bipolar (e.g., ±5 V). For the board shown, this gives a unit gain input resolution ($E_{FSR} = 10$ V) of $Q = 152.6$ µV/bit. The selectable gain onboard amplifier extends the analog input range from ±10V down to ± 0.2 V. The module also contains two 16-bit analog output channels driven by D/A converters. There are four digital I/O channels and two 32-bit counters for instrument control applications and external triggering. This USB-powered device is comparable in size to a pocket-sized camera, a small size that can still accommodate a row of screw terminal wire connectors to attach transducers and devices.

Data acquisition technology is rapidly changing to bring new capabilities to the marketplace. A full-featured, self-contained DAS module is shown in Figure 7.21. This particular system is powered by an onboard dual core microprocessor with an operating system enabling standalone programmable real-time data processing and control. Data acquisition and control capabilities are retained: 10 single-ended and differential-ended analog inputs, six analog outputs, 40 digital I/O, and three-axis accelerometer along with the needed I/O connector terminals. It can interface with other devices, such as tablets and external computers, either through a USB port or its onboard

Photograph of NI USB-621x; Courtesy of National Instruments.

**FIGURE 7.20**     Photograph of a multipurpose USB-based data acquisition module.

Photograph of NI myRIO; Courtesy of National Instruments.

**FIGURE 7.21**     Example of a self-contained, portable multipurpose data acquisition system.

WiFi communication. Aside from the portability offered by the small package, the design is also attractive for data acquisition needs in moving (e.g., vehicles, robotics) and rotating systems, because there is no need for connection cables to computers and power sources.

## Single- and Differential-Ended Connections

Analog signal input connections to a DAS board may be single- or differential-ended. *Single-ended connections* use only one signal line (+ or HIGH) that is measured relative to ground (GRD), as shown in Figure 7.22. The return line (− or LOW) and ground are connected together. There is a common external ground point, usually through the DAS board. Multiple single-ended connections to a DAS board are shown in Figure 7.23.

Single-ended connecting wires should never be long (>1 m), as they are susceptible to EMI (electromagnetic interference) noise, particularly at high speed transfer rates. Because of noise, the signal should be large compared to the DAS resolution. Single-ended connections are suitable only when all the analog signals can be made relative to the common ground point.

Why the concern over ground points? Electrical grounds are not all at the same voltage value (see Ground and Ground Loops, in Chapter 6). So if a signal is grounded at two different points, such as at its source and at the DAS board, the grounds could be at different voltage levels. When a grounded source is wired as a single-ended connection, the difference between the source ground voltage and the board ground voltage gives rise to a *common-mode voltage* (CMV). The CMV

**(a)**                      **(b)**

**FIGURE 7.22**  (a) Single-ended connection. (b) Differential-ended connection.

**FIGURE 7.23**
Multiple single-ended analog connections to a data acquisition board.

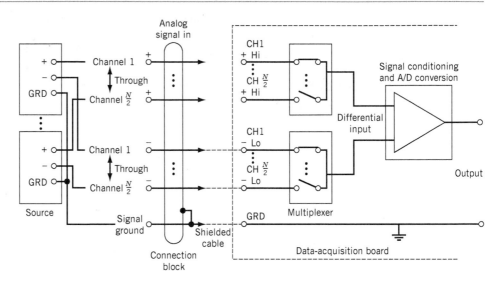

**FIGURE 7.24**
Multiple differential-ended
analog connections to a data
acquisition board.

combines with the input signal superimposing interference and noise on the signal. This effect is referred to as a ground loop. The measured signal is unreliable.

*Differential-ended connections* allow the voltage difference between two distinct input signals to be measured. Here the signal input (+ or HIGH) line is paired with a signal return (− or LOW) line, which is isolated from ground (GRD), as shown in Figure 7.22. By using twisted pairs, the effects of noise are greatly reduced. A differential-ended connection to a DAS board is shown in Figure 7.24. The measured voltage is the voltage difference between these two (+ and −) lines for each channel.

Differential-ended connections are the preferred way to measure low-level signals. However, for low-level measurements, a 10-kΩ to 100-kΩ resistor should be connected between the signal return (− or LOW) line and ground (GRD) at the DAS board. The differential-ended connection is less prone to common-mode voltage errors. But if the measured voltage exceeds the board's CMV range limit specification, it can be damaged.

### Special Signal-Conditioning Modules

Signal-conditioning modules exist for different transducer applications [5]. They connect between the transducer and the data acquisition module. For example, resistance bridge modules allow the direct interfacing of strain gauges or other resistance sensors through an on-board Wheatstone bridge, such as depicted in Figure 7.25. There are temperature modules to allow for electronic thermocouple cold junction compensation that can provide for signal linearization for reasonably accurate (down to 0.5°C) temperature measurements.

### Data Acquisition Triggering

With DAS boards, data acquisition or equipment control can be triggered by software command, external pulse, or on-board clock. Software-controlled triggers can send TTL-level pulses to external devices for synchronization, activation, and control.

## 7.9 Digital Input–Output Communication

Certain standards exist for the manner in which digital information is communicated between digital devices. Serial communication methods transmit data bit by bit. Parallel communication methods transmit data in simultaneous groups of bits—for example, byte by byte. Both methods use

**FIGURE 7.25**   A strain-gauge interface. Gauges are connected by wires to the interface.

a *handshake*, an interface protocol that initiates, allows, and terminates the data transfer between devices and TTL-level signals. Most lab equipment can be equipped to communicate by at least one of these means, and standards have been defined to codify communications between devices of different manufacturers. Standards continuously evolve, so the present discussion focuses on concepts and application of representative standards.

## Data Transmission

Serial communication implies that data are sent in successive streams of information 1 bit at a time. The value of each bit is represented by an analog voltage pulse with a 1 and 0 distinguished by two equal voltages of opposite polarity in the 3- to 25-V range. Communication rates are measured in baud, which refers to the number of signal pulses per second.

A typical asynchronous transmission is 10 serial bits composed of a start bit followed by a 7- or 8-bit data stream, either one or no parity bit, and terminated by 1 or 2 stop bits. The start and stop bits form the serial "handshake." *Asynchronous transmission* means that information may be sent at random intervals. So the start and stop bits are signals used to initiate and to end the transmission of each byte of data transmitted, hence the analogy of a handshake. The start bit allows for synchronization of the clocks of the two communicating devices. The parity bit allows for limited error checking. *Parity* involves counting the number of 1s in a byte. In 1 byte of data, there will be an even or odd number of bits with a value of 1. An additional bit added to each byte to make the number of 1 bits a predetermined even or odd number is called a parity bit. The receiving device counts the number of transmitted bits, checking for the predetermined even (even parity) or odd (odd parity) number. In *synchronous transmission*, the two devices initialize and synchronize communication with each other. Even when there are no data to send, the devices transmit characters just to maintain synchronization. Stop and start bits are then unnecessary, allowing for higher data transfer rates.

Devices that employ data buffers, a region of RAM that serves as a data holding area, use a software handshaking protocol such as XON/XOFF. With this, the receiving device transmits an XOFF signal to halt transmission as the buffer nears full and an XON signal when it has emptied and is again ready to receive.

## Universal Serial Bus

The Universal Serial Bus (USB) permits peripheral expansion for up to 127 devices at low- to high-speed serial data transfer rates. The transfer rate of the USB depends on the mode used. The original USB (USB 1.0/1.1) supports transfer rates from 1.5 Mbs up to 12 Mbs, and the high-speed USB (USB 2.0) supports rates up to 480 Mbs. The USB 3.2 standard boasts data rates of up to 20 Gbit/s. Notably, the maximum current that a USB 3.x device can draw is 900 mA, an 80% increase over USB 2.0. This bus connects USB devices to a single computer host through a USB root hub. The USB physical interconnect is a tiered star topology, as illustrated in Figure 7.26a. In this setup, the root hub permits one to four attachments, which can be a combination of USB peripheral devices and additional USB hubs. Each successive hub can in turn support up to four devices or hubs. Cable length between a device and hub or between two hubs is limited to 5 m for USB 2.0 and, while unspecified, a practical limit of 3 m for USB 3.0. The connecting cable (Figure 7.26b) is a four-line wire consisting of a hub power line, two signal lines (+ and −), and a ground (GRD) line.

## Bluetooth Communications

Bluetooth is a standardized specification permitting wireless communication between devices over short-to-moderate distances. The specification is based on the use of radio waves to transmit information, allowing signals to pass through enclosures, walls, and clothes. Devices are independent of aiming position. As such, the specification overcomes problems inherent with popular infrared port communication devices—for example, those used on common electronic remote controls, which require line-of-sight interaction. However, Bluetooth is not intended or written to be a wireless LAN network protocol, such as is IEEE 802.11. Instead, it does permit communication between peripheral devices that use their own protocol by setting up a personal

**FIGURE 7.26**
General purpose interface bus (GPIB) assignments to a 25-pin connector.

(a) Star topology

(b) Wiring connection

area network (PAN). Up to eight devices can communicate simultaneously within the 2.402- and 2.480-GHz radio spectrum on a PAN. Bluetooth 5.0 can manage data transfer rates up to 50 Mbps.

## Other Serial Communications: RS-232C

The RS-232C protocol, initially set up to translate signals between telephone lines and computers via a modem (modulator–demodulator), remains a well-used interface for communication between a computer and any serial device. Basically, this protocol allows two-way communication using two single-ended signal (+) wires, noted as TRANSMIT and RECEIVE, between data communications equipment (DCE), such as an instrument, and data terminal equipment (DTE), such as the DAS or computer. These two signals are analogous to a telephone's mouthpiece and earpiece signals. A signal GROUND wire allows signal return (−) paths. Many lab computers have an RS-232C–compatible I/O port. The popularity of this interface comes from the wide range of equipment that can use it. Either a 9-pin or 25-pin connector can be used.

## Parallel Communications

The *general purpose interface bus* (*GPIB*) is a classic example of a high-speed parallel interface. Originally developed by Hewlett-Packard, it is sometimes referred to as the HP-IB. The GPIB is usually operated under the IEEE-488 communication standard. In 2004, the international and IEEE standards were combined and the current standard is designated IEC-60488-1. This 8-bit parallel bus allows for the control of other devices through a central controller, usually a lab computer, and it allows devices to receive/transmit information from/to the controller. This standard is well defined to interface communication between computers and printers and scientific instrumentation. The standard for the GPIB operates from a 16-wire bus. A 25-pin connector is standard. Bit-parallel, byte-serial communication at data rates up to 8 Mbytes/s is possible.

---

**EXAMPLE 7.7**

A strain transducer has a static sensitivity of 2.5 V/unit strain (2.5 μV/μɛ) and requires a supply voltage of 5 VDC. It is to be connected to a DAS having a ±5 V, 12-bit A/D converter and its signal measured at 1,000 Hz. The transducer signal is to be amplified and filtered. For an expected measurement range of 1 to 500 μɛ, specify appropriate values for amplifier gain, filter type, and cutoff frequency, and show a signal flow diagram for the connections.

**SOLUTION**

The signal flow diagram is shown in Figure 7.27. Power is drawn off the data acquisition board and routed to the transducer. Transducer signal wires are shielded and routed through the amplifier, filter, and connected to the data acquisition board connector block using twisted pairs to channel 0, as shown. The differential-ended connection at the board reduces noise.

The amplifier gain is determined by considering the minimum and maximum signal magnitudes expected, and the quantization error of the DAS. The nominal signal magnitude ranges from 2.5 μV to 1.25 mV. Using an amplifier gain of $G = 1,000$ would boost these signal levels to 2.5 mV to 1.25 V. This lower value is on the order of the quantization error of the 12-bit converter. Using a gain of $G = 3,000$ would raise the low end of the signal out of quantization noise while keeping the high end out of saturation achieving an input signal range of 7.5 mV to 3.75 V.

With a sample rate of $f_s = 1,000$ Hz, an anti-alias, low-pass Butterworth filter with 12 dB/octave roll-off and a cutoff frequency set at $f_c = f_N = 500$ Hz would meet the task.

**FIGURE 7.27**   Line connections for Example 7.7.

### EXAMPLE 7.8

The output from an analog device (nominal output impedance of 600 $\Omega$) is the input to the 12-bit A/D converter (nominal input impedance of 1 M$\Omega$) of a DAS. For a 2-V signal, will interstage loading be a problem?

**KNOWN** $Z_1 = 600\ \Omega$.
$\qquad$ $Z_m = 1\ M\Omega$.
$\qquad$ $E_1 = 2\ V$.

**FIND** $e_I$.

### SOLUTION

The loading error is given by $e_I = E_m - E_I$, where $E_1$ is the true voltage and $E_m$ is the measured voltage. From Equation 6.39,

$$e_I = E_1 \left( \frac{1}{1 + Z_1/Z_m} - 1 \right) = -1.2\ mV$$

The interstage loading error at 2-V input actually is less than the $\pm 2.4$-mV quantization error of the 12-bit device.

### EXAMPLE 7.9

An analog signal is to be sampled at a rate of 200 Hz with a 12-bit A/D converter that has an input range of $-10$ to 10 V. The signal contains a 200-Hz component, $f_1$, with an amplitude of 100 mV. Specify a suitable LC filter that attenuates the 200-Hz component down to the A/D converter quantization resolution level and acts as an anti-alias filter for quantization.

**KNOWN** $fs = 200\ Hz$.
$\qquad$ $f_1 = 200\ Hz$.
$\qquad$ $A_1 = 100\ mV$.
$\qquad$ $E_{FSR} = 10V$.
$\qquad$ $M = 12$.

**FIND** Specify a suitable filter.

**SOLUTION**
For $f_s = 200$ Hz, the Nyquist frequency $f_N$ is 100 Hz. So an appropriate design for an anti-alias filter might have the properties of $f_c = 100$ Hz and $M$ (100 Hz) $= -3$ dB and a maximally flat passband (Butterworth). The A/D converter quantization resolution level is

$$Q = \frac{E_{FSR}}{2^M} = \frac{20 \text{ V}}{4096} = 4.88 \text{ mV/bit}$$

For a 100-mV signal at 200 Hz, this requires a $k$-stage, low-pass Butterworth filter with an attenuation of

$$M(200 \text{ HZ}) = \frac{4.88 \text{ mV}}{100 \text{ mV}} = 0.0488 \text{ (or } -26 \text{ dB)}$$
$$= [1 + (f/f_c)^{2k}]^{-1/2} = [1 + (200/100)^{2k}]^{-1/2}$$

so $k = 4.3 \approx 5$. Appropriate values for $L$ and $C$ can be set from Table 6.1 and Figure 6.30 and scaled.

---

**EXAMPLE 7.10    Digital Signal Processors**

We enjoy our personal mobile devices to store audio (e.g., music, books) and use them for playback as we go about our normal activities. At some point, this audio information was converted from an analog to a digital signal by passing it through an A/D converter and subsequently storing it in digital form, often compressed to minimize memory needs. This would be the form of a file that you might download from a website.

Digital signal processors (DSP) are devices that process digital information [5]. They use simple mathematical algorithms to manipulate information quickly. A common mobile audio device (schematic in Figure 7.28) uses a DSP to encode incoming digital information, such as from a download, and save it to the audio device's memory. In storing the information, the user interacts directly with the DSP through a user interface, assigning it a filename or organizing the order of the files. To play back some particular information, the user again interacts with the DSP so that the DSP pulls the correct digital file from memory and decodes, expands, and converts it back to its original analog signal by passing it through a D/A converter. The now analog signal passes through an amplifier before the signal is coupled with the headset earpiece or external speakers.

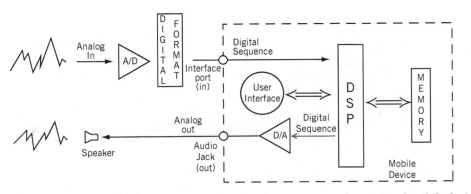

**FIGURE 7.28**   An internal digital signal processor (DSP) is common within personal mobile devices to handle digital data processing, conditioning, and handling.

A DSP is programmable and works with any type of physical signal that has been digitized. More and more, we interact with DSPs without realizing it. For example, a cell phone uses a DSP to provide echo cancellation during conversations, and your television or video player uses a DSP to expand the movie you download whose file had been previously compressed from the original signal to manage resource usage. Furthermore, that movie's audio format is decoded by the DSPs within both the video player and audio amplifier to create an impressive home theater 7.1 sound.

## 7.10  Digital Image Acquisition and Processing

Digital images are widely used in applications as varied as inspection, medical imaging, quality assurance, and sorting. This section serves as a brief introduction to acquiring and processing digital images.

### Image Acquisition

Consider the characteristics of a digital grayscale image, such as the one in Figure 7.29. A digital image is composed of an $m \times n$ array of pixels, as illustrated in Figure 7.30. Here we have zoomed in on the edge of one of the coins in Figure 7.29, and we see that the image is composed of pixels having various levels of gray shading. The grayscale images in Figures 7.29 and 7.30 are 8-bit digital images; each pixel is assigned a number between 0 and 256, with zero corresponding to pure black.

The device that enables digital imaging is termed a charge-coupled device (CCD).[2] A CCD image sensor is essentially an array of capacitors that gain charge when exposed to light. Each capacitor represents one pixel in the resulting digital image. A digital camera may have a physical

Donald Beasley

**FIGURE 7.29**   Characteristics of a digital grayscale image.

---

[2]The CCD was invented in 1969 at AT&T Bell Labs by Willard Boyle and George Smith (*Bell System Technical Journal* 49(4), 1970). In 2009 they shared the Nobel Prize in physics for this invention.

Donald Beasley

**FIGURE 7.30** Digital image of an $m \times n$ pixel array.

or an electronic shutter, but in either case the CCD array is active for an exposure time exactly as film cameras function. Once the exposure is over, the CCD accumulated charge is converted to a digital representation of the image. The resolution of the image is determined by the number of pixels. Common cameras may have resolutions that range from 0.3 million pixels (MP) to 21 MP, but higher resolutions are available.

In many applications for digital imaging, the images are transferred directly from the camera to a computer, with image acquisition rates as fast as 200,000 frames per second. The term *framegrabber* is used to describe the computer hardware that acquires the images from a CCD camera. Most framegrabber hardware is in the form of a plug-in board or an external module for a computer. As with any photography, very short exposure times require high-intensity lighting.

## Image Processing

A basic treatment of the fundamentals and applications of digital image processing would easily require an entire book. So our purpose here is to describe one basic issue as illustrative of the impact of digital image processing. Let's consider again Figure 7.29, which shows five coins. Two tasks that might be reasonable to ask an imaging system to accomplish are to locate the coins and determine their diameter, sensing tasks with parallels in automated manufacturing.

Both these tasks can be accomplished if we can locate the outer edges of the coins. In principle, an edge occurs where the gradient of the pixel intensity is a maximum. So, some numerical scheme for finding the gradient is required. In a two-dimensional image, the gradient has both magnitude and direction. For the two-dimensional array that comprises a grayscale image, two gradients are calculated, one in the $x$-direction and one in the $y$-direction (horizontal and vertical in the image). Using these two numerical estimates of the gradient, a magnitude and direction can be determined. We will explore two gradient-based methods, *Sobel* and *Canny*, as they are implemented in Matlab. The basic differences in these two methods are the degree of smoothing and the criteria for defining an edge. The *Canny* method smoothes the image to suppress noise and uses two threshold values in determining the location of an edge. Gradient values that are below the low threshold result in a pixel being assigned as a nonedge. Pixels where gradient values are above the high threshold are set as edges pixels. Gradient values between the two thresholds are examined to see if adjacent pixels are an edge, and if there is a direct path to an edge pixel that pixel is included as an edge.

The implementation of edge detection in Matlab is a very straightforward process for grayscale images. The function *imread* creates the $m \times n$ matrix of grayscale values. The function *edge* identifies the edges with the method selected. The function *imshow* views the image.

Figure 7.31 shows the results of the two edge detection methods for our image of coins. Clearly the grayscale image contains information that causes the edge detection methods to find many edges that have nothing to do with our coins. One method to improve the chances of finding the edges of the coins is to subject the image to a process called *thresholding* whereby all pixel values above a certain level are set to pure white and all below to pure black. Let's threshold our image at level 127. The resulting image is shown in Figure 7.32. Applying the same edge detection methods used on the grayscale image gives the image shown in Figure 7.33. This is a step closer to finding the edges of the coins and sizing their diameters.

Donald Beasley                    (a)

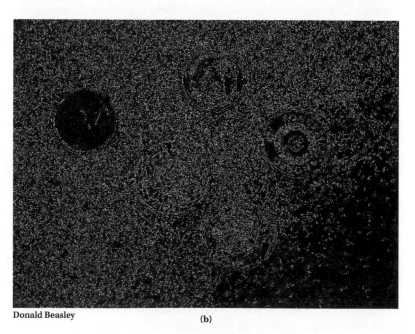

Donald Beasley                    (b)

**FIGURE 7.31**    Edge detection methods. (a) Sobel method of edge detection. (b) Canny method of edge detection.

Donald Beasley

**FIGURE 7.32** Improved effect of using threshold values to eliminate noise.

Donald Beasley (a)

**FIGURE 7.33** Effects of using threshold values on a grayscale image. (a) Sobel method of edge detection. (b) Canny method of edge detection.

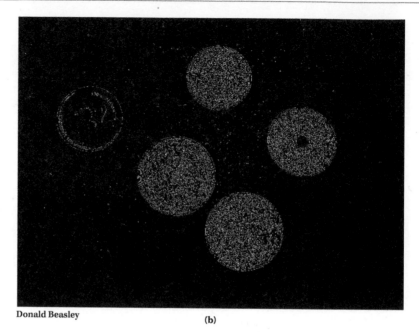

Donald Beasley

**(b)**

**FIGURE 7.33**    (*Continued*)

## SUMMARY

This chapter has focused on sampling concepts, the interfacing of analog and digital devices, and data acquisition systems. Despite the advantages of digital systems, they must interact with an analog world, and the back-and-forth exchange between an analog and digital signal has limitations. With this in mind, a thorough discussion has been provided on the selection of sample rate and the number of measurements required to reconstruct a continuous process variable from a discrete representation. Equations 7.2 and 7.10 explain the criteria by which a periodic waveform can be accurately represented by such a discrete time series. The limitations resulting from improper sampling and the discrete representation include frequency alias and leakage, as well as amplitude ambiguity.

The fact that practically any electrical instrument can be interfaced with a microcontroller or a DAS is significant. The working mechanics of A/D and D/A converters is basic to interfacing such analog and digital devices with DAS. The limitations imposed by the resolution and range of these devices mandate certain signal-conditioning requirements in terms of signal gain and filtering prior to acquisition. These require analog amplifiers and filters, devices discussed previously. Communication between digital devices surrounds us. But their proliferation means that standards for communication must be set. These standards are evaluated and replaced as newer devices demand faster data transfer.

## REFERENCES

1. Bendat, J., and A. Piersol, *Random Data: Analysis and Measurement Procedures*, 4th ed., Wiley, New York, 2010.

2. Proakis, J. G., and D. K. Manolakis, *Digital Signal Processing*, 4th ed., Prentice-Hall, Pearson Education, Upper Saddle River, NJ, 2007.

3. Lyons, R., *Understanding Digital Signal Processing*, 3rd ed., Pearson Education, Upper Saddle River, NJ, 2010.

4. Swedlin, E. G., and D. L. Ferro, *Computers: The Life Story of a Technology*, Johns Hopkins University Press, Baltimore, MD, 2007.

5. Di Paolo, M., *Data-Acquisition Systems: From Fundamentals to Applied Design*, Springer, New York, 2013.

## NOMENCLATURE

| | | | |
|---|---|---|---|
| $e$ | error | $f_a$ | alias frequency $(t^{-1})$ |
| $e_Q$ | quantization error | $f_c$ | filter cutoff frequency $(t^{-1})$ |
| $f$ | frequency $(t^{-1})$ | $f_m$ | maximum analog signal frequency $(t^{-1})$ |

| | | | | |
|---|---|---|---|---|
| $f_N$ | Nyquist frequency ($t^{-1}$) | | $I$ | electric current (A) |
| $f_s$ | sample rate frequency ($t^{-1}$) | | $M$ | number of bits |
| $k$ | cascaded filter stage number | | $M(f)$ | magnitude ratio at frequency $f$ |
| $t$ | time ($t$) | | $N$ | data set size |
| $y(t)$ | analog signal, $y_i$ | | $N\delta t$ | total digital sample period ($t$) |
| $y(r\delta t)$ | discrete values of a signal, $y(t)$ | | $R$ | resistance ($\Omega$) |
| $\{y(r\delta t)\}$ | complete discrete time signal of $y(t)$ | | $Q$ | A/D converter resolution (V) |
| $A$ | amplifier open loop gain; also, constant | | $\delta f$ | frequency resolution of DFT ($t^{-1}$) |
| $E$ | voltage (V) | | $\delta R$ | change in resistance ($\Omega$) |
| $E_i$ | input voltage (V) | | $\delta t$ | sample time increment ($t$) |
| $E_o$ | output voltage (V) | | $\tau$ | time constant ($t$) |
| $E_{FSR}$ | full-scale analog voltage range | | $\phi(f)$ | phase shift at frequency $f$ |
| $G$ | amplifier gain | | | |

# 8

# Temperature Measurements

## 8.1 Introduction

Temperature is one of the most commonly used and measured engineering variables. Although much of our lives is affected by the diurnal and seasonal variations in ambient temperature, the fundamental scientific definition of temperature and a scale for the measurement of temperature are not commonly understood. This chapter explores the establishment of a practical temperature scale and common methods of temperature measurement. In addition, errors associated with the design and installation of temperature sensors are discussed.

Upon completion of this chapter, the reader will be able to:

- Describe the primary standards for temperature

- State the role of fixed point calibration and the necessity for an interpolation method in establishing a temperature standard

- State the physical principle underlying electrical resistance thermometry

- Employ standard relationships to determine temperature from resistance devices

- Analyze thermoelectric circuits designed to measure temperature

- State the principles employed in radiation temperature measurements

- Estimate the impact of loading errors in temperature measurement

### Historical Background

Guillaume Amontons (1663–1705), a French scientist, was one of the first to explore the thermodynamic nature of temperature. His efforts examined the behavior of a constant volume of air that was subject to temperature changes. The modern liquid-in-glass bulb thermometer traces its origin to Galileo (1565–1642), who attempted to use the volumetric expansion of liquids in tubes as a relative measure of temperature. Unfortunately, this open tube device was sensitive to both barometric pressure and temperature changes. A major advance in temperature measurement occurred in 1630 as a result of a seemingly unrelated event: the development of the technology to manufacture capillary glass tubes. These tubes were then used with water and alcohol in a thermometric device resembling a bulb thermometer, and these devices eventually led to the development of a practical temperature measuring instrument.

A temperature scale proposed by Gabriel D. Fahrenheit, a German physicist (1686–1736), in 1715 attempted to incorporate body temperature as the median point on a scale having 180 divisions between the freezing point and the boiling point of water. Fahrenheit also successfully used mercury as the liquid in a bulb thermometer, making significant improvements over the attempts of Ismael Boulliau in 1659. In 1742, the Swedish astronomer Anders Celsius[1] (1701–1744) described a temperature scale that divided the interval between the boiling and freezing points of water at 1 atm pressure into 100 equal parts. The boiling point of water was fixed as 0 and the freezing point of water as 100. Shortly after Celsius's death, Carolus Linnaeus (1707–1778) reversed the scale so that the 0 point corresponded to the freezing point of water at 1 atm. Even though this scale may not have been originated by Celsius [1], in 1948 the change from degrees centigrade to degrees Celsius was officially adopted.

As stated by H. A. Klein in *The Science of Measurement: A Historical Survey* [2],

> *From the original thermoscopes of Galileo and some of his contemporaries, the measurement of temperature has pursued paths of increasing ingenuity, sophistication and complexity. Yet temperature remains in its innermost essence the average molecular or atomic energy of the least bits making up matter, in their endless dance. Matter without motion is unthinkable. Temperature is the most meaningful physical variable for dealing with the effects of those infinitesimal, incessant internal motions of matter.*

## 8.2 Temperature Standards and Definition

Temperature can be loosely described as the property of an object that describes its hotness or coldness, concepts that are clearly relative. Our experiences indicate that heat transfer tends to equalize temperature, or more precisely, systems that are in thermal communication eventually have equal temperatures. The *zeroth law of thermodynamics* states that two systems in thermal equilibrium with a third system are in thermal equilibrium with each other. Thermal equilibrium implies that no heat transfer occurs between the systems, defining the equality of temperature. Although the zeroth law of thermodynamics essentially provides the definition of the equality of temperature, it provides no means for defining a temperature scale.

A temperature scale provides for three essential aspects of temperature measurement: (1) the definition of the size of the degree, (2) fixed reference points for establishing known temperatures, and (3) a means for interpolating between these fixed temperature points. These provisions are consistent with the requirements for any standard, as described in Chapter 1.

### Fixed Point Temperatures and Interpolation

To begin, consider the definition of the triple point of water as having a value of 0.01 for our temperature scale, as is done for the Celsius scale (0.01°C). This provides for an arbitrary starting, or fixed, point for a temperature scale; in fact, the numeric value assigned to this temperature could be anything. For example, on the Fahrenheit temperature scale, it has a value very close to 32. Fixed points are typically defined by melting or boiling point temperatures, or the triple point, of a pure substance. The point at which pure water boils at one standard atmosphere pressure is an easily reproducible fixed point temperature. For our purposes, let's assign this fixed point a numerical value of 100.

The next problem is to define the size of the degree. Because we have two fixed points on our temperature scale, we can see that the degree is 1/100th of the temperature difference between the ice point and the boiling point of water at atmospheric pressure.

Conceptually, this defines a workable scale for the measurement of temperature, but as yet we have made no provision for interpolating between the two fixed point temperatures.

---

[1]It is interesting to note that in addition to his work in thermometry, Celsius published significant papers on the aurora borealis and the falling level of the Baltic Sea.

**FIGURE 8.1**    Calibration and interpolation for a liquid-in-glass thermometer.

## Interpolation

The calibration of a temperature measurement device entails not only the establishment of fixed temperature points but also the indication of any temperature between fixed points. The operation of a liquid-in-glass thermometer is based on the thermal expansion of a liquid, often mercury or alcohol, contained in a glass capillary. The level of the liquid is read as an indication of the temperature. Imagine that we submerged the thermometer in water at the ice point, made a mark on the glass at the height of the column of liquid, and labeled it 0°C, as illustrated in Figure 8.1. Next, we submerged the thermometer in boiling water and again marked the level of the liquid, this time labeling it 100°C.

Using reproducible fixed temperature points, we have calibrated our thermometer at two points, but we want to be able to measure temperatures other than these two fixed points. How can we determine the appropriate place on the thermometer to mark, say, 50°C?

The process of establishing 50°C without a fixed point calibration is called interpolation. The simplest option would be to divide the distance on the thermometer between the marks representing 0 and 100 into equally spaced degree divisions. This places 50°C as shown in Figure 8.1. What assumption is implicit in this method of interpolation? It is obvious that we do not have enough information to appropriately divide the interval between 0 and 100 on the thermometer into degrees. A theory of the behavior of the liquid in the thermometer or many fixed points for calibration is necessary to resolve our dilemma.

Even by the late eighteenth century, there was no standard for interpolating between fixed points on the temperature scale; the result was that different thermometers indicated different temperatures away from fixed points, sometimes with surprisingly large errors.

## Temperature Scales and Standards

It is necessary to reconcile an arbitrary temperature scale with the idea of absolute temperature. Thermodynamics defines a temperature scale that has an absolute reference and defines an absolute zero for temperature. For example, this absolute temperature governs the energy behavior of an ideal gas and is used in the ideal gas equation of state. The behavior of real gases at very low pressure may be used as a temperature standard to define a practical measure of temperature that approximates the thermodynamic temperature. The unit of degrees Celsius (°C) is a practical scale related to the absolute Kelvin scale as °C = K − 273.15.

The modern engineering definition of the temperature scale is provided by a standard called the International Temperature Scale of 1990 (ITS-90) [3]. This standard establishes fixed points

**Table 8.1  Temperature Fixed Points as Defined by ITS-90**

| Defining State | Temperature[a] | |
|---|---|---|
| | K | °C |
| Triple point of hydrogen | 13.8033 | −259.3467 |
| Liquid–vapor equilibrium for hydrogen at 25/76 atm | ~17 | ~−256.15 |
| Liquid–vapor equilibrium for hydrogen at 1 atm | ~20.3 | ~−252.87 |
| Triple point of neon | 24.5561 | −248.5939 |
| Triple point of oxygen | 54.3584 | −218.7916 |
| Triple point of argon | 83.8058 | −189.3442 |
| Triple point of water | 273.16 | 0.01 |
| Solid–liquid equilibrium for gallium at 1 atm | 302.9146 | 29.7646 |
| Solid–liquid equilibrium for tin at 1 atm | 505.078 | 231.928 |
| Solid–liquid equilibrium for zinc at 1 atm | 692.677 | 419.527 |
| Solid–liquid equilibrium for silver at 1 atm | 1,234.93 | 961.78 |
| Solid–liquid equilibrium for gold at 1 atm | 1,337.33 | 1,064.18 |
| Solid–liquid equilibrium for copper at 1 atm | 1,357.77 | 1,084.62 |

[a]Significant digits shown are as provided in ITS-90.

for temperature and provides standard procedures and devices for interpolating between fixed points. It establishes the Kelvin (K) as the unit for the fundamental increment in temperature. Temperatures established according to ITS-90 do not deviate from the thermodynamic temperature scale by more than the uncertainty in the thermodynamic temperature at the time of adoption of ITS-90. The primary fixed points from ITS-90 are shown in Table 8.1. In addition to these fixed points, other fixed points of secondary importance are available in ITS-90.

### Standards for Interpolation

Along with the fixed temperature points established by ITS-90 [3], a standard for interpolation between these fixed points is necessary. Standards for acceptable thermometers and interpolating equations are provided in ITS-90. For temperatures ranging from 13.8033 to 1,234.93 K, ITS-90 establishes a platinum resistance thermometer as the standard interpolating instrument and establishes interpolating equations that relate temperature to resistance. Above 1,234.93 K, temperature is defined in terms of blackbody radiation, without specifying an instrument for interpolation.

In summary, temperature measurement, a practical temperature scale, and standards for fixed points and interpolation have evolved over a period of about two centuries. Present standards for fixed point temperatures and interpolation allow for practical and accurate measurements of temperature. In the United States, the National Institute of Standards and Technology (NIST) provides for a means to obtain accurately calibrated platinum resistance thermometers for use as secondary standards in the calibration of a temperature measuring system to any practical level of uncertainty.

## 8.3 Thermometry Based on Thermal Expansion

Most materials exhibit a change in size with changes in temperature. Since this physical phenomenon is well-defined and repeatable, it is useful for temperature measurement. The liquid-in-glass thermometer and the bimetallic thermometer are based on this phenomenon.

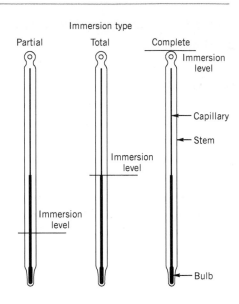

**FIGURE 8.2**    Liquid-in-glass thermometers.

## Liquid-in-Glass Thermometers

A liquid-in-glass thermometer measures temperature by virtue of thermal expansion of a liquid. The construction of a liquid-in-glass thermometer is shown in Figure 8.2. The liquid is contained in a glass structure that consists of a bulb and a stem. The bulb serves as a reservoir and provides sufficient fluid for the total volume change of the fluid to cause a detectable rise of the liquid in the stem of the thermometer. The stem contains a capillary tube, and the difference in thermal expansion between the liquid and the glass produces a detectable change in the level of the liquid in the glass capillary. Principles and practices of temperature measurement using liquid-in-glass thermometers are described elsewhere [4].

During calibration, such a thermometer is subject to one of three measuring environments:

1. For a *complete immersion thermometer*, the entire thermometer is immersed in the calibrating temperature environment or fluid.

2. For a *total immersion thermometer*, the thermometer is immersed in the calibrating temperature environment up to the liquid level in the capillary.

3. For a *partial immersion thermometer*, the thermometer is immersed to a predetermined level in the calibrating environment.

For the most accurate temperature measurements, the thermometer should be immersed in the same manner in use as it was during calibration. Liquid-in-glass thermometers have limited engineering applications but do provide reliable, accurate temperature measurement. Thus, they may be used as a local standard for calibration of other temperature sensors.

## Bimetallic Thermometers

The physical phenomenon employed in a bimetallic temperature sensor is the differential thermal expansion of two metals. Figure 8.3 shows the construction and response of bimetallic sensors to a change in temperature. The sensor is constructed by bonding two strips of different metals, *A* and *B*. The resulting bimetallic strip may be in a variety of shapes, depending on the particular application. Consider the simple linear construction shown in Figure 8.3. At the assembly temperature, $T_1$, the bimetallic strip is straight, but for temperatures other than $T_1$ the strip has a

**FIGURE 8.3**  Expansion thermometry using bimetallic materials: strip, spiral, and helix forms.

curvature. The physical basis for the relationship between the radius of curvature and temperature is given as

$$r_c \propto \frac{d}{[(C_\alpha)_A - (C_\alpha)_B](T_2 - T_1)} \tag{8.1}$$

where $r_c$ is the radius of curvature, $C_\alpha$ is the material thermal expansion coefficient, $T$ is temperature, and $d$ is the thickness of the bimetallic strip.

Bimetallic strips employ one metal having a high coefficient of thermal expansion with another having a low coefficient, providing increased sensitivity. Invar is often used as one of the metals, because for this material $C_\alpha = 1.7 \times 10^{-8}$ m/m-°C as compared to typical values for other metals, such as steels, which range from approximately $2 \times 10^{-5}$ to $20 \times 10^{-5}$ m/m-°C.

The bimetallic sensor is the primary element in most dial thermometers and many thermostats. The geometries shown in Figure 8.3 serve to provide the desired deflection in the bimetallic strip for a given application. Dial thermometers using a bimetallic strip as their sensing element typically provide temperature measurements with uncertainties of $\pm 1$°C.

## 8.4 Electrical Resistance Thermometry

As a result of the physical nature of the conduction of electricity, electrical resistance of a conductor or semiconductor varies with temperature. Using this behavior as the basis for temperature measurement is extremely simple in principle and leads to two basic classes of resistance thermometers: resistance temperature detectors (conductors) and thermistors (semiconductors). Resistance temperature detectors (RTDs) may be formed from a solid metal wire that exhibits a change in electrical resistance with temperature. Depending on the material selected, the resistance may increase or decrease with temperature. A thermistor may have a positive temperature coefficient (PTC) or a negative temperature coefficient (NTC). Figure 8.4 shows resistance as a function of temperature for a variety of conductor and semiconductor materials used to measure temperature. The PTC materials are metals or alloys and the NTC materials are semiconductors. Cryogenic temperatures are included in this figure, and germanium is clearly an excellent choice for low temperature measurement because of its large sensitivity.

**FIGURE 8.4**    Resistance as a function of temperature for selected materials used as temperature sensors.
(©2001, IEEE. Reprinted, with permission, from Yeager, C. J., and S. S. Courts, A Review of Cryogenic Thermometry and Common Temperature Sensors, *IEEE Sensors Journal*, 1(4), 2001.)

## Resistance Temperature Detectors

In the case of an RTD,[2] the sensor is generally constructed by mounting a metal wire on an insulating support structure to eliminate mechanical strains and by encasing the wire to prevent changes in resistance resulting from influences from the sensor's environment, such as corrosion. Figure 8.5 shows such a typical RTD construction.

Mechanical strain significantly changes a conductor's resistance and must be eliminated if accurate temperature measurements are to be made. Such mechanical stresses and resulting strains can be created by thermal expansion. Thus, provision for strain-free expansion of the conductor as its temperature changes is essential in the construction of an RTD. The support structure also expands as the temperature of the RTD increases, and the construction allows for strain-free differential expansion.

The physical basis for the relationship between resistance and temperature is the temperature dependence of the resistivity, $\rho_e$, of a material. The resistance of a conductor of length $l$ and cross-sectional area $A_c$ may be expressed in terms of the resistivity $\rho_e$ as

$$R = \frac{\rho_e l}{A_c} \tag{8.2}$$

The relationship between the resistance of a metal conductor and its temperature may also be expressed by the polynomial expansion:

$$R = R_0[1 + \alpha(T - T_0) + \beta(T - T_0)^2 + \cdots] \tag{8.3}$$

[2]The term RTD in this context refers to metallic PTC resistance sensors.

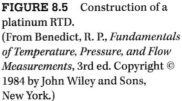

**FIGURE 8.5** Construction of a platinum RTD.
(From Benedict, R. P., *Fundamentals of Temperature, Pressure, and Flow Measurements*, 3rd ed. Copyright © 1984 by John Wiley and Sons, New York.)

**FIGURE 8.6** Relative resistance of three pure metals ($R_0$ at 0°C).

where $R_0$ is a reference resistance measured at a reference temperature $T_0$. The coefficients $\alpha$, $\beta$, ... are material constants. Figure 8.6 shows the relative relation between resistance and temperature for three common metals. This figure provides evidence that the relationship between temperature and resistance over specific small temperature ranges is linear. This approximation can be expressed as

$$R = R_0[1 + \alpha(T - T_0)] \tag{8.4}$$

where $\alpha$ is termed the temperature coefficient of resistivity. For example, for platinum conductors the linear approximation is accurate to within an uncertainty of ±0.3% over the range 0–200°C and ±1.2% over the range 200–800°C. Table 8.2 lists a number of temperature coefficients of resistivity $\alpha$ for materials at 20°C.

Table 8.2    **Temperature Coefficient of Resistivity for Selected Materials at 20°C**

| Substance | $\alpha$ [°C$^{-1}$] |
| --- | --- |
| Aluminum (Al) | 0.00429 |
| Carbon (C) | −0.0007 |
| Copper (Cu) | 0.0043 |
| Gold (Au) | 0.004 |
| Iron (Fe) | 0.00651 |
| Lead (Pb) | 0.0042 |
| Nickel (Ni) | 0.0067 |
| Nichrome | 0.00017 |
| Platinum (Pt) | 0.003927 |
| Tungsten (W) | 0.0048 |

### Platinum Resistance Temperature Device (RTD)

Platinum is the most common material chosen for the construction of RTDs. The principle of operation is quite simple: platinum exhibits a predictable and reproducible change in electrical resistance with temperature, which can be calibrated and interpolated to a high degree of accuracy. The linear approximation for the relationship between temperature and resistance is valid over a wide temperature range, and platinum is highly stable. To be suitable for use as a secondary temperature standard, a platinum resistance thermometer should have a value of $\alpha$ not less than 0.003925°C$^{-1}$. This minimum value is an indication of the purity of the platinum. In general, RTDs may be used for the measurement of temperatures ranging from cryogenic to approximately 650°C. By properly constructing an RTD and correctly measuring its resistance, an uncertainty in temperature measurement as low as ±0.005°C is possible. Because of this potential for low uncertainties and the predictable and stable behavior of platinum, the platinum RTD is widely used as a local standard

### Resistance Temperature Device Resistance Measurement

The resistance of an RTD may be measured by a number of means, and the choice of an appropriate resistance measuring device must be made based on the required level of uncertainty in the final temperature measurement. Conventional ohmmeters cause a small current to flow during resistance measurements, creating self-heating in the RTD. An appreciable temperature change of the sensor may be caused by this current, in effect causing a loading error. This is an important consideration for RTDs.

Bridge circuits, as described in Chapter 6, are used to measure the resistance of RTDs to provide low uncertainties in measured resistance values. Wheatstone bridge circuits are commonly used for these measurements. However, the basic Wheatstone bridge circuit does not compensate for the resistance of the leads in measuring the resistance of an RTD, which is a major source of error in electrical resistance thermometers. When greater accuracies are required, three-wire and four-wire bridge circuits are used. Figure 8.7a shows a three-wire Callender-Griffiths bridge circuit. The lead wires numbered 1, 2, and 3 have resistances $r_1$, $r_2$, and $r_3$, respectively. At balanced conditions neglecting lead wire effects,

$$\frac{R_1}{R_2} = \frac{R_3}{R_{\text{RTD}}} \tag{8.5}$$

**FIGURE 8.7** Bridge circuits. (a) Callender-Griffiths 3-wire bridge; (b) and (c) Mueller 4-wire bridge. An average of the readings in (b) and (c) eliminates the effect of lead wire resistances.

but with the lead wire resistances included in the circuit analysis,

$$\frac{R_1}{R_2} = \frac{R_3 + r_1}{R_{\text{RTD}} + r_3} \tag{8.6}$$

and with $R_1 = R_2$, the resistance of the RTD, $R_{\text{RTD}}$, can be found as

$$R_{\text{RTD}} = R_3 + r_1 - r_3 \tag{8.7}$$

If $r_1 = r_3$, the effect of these lead wires is eliminated from the determination of the RTD resistance by this bridge circuit. Note that the resistance of lead wire 2 does not contribute to any error in the measurement at balanced conditions, because no current flows through the galvanometer $G$.

The four-wire Mueller bridge, as shown in Figure 8.7b,c, provides increased compensation for lead-wire resistances compared to the Callender-Griffiths bridge and is used with four-wire RTDs. The four-wire Mueller bridge is typically used when low uncertainties are desired, as in cases where the RTD is used as a laboratory standard. A circuit analysis of the bridge circuit in the first measurement configuration, Figure 8.7b, yields

$$R_{\text{RTD}} + r_3 = R_3 + r_1 \tag{8.8}$$

and in the second measurement configuration, Figure 8.7c, yields

$$R_{\text{RTD}} + r_1 = R_3' + r_3 \tag{8.9}$$

where $R_3$ and $R_3'$ represent the indicated values of resistance in the first and second configurations, respectively. Adding Equations 8.8 and 8.9 results in an expression for the resistance of the RTD in terms of the indicated values for the two measurements:

$$R_{\text{RTD}} = \frac{R_3 + R_3'}{2} \tag{8.10}$$

With this approach, the effect of variations in lead wire resistances is essentially eliminated.

### EXAMPLE 8.1

An RTD forms one arm of an equal-arm Wheatstone bridge, as shown in Figure 8.8. The fixed resistances $R_2$ and $R_3$ are equal to 25 $\Omega$. The RTD has a resistance of 25 $\Omega$ at a temperature of 0°C and is used to measure a temperature that is steady in time.

**FIGURE 8.8**    RTD Wheatstone bridge arrangement.

The resistance of the RTD over a small temperature range may be expressed as in Equation 8.4:

$$R_{\text{RTD}} = R_0[1 + \alpha(T - T_0)]$$

Suppose the coefficient of resistance for this RTD is 0.003925°C$^{-1}$. A temperature measurement is made by placing the RTD in the measuring environment and balancing the bridge by adjusting $R_1$. The value of $R_1$ required to balance the bridge is 37.36 $\Omega$. Determine the temperature of the RTD.

**KNOWN** $R(0°C) = 25 \ \Omega$

$\qquad\qquad \alpha = 0.003925°C^{-1}$

$\qquad\qquad R_1 = 37.36 \ \Omega$ (when bridge is balanced)

**FIND**   The temperature of the RTD.

### SOLUTION
The resistance of the RTD is measured by balancing the bridge; recall that in a balanced condition

$$R_{\text{RTD}} = R_1 \frac{R_3}{R_2}$$

The resistance of the RTD is measured to be 37.36 $\Omega$. With $R_0 = 25 \ \Omega$ at $T = 0°C$ and $\alpha = 0.003925°C^{-1}$, Equation 8.4 becomes

$$37.36 \ \Omega = 25(1 + \alpha T) \ \Omega$$

The temperature of the RTD is 126°C.

## EXAMPLE 8.2

Consider the bridge circuit and RTD of Example 8.1. To select or design a bridge circuit for measuring the resistance of the RTD in this example, the required uncertainty in temperature would be specified. If the required uncertainty in the measured temperature is $\leq 0.5°C$, would a 1% total uncertainty in each of the resistors that make up the bridge be acceptable? Neglect the effects of lead wire resistances for this example.

**KNOWN** A required uncertainty in temperature of $\pm 0.5°C$, measured with the RTD and bridge circuit of Example 8.1.

**FIND** The uncertainty in the measured temperature for a 1% total uncertainty in each of the resistors that make up the bridge circuit.

**ASSUMPTION** All uncertainties are provided and evaluated at the 95% confidence level.

## SOLUTION

Perform a design-stage uncertainty analysis. Assuming at the design stage that the total uncertainty in the resistances is 1%, then with initial values of the resistances in the bridge equal to 25 $\Omega$, the design-stage uncertainties are set at

$$u_{R_1} = u_{R_2} = u_{R_3} = (0.01)(25) = 0.25\,\Omega$$

The root-sum-squares (RSS) method is used to estimate the propagation of uncertainty in each resistor to the uncertainty in determining the RTD resistance, $u_{RTD}$, by

$$u_{RTD} = \sqrt{\left[\frac{\partial R}{\partial R_1}u_{R_1}\right]^2 + \left[\frac{\partial R}{\partial R_2}u_{R_2}\right]^2 \left[\frac{\partial R}{\partial R_3}u_{R_3}\right]^2}$$

where

$$R = R_{RTD} = \frac{R_1 R_3}{R_2}$$

and we assume that the uncertainties are not correlated. Then the design-stage uncertainty in the resistance of the RTD is

$$u_{RTD} = \sqrt{\left[\frac{R_3}{R_2}u_{R_1}\right]^2 + \left[\frac{-R_1 R_3}{R_2^2}u_{R_2}\right]^2 + \left[\frac{R_1}{R_2}u_{R_3}\right]^2}$$

$$u_{RTD} = \sqrt{(1 \times 0.25)^2 + (1 \times -0.25)^2 + (1 \times 0.25)^2} = 0.433\ \Omega$$

To determine the uncertainty in temperature, we know

$$R = R_{RTD} = R_0\left[1 + \alpha(T - T_0)\right]$$

and

$$u_T = \sqrt{\left(\frac{\partial T}{\partial R}u_{RTD}\right)^2}$$

Setting $T_0 = 0°C$ with $R_0 = 25\ \Omega$, and neglecting uncertainties in $T_0$, $\alpha$, and $R_0$, we have

$$\frac{\partial T}{\partial R} = \frac{1}{\alpha R_0}$$

$$\frac{1}{\alpha R_0} = \frac{1}{(0.003925°C^{-1})(25\,\Omega)}$$

Then the design-stage uncertainty in temperature is

$$u_T = u_{RTD} \left( \frac{\partial T}{\partial R} \right) = \frac{0.433 \ \Omega}{0.098 \ \Omega/°C} = 4.4°C$$

The desired uncertainty in temperature is not achieved with the specified levels of uncertainty in the pertinent variables.

**COMMENT** Uncertainty analysis, in this case, would have prevented performing a measurement that would not provide acceptable results.

---

### EXAMPLE 8.3

Suppose the total uncertainty in the bridge resistances of Example 8.1 was reduced to 0.1%. Would the required level of uncertainty in temperature be achieved?

**KNOWN** The uncertainty in each of the resistors in the bridge circuit for temperature measurement from Example 8.1 is $\pm 0.1\%$.

**FIND** The resulting uncertainty in temperature.

**SOLUTION**
The uncertainty analysis from the previous example may be directly applied, with the uncertainty values for the resistances appropriately reduced. The uncertainties for the resistances are reduced from 0.25 to 0.025, yielding

$$u_{RTD} = \sqrt{(1 \times 0.025)^2 + (1 \times -0.025)^2 + (1 \times 0.025)^2} = 0.0433 \ \Omega$$

and the resulting 95% uncertainty interval in temperature is $\pm 0.44°C$, which satisfies the design constraint.

**COMMENT** This result provides confidence that the effects of the resistors' uncertainties will not cause the uncertainty in temperature to exceed the target value. However, the uncertainty in temperature also depends on other aspects of the measurement system. The design-stage uncertainty analysis performed in this example may be viewed as ensuring that the factors considered do not produce a higher-than-acceptable uncertainty level.

---

## Practical Considerations

The transient thermal response of typical commercial RTDs is generally quite slow compared with other temperature sensors, and for transient measurements bridge circuits must be operated in a deflection mode or use auto-balancing circuits. For these reasons, RTDs are not generally chosen for transient temperature measurements. A notable exception is the use of very small platinum wires and films for temperature measurements in noncorrosive flowing gases. In such an application, wires having diameters on the order of 10 μm can have frequency responses higher than any other temperature sensor because of their extremely low thermal capacitance. Obviously, the smallest impact would destroy this sensor. Other resistance sensors in the form of thin metallic films provide fast transient response temperature measurements, often in conjunction with anemometry or heat flux measurements. Such metallic films are constructed by depositing a film, commonly of platinum, onto a substrate and coating the film with a ceramic glass for mechanical protection [5]. Typical film thickness ranges from 1 to 2 μm, with a 10-μm protective coating. Continuous exposure at temperatures of 600°C is possible with this construction. Some practical uses for film sensors include temperature control circuits for heating systems and cooking devices

**FIGURE 8.9** Electronic Circuit for RTD resistance measurement. (Adapted from Texas Instruments SLAU 520A, 2013)

**FIGURE 8.10** Electronic circuit for RTD resistance measurement. (Adapted from Texas Instruments SLAU 520A, 2013)

and surface temperature monitoring on electronic components subject to overheating. Uncertainty levels range from about ±0.1 to 2°C.

An alternative to a true resistance bridge for measuring the resistance of an RTD is the electronic circuit shown in Figure 8.9. Here, a known current flows through the RTD and the voltage drop across the RTD is measured. The signal conditioning stage will normally include a gain, followed by A/D conversion, thus forming a measuring system. As with true bridge circuits, there is a current flow through the RTD, and some error is introduced by the resulting self-heating.

As discussed in association with bridge measurements, it is important to compensate for lead resistance in measuring the resistance of the RTD. Figure 8.10 illustrates a circuit employing two matched current sources and a reference resistor. If the current sources are identical,

$$R_{RTD} = V_{DIFF}/I$$

## Thermistors

Thermistors (from *thermally* sensitive *resistors*) are ceramic-like semiconductor devices. The most common thermistors are NTC, and the resistance of these thermistors decreases rapidly with temperature, in contrast to the small increases of resistance with temperature for RTDs.

An accurate functional relationship between resistance and temperature for a thermistor is generally assumed to be of the form

$$R = R_0 e^{\beta(1/T - 1/T_0)} \tag{8.11}$$

The parameter $\beta$ ranges from 3,500 to 4,600 K, depending on the material, temperature, and individual construction for each sensor, and thus must be determined for each thermistor. Figure 8.11 shows the variation of resistance with temperature for two common thermistor materials; the ordinate is the ratio of the resistance to the resistance at 25°C. Thermistors

**FIGURE 8.11**  Representative thermistor resistance variations with temperature.

exhibit large resistance changes with temperature in comparison to a typical RTD as indicated by comparison of Figures 8.6 and 8.11. Equation 8.11 is not accurate over a wide range of temperature unless $\beta$ is taken to be a function of temperature; typically, the value of $\beta$ specified by a manufacturer for a sensor is assumed to be constant over a limited temperature range. A simple calibration is possible for determining $\beta$ as a function of temperature, as illustrated in the circuits shown in Figure 8.12. Other circuits and a more complete discussion of measuring $\beta$ may be found in the Electronic Industries Association Standard Thermistor Definitions and Test Methods [6].

Thermistors are generally used when high sensitivity, ruggedness, or fast response times are required [7]. Thermistors are often encapsulated in glass and thus can be used in corrosive or abrasive environments. The resistance characteristics of the semiconductor material may change at elevated temperatures, and some aging of a thermistor occurs at temperatures above 200°C. The high resistance of a thermistor, compared to that of an RTD, eliminates the problems of lead wire resistance compensation.

A commonly reported specification of a thermistor is the zero-power resistance and dissipation constant. The zero-power resistance of a thermistor is the resistance value of the thermistor with

**FIGURE 8.12**  Circuits for determining $\beta$ for thermistors. (a) Voltage divider method: $R_T = R_1 (E_i / E_1 - 1)$. Note: Both $R_1$ and $E_i$ must be known values. The value of $R_1$ may be varied to achieve appropriate values of thermistor current. (b) Volt-ammeter method. Note: Both current and voltage are measured.

no flow of electric current. The zero power resistance should be measured such that a decrease in the current flow to the thermistor results in not more than a 0.1% change in resistance. The dissipation constant for a thermistor is defined at a given ambient temperature as

$$\delta = \frac{P}{T - T_\infty} \tag{8.12}$$

where

$\delta$ = dissipation constant

$P$ = power supplied to the thermistor

$T, T_\infty$ = thermistor and ambient temperatures, respectively

## EXAMPLE 8.4

The output of a thermistor is highly nonlinear with temperature, and there is often a benefit to linearizing the output through appropriate circuit, whether active or passive. In this example, we examine the output of an initially balanced bridge circuit in which one of the arms contains a thermistor. Consider a Wheatstone bridge as shown in Figure 8.8, but replace the RTD with a thermistor having a value of $R_0 = 10,000\ \Omega$ with $\beta = 3680$ K. Here, we examine the output of the circuit over two temperature ranges: (a) 25–325°C and (b) 25–75°C.

**KNOWN** A Wheatstone bridge where $R_2 = R_3 = R_4 = 10,000\ \Omega$ and where $R_1$ is a thermistor.

**FIND** The output of the bridge circuit as a function of temperature.

**SOLUTION**
The fundamental relationship between resistances in a Wheatstone bridge and the normalized output voltage is provided in Equation 6.11:

$$\frac{E_o}{E_i} = \left( \frac{R_1}{R_1 + R_2} - \frac{R_3}{R_3 + R_4} \right) \tag{6.11}$$

and the resistance of the thermistor is

$$R = R_o e^{\beta(1/T - 1/T_o)}$$

Substituting in Equation 6.11 for $R_1$ yields

$$\frac{E_o}{E_i} = \left( \frac{R_o e^{\beta(1/T - 1/T_o)}}{R_o e^{\beta(1/T - 1/T_o)} + R_2} - \frac{R_3}{R_3 + R_4} \right)$$

Figure 8.13a is a plot of this function over the range 25 to 325°C. Clearly, the sensitivity of the circuit to changes in temperature greatly diminishes as the temperature increases above 100°C, with an asymptotic value of 0.5. Figure 8.13b shows the behavior over the restricted range 25 to 75°C; a linear curve fit is also shown for comparison. Over this small range of temperature, the assumption that a linear relationship between normalized output and temperature would be suitable for many applications.

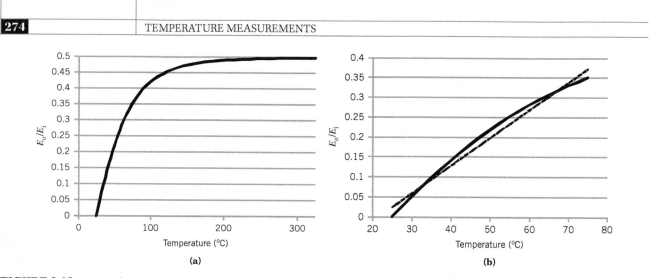

**FIGURE 8.13**    Normalized bridge output voltage as a function of temperature with a thermistor as the temperature sensor: (a) 25 to 325°C, (b) 25 to 75°C.

## EXAMPLE 8.5

The material constant $\beta$ is to be determined for a particular thermistor using the circuit shown in Figure 8.12a. The thermistor has a resistance of 60 k$\Omega$ at 25°C. The reference resistor in the circuit, $R_1$, has a resistance of 130.5 k$\Omega$. The dissipation constant $\delta$ is 0.09 mW/°C. The voltage source used for the measurement is constant at 1.564 V. The thermistor is to be used at temperatures ranging from 100 to 150°C. Determine the value of $\beta$.

**KNOWN** The temperature range of interest is from 100 to 150°C.

$$R_0 = 60,000 \ \Omega, \quad T_0 = 25°C$$
$$E_i = 1.564 \ \text{V}, \quad \delta = 0.09 \ \text{mW/°C}, \ R_1 = 130.5 \ \text{k}\Omega$$

**FIND** The value of $\beta$ over the temperature range from 100 to 150°C.

**SOLUTION**
The voltage drop across the fixed resistor is measured for three known values of thermistor temperature. The thermistor temperature is controlled and determined by placing the thermistor in a laboratory oven and measuring the temperature of the oven. For each measured voltage across the reference resistor, the thermistor resistance $R_T$ is determined from

$$R_T = R_1 \left( \frac{E_i}{E_1} - 1 \right)$$

The results of these measurements are as follows:

| Temperature (°C) | $E_1$ Voltage (V) | $R_T$ ($\Omega$) |
|---|---|---|
| 100 | 1.501 | 5,477.4 |
| 125 | 1.531 | 2,812.9 |
| 150 | 1.545 | 1,604.9 |

Equation 8.11 can be expressed in the form of a linear equation as

$$\ln \frac{R_T}{R_0} = \beta \left( \frac{1}{T} - \frac{1}{T_0} \right)$$

(8.13)

Applying this equation to the measured data, with $R_0 = 60{,}000\ \Omega$, the three data points above yield the following:

| $\ln(R_T/R_0)$ | $(1/T - 1/T_0)[\mathrm{K}^{-1}]$ | $\beta(\mathrm{K})$ |
|---|---|---|
| −2.394 | −6.75 × 10⁻⁴ | 3,546.7 |
| −3.060 | −8.43 × 10⁻⁴ | 3,629.9 |
| −3.621 | −9.92 × 10⁻⁴ | 3,650.2 |

**COMMENT** These results are for constant $\beta$ and are based on the behavior described by Equation 8.11 over the temperature range from $T_0$ to the temperature $T$. The significance of the measured differences in $\beta$ is examined further.

The measured values of $\beta$ in Example 8.4 are different at each value of temperature. If $\beta$ were truly a temperature-independent constant and these measurements had negligible uncertainty, then all three measurements would yield the same value of $\beta$. The variation in $\beta$ may result from a physical effect of temperature or may be attributable to the uncertainty in the measured values.

Are the measured differences significant? If so, what value of $\beta$ best represents the behavior of the thermistor over this temperature range? To perform the necessary uncertainty analysis, additional information must be provided concerning the instruments and procedures used in the measurement.

The original results of the measured values of $\beta$ must now be reexamined. The results from Table 8.3 are plotted as a function of temperature in Figure 8.14, with 95% uncertainty limits on

**Table 8.3   Uncertainties in $\beta$**

| T [°C] | Uncertainty | | |
|---|---|---|---|
| | Random $s_{\bar{\beta}}$ [K] | Systematic $b_\beta$ [K] | Total $\pm u_\beta$ 95%[K] |
| 100 | 13.3 | 37.4 | 79.4 |
| 125 | 10.6 | 53.9 | 109.9 |
| 150 | 8.8 | 78.4 | 157.8 |

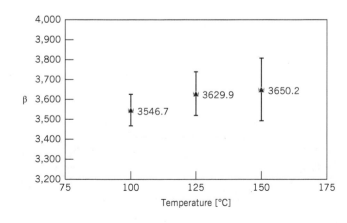

**FIGURE 8.14**   Measured values of $\beta$ and associated uncertainties for three temperatures. Each data point represents $\bar{\beta} \pm u_\beta$.

each data point. Clearly, there is no justification for assuming that the measured values indicate a trend of changes with temperature, and it would be appropriate to use either the average value of $\beta$ or the value determined from the linear least-squares curve fit.

## 8.5 Thermoelectric Temperature Measurement

A common method of measuring and controlling temperature uses an electrical circuit called a thermocouple. A *thermocouple* consists of two electrical conductors that are made of dissimilar metals and have at least one electrical connection. This electrical connection is referred to as a *junction*. A thermocouple junction may be created by welding, by soldering, or by any method that provides good electrical contact between the two conductors, such as twisting the wires around one another. The output of a thermocouple circuit is a voltage, and there is a definite relationship between this voltage and the temperatures of the junctions that make up the thermocouple circuit. We will examine the causes of this voltage and develop the basis for using thermocouples to make engineering measurements of temperature.

Consider the thermocouple circuit shown in Figure 8.15. The junction labeled 1 is at a temperature $T_1$ and the junction labeled 2 is at a temperature $T_2$. This thermocouple circuit measures the difference between $T_1$ and $T_2$. If $T_1$ and $T_2$ are not equal, a finite open-circuit voltage, $V_o$, is measured. The magnitude of the voltage depends on the difference in the temperatures and the particular metals used in the thermocouple circuit.

A thermocouple junction is the source of a voltage that gives rise to the total voltage difference in a thermocouple circuit. It is the basis for temperature measurement using thermocouples. The circuit shown in Figure 8.15 is the most common form of a thermocouple circuit used for measuring temperature.

It is our goal to understand the origin of thermoelectric phenomena and the requirements for providing accurate temperature measurements using thermocouples. In an electrical conductor that is subject to a temperature gradient, there will be both a flow of thermal energy and a flow of electricity. These phenomena are closely tied to the behavior of the free electrons in a metal; it is no coincidence that good electrical conductors are, in general, good thermal conductors. The characteristic behavior of these free electrons in an electrical circuit composed of dissimilar metals results in a useful relationship between temperature and thermocouple voltage. There are three basic phenomena that can occur in a thermocouple circuit: (1) the *Seebeck effect*, (2) the *Peltier effect*, and (3) the *Thomson effect*.

The open-circuit voltage, $V_o$, generated by a thermocouple circuit is the result of the Seebeck effect only. Under measurement conditions with no loading errors, the measured voltage would equal the open-circuit voltage.

### Seebeck Effect

The Seebeck effect, named for Thomas Johann Seebeck (1770–1831), refers to the generation of a voltage difference in an open thermocouple circuit due to a difference in temperature between junctions in the circuit. The Seebeck effect refers to the case when there is no current flow in the

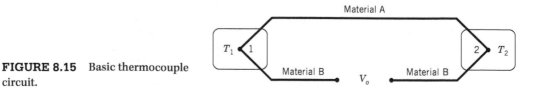

**FIGURE 8.15**  Basic thermocouple circuit.

**FIGURE 8.16**  Peltier effect due to current flow across a junction of dissimilar metals.

circuit. There is a fixed, reproducible relationship between the voltage and the junction temperatures $T_1$ and $T_2$ (Figure 8.13). This relationship is expressed by the Seebeck coefficient, $\alpha_{AB}$, defined as

$$\alpha_{AB} = \left[ \frac{\partial(V)}{\partial T} \right]_{\text{open circuit}} \tag{8.14}$$

where $A$ and $B$ refer to the two materials that comprise the thermocouple. Since the Seebeck coefficient specifies the rate of change of voltage with temperature for the materials $A$ and $B$, it is equal to the static sensitivity of the open-circuit thermocouple.

## Peltier Effect

A familiar concept is that of $I^2R$ or joule heating in a conductor through which an electrical current flows. Consider the two conductors having a common junction, shown in Figure 8.16, through which an electrical current $I$ flows due to an externally applied voltage. For any portion of either of the conductors, the energy removal rate required to maintain a constant temperature is $I^2R$, where $R$ is the resistance to a current flow and is determined by the resistivity and size of the conductor. However, at the junction of the two dissimilar metals, the removal of a quantity of energy different than $I^2R$ is required to maintain a constant temperature. The difference in $I^2R$ and the amount of energy generated by the current flowing through the junction is due to the Peltier effect. The Peltier effect is a thermodynamically reversible conversion of energy as current flows across the junction, in contrast to the irreversible dissipation of energy associated with $I^2R$ losses. The Peltier heat is the quantity of heat in addition to the quantity $I^2R$ that must be removed from the junction to maintain the junction at a constant temperature. This amount of energy is proportional to the current flowing through the junction; the proportionality constant is the Peltier coefficient $\pi_{AB}$, and the heat transfer required to maintain a constant temperature is

$$Q_\pi = \pi_{AB} I \tag{8.15}$$

caused by the Peltier effect alone. This behavior was discovered by Jean Charles Athanase Peltier (1785–1845) during experiments using Seebeck's thermocouple. He observed that passing a current through a thermocouple circuit having two junctions, as in Figure 8.15, raised the temperature at one junction while lowering the temperature at the other junction. This effect forms the basis of a device known as a Peltier refrigerator, which provides cooling without moving parts. An assessment of the prospects of thermoelectric cooling technologies is provided in [8].

## Thomson Effect

In addition to the Seebeck effect and the Peltier effect, there is a third phenomenon that occurs in thermoelectric circuits. Consider the conductor shown in Figure 8.17, which is subject to a longitudinal temperature gradient and also to a voltage difference, such that there is a flow of

**FIGURE 8.17**    Thomson effect due to simultaneous flows of current and heat.

$q_1$ Energy flow as a result of a temperature gradient
$q_2$ Heat transfer to maintain constant temperature

current and heat in the conductor. Again, to maintain a constant temperature in the conductor, it is found that a quantity of energy different than the joule heat, $I^2R$, must be removed from the conductor. First noted by William Thomson (1824–1907, known as Lord Kelvin from 1892) in 1851, this energy is expressed in terms of the Thomson coefficient, $\sigma$, as

$$Q_\sigma = \sigma I(T_1 - T_2) \tag{8.16}$$

For a thermocouple circuit, all three of these effects may be present and may contribute to the overall voltage difference in the circuit.

## Fundamental Thermocouple Laws

The basic thermocouple circuit shown in Figure 8.18 can be used to measure the difference between the two temperatures $T_1$ and $T_2$. For practical temperature measurements, one of these junctions becomes a reference junction and is maintained at some known, constant reference temperature—say, $T_2$. The other junction then becomes the measuring junction, and the voltage existing in the circuit provides a direct indication of the temperature of the measuring junction $T_1$.

The use of thermocouple circuits to measure temperature is based on observed behaviors of carefully controlled thermocouple materials and circuits. The following laws provide the basis necessary for temperature measurement with thermocouples:

1. **Law of homogeneous materials:**    *A thermoelectric current cannot be sustained in a circuit of a single homogeneous material by the application of heat alone, regardless of how it might vary in cross section.* Simply stated, this law requires that at least two materials be used to construct a thermocouple circuit for measuring temperature. It is interesting to note that a current may occur in an inhomogeneous wire that is nonuniformly heated, but this is neither useful nor desirable in a thermocouple.

2. **Law of intermediate materials:**    *The algebraic sum of the voltage differences in a circuit composed of any number of dissimilar materials is zero if all of the circuit is at a uniform*

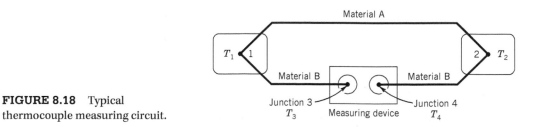

**FIGURE 8.18**    Typical thermocouple measuring circuit.

*temperature*. This law allows a material other than the thermocouple materials to be inserted into a thermocouple circuit without changing the output voltage of the circuit. As an example, consider the thermocouple circuit shown in Figure 8.18, where the junctions of the measuring device are made of copper and material B is an alloy (not pure copper). The electrical connection between the measuring device and the thermocouple circuit forms yet another thermocouple junction. The law of intermediate materials, in this case, provides that the measured voltage will be unchanged from the open-circuit voltage, which corresponds to the temperature difference between $T_1$ and $T_2$, if $T_3 = T_4$. Another practical consequence of this law is that copper extension wires may be used to transmit thermocouple voltages to a measuring device.

3. ***Law of successive or intermediate temperatures:*** *If two dissimilar homogeneous materials that form a thermocouple circuit produce $V_1$ when the junctions are at $T_1$ and $T_2$ and produce $V_2$ when the junctions are at $T_2$ and $T_3$, the voltage generated when the junctions are at $T_1$ and $T_3$ will be $V_1 + V_2$.* This law allows a thermocouple calibrated for one reference temperature, say, $T_2$, to be used at another reference temperature, such as $T_3$, to determine temperature $T_1$.

## Basic Temperature Measurement with Thermocouples

Let's first examine a historically significant method of using a thermocouple circuit to measure temperature. Figure 8.19 shows two basic thermocouple circuits using a chromel–constantan thermocouple and a reference temperature. In Figure 8.19a, the thermocouple wires are connected directly to a voltmeter[3] to measure the voltage. In Figure 8.19b, copper extension wires are employed, creating two reference junctions. The law of intermediate materials ensures that neither the voltmeter nor the extension wires will change the output voltage of the circuit so long

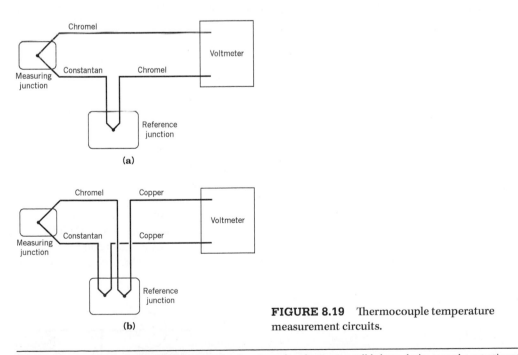

**(a)**

**(b)**

**FIGURE 8.19** Thermocouple temperature measurement circuits.

---

[3]Historically, the most accurate method of measuring thermocouple voltage was a null-balance device termed a potentiometer. However, modern nano-voltmeters have replaced potentiometers for accurate voltage measurement with essentially zero loading error.

as the two connecting junctions at the voltmeter and the two at the reference temperature experience no temperature difference. All that is required to be able to measure temperature with this circuit is to know the relationship between the output voltage and the temperature of the measuring junction, for the particular reference temperature. One method of determining this relationship is to calibrate the thermocouple. However, we shall see that for reasonable levels of uncertainty for temperature measurement, standard materials and procedures allow thermocouples to be accurate temperature-measuring devices without the necessity of calibration.

### Reference Junction

The provisions for a reference junction should provide a temperature that is accurately known, stable, and reproducible. A reference junction temperature could be provided by the ice point, 0°C, because of the relative ease with which it could historically be obtained. An ice bath can provide the reference junction temperature. An ice bath is typically made by filling a vacuum flask, or Dewar, with finely crushed ice, and adding just enough water to create a transparent slush. Ice baths can be constructed to provide a reference junction temperature to an uncertainty within ±0.01°C.

Modern reference junction temperatures are created electronically. Termed *electronic reference junction compensation*, these dedicated circuits provide a convenient means of the measurement of temperature without the necessity to construct an ice bath. Numerous manufacturers produce commercial temperature measuring devices with built-in reference junction compensation, and many digital data acquisition cards include reference junction compensation. The electronics generally rely on a thermistor, a temperature-sensitive integrated circuit, or an RTD to determine the reference junction temperature, as shown for a thermistor in Figure 8.20. Uncertainties for the reference junction temperature in this case are on the order of ±0.1°C, with ±0.5°C as typical. In addition, data acquisition systems (DASs) may include a microprocessor that converts compensated voltage readings directly into temperature output based on the standard thermocouple voltages. Combining a thermocouple, reference junction compensation, and a microprocessor results in a device termed a "smart sensor."

## Thermocouple Standards

The National Institute of Standards and Technology (NIST) provides specifications for the materials and construction of standard thermocouple circuits for temperature measurement [9]. Many material combinations exist for thermocouples; these material combinations are identified by a thermocouple type and denoted by a letter. Table 8.4 shows the letter designations and the polarity of common thermocouples, along with some basic application information for each type. The choice of a type of thermocouple depends on the temperature range to be measured, the particular application, and the desired uncertainty level.

To determine the voltage output of a particular material combination, a thermocouple is formed from a candidate material and a standard platinum alloy to form a thermocouple circuit having a 0°C reference temperature. Figure 8.21 shows the output of various materials in combination with platinum-67. The notation indicates the thermocouple type. The law of intermediate temperatures

**FIGURE 8.20**   Basic thermistor circuit for thermocouple reference junction compensation.

Table 8.4    **Thermocouple Designations**

|      | Material combination | | Applications |
|------|------------------------|-----------------|--------------|
| **Type** | **Positive** | **Negative** | **Applications** |
| E | Chromel(+) | Constantan(−) | Highest sensitivity (<1,000°C) |
| J | Iron(+) | Constantan(−) | Nonoxidizing environment (<760°C) |
| K | Chromel(+) | Alumel(−) | High temperature (<1,372°C) |
| S | Platinum/10% rhodium | Platinum(−) | Long-term stability high temperature (<1,768°C) |
| T | Copper(+) | Constantan(−) | Reducing or vacuum environments (<400°C) |

**FIGURE 8.21**   Thermal voltage of thermocouple materials relative to platinum-67. Note: For example, JP indicates the positive leg of a J thermocouple, or iron. JN indicates the negative leg of a J thermocouple, or constantan. All other notations are similar for each type of thermocouple. (Adapted from Benedict, R. P., *Fundamentals of Temperature, Pressure, and Flow Measurements*, 3rd ed. Copyright © 1984 by John Wiley and Sons, New York. Reprinted by permission.)

**FIGURE 8.22** Thermocouple voltage output as a function of temperature for some common thermocouple materials. Reference junction is at 32 °F (0 °C).
(Adapted from Benedict, R. P., *Fundamentals of Temperature, Pressure and Flow Measurements*, 3rd ed. Copyright © 1984 by John Wiley and Sons, New York. Reprinted by permission.)

then allows the voltage of any two materials, whose voltage relative to platinum is known, to be determined. Figure 8.22 shows a plot of the voltage as a function of temperature for some common thermocouple material combinations. The slope of the curves in this figure corresponds to the static sensitivity of the thermocouple measuring circuit.

### Standard Thermocouple Voltage

Table 8.5 provides the standard composition of thermocouple materials, along with standard limits of error for the various material combinations. These limits specify the expected maximum errors resulting from the thermocouple materials. The NIST uses high-purity materials to establish the standard value of voltage output for a thermocouple composed of two specific materials. This results in standard tables or equations used to determine a measured temperature from a measured output voltage [9]. An example of such a table is provided in Table 8.6 for an iron/constantan thermocouple, usually referred to as a J-type thermocouple. Table 8.7 provides the corresponding polynomial equations that relate output voltage and temperature for standard thermocouples.

Because of the widespread need to measure temperature, an industry has grown up to supply high-grade thermocouple wire. Manufacturers can also provide thermocouples having special tolerance limits relative to the NIST standard voltages with uncertainties in temperature ranging

## Table 8.5    Standard Thermocouple Compositions[a]

| Type | Wire Positive | Negative | Expected systematic uncertainty[b] |
|---|---|---|---|
| S | Platinum | Platinum/10% rhodium | ±1.5°C or 0.25% |
| R | Platinum | Platinum/13% rhodium | ±1.5°C |
| B | Platinum/30% rhodium | Platinum/6% rhodium | ±0.5% |
| T | Copper | Constantan | ±1.0°C or 0.75% |
| J | Iron | Constantan | ±2.2°C or 0.75% |
| K | Chromel | Alumel | ±2.2°C or 0.75% |
| E | Chromel | Constantan | ±1.7°C or 0.5% |

*Alloy Designations*

Constantan: 55% copper with 45% nickel

Chromel: 90% nickel with 10% chromium

Alumel: 94% nickel with 3% manganese, 2% aluminum, and 1% silicon

[a]From Temperature Measurements ANSI PTC 19.3 1974.
[b]Use greater value; these limits of error do not include installation errors.

## Table 8.6    Thermocouple Reference Table for Type-J Thermocouple[a]

| | Temperature (°C) | | | | | | | Thermocouple voltage (mV) | | |
|---|---|---|---|---|---|---|---|---|---|---|
| | 0 | −1 | −2 | −3 | −4 | −5 | −6 | −7 | −8 | −9 |
| −210 | −8.095 | | | | | | | | | |
| −200 | −7.890 | −7.912 | −7.934 | −7.955 | 7.976 | −7.996 | −8.017 | −8.037 | −8.057 | −8.076 |
| −190 | −7.659 | −7.683 | −7.707 | −7.731 | −7.755 | −7.778 | −7.801 | −7.824 | −7.846 | −7.868 |
| −180 | −7.403 | −7.429 | 7.456 | −7.482 | −7.508 | −7.534 | −7.559 | −7.585 | −7.610 | −7.634 |
| −170 | −7.123 | −7.152 | −7.181 | −7.209 | −7.237 | −7.265 | −7.293 | −7.321 | −7.348 | −7.376 |
| −160 | −6.821 | −6.853 | −6.883 | −6.914 | −6.944 | −6.975 | −7.005 | −7.035 | −7.064 | −7.094 |
| −150 | −6.500 | −6.533 | −6.566 | −6.598 | −6.631 | −6.663 | −6.695 | −6.727 | −6.759 | −6.790 |
| −140 | −6.159 | −6.194 | 6.229 | −6.263 | −6.298 | −6.332 | −6.366 | −6.400 | −6.433 | −6.467 |
| −130 | −5.801 | −5.838 | −5.874 | −5.910 | −5.946 | −5.982 | −6.018 | −6.054 | −6.089 | −6.124 |
| −120 | −5.426 | −5.465 | −5.503 | −5.541 | −5.578 | −5.616 | −5.653 | −5.690 | −5.727 | −5.764 |
| −110 | −5.037 | −5.076 | −5.116 | −5.155 | −5.194 | −5.233 | −5.272 | −5.311 | −5.350 | −5.388 |
| −100 | −4.633 | −4.674 | −4.714 | −4.755 | −4.796 | −4.836 | −4.877 | −4.917 | −4.957 | −4.997 |
| −90 | −4.215 | −4.257 | −4.300 | −4.342 | −4.384 | −4.425 | −4.467 | −4.509 | −4.550 | −4.591 |
| −80 | −3.786 | −3.829 | −3.872 | −3.916 | −3.959 | −4.002 | −4.045 | −4.088 | −4.130 | −4.173 |
| −70 | −3.344 | −3.389 | −3.434 | −3.478 | −3.522 | −3.566 | −3.610 | −3.654 | −3.698 | −3.742 |
| −60 | −2.893 | −2.938 | −2.984 | −3.029 | −3.075 | −3.120 | −3.165 | −3.210 | −3.255 | −3.300 |
| −50 | −2.431 | −2.478 | −2.524 | −2.571 | −2.617 | −2.663 | −2.709 | −2.755 | −2.801 | −2.847 |
| −40 | −1.961 | −2.008 | −2.055 | −2.103 | −2.150 | −2.197 | −2.244 | −2.291 | −2.338 | −2.385 |
| −30 | −1.482 | −1.530 | −1.578 | −1.626 | −1.674 | −1.722 | −1.770 | −1.818 | −1.865 | −1.913 |
| −20 | −0.995 | −1.044 | −1.093 | −1.142 | −1.190 | −1.239 | −1.288 | −1.336 | −1.385 | −1.433 |
| −10 | −0.501 | −0.550 | −0.600 | −0.650 | −0.699 | −0.749 | −0.798 | −0.847 | −0.896 | −0.946 |
| 0 | 0.000 | −0.050 | −0.101 | −0.151 | −0.201 | −0.251 | −0.301 | −0.351 | −0.401 | −0.451 |

*(continued)*

## Table 8.6    (Continued)

| Temperature (°C) | | | | | | | | Thermocouple voltage (mV) | | |
|---|---|---|---|---|---|---|---|---|---|---|
| | **0** | **+1** | **+2** | **+3** | **+4** | **+5** | **+6** | **+7** | **+8** | **+9** |
| 0 | 0.000 | 0.050 | 0.101 | 0.151 | 0.202 | 0.253 | 0.303 | 0.354 | 0.405 | 0.451 |
| 10 | 0.507 | 0.558 | 0.609 | 0.660 | 0.711 | 0.762 | 0.814 | 0.865 | 0.916 | 0.968 |
| 20 | 1.019 | 1.071 | 1.122 | 1.174 | 1.226 | 1.277 | 1.329 | 1.381 | 1.433 | 1.485 |
| 30 | 1.537 | 1.589 | 1.641 | 1.693 | 1.745 | 1.797 | 1.849 | 1.902 | 1.954 | 2.006 |
| 40 | 2.059 | 2.111 | 2.164 | 2.216 | 2.269 | 2.322 | 2.374 | 2.427 | 2.480 | 2.532 |
| 50 | 2.585 | 2.638 | 2.691 | 2.744 | 2.797 | 2.850 | 2.903 | 2.956 | 3.009 | 3.062 |
| 60 | 3.116 | 3.169 | 3.222 | 3.275 | 3.329 | 3.382 | 3.436 | 3.489 | 3.543 | 3.596 |
| 70 | 3.650 | 3.703 | 3.757 | 3.810 | 3.864 | 3.918 | 3.971 | 4.025 | 4.079 | 4.133 |
| 80 | 4.187 | 4.240 | 4.294 | 4.348 | 4.402 | 4.456 | 4.510 | 4.564 | 4.618 | 4.672 |
| 90 | 4.726 | 4.781 | 4.835 | 4.889 | 4.943 | 4.997 | 5.052 | 5.106 | 5.160 | 5.215 |
| 100 | 5.269 | 5.323 | 5.378 | 5.432 | 5.487 | 5.541 | 5.595 | 5.650 | 5.705 | 5.759 |
| 110 | 5.814 | 5.868 | 5.923 | 5.977 | 6.032 | 6.087 | 6.141 | 6.196 | 6.251 | 6.306 |
| 120 | 6.360 | 6.415 | 6.470 | 6.525 | 6.579 | 6.634 | 6.689 | 6.744 | 6.799 | 6.854 |
| 130 | 6.909 | 6.964 | 7.019 | 7.074 | 7.129 | 7.184 | 7.239 | 7.294 | 7.349 | 7.404 |
| 140 | 7.459 | 7.514 | 7.569 | 7.624 | 7.679 | 7.734 | 7.789 | 7.844 | 7.900 | 7.955 |
| 150 | 8.010 | 8.065 | 8.120 | 8.175 | 8.231 | 8.286 | 8.341 | 8.396 | 8.452 | 8.507 |
| 160 | 8.562 | 8.618 | 8.673 | 8.728 | 8.783 | 8.839 | 8.894 | 8.949 | 9.005 | 9.060 |
| 170 | 9.115 | 9.171 | 9.226 | 9.282 | 9.337 | 9.392 | 9.448 | 9.503 | 9.559 | 9.614 |
| 180 | 9.669 | 9.725 | 9.780 | 9.836 | 9.891 | 9.947 | 10.002 | 10.057 | 10.113 | 10.168 |
| 190 | 10.224 | 10.279 | 10.335 | 10.390 | 10.446 | 10.501 | 10.557 | 10.612 | 10.668 | 10.723 |
| 200 | 10.779 | 10.834 | 10.890 | 10.945 | 11.001 | 11.056 | 11.112 | 11.167 | 11.223 | 11.278 |
| 210 | 11.334 | 11.389 | 11.445 | 11.501 | 11.556 | 11.612 | 11.667 | 11.723 | 11.778 | 11.8.34 |
| 220 | 11.889 | 11.945 | 12.000 | 12.056 | 12.111 | 12.167 | 12.222 | 12.278 | 12.334 | 12.389 |
| 230 | 12.445 | 12.500 | 12.556 | 12.611 | 12.667 | 12.722 | 12.778 | 12.833 | 12.889 | 12.944 |
| 240 | 13.000 | 13.056 | 13.111 | 13.167 | 13.222 | 13.278 | 13.333 | 13.389 | 13.444 | 13.500 |
| 250 | 13.555 | 13.611 | 13.666 | 13.722 | 13.777 | 13.833 | 13.888 | 13.944 | 13.999 | 14.055 |
| 260 | 14.110 | 14.166 | 14.221 | 14.277 | 14.332 | 14.388 | 14.443 | 14.499 | 14.554 | 14.609 |
| 270 | 14.665 | 14.720 | 14.776 | 14.831 | 14.887 | 14.942 | 14.998 | 15.053 | 15.109 | 15.164 |
| 280 | 15.219 | 15.275 | 15.330 | 15.386 | 15.441 | 15.496 | 15.552 | 15.607 | 15.663 | 15.718 |
| 290 | 15.773 | 15.829 | 15.884 | 15.940 | 15.995 | 16.050 | 16.106 | 16.161 | 16.216 | 16.272 |
| 300 | 16.327 | 16.383 | 16.438 | 16.493 | 16.549 | 16.604 | 16.659 | 16.715 | 16.770 | 16.825 |
| 310 | 16.881 | 16.936 | 16.991 | 17.046 | 17.102 | 17.157 | 17.212 | 17.268 | 17.323 | 17.378 |
| 320 | 17.434 | 17.489 | 17.544 | 17.599 | 17.655 | 17.710 | 17.765 | 17.820 | 17.876 | 17.931 |
| 330 | 17.986 | 18.041 | 18.097 | 18.152 | 18.207 | 18.262 | 18.318 | 18.373 | 18.428 | 18.483 |
| 340 | 18.538 | 18.594 | 18.649 | 18.704 | 18.759 | 18.814 | 18.870 | 18.925 | 18.980 | 19.035 |
| 350 | 19.090 | 19.146 | 19.201 | 19.256 | 19.311 | 19.366 | 19.422 | 19.477 | 19.532 | 19.587 |
| 360 | 19.642 | 19.697 | 19.753 | 19.808 | 19.863 | 19.918 | 19.973 | 20.028 | 20.083 | 20.139 |
| 370 | 20.194 | 20.249 | 20.304 | 20.359 | 20.414 | 20.469 | 20.525 | 20.580 | 20.635 | 20.690 |
| 380 | 20.745 | 20.800 | 20.855 | 20.911 | 20.966 | 21.021 | 21.076 | 21.131 | 21.186 | 21.241 |
| 390 | 21.297 | 21.352 | 21.407 | 21.462 | 21.517 | 21.572 | 21.627 | 21.683 | 21.738 | 21.793 |

Table 8.6    (Continued)

| | Temperature (°C) | | | | | | | Thermocouple voltage (mV) | | |
|---|---|---|---|---|---|---|---|---|---|---|
| | **0** | **+1** | **+2** | **+3** | **+4** | **+5** | **+6** | **+7** | **+8** | **+9** |
| 400 | 21.848 | 21.903 | 21.958 | 22.014 | 22.069 | 22.124 | 22.179 | 22.234 | 22.289 | 22.345 |
| 410 | 22.400 | 22.455 | 22.510 | 22.565 | 22.620 | 22.676 | 22.731 | 22.786 | 22.841 | 22.896 |
| 420 | 22.952 | 23.007 | 23.062 | 23.117 | 23.172 | 23.228 | 23.283 | 23.338 | 23.393 | 23.449 |
| 430 | 23.504 | 23.559 | 23.614 | 23.670 | 23.725 | 23.780 | 23.835 | 23.891 | 23.946 | 24.001 |
| 440 | 24.057 | 24.112 | 24.167 | 24.223 | 24.278 | 24.333 | 24.389 | 24.444 | 24.499 | 24.555 |
| 450 | 24.610 | 24.665 | 24.721 | 24.776 | 24.832 | 24.887 | 24.943 | 24.998 | 25.053 | 25.109 |
| 460 | 25.164 | 25.220 | 25.275 | 25.331 | 25.386 | 25.442 | 25.497 | 25.553 | 25.608 | 25.664 |
| 470 | 25.720 | 25.775 | 25.831 | 25.886 | 25.942 | 25.998 | 26.053 | 26.109 | 26.165 | 26.220 |
| 480 | 26.276 | 26.332 | 26.387 | 26.443 | 26.499 | 26.555 | 26.610 | 26.666 | 26.722 | 26.778 |
| 490 | 26.834 | 26.889 | 26.945 | 27.001 | 27.057 | 27.113 | 27.169 | 27.225 | 27.281 | 27.337 |
| 500 | 27.393 | 27.449 | 27.505 | 27.561 | 27.617 | 27.673 | 27.729 | 27.785 | 27.841 | 27.897 |
| 510 | 27.953 | 28.010 | 28.066 | 28.122 | 28.178 | 28.234 | 28.291 | 28.347 | 28.403 | 28.460 |
| 520 | 28.516 | 28.572 | 28.629 | 28.685 | 28.741 | 28.798 | 28.854 | 28.911 | 28.967 | 29.024 |
| 530 | 29.080 | 29.137 | 29.194 | 29.250 | 29.307 | 29.363 | 29.420 | 29.477 | 29.534 | 29.590 |
| 540 | 29.647 | 29.704 | 29.761 | 29.818 | 29.874 | 29.931 | 29.988 | 30.045 | 30.102 | 30.159 |
| 550 | 30.216 | 30.273 | 30.330 | 30.387 | 30.444 | 30.502 | 30.559 | 30.616 | 30.673 | 30.730 |
| 560 | 30.788 | 30.845 | 30.902 | 30.960 | 31.017 | 31.074 | 31.132 | 31.189 | 31.247 | 31.304 |
| 570 | 31.362 | 31.419 | 31.477 | 31.535 | 31.592 | 31.650 | 31.708 | 31.766 | 31.823 | 31.881 |
| 580 | 31.939 | 31.997 | 32.055 | 32.113 | 32.171 | 32.229 | 32.287 | 32.345 | 32.403 | 32.461 |
| 590 | 32.519 | 32.577 | 32.636 | 32.694 | 32.752 | 32.810 | 32.869 | 32.927 | 32.985 | 33.044 |
| 600 | 33.102 | 33.161 | 33.219 | 33.278 | 33.337 | 33.395 | 33.454 | 33.513 | 33.571 | 33.630 |
| 610 | 33.689 | 33.748 | 33.807 | 33.866 | 33.925 | 33.984 | 34.043 | 34.102 | 34.161 | 34.220 |
| 620 | 34.279 | 34.338 | 34.397 | 34.457 | 34.516 | 34.575 | 34.635 | 34.694 | 34.754 | 34.813 |
| 630 | 34.873 | 34.932 | 34.992 | 35.051 | 35.111 | 35.171 | 35.2.30 | 35.290 | 35.350 | 35.410 |
| 640 | 35.470 | 35.530 | 35.590 | 35.650 | 35.710 | 35.770 | 35.830 | 35.890 | 35.950 | 36.010 |
| 650 | 36.071 | 36.131 | 36.191 | 36.191 | 36.252 | 36.373 | 36.433 | 36.494 | 36.554 | 36.615 |
| 660 | 36.675 | 36.736 | 36.797 | 36.858 | 36.918 | 36.979 | 37.040 | 37.101 | 37.162 | 37.22.3 |
| 670 | 37.284 | 37.345 | 37.406 | 37.467 | 37.528 | 37.590 | 37.651 | 37.712 | 37.773 | 37.835 |
| 680 | 37.896 | 37.958 | 38.019 | 38.081 | 38.142 | 38.204 | 38.265 | 38.327 | 38.389 | 38.450 |
| 690 | 38.512 | 38.574 | 38.636 | 38.698 | 38.760 | 38.822 | 38.884 | 38.946 | 39.008 | 39.070 |
| 700 | 39.132 | 39.194 | 39.256 | 39.318 | 39.381 | 39.443 | 39.505 | 39.568 | 39.630 | 39.693 |
| 710 | 39.755 | 39.818 | 39.880 | 39.943 | 40.005 | 40.068 | 40.131 | 40.193 | 40.256 | 40.319 |
| 720 | 40.382 | 40.445 | 40.508 | 40.570 | 40.633 | 40.696 | 40.759 | 40.822 | 40.886 | 40.949 |
| 730 | 41.012 | 41.075 | 41.138 | 41.201 | 41.265 | 41.328 | 41.391 | 41.455 | 41.518 | 41.581 |
| 740 | 41.645 | 41.708 | 41.772 | 41.835 | 41.899 | 41.962 | 42.026 | 42.090 | 42.153 | 42.217 |
| 750 | 42.281 | 42.344 | 42.408 | 42.472 | 42.536 | 42.599 | 42.663 | 42.727 | 42.791 | 42.855 |
| 760 | 42.919 | 42.983 | 43.047 | 43.110 | 43.174 | 43.238 | 43.303 | 43.367 | 43.431 | 43.495 |

[a]Reference junction at 0°C.

**Table 8.7**    **Reference Functions for Selected Letter Designated Thermocouples**

The relationship between voltage and temperature is provided in the form of a polynomial in temperature [9]

$$E = \sum_{i=0}^{n} c_i T^i$$

where $E$ is in $\mu V$ and $T$ is in °C. Constants are provided below.

| Thermocouple type | Temperature range | Constants |
|---|---|---|
| J-type | −210 to 760°C | $c_0 = 0.000\ 000\ 000\ 0$ |
| | | $c_1 = 5.038\ 118\ 7815 \times 10^1$ |
| | | $c_2 = 3.047\ 583\ 693\ 0 \times 10^{-2}$ |
| | | $c_3 = -8.568\ 106\ 572\ 0 \times 10^{-5}$ |
| | | $c_4 = 1.322\ 819\ 529\ 5 \times 10^{-7}$ |
| | | $c_5 = -1.705\ 295\ 833\ 7 \times 10^{-10}$ |
| | | $c_6 = 2.094\ 809\ 069\ 7 \times 10^{-13}$ |
| | | $c_7 = -1.253\ 839\ 533\ 6 \times 10^{-16}$ |
| | | $c_8 = 1.563\ 172\ 569\ 7 \times 10^{-20}$ |
| T-type | −270 to 0°C | $c_0 = 0.000\ 000\ 000\ 0$ |
| | | $c_1 = 3.874\ 810\ 6364 \times 10^1$ |
| | | $c_2 = 4.419\ 443\ 434\ 7 \times 10^{-2}$ |
| | | $c_3 = 1.184\ 432\ 310\ 5 \times 10^{-4}$ |
| | | $c_4 = 2.003\ 297\ 355\ 4 \times 10^{-5}$ |
| | | $c_5 = 9.013\ 801\ 955\ 9 \times 10^{-7}$ |
| | | $c_6 = 2.265\ 115\ 659\ 3 \times 10^{-8}$ |
| | | $c_7 = 3.607\ 115\ 420\ 5 \times 10^{-10}$ |
| | | $c_8 = 3.849\ 393\ 988\ 3 \times 10^{-12}$ |
| | | $c_9 = 2.821\ 352\ 192\ 5 \times 10^{-14}$ |
| | | $c_{10} = 1.425\ 159\ 477\ 9 \times 10^{-16}$ |
| | | $c_{11} = 4.876\ 866\ 228\ 6 \times 10^{-19}$ |
| | | $c_{12} = 1.079\ 553\ 927\ 0 \times 10^{-21}$ |
| | | $c_{13} = 1.394\ 502\ 706\ 2 \times 10^{-24}$ |
| | | $c_{14} = 7.979\ 515\ 392\ 7 \times 10^{-28}$ |
| T-type | 0 to 400°C | $c_0 = 0.000\ 000\ 000\ 0$ |
| | | $c_1 = 3.874\ 810\ 636\ 4 \times 10^1$ |
| | | $c_2 = 3.329\ 222\ 788\ 0 \times 10^{-2}$ |
| | | $c_3 = 2.061\ 824\ 340\ 4 \times 10^{-4}$ |
| | | $c_4 = -2.188\ 225\ 684\ 6 \times 10^{-6}$ |
| | | $c_5 = 1.099\ 688\ 092\ 8 \times 10^{-8}$ |
| | | $c_6 = -3.081\ 575\ 877\ 2 \times 10^{-11}$ |
| | | $c_7 = 4.547\ 913\ 529\ 0 \times 10^{-14}$ |
| | | $c_8 = -2.751\ 290\ 167\ 3 \times 10^{-17}$ |

from $\pm 1.0°C$ to perhaps $\pm 0.1°C$. Thermocouples constructed of standard thermocouple wire do not require calibration to provide measurement of temperature within the tolerance limits given in Table 8.5.

## Thermocouple Voltage Measurement

The Seebeck voltage for a thermocouple circuit is measured with no current flow in the circuit. From our discussion of the Thomson and Peltier effects, it is clear that the voltage is different from the open-circuit value when there is a current flow in the thermocouple circuit. As such, the best method for the measurement of thermocouple voltages is a device that minimizes current flow. For many years, the potentiometer was the laboratory standard for voltage measurement in thermocouple circuits. A potentiometer, as described in Chapter 6, has nearly zero loading error at a balanced condition. However, modern voltage measuring devices, such as digital voltmeters or data acquisition cards, have sufficiently large values of input impedance that they can be used with minimal loading error. These devices can also be used in either static or dynamic measuring situations where the loading error created by the measurement device is acceptable for the particular application. For such needs, high-impedance voltmeters have been incorporated into commercially available temperature indicators, temperature controllers, and digital DASs.

### EXAMPLE 8.6

The thermocouple circuit shown in Figure 8.23 is used to measure the temperature $T_1$. The thermocouple reference junction labeled 2 is at a temperature of $0°C$. The voltage output is measured and found to be 9.669 mV. What is $T_1$?

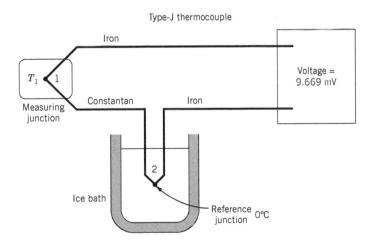

**FIGURE 8.23**    Thermocouple circuit for Example 8.7.

**KNOWN**   A thermocouple circuit having one junction at $0°C$ and a second junction at an unknown temperature. The circuit produces a voltage of 9.669 mV.

**FIND**   The temperature $T_1$.

**ASSUMPTION**   Thermocouple follows NIST standard voltage behavior.

**SOLUTION**
Standard thermocouple tables such as Table 8.6 are referenced to 0°C. The temperature of the reference junction for this case is 0°C. Therefore, the temperature corresponding to the output voltage may simply be determined from Table 8.6, in this case as 180.0°C.

**COMMENT** Because of the law of intermediate metals, the junctions formed at the voltage measuring device do not affect the voltage measured for the thermocouple circuit, and the voltage output reflects accurately the temperature difference between junctions 1 and 2.

---

**EXAMPLE 8.7**

Suppose the thermocouple circuit in Example 8.6 now has junction 2 maintained at a temperature of 30°C and produces an output voltage of 8.132 mV. What temperature is sensed by the measuring junction?

**KNOWN** The value of $T_2$ is 30°C, and the output is 8.132 mV.

**ASSUMPTION** Thermocouple follows NIST standard voltage behavior.

**FIND** The temperature of the measuring junction.

**SOLUTION**
By the law of intermediate temperatures, the output voltage for a thermocouple circuit having two junctions, one at 0°C and the other at $T_1$, would be the sum of the voltages for a thermocouple circuit between 0 and 30°C and between 30°C and $T_1$. Thus,

$$V_{0-30} + V_{30-T_1} = V_{0-T_1}$$

This relationship allows the voltage reading from the nonstandard reference temperature to be converted into a 0°C reference temperature by adding $V_{0-30} = 1.537$ to the existing reading. This results in an equivalent output voltage, referenced to 0°C as

$$1.537 + 8.132 = 9.669 \text{ mV}$$

Clearly, this thermocouple is sensing the same temperature as in the previous example, 180.0°C. This value is determined from Table 8.6.

**COMMENT** Note that the effect of raising the reference junction temperature is to lower the output voltage of the thermocouple circuit. Negative values of voltage, as compared with the polarity listed in Table 8.4, indicate that the measured temperature is less than the reference junction temperature.

---

**EXAMPLE 8.8**

A J-type thermocouple measures a temperature of 100°C and is referenced to 0°C. The thermocouple is AWG 30 (American wire gauge [AWG]; AWG 30 is 0.010-in. wire diameter) and is arranged in a circuit as shown in Figure 8.19a. The length of the thermocouple wire is 10 ft in order to run from the measurement point to the ice bath and to a voltmeter. The resolution of the voltmeter is 0.005 mV. If the thermocouple wire has a resistance per unit length, as specified by the manufacturer, of 5.6 Ω/ft, estimate the residual current in the thermocouple. Assume that the voltmeter has an infinite input impedance.

**KNOWN**  A voltmeter having a resolution of 0.005 mV is used to measure the output voltage of a J-type thermocouple that is 10 ft long.

**FIND**  The residual current in the thermocouple circuit.

**SOLUTION**
The total resistance of the thermocouple circuit is 56 $\Omega$ for 10 ft of thermocouple wire. The residual current is then found from Ohm's law as

$$I = \frac{E}{R} = \frac{0.005 \text{ mV}}{56 \ \Omega} = 8.9 \times 10^{-8} \text{A}$$

**COMMENT**  The loading error due to this current flow is $\sim 0.005$ mV $/ 54.3$ $\mu$V $/ \, ^\circ$C $\approx 0.09\,^\circ$C.

---

### EXAMPLE 8.9

Suppose that instead of assuming an infinite input impedance for the voltmeter, we assume that a high-impedance voltmeter is used. Determine the minimum input impedance required for the voltmeter that will limit the loading error to the same level as the infinite impedance case.

**KNOWN**  Loading error should be less than $8.9 \times 10^{-8}$ A.

**FIND**  Input impedance for a voltmeter that would produce the same current flow or loading error as in Example 8.8.

**SOLUTION**
At 100°C a J-type thermocouple referenced to 0°C has a Seebeck voltage of $E_s = 5.269$ mV. At this temperature, the required voltmeter impedance to limit the current flow to $8.9 \times 10^{-8}$ A is found from Ohm's law:

$$\frac{E_s}{I} = 5.269 \times 10^{-3} \text{ V}/8.9 \times 10^{-8} \text{ A} = 59.2 \text{ k}\Omega$$

**COMMENT**  This input impedance is not at all high for a modern nanovoltmeter, and such a voltmeter would be a reasonable choice in this situation. As always, the allowable loading error should be determined based on the required uncertainty in the measured temperature.

---

## Multiple-Junction Thermocouple Circuits

A thermocouple circuit composed of two junctions of dissimilar metals produces an open-circuit voltage that is related to the temperature difference between the two junctions. More than two junctions can be employed in a thermocouple circuit, and thermocouple circuits can be devised to measure temperature differences, or average temperature, or to amplify the output voltage of a thermocouple circuit.

### Thermopiles

*Thermopile* is the term used to describe a multiple-junction thermocouple circuit that is designed to amplify the output of the circuit. Because thermocouple voltage outputs are typically in the millivolt range, increasing the voltage output may be a key element in reducing the uncertainty in the temperature measurement or may be necessary to allow transmission of the thermocouple signal to the recording device. Figure 8.24 shows a thermopile for providing an amplified output

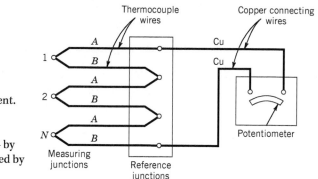

**FIGURE 8.24**  Thermopile arrangement. (From Benedict, R. P. *Fundamentals of Temperature, Pressure and Flow Measurements*, 3rd ed. Copyright © 1984 by John Wiley and Sons, New York. Reprinted by permission.)

signal; in this case, the output voltage would be $N$ times the single thermocouple output, where $N$ is the number of measuring junctions in the circuit. The average output voltage corresponds to the average temperature level sensed by the $N$ junctions. Thus, local point measurements entail consideration of the physical size of a thermopile, as compared to a single thermocouple. In transient measurements, a thermopile may have a more limited frequency range than a single thermocouple owing to its increased thermal capacitance. Thermopiles are particularly useful for reducing the uncertainty in measuring small temperature differences between the measuring and reference junctions. The principle has also been used to generate small amounts of power in spacecraft and to provide thermoelectric cooling using the Peltier effect.

Figure 8.25 shows a series arrangement of thermocouple junctions designed to measure the average temperature difference between junctions. This thermocouple circuit could also be used to measure deviations from a uniform temperature. In that case, any voltage output would indicate that a temperature difference existed between two of the thermocouple junctions. Alternatively, junctions 1, 2,..., $N$ could be located at one physical location, whereas junctions 1′, 2′,..., $N′$ could be located at another physical location. Such a circuit might be used to measure the heat flux through a solid.

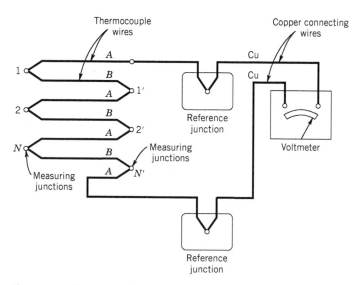

**FIGURE 8.25**  Thermocouples arranged to sense temperature differences. (From Benedict, R. P. *Fundamentals of Temperature, Pressure and Flow Measurements*, 3rd ed. Copyright © 1984 by John Wiley and Sons, New York. Reprinted by permission.)

**FIGURE 8.26**   Parallel arrangement of thermocouples for sensing the average temperature of the measuring junctions. (From Benedict, R. P. *Fundamentals of Temperature, Pressure and Flow Measurements*, 3rd ed. Copyright © 1984 by John Wiley and Sons, New York. Reprinted by permission.)

## Thermocouples in Parallel

When a spatially averaged temperature is desired, multiple thermocouple junctions can be arranged, as shown in Figure 8.26. In such an arrangement of $N$ junctions, a mean voltage is produced, given by

$$\overline{V} = \frac{1}{N} \sum_{i-1}^{N} (V)_i \tag{8.17}$$

The mean voltage is indicative of a mean temperature

$$\overline{T} = \frac{1}{N} \sum_{i-1}^{N} T_i \tag{8.18}$$

# Applications for Thermoelectric Temperature Measurement: Heat Flux

Now that we understand measuring temperature using thermocouples and thermopiles, we will consider an important application for the use of these devices. Figure 8.27 illustrates conceptually how we might construct a *heat flux sensor*. We wish to measure the heat flux created by the temperature difference indicated in the figure, which flows from surface 1 to surface 2.

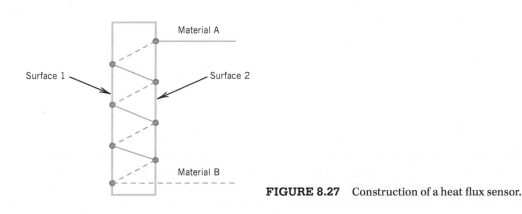

**FIGURE 8.27**   Construction of a heat flux sensor.

**FIGURE 8.28**   Thin-film heat flux sensor.

In this arrangement, thermopile junctions are placed on each side of a solid material, sometimes termed a barrier material. The thermal conductivity, $k$, of the barrier material is precisely known. We may compute the averaged heat flux through the barrier material as

$$|\dot{q}| = k\frac{|T_1 - T_2|}{L} \tag{8.19}$$

where
    $|\dot{q}|$ = absolute value of heat flux (W/m$^2$)
    $k$ = thermal conductivity of the barrier material (W/m-K)
    $L$ = thickness of the barrier material (m)
    $|T_1 - T_2|$ = absolute value of the temperature difference across the barrier material (K)

One embodiment of this concept is a thin-film heat flux sensor as illustrated in Figure 8.28. This sensor consists of thin-film thermocouples bonded to a flexible barrier material, such as Kapton. When properly mounted on a surface, the heat flux through the sensor represents the heat flux through the surface. A typical sensitivity of a thin-film heat flux sensor is 3 μV/(W / m$^2$).

Consider measuring the heat flux through the wall of a residence by mounting the heat flux sensor on the inside dry wall. Under these circumstances, the most accurate heat flux measurement would result if the sensor had the same surface characteristics as the dry wall. For example, so that the radiative component of heat transfer would match that of the dry wall, the sensor could be painted to match the wall.

### EXAMPLE 8.10

Consider using a thermopile arranged to measure temperature difference to determine the heat flux across a single-pane glass window. The thermal conductivity of the glass is 1.5 W/m-K and the window is 5 mm thick. The thermopile is constructed of J-type thermocouples, and there are 10 active junctions. Determine the heat flux if the output of the thermopile is 4.61 mV.

**KNOWN** For the thermopile, $N = 10$ and it is J-type with an output of 4.61 mV. Glass window has $L = 0.005$ m $k = 1.5$ W/m-k.

**FIND** The heat flux, $\dot{q}$, through the glass.

**SOLUTION**
The heat flux is given by

$$|q| = k\frac{|T_1 - T_2|}{L} \tag{8.19}$$

The temperature difference can be determined from the thermopile output and Table 8.6. With 10 active junctions, the single junction equivalent output voltage would be 4.61 mV/10 = 0.461 mV. At temperatures near room temperature, the sensitivity of a J-type thermocouple is approximately 0.0512 mV/°C. Therefore, the temperature difference may be determined as

$$T_2 - T_1 = 0.461 \text{ mV}/0.0512 \text{ mV}/°C = 9°C$$

And the resulting heat flux is

$$|q| = k\frac{|T_1 - T_2|}{L} = (1.5 \text{ W/m-K})\frac{9°C}{0.005 \text{ m}} = 2700 \text{ W}/m^2$$

**COMMENT** Note that the size of the degree, whether Kelvin or Celsius, is the same, and no unit conversion is required.

---

**EXAMPLE 8.11    Heat Flux Case Study**

Phase-change gel encapsulated within beads is being embedded in various bedding products to better regulate temperature. A person's body heat is absorbed by a mattress, warming the surface. However, a person can feel thermal discomfort if this heat is not conducted away sufficiently, in effect insulating the sleeper. The most common targets are mattresses made of foam known to cause thermal discomfort for many users. The idea is to absorb body energy within the gel more rapidly so as to maintain a cooler body feel longer. Supplemental air gaps in the foam help to move trapped energy away out of the gel and from the sleeper's body.

To quantify the effect of encapsulated gel on the cooling effect for a mattress, a major supplier of medical mattresses conducted a test using three different gel materials. A square sample of a gel mattress approximately 0.4 m square was used for testing. The top surface of the mattress was instrumented with a thin-film heat flux gauge, and a weighted constant-temperature source of the same size as the mattress sample was placed on top of the mattress sample. The conditions were designed to simulate a person's torso and the step-change in heat flux associated with lying down onto the mattress. The heat flux sensor output was appropriately amplified and recorded. Figure 8.29 shows the results of this test for three different gel types. The larger the heat flux, the greater the cooling effect. It is clear that the choice of particular gel material has a significant effect on the heat flux, both during early times and in steady state. Within 10 minutes, the heat flux is approaching a steady-state value, which varies between samples. The experience of lying down on a mattress containing encapsulated phase-change gel creates a perception that the surface is initially quite cool, but within a short period the perceived level of cooling is significantly reduced. The test shows that some designs maintain the cooler perception longer, supposedly an aid to better sleeping. Tests such as this are intended to test the effectiveness of preliminary designs as an aide to making informed decisions to improve them.

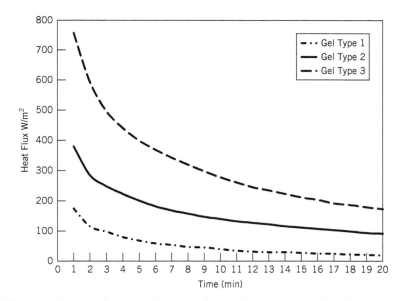

**FIGURE 8.29**    Heat flux as a function of time for three gel types employed in a foam mattress.

## Data Acquisition Considerations

Measuring temperatures by using thermocouples connected to a DAS is common practice. However, the characteristics of the thermocouple, including the need for a reference or cold junction and the low signal voltages produced, complicate its use. Nevertheless, with a little attention and realistic expectation of achievable accuracy, the systems are quite acceptable for most monitoring and moderate accuracy measurements.

Once the appropriate thermocouple type is selected, attention must be given to the cold junction compensation method. The two connection points of the thermocouple to the DAS board form two new thermocouple connections. Use of external cold junction methods between the thermocouple and the board eliminates this problem, but more frequently the thermocouple is connected directly to the board and uses built-in electronic cold junction compensation. This is usually accomplished by using a separate thermistor sensor, which measures the temperature at the system connection point to determine the cold junction error, and by providing an appropriate bias voltage correction either directly or through software. An important consideration is that the internal correction method has a typical uncertainty of the order of 0.5 to 1.5°C and, as a systematic error, this error is directly passed on to the measurement as an offset.

These boards also may use internal polynomial interpolation for converting measured voltage into temperature. If not, this can be programmed into the data reduction scheme by using, for example, the information of Table 8.7. Nonetheless, this introduces a "linearization" error, which is a function of thermocouple material and temperature range and typically specified with the DAS board.

Thermocouples are often used in harsh industrial environments with significant electromagnetic interference (EMI) and radiofrequency (rf) noise sources. Thermocouple wire pairs should be twisted to reduce noise. Also, a differential-ended connection is preferred between the thermocouple and the DAS board. In this arrangement, though, the thermocouple becomes an isolated voltage source, meaning that there is no longer a direct ground path keeping the input within its common mode range. As a consequence, a common complaint is that the measured signal may drift or suddenly jump in the level of its output. Usually, this interference behavior is eliminated by placing a 10- to 100-k$\Omega$ resistor between the low terminal of the input and low-level ground.

Because most DAS boards use analog-to-digital (A/D) converters having a $\pm 5$ V full scale, the signal must be conditioned using an amplifier. Usually, a gain of 100 to 500 is sufficient. High-gain, very-low-noise amplifiers are important for accurate measurements. Consider a J-type thermocouple using a 12-bit A/D converter with a signal conditioning gain of 100. This allows for a full-scale input range of 100 mV, which is a suitable range for most measurements. Then, the A/D conversion resolution is

$$\frac{E_{FSR}}{(G)(2^M)} = \frac{10 \text{ V}}{(100)(1^{12})} = 24.4 \text{ } \mu\text{V/bit}$$

A J-type thermocouple has a sensitivity of $\sim 55$ $\mu$V/°C. Thus, the measurement resolution becomes

$$(24.4 \text{ } \mu\text{V/bit})/(55 \text{ } \mu\text{V/°C}) = 0.44\text{°C/bit}$$

Thermocouples tend to have long time constants relative to the typical sample rate capabilities of general-purpose DAS boards. If temperature measurements show greater-than-expected fluctuations, high-frequency sampling noise is a likely cause. Reducing the sample rate and using a smoothing filter are simple solutions. The period of averaging should be on the scale of the time constant of the thermocouple.

**EXAMPLE 8.12**

It is desired to create an off-the-shelf temperature measuring system for a personal computer (PC)–based control application. The proposed temperature measuring system is illustrated schematically in Figure 8.30. The temperature measurement system consists of the following:

- A PC-based DAS composed of a data acquisition board, a computer, and appropriate software to allow measurement of analog input voltage signals.

- A J-type thermocouple and reference junction compensator. The reference junction compensator provides an output voltage from the thermocouple equivalent to the value that would exist between the measuring junction and a reference temperature of 0°C.

- A J-type thermocouple, which is uncalibrated but meets NIST standard limits of error.

The system is designed to measure and control a process temperature that varies slowly in time compared to sampling rate of the DAS. The process nominally operates at 185°C.

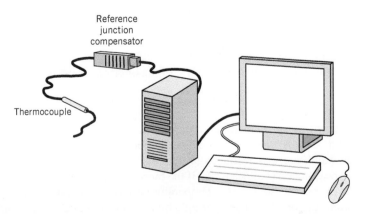

**FIGURE 8.30**  Personal computer (PC)–based temperature measurement system.

The following specifications are applicable to the measurement system components:

| Component | Characteristics | Accuracy specifications |
|---|---|---|
| Data acquisition board | Analog voltage input range: 0 to 0.1 V | 12-bit A/D converter accuracy: $\pm 0.01\%$ of reading |
| Reference junction compensator | J-type compensation range from 0 to 50°C | $\pm 0.5$°C over the range 20 to 36°C |
| Thermocouple (J-type) | Stainless steel sheathed ungrounded junction | Accuracy: $\pm 1.0$°C based on NIST standard limits of error |

The purpose of the measurement system requires that the temperature measurement have a total uncertainty of less than 1.5°C. Based on a design-stage uncertainty analysis, does this measurement system meet the overall accuracy requirement?

## SOLUTION

The design-stage uncertainty for this measurement system is determined by expressing the uncertainty of each system component as an equivalent uncertainty in temperature and then combining these design-stage uncertainties.

The 12-bit A/D converter divides the full-scale voltage range into $2^{12}$, or 4,096, equal-sized intervals. Thus, the resolution (quantization error) of the A/D in measuring voltage is

$$\frac{0.1 \text{ V}}{4096 \text{ intervals}} = 0.0244 \text{ mV}$$

The uncertainty of the DAS is specified as 0.01% of the reading. A nominal value for the thermocouple voltage must be known or established. In the present case, the nominal process temperature is 185°C, which corresponds to a thermocouple voltage of approximately 10 mV. Thus, the calibration uncertainty of the DAS is 0.001 mV.

The contribution to the total uncertainty of the temperature measurement system from the DAS can now be determined. First, combine the resolution and calibration uncertainties as

$$u_{DAS} = \sqrt{(0.0244)^2 + (0.001)^2} = 0.0244 \text{ mV}$$

The relationship between uncertainty in voltage and temperature is provided by the static sensitivity, which can be estimated from Table 8.6 at 185°C as 0.055 mV/°C. Thus, an uncertainty of 0.0244 mV corresponds to an uncertainty in temperature of

$$\frac{0.0244 \text{ mV}}{0.055 \text{ mV}/°C} = 0.444°C$$

This uncertainty can now be combined directly with the reference junction uncertainty and the uncertainty interval associated with the standard limits of error for the thermocouple, as

$$u_T = \sqrt{(0.44)^2 + (1.0)^2 + (0.5)^2} = \pm 1.2°C \quad (95\%)$$

**COMMENT** It might be appropriate in certain cases to calibrate the thermocouple against a laboratory standard, such as an RTD, which has a calibration traceable to NIST standards. With reasonable expense and care, an uncertainty level in the thermocouple of $\pm 0.1$°C can be achieved, compared to the uncalibrated value of $\pm 1$°C. If the thermocouple were so calibrated, the resulting uncertainty in the overall measurement of temperature is reduced to $\pm 0.68$°C, so that by reducing the uncertainty contribution of the thermocouple by a factor of 10, the system uncertainty would be reduced by a factor of 2.

# 8.6 Radiative Temperature Measurements

There is a distinct advantage to measuring temperature by detecting thermal radiation. The sensor for thermal radiation need not be in contact with the surface to be measured, making this method attractive for a wide variety of applications. The basic operation of a radiation thermometer is predicated upon some knowledge of the radiation characteristics of the surface whose temperature is being measured, relative to the calibration of the thermometer. The spectral characteristics of radiative measurements of temperature are beyond the scope of the present discussion; an excellent source for further information is Dewitt and Nutter [10].

## Radiation Fundamentals

Radiation refers to the emission of electromagnetic waves from the surface of an object. This radiation has characteristics of both waves and particles, which leads to a description of the radiation as being composed of photons. The photons generally travel in straight lines from points of emission to another surface, where they are absorbed, reflected, or transmitted. This electromagnetic radiation exists over a large range of wavelengths that includes x-rays, ultraviolet radiation, visible light, and infrared or thermal radiation, as shown in Figure 8.31. The thermal radiation emitted from an object is related to its temperature and has wavelengths ranging from approximately $10^{-7}$ to $10^{-3}$ m. It is necessary to understand two key aspects of radiative heat transfer in relation to temperature measurements. First, the radiation emitted by an object is proportional to the fourth power of its temperature. In the ideal case, this may be expressed as

$$E_b = \sigma T^4 \tag{8.20}$$

where $E_b$ is the flux of energy radiating from an ideal surface, or the blackbody emissive power. The emissive power of a body is the energy emitted per unit area and per unit time. The term "blackbody" implies a surface that absorbs all incident radiation and as a result emits radiation in an "ideal" manner.

Second, the emissive power is a direct measure of the total radiation emitted by an object. However, energy is emitted by an ideal radiator over a range of wavelengths, and at any given temperature the distribution of the energy emitted as a function of wavelength is unique.

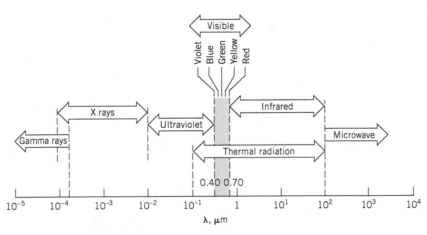

**FIGURE 8.31** The electromagnetic spectrum.
(From Incropera, F. P., and D. P. DeWitt, *Fundamentals of Heat and Mass Transfer*, 2nd ed. Copyright © 1985 by John Wiley & Sons, New York. Reprinted by permission.)

Max Planck (1858–1947) developed the basis for the theory of quantum mechanics in 1900 as a result of examining the wavelength distribution of radiation. He proposed the following equation to describe the wavelength distribution of thermal radiation for an ideal or blackbody radiator:

$$E_{b\lambda} = \frac{2\pi h_p c^2}{\lambda^5 [\exp(h_p c / k_B \lambda T) - 1]} \tag{8.21}$$

where

$E_{b\lambda}$ = total emissive power at the wavelength $\lambda$

$\lambda$ = wavelength

$c$ = speed of light in a vacuum = $2.998 \times 10^8$ m/s

$h_p$ = Planck's constant = $6.6256 \times 10^{-34}$ Js

$k_B$ = Boltzmann's constant = $1.3805 \times 10^{-23}$ J/K

Figure 8.32 is a plot of this wavelength distribution for various temperatures. For the purposes of radiative temperature measurements, it is crucial to note that the maximum energy emission shifts to shorter wavelengths at higher temperatures. Our experiences confirm this behavior through observation of color changes as a surface is heated.

Consider an electrical heating element such as that can be found in an electric oven. With no electric current flow through the element, it appears almost black, its room temperature color. With a current flow, the element temperature rises and it appears to change color to a dull red,

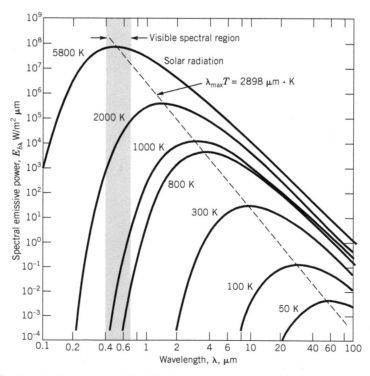

**FIGURE 8.32** Planck distribution of blackbody emissive power as a function of wavelength. (From Incropera, F. P., and D. P. DeWitt, *Fundamentals of Heat and Mass Transfer*, 2nd ed. Copyright © 1985 by John Wiley & Sons, New York. Reprinted by permission.)

**FIGURE 8.33** Schematic of a basic radiometer: (1) lens; (2) focusing mirror; (3) detector (thermopile or thermistor).

perhaps to a reddish orange. If its temperature continued to increase, eventually it would appear white. This change in color signifies a shift in the maximum intensity of the emitted radiation to shorter wavelengths, out of the infrared and into the visible. There is also an increase in total emitted energy. The Planck distribution provides a basis for the measurement of temperature through color comparison.

## Radiation Detectors

Radiative energy flux can be detected in a sensor by two basic techniques. The detector is subject to radiant energy from the source whose temperature is to be measured. The first technique involves a thermal detector in which absorbed radiative energy elevates the detector temperature, as shown in Figure 8.33. These thermal detectors are certainly the oldest sensors for radiation, and the first such a detector can probably be credited to Sir William Herschel, who verified the presence of infrared radiation using a thermometer and a prism. The equilibrium temperature of the detector is a direct measure of the amount of radiation absorbed. The resulting rise in temperature must then be measured. Thermopile detectors provide a thermoelectric power resulting from a change in temperature. A thermistor can also be used as the detector and results in a change in resistance with temperature.

A second basic type of detector relies on the interaction of a photon with an electron, resulting in an electric current. In a photomultiplier tube, the emitted electrons are accelerated and used to create an amplified current, which is measured. Photovoltaic cells may be employed as radiation detectors. The photovoltaic effect results from the generation of a potential across a positive–negative junction in a semiconductor when it is subjected to a flux of photons. Electron–hole pairs are formed if the incident photon has an energy level of sufficient magnitude. This process results in the direct conversion of radiation into electrical energy, in high sensitivity, and in a fast response time when used as a detector. In general, the photon detectors tend to be spectrally selective, so that the relative sensitivity of the detector tends to change with the wavelength of the measured radiation.

Many considerations enter into the choice of a detector for radiative measurements. If time response is important, photon detectors are significantly faster than thermopile detectors or thermistor detectors and thus have a much wider frequency response. Photodetectors saturate, whereas thermopile sensors may slowly change their characteristics over time. Some instruments have variations of sensitivity with the incident angle of the incoming radiation; this factor may be important for solar insolation measurements. Other considerations include wavelength sensitivity, cost, and allowable operating temperatures.

## Radiometer

Perhaps the simplest form of radiation detector is a radiometer that measures a source temperature by measuring the voltage output from a thermopile detector. A schematic of such a device is shown in Figure 8.33. The increase in temperature of the thermopile is a direct indication of the temperature of the radiation source. One application of this principle is in the measurement of total

**FIGURE 8.34**   Pyranometer construction.

**FIGURE 8.35**   Industrial infrared (IR) thermopile sensor.

solar radiation incident upon a surface. Figure 8.34 shows a schematic of a pyranometer, used to measure global solar irradiance. It would have a hemispherical field of view and measures both the direct or beam radiation and diffuse radiation. The diffuse and beam components of radiation can be separated by shading the pyranometer from the direct solar radiation, thereby measuring the diffuse component.

Figure 8.35 shows infrared (IR) thermopile sensors that are manufactured using micromachining and advanced semiconductor processing methods. The hot junction of the thermopile is placed under the IR filter and the cold junctions under the IR mask. Manufacturing methods allow hundreds of junctions to be created in a microdevice. These sensors form the basis for many practical applications, ranging from tympanic thermometers to automatic climate control systems for automotive applications.

## Pyrometry

Optical pyrometry identifies the temperature of a surface by its color, or more precisely, the color of the radiation it emits. A schematic of an optical pyrometer is shown in Figure 8.36. A standard lamp is calibrated so that the current flow through its filament is controlled and calibrated in terms of the filament temperature. Comparison is made optically between the color of this filament and the surface of the object whose temperature is being measured. The comparator can be the human eye. Uncertainties in the measurement may be reduced by appropriately filtering the incoming light. Corrections must be applied for surface emissivity associated with the measured radiation; uncertainties vary with the skill of the user and generally are on the order of 5°C. Replacing the

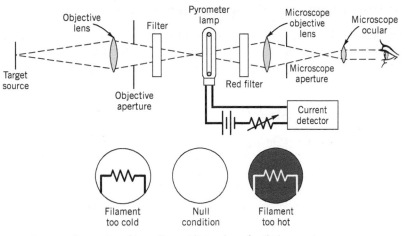

Appearance of lamp filament in eyepiece of optical pyrometer.

**FIGURE 8.36** Schematic diagram of a disappearing-filament optical pyrometer.

human eye with a different detector extends the range of useful temperature measurement and reduces the random uncertainty.

The major advantage of an optical pyrometer lies in its ability to measure high temperatures remotely up to 4500°C. For example, it could be used to measure the temperature of a furnace without having any sensor in the furnace itself. For many applications, this provides a safe and economical means of measuring high temperatures.

## Optical Fiber Thermometers

The optical fiber thermometer is based on the creation of an ideal radiator that is optically coupled to a fiber-optic transmission system [11, 12], as shown in Figure 8.37. The temperature sensor in this system is a thin, single-crystal aluminum oxide (sapphire) fiber; a metallic coating on the tip of the fiber forms a blackbody radiating cavity, which radiates directly along the sapphire crystal fiber. The single-crystal sapphire fiber is necessary because of the high-temperature operation of the thermometer. The operating range of this thermometer is 300 to 1,900°C. Signal transmission is accomplished using standard, low-temperature fiber optics. A specific wavelength band of the transmitted radiation is detected and measured, and these raw data are reduced to yield the temperature of the blackbody sensor.

The absence of electrical signals associated with the sensor signal provides excellent immunity from electromagnetic and radiofrequency interference. The measurement system has superior frequency response and sensitivity. The system has been employed for measurement in combustion applications. Temperature resolution of 0.1 mK is possible.

Optical path to photodetector

**FIGURE 8.37** Optical fiber thermometer: (1) blackbody cavity (iridium film); (2) sapphire fiber (single crystal); (3) protective coating ($Al_2O_3$).

## Narrow-Band Infrared Temperature Measurement

Infrared temperature measurement allows noncontact point or full-field temperature measurements. Advances in detector technology and optical filters have allowed affordable, advanced infrared thermometers and cameras to be developed at reasonable cost.

## Fundamental Principles

Consider the infrared emission, in wavelengths from 0.7 to 100 μm, from a blackbody whose spectral emission curve is defined by Equation 8.21. Figure 8.38 shows two blackbody emission curves, for $T = 800$ and 2,000 K. At a wavelength of 2 μm, we see that a blackbody at 2,000 K emits $3.29 \times 10^5$ W/m²-μm and at 800 K a blackbody emits $1.45 \times 10^3$ W/m²-μm. Thus, in principle, measuring the radiative power emitted at one wavelength would allow the determination of the temperature of a blackbody. In fact, appropriate construction of a cavity to simulate a blackbody can be used as a calibration standard for infrared thermometers [13].

A blackbody emits the maximum possible radiation at a given temperature. Therefore, all real surfaces emit less energy than a blackbody. One model of real surfaces is termed a gray surface; it is simply a scaled version of the blackbody with the scale factor being the total hemispherical emissivity. Figure 8.38 shows a blackbody emission curve and a gray surface emission curve at 800 K; the emissivity of the gray surface, ε, is 0.6. Thus, for real surfaces it is necessary to either know the emissivity or develop some means of eliminating the emissivity from the temperature measurement. Modern infrared thermometers have user input for surface emissivity and make corrections to the reported temperature based on this input.

The practical application of infrared thermometry requires an understanding of the conditions under which the measurement is being made and the surface characteristics of the object for which the temperature is being measured. An infrared thermometer gathers all radiation from the surface being measured, both emitted and reflected. Thus, for a surface having a significant reflectivity, the temperature of the surroundings can affect the measured value. Absorption in the intervening medium can also affect the temperature measurement. For example, $CO_2$ and $H_2O$

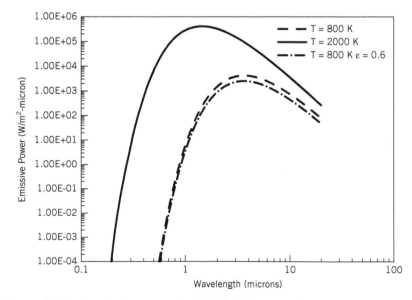

**FIGURE 8.38**    Blackbody emission curves for 800 K and 2,000 K and gray surface emission curve for 800 K and an emissivity of 0.6.

have several absorption bands between 1.5 and 3 μm; these particular wavelengths should not be selected for temperature measurement. The selection of a wavelength band depends on the expected temperature and the particular application.

## Two-Color Thermometry

Two-color infrared thermometry refers to a technique that uses measurement of radiative emissive power at two different narrow wavelength bands. Equations 8.22 show the theoretical emissive power for two different wavelengths where we incorporated spectral and temperature-dependent hemispherical emissivity.

$$E_{\lambda_1} = \varepsilon(\lambda_1, T) \frac{2\pi h_p c^2}{\lambda_1^5 [\exp(h_p c / k_B \lambda_1 T) - 1]}$$
$$E_{\lambda_2} = \varepsilon(\lambda_2, T) \frac{2\pi h_p c^2}{\lambda_2^5 [\exp(h_p c / k_B \lambda_2 T) - 1]} \tag{8.22}$$

For opaque diffuse surfaces, an excellent engineering model for emissivity is the gray surface assumption. Under this assumption, the emissivity is not a function of wavelength or temperature. Thus, if we take the ratio of the two equations in Equations 8.22, the result is independent of the emissivity:

$$\frac{E_{\lambda_1}}{E_{\lambda_2}} = \frac{\lambda_2^5 [\exp(h_p c / k_B \lambda_2 T) - 1]}{\lambda_1^5 [\exp(h_p c / k_B \lambda_1 T) - 1]} \tag{8.23}$$

In practice, two independent measurements of emissive power at two different narrow wavelength bands are made, and the signal ratio is a direct indicator of the surface temperature without knowing the surface emissivity. Figure 8.39 shows a conceptual arrangement of the optics and detectors required to implement the two-color measurement method.

## Full-Field IR Imaging

Advances in CCD detector and microprocessor technology have allowed infrared cameras to gain wide usage in industry and for research purposes. Applications include assessing the effectiveness of insulation in buildings, finding victims in a fire, and mapping body surface temperature for a

**FIGURE 8.39** Schematic diagram of a two-color infrared thermometer. (1) surface whose temperature is being measured, (2) lens, (3) narrow band optical filter, (4) detector.

variety of medical purposes [14]. Other applications include night vision, including both security and military uses, and maritime night navigation.

**Example 8.13** | **Infrared Imaging of Environmental Water Systems to Visualize Fluid Motion Patterns**

A variety of situations exist where it may be useful to monitor the discharge of warm water into an environmental water system, such as that occurs at power plants that use lakes for a cooling water source. Figure 8.40 is an infrared image obtained using an IR camera sensitive to the 3 to 5 μm wavelength band [15]. The image is essentially the temperature field of a warm water surface that is evaporating into still air. The evaporation of the water that occurs at the interface between the water and air cools the water at the interface and creates a buoyancy-driven flow in the water. In Figure 8.40, the darker colored pixels represent cooler temperatures. The complex pattern of cool and warm regions in this image reveals the underlying pattern of liquid motion created by the buoyancy forces resulting from the temperature difference between the fluid at the interface and the underlying warmer fluid.

**FIGURE 8.40** Infrared image of water surface under conditions where water is evaporating into still air.
(Reprinted from Bower, S. M., and Saylor, J. R., The Effects of Surfactant Monolayers on Free Surface Natural Convection, *International Journal of Heat and Mass Transfer*, 54: 5348–5358, (2011). With permission from Elsevier.)

# 8.7 Physical Errors in Temperature Measurement

In general, errors in temperature measurement derive from two fundamental sources. The first source of errors derives from uncertain information about the temperature of the sensor itself. Such uncertainties can result from random interpolation errors, calibration systematic errors, or a host of other error sources. Instrument and procedural uncertainty in the sensor temperature can be reduced by improved calibration or by changes in the measuring and recording instruments. However, errors in temperature measurement can still occur even if the temperature of the sensor was measured exactly. In such cases, the probe does not sense accurately the temperature it was intended to measure.

A list of typical errors associated with the use of temperature sensors is provided in Table 8.8. Random errors in temperature measurements are a result of resolution limits of measuring and recording equipment, time variations in extraneous variables, and other sources of variation. Thermocouples have some characteristics that can lead to both systematic and random errors, such as the effect of extension wires and connectors. Another major source of error for thermocouples

**Table 8.8    Measuring Errors Associated with Temperature Sensors**

**RANDOM ERRORS**
1. Imprecision of readings
2. Time and spatial variations

**SYSTEMATIC ERRORS**
1. Insertion errors, heating or cooling of junctions
   a. Conduction errors
   b. Radiation errors
   c. Recovery errors
2. Effects of plugs and extension wires
   a. Nonisothermal connections
   b. Loading errors
3. Ignorance of materials or material changes during measurements
   a. Aging following calibration
   b. Annealing effects
   c. Cold work hardening
4. Ground loops
5. Magnetic field effects
6. Galvanic error
7. Reference junction inaccuracies

involves the accuracy of the reference junction. Ground loops can lead to spurious readings, especially when thermocouple outputs are amplified for control or data acquisition purposes. As with any measurement system, calibration of the entire measurement system, in place if possible, is the best means of identifying error sources and reducing the resulting measurement uncertainty to acceptable limits.

# Insertion Errors

This discussion focuses on ensuring that a sensor output accurately represents the temperature it is intended to measure. For example, suppose it is desired to measure the outdoor temperature. This measurement could employ a large dial thermometer, which might be placed on a football field or a tennis court in the direct sunlight, and assumed to represent "the temperature," perhaps as high as 50°C (120°F). But what temperature is being indicated by this thermometer? Certainly, the thermometer is not measuring the air temperature, nor is it measuring the temperature of the field or the court. The thermometer is subject to the very sources of error we wish to describe and analyze. *The thermometer, very simply, indicates its own temperature!*

In our example, the temperature of the thermometer is the thermodynamic equilibrium temperature that results from the radiant energy gained from the sun, convective exchange with the air, and conduction heat transfer with the surface on which it is resting. Considering the fact that these thermometers typically have a glass cover that ensures a greenhouse effect, it is very likely that the thermometer temperature is significantly higher than the air temperature.

The physical mechanisms that may cause a temperature probe to indicate a temperature different from that intended include conduction, radiation, and velocity recovery errors. In any real measurement system, their effects could be coupled and, therefore, should not be considered independently. However, for simplicity, we will consider each separately because our purpose is to provide only estimates of the errors and not to provide predictive techniques for correcting measured temperatures. The goal of the measurement engineer should be to minimize these errors, as far as possible, through the careful installation and design of temperature probes.

## Conduction Errors

Errors that result from conduction heat transfer between the measuring environment and the ambient are often called immersion errors. Consider the temperature probe shown in Figure 8.41. In many circumstances, a temperature probe extends from the measuring environment through a wall into the ambient environment, where indicating or recording systems are located. The probe and the electrical leads form a path for the conduction of energy from the measuring environment to the ambient. The fundamental nature of the error created by conduction in measured temperatures can be illustrated by the model of a temperature probe shown in Figure 8.42. The essential physics of immersion errors associated with conduction can be discerned by modeling the temperature probe as a fin. Suppose we assume that the measured temperature is higher than the ambient temperature. If we consider a differential element of the fin, as shown in Figure 8.42b, at steady state there is energy conducted along the fin and transferred by convection from its surface. The surface area for convection is $P dx$, where $P$ is the perimeter or circumference. Applying the first law of thermodynamics to this differential element yields

$$q_{x+dx} - q_x = hP \, dx[T(x) - T_\infty] \tag{8.24}$$

where $h$ is the convection coefficient. If $q$ is expanded in a Taylor series about the point $x$ and the substitutions

$$\theta = T - T_\infty \qquad q = -kA\frac{dT}{dx} \qquad m = \sqrt{\frac{hp}{kA}} \tag{8.25}$$

are made, then the governing differential equation becomes

$$\frac{d^2\theta}{dx^2} - m^2\theta = 0 \tag{8.26}$$

**FIGURE 8.41**   Temperature probe inserted into a measuring environment.

(a)                                        (b)

**FIGURE 8.42**   Model of a temperature probe as a one-dimensional fin.

Here $k$ is the effective thermal conductivity of the temperature probe. The solution to this differential equation for the boundary conditions that the wall has a temperature $T_w$, or a normalized value $\theta_w = T_w - T_\infty$, and the end of the fin is small in surface area, is

$$\frac{\theta(x)}{\theta_w} = \frac{\cosh mx}{\cosh mL} \tag{8.27}$$

The point $x = 0$ is the location where the temperature is assumed to be measured, and so the solution is evaluated at $x = 0$ as

$$\frac{\theta(0)}{\theta_w} = \frac{T(0) - T_\infty}{T_w - T_\infty} = \frac{1}{\cosh mL} \tag{8.28}$$

From this analysis, the error arising from conduction, $e_c$, can be estimated. An ideal sensor would indicate the fluid temperature $T_\infty$; thus, if the sensor temperature is $T_p = T(0)$, then the conduction error is

$$e_c = T_p - T_\infty = \frac{T_w - T_\infty}{\cosh mL} \tag{8.29}$$

Normally, the uncertainty due to conduction error is set at $u = e_c$. The uncertainty interval may not be symmetric.

## Probe Design

The purpose of the preceding analysis is to gain some physical understanding of ways to minimize conduction errors (not to correct inaccurate measurements). The behavior of this solution is such that the ideal temperature probe would have $T_p = T_\infty$, or $\theta(0) = 0$, implying that $e_c = 0$. Equation 8.29 shows that a value of $\theta(0) \neq 0$ results from a nonzero value of $\theta_w$ and a finite value of $\cosh mL$. The difference between the fluid temperature being measured and the wall temperature should be as small as possible; clearly, this implies that the wall should be insulated to minimize this temperature difference and the resulting conduction error.

The term "$\cosh mL$" should be as large as possible. The behavior of the cosh function is shown in Figure 8.43. Since the hyperbolic cosine monotonically increases for increasing values of the argument, the goal of a probe design should be to maximize the value of the product $mL$, or $(hP/kA)^{1/2}L$. In general, the thermal conductivity of a temperature probe and the convection coefficient are not design parameters. Thus, two important conclusions are that the probe should be as small in diameter as possible and should be inserted as far as possible into the measuring environment away from the bounding surface to make $L$ large. A small diameter increases the ratio of the perimeter $P$ to the cross-sectional area $A$. For a circular cross section, this ratio is $4/D$, where $D$ is the diameter. A good rule of thumb based on Equation 8.29 is to have an $L/D > 50$ for negligible conduction errors.

Although this analysis clearly indicates the fundamental aspects of conduction errors in temperature measurements, it does not provide the capability to correct measured temperatures for

**FIGURE 8.43**   Behavior of the hyperbolic cosine.

conduction errors. Conduction errors should be minimized through appropriate design and installation of temperature probes. Usually, the physical situation is sufficiently complex to preclude accurate mathematical description of the measurement errors. Additional information on modeling conduction errors may be found in Sparrow [16].

## Radiation Errors

Consider a temperature probe used to measure a gas temperature. In the presence of significant radiation heat transfer, the equilibrium temperature of a temperature probe may be different from the fluid temperature being measured. Because radiation heat transfer is proportional to the fourth power of temperature, the importance of radiation effects increases as the absolute temperature of the measuring environment increases. The error due to radiation can be estimated by considering steady-state thermodynamic equilibrium conditions for a temperature sensor. Consider the case where energy is transferred to/from a sensor by convection from the environment and from/to the sensor by radiation to a body at a different temperature, such as a pipe or furnace wall. For this analysis, conduction is neglected. A first law analysis of a system containing the probe, at steady-state conditions, yields

$$q_c + q_r = 0 \tag{8.30}$$

where

$q_c$ = convective heat transfer

$q_r$ = radiative heat transfer

The heat transfer components can then be expressed in terms of the appropriate fundamental relations as

$$q_c = hA_s(T_\infty - T) \quad q_r = FA_s\varepsilon\sigma(T_w^4 - T^4) \tag{8.31}$$

Assuming that the surroundings may be treated as a blackbody, the first law for a system consisting of the temperature probe is

$$hA_s(T_\infty - T_p) = FA_s\varepsilon\sigma(T_p^4 - T_w^4) \tag{8.32}$$

A temperature probe is generally small compared to its surroundings, which justifies the assumption that the surroundings may be treated as black. The radiation error $e_r$ is estimated by

$$e_r = (T_p - T_\infty) = \frac{F\varepsilon\sigma}{h}(T_w^4 - T_p^4) \tag{8.33}$$

where

$\sigma$ = the Stefan–Boltzmann constant ($\sigma = 5.669 \times 10^{-8}$ W/m$^2$K$^4$)

$\varepsilon$ = emissivity of the sensor

$F$ = radiation view factor

$T_p$ = probe temperature

$T_w$ = temperature of the surrounding walls

$T_\infty$ = fluid temperature being measured

If the sensor is small compared to the scale of the surroundings, the view factor from the sensor to the surroundings may be taken as 1. Normally, the uncertainty due to radiation error is set as

$u = e_r$. The uncertainty interval is not symmetrical and might be modeled as a uniform (rectangular) distribution with bounds of 0 and $e_r$.

### EXAMPLE 8.14

A typical situation where radiation would be important occurs in measuring the temperature of a furnace. Figure 8.44 shows a small temperature probe for which conduction errors are negligible, which is placed in a high-temperature enclosure, where the fluid temperature is $T_\infty$ and the walls of the enclosure are at $T_w$.

Convection and radiation are assumed to be the only contributing heat transfer modes at steady state. Develop an expression for the equilibrium temperature of the probe. Also determine the equilibrium temperature of the probe and the radiation error in the case where $T_\infty = 800°C$, $T_w = 500°C$, and the emissivity of the probe is 0.8. The convective heat transfer coefficient is 100 W/m²°C.

**KNOWN**  Temperatures $T_\infty = 800°C$ and $T_w = 500°C$, with $h = 100$ W/m²°C. The emissivity of the probe is 0.8.

**FIND**  An expression for the equilibrium temperature of the probe, and the resulting probe temperature for the stated conditions.

**ASSUMPTIONS**  The surroundings may be treated as black, and conduction heat transfer is neglected.

**SOLUTION**
The probe is modeled as a small, spherical body within the enclosed furnace; the view factor from the probe to the furnace is 1. At steady state, an equilibrium temperature may be found from an energy balance. The first law for a system consisting of the temperature probe from Equation 8.32 is

$$h(T_\infty - T_p) + \sigma\varepsilon(T_w^4 - T_p^4) = 0 \tag{8.34}$$

The probe is attempting to measure $T_\infty$. Equation 8.34 yields an equilibrium temperature for the probe of $T_p = 642.5°C$. The predicted radiation error in this case can be as large as (from Equation 8.33) $-157.5°C$. This result indicates that radiative heat transfer can create significant measurement errors at elevated temperatures.

**COMMENT**  Because the radiation error in this case can only lower the indicated temperature, it is an example of a nonsymmetrical error (see Section 5.10). We can use this predicted value as an estimate of the uncertainty resulting from this effect, or we can use the predicted value to correct the indicated value and assign a smaller uncertainty based on the uncertainty in the correction.

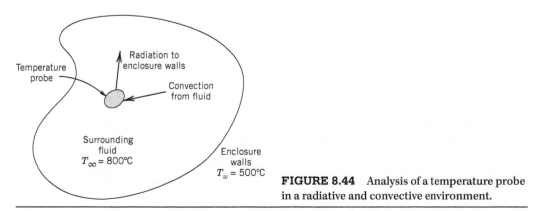

**FIGURE 8.44**  Analysis of a temperature probe in a radiative and convective environment.

# Radiation Shielding

Radiation shielding is a key concept in controlling radiative heat transfer; shielding for radiation is analogous to insulation to reduce conduction heat transfer. A radiation shield is an opaque surface interposed between a temperature sensor and its radiative surroundings so as to reduce electromagnetic wave interchange. In principle, the shield attains an equilibrium temperature closer to the fluid temperature than the surroundings. Because the probe can no longer "see" the surroundings, with the radiation shield in place, the probe temperature is closer to the fluid temperature. Additional information on radiation error in temperature measurements may be found in Sparrow [16]. The following example serves to demonstrate radiation errors and the effect of shielding.

### EXAMPLE 8.15

Consider again Example 8.14 where the oven is maintained at a temperature of 800°C. Because of energy losses, the walls of the oven are cooler, having a temperature of 500°C. For the present case, consider the temperature probe as a small, spherical object located in the oven, having no thermal conduction path to the ambient. (All energy exchange is through convection and radiation.) Under these conditions, the probe temperature is 642.5°C, as found in Example 8.14.

Suppose a radiation shield is placed between the temperature probe and the walls of the furnace, which blocks the path for radiative energy transfer. Examine the effect of adding a radiation shield on the probe temperature.

**KNOWN** A radiation shield is added to a temperature probe in an environment with $T_\infty = 800°C$ and $T_w = 500°C$.

**FIND** The radiation error in the presence of the shield.

**ASSUMPTIONS** The radiation shield completely surrounds the probe, and the surroundings may be treated as a blackbody.

**SOLUTION**
The shield equilibrium temperature is higher than the wall temperature by virtue of convection with the fluid. As a result, the probe "sees" a higher temperature surface, and the probe temperature is closer to the fluid temperature, resulting in less measurement error.

For a single radiation shield placed so that it completely surrounds the probe, which is small compared to the size of the enclosure and has an emissivity of 1, the equilibrium temperature of the shield can be determined from Equation 8.32. The temperature of the shield is found to be 628°C. Because the sensor now "sees" the shield rather than the wall, the temperature measured by the probe is 697°C, which is also determined from Equation 8.32.

One shield with an emissivity of 1 provides for an improvement over the case of no shields, but a better choice of the surface characteristics of the shield material can result in much better performance. If the shield has an emissivity of 0.1, the shield temperature rises to 756°C and the probe temperature to 771°C.

**COMMENT** Shielding provides improved temperature measurements by reducing radiative heat transfer. Another area for improvement in this temperature measurement could be the elevation of the wall temperature through insulation. This discussion of radiation shielding serves to demonstrate the usefulness of shielding as a means of improving temperature measurements in radiative environments. As with conduction errors, the development should be used to guide the design and installation of temperature sensors rather than to correct measured temperatures.

## Recovery Errors in Temperature Measurement

The kinetic energy of a gas moving at high velocity can be converted into sensible energy by reversibly and adiabatically bringing the flow to rest at a point. The temperature resulting from this process is called the stagnation or total temperature $T_t$. On the other hand, the static temperature of the gas, $T_\infty$, is the temperature that would be measured by an instrument moving at the local fluid velocity. From a molecular point of view, the static temperature measures the magnitude of the random kinetic energy of the molecules that comprise the gas, while the stagnation temperature includes both the directed and random components of kinetic energy. Generally, the engineer would be content with knowledge of either temperature, but in high-speed gas flows the sensor indicates neither temperature.

For negligible changes in potential energy, and in the absence of heat transfer or work, the energy equation for a flow may be written in terms of enthalpy and kinetic energy as

$$h_1 + \frac{U^2}{2} = h_2 \tag{8.35}$$

where state 2 refers to the stagnation condition and state 1 a condition whereby the gas is flowing with the velocity $U$. Assuming ideal gas behavior, the enthalpy difference $h_2 - h_1$ may be expressed as $c_p(T_2 - T_1)$, or in terms of static and stagnation temperatures

$$\frac{U^2}{2c_p} = T_t - T_\infty \tag{8.36}$$

The term $U^2/2c_p$ is called the *dynamic temperature*.

What implication does this have for the measurement of temperature in a flowing gas stream? The physical nature of gases at normal pressures and temperatures is such that the velocity of the gas on a solid surface is zero, because of the effects of viscosity. Thus, when a temperature probe is placed in a moving fluid, the fluid is brought to rest on the surface of the probe. Deceleration of the flow by the probe converts some portion of the directed kinetic energy of the flow to thermal energy and elevates the temperature of the probe above the static temperature of the gas. The fraction of the kinetic energy recovered as thermal energy is called the recovery factor, $r$, defined as

$$r \equiv \frac{T_p - T_\infty}{U^2/2c_p} \tag{8.37}$$

where $T_p$ represents the equilibrium temperature of the stationary (with respect to the flow) real temperature probe. In general, $r$ may be a function of the velocity of the flow, or, more precisely, the Mach number and the Reynolds number of the flow, and the shape and orientation of the temperature probe. For thermocouple junctions of round wire, Moffat [17] reports values of

$$r = 0.68 \pm 0.07 \quad (95\%) \quad \text{for wires normal to the flow}$$
$$r = 0.86 \pm 0.09 \quad (95\%) \quad \text{for wires parallel to the flow}$$

These recovery factor values tend to be constant at velocities for which temperature errors are significant, usually flows where the Mach number is greater than 0.1. For thermocouples having a welded junction, a spherical weld bead significantly larger than the wire diameter tends to a value of the recovery factor of 0.75, for the wires parallel or normal to the flow. The relationships between temperature and velocity for temperature probes with known recovery factors are

$$T_p = T_\infty + \frac{rU^2}{2c_p} \tag{8.38}$$

or in terms of the recovery error, $e_U$,

$$e_U = T_p - T_\infty = \frac{rU^2}{2c_p} \tag{8.39}$$

The probe temperature is related to the stagnation temperature by

$$T_p = T_t - \frac{(1-r)U^2}{2c_p} \tag{8.40}$$

Fundamentally, in liquids the stagnation and static temperatures are essentially equal, and the recovery error may generally be taken as zero for liquid flows. In any case, high-velocity flows are rarely encountered in liquids.

Normally, the uncertainty assigned to the recovery error is set at $u = e_U$. The uncertainty interval is often not symmetrical.

### EXAMPLE 8.16

A temperature probe having a recovery factor of 0.86 is to be used to measure a flow of air at velocities up to the sonic velocity, at a pressure of 1 atm and a static temperature of 30°C. Calculate the value of the recovery error in the temperature measurement as the velocity of the air flow increases, from 0 to the speed of sound, using Equation 8.39.

**KNOWN** $r = 0.86$ $p_\infty = 1$ atm abs $= 101$ kPa abs
$\qquad M \leq 1$ $T_\infty = 30\,°C = 303$ K

**FIND** The recovery error as a function of air velocity.

**ASSUMPTION** Air behaves as an ideal gas.

**SOLUTION**
Assuming that air behaves as an ideal gas, the speed of sound is expressed as

$$a = \sqrt{kRT} \tag{8.41}$$

For air at 101 kPa and 303 K, with $R = 0.287$ kJ/kg-K, the speed of sound is approximately 349 m/s.

Figure 8.45 shows the error in temperature measurement as a function of the Mach number for this temperature probe. The *Mach number* is defined as the ratio of the flow velocity to the speed of sound.

**COMMENT** Typically, the static and total temperatures of the flowing fluid stream are to be determined from the measured probe temperature. In this case, a second independent measurement of the velocity is necessary.

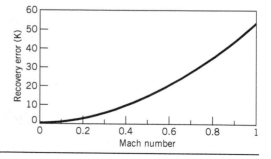

**FIGURE 8.45** Behavior of recovery error as a function of the Mach number.

# SUMMARY

Temperature is a fundamentally important quantity in science and engineering, both in concept and practice. Thus, temperature is one of the most widely measured engineering variables, providing the basis for a variety of control and safety systems. This chapter provides the basis for the selection and installation of temperature sensors.

Temperature is defined for practical purposes through the establishment of a temperature scale, such as the Kelvin scale, that encompasses fixed reference points and interpolation standards. The International Temperature Scale 1990 is the accepted standard for temperature measurement.

The two most common methods of temperature measurement use thermocouples and RTDs. Standards for the construction and use of these temperature measuring devices have been established and provide the basis for selection and installation of commercially available sensors and measuring systems.

Installation effects on the accuracy of temperature measurements are a direct result of the influence of radiation, conduction, and convection heat transfer on the equilibrium temperature of a temperature sensor. The installation of a temperature probe into a measuring environment can be accomplished in such a way as to minimize the uncertainty in the resulting temperature measurement.

# REFERENCES

1. Patterson, E. C., Eponyms: Why Celsius? *American Scientist* 77(4): 413, 1989.

2. Klein, H. A., *The Science of Measurement: A Historical Survey*, Dover, Mineola, NY, 1988.

3. Committee Report, the International Temperature Scale of 1990,[4] *Metrologia* 27(1): 3–10, 1990. (An Erratum appears in *Metrologia* 27(2): 107, 1990.)

4. *Temperature Measurement*, Supplement to American Society of Mechanical Engineers PTC 19.3, 1974.

5. Diehl, W., Thin-film PRTD, *Measurements and Control*, 155–159, December 1982.

6. Thermistor Definitions and Test Methods, Electronic Industries Association Standard RS-275-A (ANSI Standard C83. 68–1972), June 1971.

7. Thermometrics, Inc., Vol. 1. NTC Thermistors, 1993.

8. Brown, D. R., N. Fernandez, J. A. Dirks, and T. B. Stout, The Prospects of Alternatives to Vapor Compression Technology for Space Cooling and Food Refrigeration Applications, Pacific Northwest National Laboratory (PNL). U.S. Department of Energy, March 2010.

9. Burns, G. W., M. G. Scroger, and G. F. Strouse, Temperature-Electromotive Force Reference Functions and Tables for the Letter-Designated Thermocouple Types Based on the ITS-90, *NIST Monograph* 175, April 1993 (supersedes NBS Monograph 125).

10. Dewitt, D. P., and G. D. Nutter, *Theory and Practice of Radiation Thermometry*, Wiley-Interscience, New York, 1988.

11. Dils, R. R., High-temperature Optical Fiber Thermometer, *Journal of Applied Physics*, 54(3): 1198–1201, 1983.

12. Optical Fiber Thermometer, *Measurements and Control*, April 1987.

13. Dalta, R. U., M. C. Croarkin, and A. C. Parr, Cryogenic Blackbody Calibrations at the NIST Low Background Infrared Calibration Facility, *Journal of Research of NIST*, 99(1): 1994.

14. Lahiri, B. B., S. Bagavathiappan, T. Jayakumar, and J. Philip, Medical Applications of Infrared Thermometry: A Review, *Infrared Physics and Technology*, 55: 221–235, 2012.

15. Bower, S. M., and J. R. Saylor, The Effects of Surfactant Monolayers on Free Surface Natural Convection, *International Journal of Heat and Mass Transfer*, 54: 5348–5358, 2011.

16. Sparrow, E. M., Error estimates in temperature measurement. In E. R. G. Eckert and R. J. Goldstein (Eds.), *Measurements in Heat Transfer*, 2nd ed., Hemisphere, Washington, DC, 1976.

17. Moffat, R. J., Gas temperature measurements. In *Temperature—Its Measurement and Control in Science and Industry*, Vol. 3, Part 2, Reinhold, New York, 1962.

# SUGGESTED READING

Benedict, R. P., *Fundamentals of Temperature, Pressure, and Flow Measurements*, 3rd ed., Wiley, New York, 1984.

---

[4]The text of the superseded International Practical Temperature Scale of 1968 appears as an appendix in the National Bureau of Standards monograph 124. An amended version was adopted in 1975, and the English text published: *Metrologia* 12: 7–17, 1976.

# NOMENCLATURE

$a$     speed of sound ($lt^{-1}$)

$b_{\bar{x}}$     systematic standard uncertainty in $x$

$c$     speed of light in a vacuum ($lt^{-1}$)

$c_p$     specific heat ($l^2 t^{-2}/°$)

$d$     thickness; diameter ($l$)

$e_c$     conduction temperature error (°)

$e_U$     recovery temperature error (°)

$e_r$     radiation temperature error (°)

$h$     convective heat transfer coefficient ($mt^{-3}/°$); enthalpy ($l^2 t^{-2}$)

$k$     thermal conductivity ($mlt^{-3}/°$)

$k$     ratio of specific heats

$l$     length ($l$)

$m$     $\sqrt{hP/kA}$ (eq. 8.23) ($l^{-2}$)

$q$     heat flux ($mt^3$) or heat transfer ($ml^2/t^2$)

$r$     recovery factor

$r_c$     radius of curvature ($l$)

$s_{\bar{x}}$     random standard uncertainty in $x$

$u$     uncertainty

$u_d$     design-stage uncertainty

$A_c$     cross-sectional area ($l^2$)

$A_s$     surface area ($l^2$)

$B$     systematic uncertainty

$C_\alpha$     coefficient of thermal expansion ($ll^{-1}/°$)

$D$     diameter ($l$)

$E_b$     blackbody emissive power ($mt^{-3}$)

$E_{b\lambda}$     spectral emissive power ($mt^{-3}$)

$E_i$     input voltage (V)

$E_1$     voltage drop across $R_1$ (V)

$E_o$     output voltage (V)

$F$     radiation view factor

$I$     current (A)

$L$     length or thickness ($l$)

$P$     perimeter ($l$)

$R$     resistance ($\Omega$); gas constant ($l^2 t^{-2}/°$)

$R_0$     reference resistance ($\Omega$)

$R_T$     thermistor resistance ($\Omega$)

$T$     temperature (°)

$T_0$     reference temperature (°)

$T_p$     probe temperature (°)

$T_w$     wall or boundary temperature (°)

$T_\infty$     fluid temperature (°)

$U$     fluid velocity ($lt^{-1}$)$V$ voltage (V)

$\alpha$     temperature coefficient of resistivity ($\Omega/°$)

$\alpha_{AB}$     Seebeck coefficient

$\beta$     material constant for thermistor resistance (°); constant in polynomial expansion

$\pi_{AB}$     Peltier coefficient ($m\,l^2 t^{-3}\,A^{-1}$)

$\sigma$     Thomson coefficient ($ml^2 t^{-3}\,A^{-1}$); Stefan–Boltzmann constant for radiation ($mt^{-3}/(°)^4$)

$\theta$     nondimensional temperature

$\theta_x$     sensitivity index for variable $x$ (uncertainty analysis)

$\rho_e$     resistivity ($\Omega, l$)

$\delta$     thermistor dissipation constant ($ml^2 t^{-3}/°$)

$\varepsilon$     emissivity

# Pressure and Velocity Measurements

# 9

## 9.1 Introduction

This chapter introduces methods for measuring pressure and the velocity within fluids. Pressure is measured in static and dynamic systems, in acoustics, and in fluid systems. Instruments and procedures for establishing known values of pressure for calibration purposes, as well as various types of transducers for pressure measurement, are discussed. We also discuss well-established methods measuring the local and full-field velocity within a moving fluid. Finally, we present practical considerations, including common error sources, for pressure and velocity measurements. Although there are various practical test standards for pressure, many of which are applied to a specific application or measuring device, the American Society of Mechanical Engineers' Performance Test Code (ASME PTC) 19.2 provides an overview of basic pressure concepts and measuring instruments and is an accepted standard [1].

Upon completion of this chapter, the reader will be able to:

- Explain absolute and gauge pressure concepts and describe the working standards that measure pressure directly

- Explain the physical principles underlying mechanical pressure measurements and the various types of transducers used to measure pressure

- Explain pressure concepts related to static systems and with moving fluids

- Analyze the dynamic behavior of pressure system response due to transmission line effects

- Understand the basics for measuring and reporting acoustic signals

- Explain the physical principles underlying various velocity measurement methods and their practical use

## 9.2 Pressure Concepts

Pressure represents a contact force per unit area. It acts inwardly (compressive) and normally to a surface. A pure vacuum, which contains no molecules, provides the limit for a primary standard for absolute zero pressure. As shown in Figure 9.1, the absolute pressure scale is quantified relative to this absolute zero pressure. The pressure under standard atmospheric conditions is defined as $1.01325 \times 10^5$ Pa absolute (where 1 Pa $= 1$ N/m$^2$) at $15\,^{\circ}$C [2]. This is equivalent to

101.325 kPa absolute

1 atm absolute

**FIGURE 9.1**    Relative pressure
scales.

14.696 lb/in.$^2$ absolute (written as psia)

1.013 bar absolute (where 1 bar = 100 kPa)

The term "absolute" might be abbreviated as "a" or "abs."

Also indicated in Figure 9.1 is a gauge pressure scale. The gauge pressure scale is measured relative to some absolute reference pressure, which is defined in a manner convenient to the measurement. The relation between an absolute pressure, $p_{abs}$, and its corresponding gauge pressure, $p_{gauge}$, is given by

$$p_{gauge} = p_{abs} - p_0 \tag{9.1}$$

where $p_0$ is a reference pressure. A commonly used reference pressure is the local absolute atmospheric pressure. Absolute pressure is a positive number. Gauge pressure can be positive or negative depending on the value of measured pressure relative to the reference pressure. A differential pressure, such as $p_1 - p_2$, is a relative measure of pressure.

Pressure can also be described in terms of the pressure exerted on a surface that is submerged in a column of fluid at depth $h$, as depicted in Figure 9.2. From hydrostatics, the pressure at any depth within a fluid of specific weight $\gamma$ can be written as

$$p_{abs}(h) = p(h_0) + \gamma h = p_0 + \gamma h \tag{9.2}$$

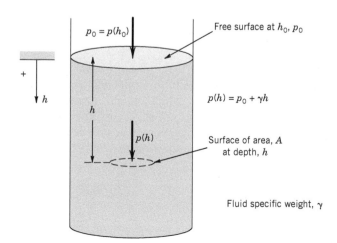

**FIGURE 9.2**
Hydrostatic-equivalent pressure head
and pressure.

where $p_0$ is the pressure at an arbitrary datum line at $h_0$, and $h$ is measured relative to $h_0$. The fluid specific weight is given by $\gamma = \rho g$ where $\rho$ is the density. When Equation 9.2 is rearranged, the equivalent head of fluid at depth $h$ becomes

$$h = \left[p_{abs}(h) - p(h)\right]/\gamma = (p_{abs} - p_0)/\gamma \tag{9.3}$$

The equivalent pressure head at one standard atmosphere ($p_0 = 0$ absolute) is

$$760 \text{ mm Hg abs} = 760 \text{ torr abs} = 1 \text{ atm abs}$$
$$= 10,350.8 \text{ mm } H_2O \text{ abs} = 29.92 \text{ in Hg abs}$$
$$= 407.513 \text{ in } H_2O \text{ abs}$$

The standard is based on mercury (Hg) with a density of $0.0135951$ kg/cm$^3$ at 0°C and water at $0.000998207$ kg/cm$^3$ at 20°C [2].

## EXAMPLE 9.1

Determine the absolute and gauge pressures and the equivalent pressure head at a depth of 10 m below the free surface of a pool of water at 20°C.

**KNOWN** $h = 10$ m, where $h_0 = 0$ is the free surface.
$T = 20°C$.
$\rho_{H_2O} = 998.207$ kg/m$^3$.
Specific gravity of mercury, $S_{Hg} = 13.57$.

**ASSUMPTION** $p(h_0) = 1.0132 \times 10^5$ Pa abs.

**FIND** $p_{abs}$, $p_{gauge}$, and $h$.

## SOLUTION
The absolute pressure is found directly from Equation 9.2. Using the pressure at the free surface as the reference pressure and the datum line for $h_0$, the absolute pressure must be

$$p_{abs}(h) = 1.0132 \times 10^5 \text{ N/m}^2 + \frac{(997.4 \text{ kg/m}^3)(9.8 \text{ m/s}^2)(10 \text{ m})}{1 \text{ kg-m/N-s}^2}$$
$$= 1.991 \times 10^5 \text{ N/m}^2 \text{ abs}$$

This is equivalent to 199.1 kPa abs or 1.96 atm abs or 28.80 lb/in.$^2$ abs or 1.99 bar abs.

The pressure can be described as a gauge pressure by referencing it to atmospheric pressure. From Equation 9.1,

$$p(h) = p_{abs} - p_0 = \gamma h$$
$$= 9.775 \times 10^4 \text{ N/m}^2$$

which is also equivalent to 97.7 kPa or 0.96 atm or 14.1 lb/in.$^2$ or 0.98 bar.

We can express this pressure as an equivalent column of liquid,

$$h = \frac{p_{abs} - p_0}{\rho g} = \frac{(1.991 \times 10^5) - (1.0132 \times 10^5)\text{N/m}^2}{(998.2 \text{ kg/m}^3)(9.8 \text{ m/s}^2)(1 \text{ N-s}^2/\text{kg-m})}$$
$$= 10 \text{ m } H_2O$$

## 9.3 Pressure Reference Instruments

The units of pressure can be defined through the standards of the fundamental dimensions of mass, length, and time. In practice, pressure transducers are calibrated by comparison against certain reference instruments, which also serve as pressure measuring instruments. This section discusses several basic pressure reference instruments that can serve either as working standards or as laboratory instruments.

### McLeod Gauge

The McLeod gauge, originally devised by Herbert McLeod in 1874 [3], is a pressure-measuring instrument and laboratory reference standard used to establish gas pressures in the subatmospheric range of 1 mm Hg abs down to 0.1 mm Hg abs. A pressure that is below atmospheric pressure is also called a vacuum pressure. One variation of this instrument is sketched in Figure 9.3a, in which the gauge is connected directly to the low-pressure source. The glass tubing is arranged so that a sample of the gas at an unknown low pressure can be trapped by inverting the gauge from the sensing position, depicted as Figure 9.3a, to that of the measuring position, depicted as Figure 9.3b. In this way, the gas trapped within the capillary is isothermally compressed by a rising column of mercury. Boyle's law is then used to relate the two pressures on either side of the mercury to the distance of travel of the mercury within the capillary. Mercury is the preferred working fluid because of its high density and very low vapor pressure.

At the equilibrium and measuring position, the capillary pressure, $p_2$, is related to the unknown gas pressure to be determined, $p_1$, by $p_2 = p_1(\forall_1/\forall_2)$ where $\forall_1$ is the gas volume of the gauge in Figure 9.3a (a constant for a gauge at any pressure) and $\forall_2$ is the capillary volume in Figure 9.3b. But $\forall_2 = Ay$, where $A$ is the known cross-sectional area of the capillary and $y$ is the vertical length of the capillary occupied by the gas. With $\gamma$ as the specific weight of the mercury, the difference in pressures is related by $p_2 - p_1 = \gamma y$ such that the unknown gas pressure is just a function of $y$:

$$p_1 = \gamma Ay^2/(\forall_1 - Ay) \tag{9.4}$$

In practice, a commercial McLeod gauge has the capillary etched and calibrated to indicate either pressure, $p_1$, or its equivalent head, $p_1/\gamma$, directly. The McLeod gauge generally does not

(a) Sensing position

(b) Indicating position

**FIGURE 9.3**
McLeod gauge.

require correction. The reference stem offsets capillary forces acting in the measuring capillary. Instrument systematic uncertainty is on the order of 0.5% (95%) at 1 mm Hg abs and increases to 3% (95%) at 0.1 mm Hg abs.

## Barometer

A barometer consists of an inverted tube containing a fluid and is used to measure atmospheric pressure. To create the barometer, the tube, which is sealed at only one end, is evacuated to zero absolute pressure. The tube is immersed with the open end down within a liquid-filled reservoir as shown in the illustration of the Fortin barometer in Figure 9.4. The reservoir is open to atmospheric pressure, which forces the liquid to rise up the tube.

From Equations 9.2 and 9.3, the resulting height of the liquid column above the reservoir free surface is a measure of the absolute atmospheric pressure in the equivalent head (Equation 9.3). Evangelista Torricelli (1608–1647), a colleague of Galileo, can be credited with developing and interpreting the working principles of the barometer in 1644.

**FIGURE 9.4** Fortin barometer.

As Figure 9.4 shows, the closed end of the tube is at the vapor pressure of the barometric liquid at room temperature. So the indicated pressure is the atmospheric pressure minus the liquid vapor pressure. Mercury is the most common liquid used, because it has a very low vapor pressure and so, for practical use, the indicated pressure can be taken as the local absolute barometric pressure. However, for very accurate work, the barometer needs to be corrected for temperature effects on the vapor pressure, for temperature and altitude effects on the weight of mercury, and for deviations from standard gravity (9.80665 m/s$^2$ or 32.17405 ft/s$^2$). Correction curves are provided by instrument manufacturers.

Barometers are used as local standards for the measurement of absolute atmospheric pressure. Under standard conditions for pressure temperature and gravity, the mercury rises 760 mm (29.92 in.) above the reservoir surface. The U.S. National Weather Service always reports a barometric pressure that has been corrected to sea-level elevation.

## Manometer

A manometer is an instrument used to measure differential pressure based on the relationship between pressure and the hydrostatic equivalent head of fluid. Several design variations are available, allowing practical measurements ranging from the order of 0.005 mm of manometer fluid to several meters.

The U-tube manometer in Figure 9.5 consists of a transparent tube filled with an indicating manometer fluid, which is a liquid of specific weight $\gamma_m$. This forms two free surfaces of the manometer fluid. The difference in pressures $p_1$ and $p_2$ applied across the two free surfaces brings about a deflection, $H$, in the level of the manometer fluid. For a measured fluid of specific weight $\gamma$, which can be either a gas or a liquid, the hydrostatic equation can be applied to the manometer of Figure 9.5 as

$$p_1 = p_2 + \gamma x + \gamma_m H - \gamma(H + x)$$

which yields the relation between the manometer deflection and applied differential pressure,

$$p_1 - p_2 = (\gamma_m - \gamma)H \tag{9.5}$$

From Equation 9.5, the static sensitivity of the U-tube manometer is given by $K = 1/(\gamma_m - \gamma)$. To maximize manometer sensitivity, choose manometer fluids that minimize the value of $(\gamma_m - \gamma)$.

**FIGURE 9.5** U-tube manometer.

**FIGURE 9.6** Micromanometer.

The manometer fluid should be selected to provide a deflection that is measurable, yet not so great that it becomes awkward to observe. From a practical standpoint, the manometer fluid must not be soluble with the measured fluid.

A variation in the U-tube manometer is the micromanometer shown in Figure 9.6. These special-purpose instruments are used to measure very small differential pressures, down to 0.005 mm $H_2O$ (0.0002 in. $H_2O$). To null the manometer, the reservoir is moved up or down until the level of the manometer fluid within the reservoir is at the same level as a set mark within a magnifying sight glass. Changes in pressure displace the fluid so that the reservoir must be moved up or down to bring the meniscus back to the set mark. The height of repositioning is equal to the change in the equivalent pressure head. The position of the reservoir is controlled by a micrometer or other calibrated displacement measuring device.

The inclined tube manometer is also used to measure small changes in pressure. It is essentially a U-tube manometer with one leg inclined at an angle θ, typically from 10 to 30 degrees relative to the horizontal. As indicated in Figure 9.7, a change in pressure equivalent to a deflection of height $H$ in a U-tube manometer would bring about a change in position of the meniscus in the

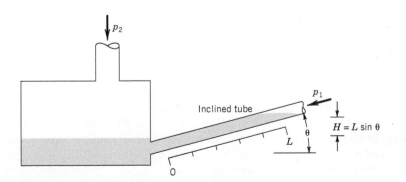

**FIGURE 9.7** Inclined tube manometer.

inclined leg of $L = H/\sin \theta$. This provides an increased sensitivity over the conventional U-tube by the factor of $1/\sin \theta$.

A number of elemental errors affect the instrument uncertainty of all types of manometers. These include scale and alignment errors, zero (null) error, temperature error, gravity error, and capillary and meniscus errors. The specific weight of the manometer fluid varies with temperature but can be corrected. For example, the temperature dependence of mercury specific weight is approximated by

$$\gamma_{Hg} = \frac{133.084}{1 + 0.00006T}[N/m^3] = \frac{848.707}{1 + 0.000101(T - 32)}[lb/ft^3]$$

with $T$ in °C or °F, respectively. A gravity correction for elevation $z$ and latitude $\phi$ corrects for gravity errors using the dimensionless correction,

$$e_1 = -(2.637 \times 10^{-3}\cos 2\phi + 9.6 \times 10^{-8}z + 5 \times 10^{-5})_{US} \tag{9.6a}$$

$$= -(2.637 \times 10^{-3}\cos 2\phi + 2.9 \times 10^{-8}z + 5 \times 10^{-5})_{metric} \tag{9.6b}$$

where $\phi$ is in degrees and $z$ is in feet for Equation 9.6a and meters in Equation 9.6b. Tube-to-liquid capillary forces can lead to the development of a meniscus in the manometer fluid. These effects can be minimized by using manometer tube bores of greater than about 6 mm (0.25 in.). In general, the instrument uncertainty in measured pressure can be as low as 0.02% to 0.2% of the reading.

## EXAMPLE 9.2    U-Tube Manometer

A high-quality U-tube manometer is a remarkably simple instrument to make. It requires only a transparent U-shaped tube, manometer fluid, and a scale to measure deflection. While a U-shaped glass tube of 6 mm or greater internal bore is the preferred choice, a length of 6-mm i.d. (inside diameter) thick-walled clear flexible tubing from the hardware store and bent to a U-shape works fine for many purposes. Water, alcohol, and mineral oil are all readily available nontoxic manometer fluids with tabulated properties. A sheet of graph paper or a ruler serves to measure meniscus deflection. There is a limit to the magnitude of pressure differential that can be measured, although use of a step stool extends this range. Tack the components to a board to produce an instrument accurate to half the resolution of the scale in terms of fluid used. Simply incline one tube leg to capture the higher sensitivity of an inclined tube manometer.

The U-tube manometer is a practical tool useful for calibrating other forms of pressure transducers in the pressure range spanning atmospheric pressure levels. As one example, the ad hoc U-tube described is convenient for calibrating surgical pressure transducers over the physiological pressure ranges.

## EXAMPLE 9.3

An inclined manometer with the inclined tube set at 30 degrees is to be used at 20°C to measure an air pressure of nominal magnitude of 100 N/m² relative to ambient. Manometer "unity" oil ($S = 1$) is to be used. The specific weight of the oil is $9,770 \pm 0.5\%$ N/m² (95%) at 20°C, the angle of inclination can be set to within 1 degree using a bubble level, and the manometer resolution is 1 mm with a manometer zero error equal to its interpolation error. Estimate the uncertainty in indicated differential pressure at the design stage.

**KNOWN** $p = 100$ N/m$^2$ (nominal).

Manometer

Resolution: 1 mm

Zero error: 0.5 mm (1/2 of resolution, 95%)

$\theta = 30 \pm 1°$ (95% assumed)

$\gamma_m = 9{,}770 \pm 0.5\%$ N/m$^3$ (95%).

**ASSUMPTIONS** Temperature and capillary effects in the manometer are negligible. The degrees of freedom in the stated uncertainties are large (see Chapter 5).

**FIND** $u_d$.

**SOLUTION**

The relation between pressure and manometer deflection is given by Equation 9.5 with $H = L \sin \theta$:

$$\Delta p = p_1 - p_2 = L(\gamma_m - \gamma) \sin \theta$$

where $p_2$ is the ambient pressure so that $\Delta p$ is the nominal pressure relative to the ambient. For a nominal $\Delta p = 100$ N/m$^2$, the nominal manometer rise $L$ would be

$$L = \frac{\Delta p}{(\gamma_m - \gamma) \sin \theta} \approx \frac{\Delta p}{\gamma_m \sin \theta} = 21 \text{ mm}$$

where $\gamma_m \gg \gamma$ and so the value for $\gamma$ and its uncertainty are neglected. For the design-stage analysis, $p = f(\gamma_m, L, \theta)$, so that the uncertainty in pressure, $\Delta p$, is estimated by

$$(u_d)_p = \pm \sqrt{\left[\frac{\partial \Delta p}{\partial \gamma_m}(u_d)_{\gamma_m}\right]^2 + \left[\frac{\partial \Delta p}{\partial L}(u_d)_L\right]^2 + \left[\frac{\partial \Delta p}{\partial \theta}(u_d)_\theta\right]^2}$$

At assumed 95% confidence levels, the manometer specific weight uncertainty and angle uncertainty are estimated from the problem statement as

$$(u_d)_{\gamma_m} = (9770 \text{ N/m}^2)(0.005) \approx 49 \text{ N/m}^3$$
$$(u_d)_\theta = 1 \text{ degree} = 0.0175 \text{ rad}$$

The uncertainty in estimating the pressure from the indicated deflection arises from both the manometer resolution, $u_o$, and the zero point offset error, which we take as its instrument error, $u_c$. Using the uncertainties associated with these errors,

$$(u_d)_L = \sqrt{u_o^2 + u_c^2} = \sqrt{(0.5 \text{ mm})^2 + (0.5 \text{ mm})^2} = 0.7 \text{ mm}$$

Evaluating the derivatives and substituting values gives a design-stage uncertainty interval of a measured $\Delta p$ of

$$(u_d)_{\Delta p} = \sqrt{(0.26)^2 + (3.42)^2 + (3.10)^2} = \pm 4.6 \text{N/m}^2 (95\%)$$

**COMMENT** At a 30-degree inclination and for this pressure, the uncertainty in pressure is affected almost equally by the instrument inclination and the deflection uncertainties. As the manometer inclination is increased to a more vertical orientation—that is, toward the U-tube manometer—inclination uncertainty becomes less important and is negligible near 90 degrees. However, for a U-tube manometer, the deflection is reduced to less than 11 mm, a 50% reduction in manometer sensitivity, with an associated design-stage uncertainty of 6.8 N/m$^2$ (95%).

# Deadweight Testers

The deadweight tester makes direct use of the fundamental definition of pressure as a force per unit area to create and to determine the pressure within a sealed chamber. These devices are a common laboratory standard for the calibration of pressure-measuring devices over the pressure range from 70 to $7 \times 10^7$ N/m$^2$ (0.01 to 10,000 psi). A deadweight tester, such as that shown in Figure 9.8, consists of an internal chamber filled with oil, a close-fitting piston and cylinder, and a set of calibrated masses. Chamber pressure acts on the end of the carefully machined piston. A static equilibrium exists when the external pressure, owing to the force exerted on the fluid by the piston acting over the equivalent area $A_e$, balances with the chamber pressure. The known weight of the piston plus the additional weight of added calibrated masses provide this external force $F$. At static equilibrium the piston floats and the chamber pressure can be deduced as

$$p = \frac{F}{A_e} + \sum \text{error corrections} \tag{9.7}$$

A pressure transducer connected to a reference port is calibrated by comparison to the chamber pressure. Except for the most rigorous calibrations, the error corrections can be ignored.

When error corrections are applied, the instrument uncertainty in the chamber pressure using a deadweight tester can be as low as 0.005% to 0.01% of the reading. A number of elemental errors contribute to Equation 9.7, including air buoyancy effects, variations in local gravity, uncertainty in the known mass of the piston and added masses, shear effects, thermal expansion of the piston area, and elastic deformation of the piston [1].

An indicated pressure, $p_i$, can be corrected for gravity effects, $e_1$, from Equation 9.6a or 9.6b, and for air buoyancy effects, $e_2$, by

$$p = p_i(1 + e_1 + e_2) \tag{9.8}$$

where

$$e_2 = -\gamma_{\text{air}}/\gamma_{\text{masses}} \tag{9.9}$$

The tester fluid lubricates the piston so that the piston is partially supported by the shear forces in the oil in the gap separating the piston and the cylinder. This error varies inversely with the tester fluid viscosity, so high-viscosity fluids are preferred. In a typical tester, the uncertainty due to this error is less than 0.01% of the reading. At high pressures, elastic deformation of the piston affects the actual piston area. For this reason, the effective area is based on the average of the piston and cylinder diameters.

**FIGURE 9.8**    Deadweight tester.

## EXAMPLE 9.4

A deadweight tester indicates 100.00 lb/in.$^2$ (i.e., 100.00 psi), at 70°F in Clemson, SC ($\phi = 34°$, $z = 841$ ft). Manufacturer specifications for the effective piston area were stated at 72°F so that thermal expansion effects remain negligible. Take $\gamma_{air} = 0.076$ lb/ft$^3$ and $\gamma_{mass} = 496$ lb/ft$^3$. Correct the indicated reading for known errors.

**KNOWN** $p_i = 100.00$ psi; $z = 841$ft; $\phi = 34°$.

**ASSUMPTION** Systematic error corrections for altitude and latitude apply.

**FIND** $p$.

**SOLUTION**
The corrected pressure is found using Equation 9.8. From Equation 9.9, the correction for buoyancy effects is

$$e_2 = -\gamma_{air}/\gamma_{masses} = -0.076/496 = -0.000154$$

The correction for gravity effects is from Equation 9.6a:

$$e_1 = -(2.637 \times 10^{-3} \cos 2\phi + 9.6 \times 10^{-8} z + 5 \times 10^{-5})$$
$$= -(0.0010 + 8 \times 10^{-5} + 5 \times 10^{-5}) = -0.001119$$

From Equation 9.8, the corrected pressure becomes

$$p = 100.00 \times (1 - 0.000154 - 0.001119)\text{lb/in.}^2 = 99.87\,\text{lb/in.}^2$$

**COMMENT** This amounts to correcting an indicated signal for known systematic errors. Here that correction is $\approx 0.13\%$.

# 9.4 Pressure Transducers

A pressure transducer is a device that converts a measured pressure into a mechanical or electrical signal. The primary sensor is usually an elastic element that deforms or deflects under the measured pressure relative to a reference pressure. Several common elastic elements used, as shown in Figure 9.9, include the Bourdon tube, bellows, and diaphragm. A transducer element converts the elastic element deflection into a readily measurable signal such as an electrical voltage or mechanical rotation of a pointer. There are many methods available to perform this transducer function, and we examine a few common ones.

General categories for pressure transducers are absolute, gauge, vacuum, and differential. These categories reflect the application and reference pressure used. Absolute transducers have a sealed reference cavity held at a pressure of absolute zero, enabling absolute pressure measurements. Gauge transducers have the reference cavity open to atmospheric pressure and measure pressure relative to this. Differential transducers measure the difference between two applied pressures. Vacuum transducers are a special form of absolute transducer for low-pressure measurements.

Pressure transducers are subject to some or all of the following elemental errors: resolution, zero error, linearity error, sensitivity error, hysteresis, noise, and drift resulting from environmental temperature changes. Electrical transducers are also subject to loading error between the transducer output and its indicating device (see Chapter 6). When this is a consideration, a voltage follower (see Chapter 6) can be inserted at the output of the transducer to isolate transducer load.

Corrugated diaphragm

**FIGURE 9.9**    Elastic elements used as pressure sensors.

## Bourdon Tube

The Bourdon tube is a curved metal tube having an elliptical cross section that mechanically deforms under pressure. In practice, one end of the tube is held fixed and the input pressure applied internally. A pressure difference between the outside of the tube and the inside of the tube brings about tube deformation and a deflection of the tube free end. This action of the tube under pressure can be likened to the action of a deflated balloon that is subsequently inflated. The magnitude of the deflection of the tube end is proportional to the magnitude of the pressure difference. Several variations exist, such as the C shape (Figure 9.9), the spiral, and the twisted tube. The exterior of the tube is usually open to atmosphere (hence, the origin of the term "gauge" pressure referring to pressure referenced to atmospheric pressure), but in some variations the tube may be placed within a sealed housing and the tube exterior exposed to some other reference pressure, allowing for absolute and differential designs.

The Bourdon tube mechanical dial gauge is a commonly used pressure transducer. A typical design is shown in Figure 9.10, in which the secondary element is a mechanical linkage that converts the tube displacement into a rotation of a pointer. Designs exist for low or high pressures, including vacuum pressures, and selections span a wide choice in range. The best Bourdon tube gauges have instrument uncertainties as low as 0.1% of the full-scale deflection of the gauge, but values of 0.5% to 2% are more common. However, the attractiveness of this device is that it is simple, portable, and robust, lasting for years of use.

## Bellows and Capsule Elements

A bellows sensing element is a thin-walled, flexible metal tube formed into deep convolutions and sealed at one end (Figure 9.9). One end is held fixed while pressure is applied internally. A difference between the internal and external pressures causes the bellows to change in length. The bellows is housed within a chamber that can be sealed and evacuated for absolute measurements, vented through a reference pressure port for differential measurements, or opened to

**FIGURE 9.10** Bourdon tube pressure gauge.

**FIGURE 9.11** Potentiometer pressure transducer.

atmosphere for gauge pressure measurements. A similar design, the capsule sensing element, is also a thin-walled, flexible metal tube whose length changes with pressure, but its shape tends to be wider in diameter as well as shorter (Figure 9.9).

A mechanical linkage is used to convert the translational displacement of the bellows or capsule sensors into a measurable form. A common transducer is the sliding arm potentiometer (voltage-divider, Chapter 6) found in the potentiometric pressure transducer shown in Figure 9.11. Another type uses a linear variable displacement transducer (LVDT; see Chapter 12) to measure the bellows or capsule displacement. The LVDT design has a high sensitivity and is commonly found in pressure transducers rated for low pressures and for small pressure ranges, such as zero to several hundred mm Hg absolute, gauge, or differential.

## Diaphragms

An effective primary pressure element is a diaphragm (Figure 9.9), which is a thin, elastic circular plate supported about its circumference. The action of a diaphragm within a pressure transducer is similar to the action of a trampoline; a pressure differential on the top and bottom diaphragm faces acts to deform it. The magnitude of the deformation is proportional to the pressure difference. Both membrane and corrugated designs are used. Membranes are made of metal or nonmetallic material, such as plastic or neoprene. The material chosen depends on the pressure range anticipated and the fluid in contact with it. Corrugated diaphragms contain a number of corrugations that serve to increase diaphragm stiffness and to increase the diaphragm effective surface area.

Diaphragm-based pressure transducers have good linearity and resolution over their useful range. An advantage of the diaphragm sensor is that the very low mass and relative stiffness of the thin diaphragm give the sensor a very high natural frequency with a small damping ratio for accurate transient or dynamic measurements. Hence, these transducers can have a very wide frequency response and very short 90% rise and settling times. The natural frequency of a circular diaphragm can be estimated by [4]

$$\omega_n(\text{rad/s}) = 10.21 \sqrt{\frac{E_m t^2}{12(1 - v_p^2)\rho r^4}} \tag{9.10}$$

where $E_m$ is the material bulk modulus (psi or N/m$^2$), $t$ the thickness (in. or m), $r$ the radius (in. or m), $\rho$ the material density (lb$_m$/in.$^3$ or kg/m$^3$), and $v_p$ the Poisson's ratio for the diaphragm material. The maximum elastic deflection of a uniformly loaded, circular diaphragm supported about its circumference occurs at its center and can be estimated by

$$y_{\text{max}} = \frac{3(p_1 - p_2)(1 - v_p^2)r^4}{16E_m t^3} \tag{9.11}$$

provided that the deflection does not exceed one-third the diaphragm thickness. Various secondary elements are available to translate this displacement of the diaphragm into a measurable signal. Several methods are discussed below.

### Strain Gauge Elements

A common method for converting diaphragm displacement into a measurable signal is to sense the strain induced on the diaphragm surface as it is displaced. Strain gauges, devices whose measurable resistance is proportional to their sensed strain (see Chapter 11), can be bonded directly onto the diaphragm, integrated within the diaphragm material itself or onto a deforming element (such as a thin beam) attached to the diaphragm so as to deform with the diaphragm and to sense strain. Metal strain gauges can be used with liquids. Strain gauge resistance is reasonably linear over a wide range of strain and can be directly related to the sensed pressure [5]. A diaphragm transducer using strain gauge detection is depicted in Figure 9.12.

By using semiconductor technology in pressure transducer construction, we have a variety of very fast, very small, highly sensitive strain gauge–diaphragm transducers for measuring both static and dynamic pressures. Silicone piezoresistive strain gauges can be diffused into a single crystal of silicone wafer, which forms the diaphragm. Semiconductor strain gauges have a static sensitivity that is 50 times greater than conventional metallic strain gauges. Because the piezoresistive gauges are integral to the diaphragm, they are relatively immune to the thermoelastic strains prevalent in conventional metallic strain gauge–diaphragm constructions. Furthermore, a silicone diaphragm does not creep with age (as does a metallic gauge), thus minimizing calibration drift over time. However, uncoated silicone does not tolerate liquids.

### Capacitance Elements

Another common method to convert diaphragm displacement to a measurable signal is a capacitance sensor. One version uses a thin metallic diaphragm as one plate of a capacitor paired with a

**(b)** Bridge–strain gauge circuit
for pressure diaphragms.

**(a)** Sensing scheme

**FIGURE 9.12** Diaphragm pressure transducer using four active resistance strain gauges.

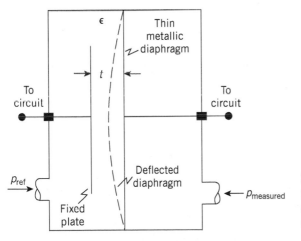

**FIGURE 9.13** Capacitance pressure transducer. In this schematic, the diaphragm is conductive and its deflection exaggerated.

fixed plate to complete the capacitor. The diaphragm is exposed to the process pressure on one side and to a reference pressure on the other or to a differential pressure. When pressure changes, this deflects the diaphragm and the gap between the plates changes, causing a change in capacitance.

To illustrate this, a transducer using this method is depicted in Figure 9.13. The capacitance $C$ developed between two parallel plates separated by average gap $t$ is determined by

$$C = c\varepsilon A/t \tag{9.12}$$

where the product $c\varepsilon$ is the permittivity of the material between the plates relative to a vacuum ($\varepsilon = 8.85 \times 10^{-12}$ F/m; $c$ = dielectric constant), and $A$ is the overlapping area of the two plates. The dielectric constant depends on the material in the gap, which for air is $c = 1$ but for water is $c = 80$. The capacitance responds to an instantaneous change in the area-averaged plate gap separation from which the time-dependent pressure is determined. However, the capacitance change is small relative to the absolute capacitance. Oscillator and bridge circuits are commonly used to operate the circuit and to measure the small capacitance change providing an output voltage $E_0$.

**EXAMPLE 9.5**

Estimate the theoretical capacitance of a sensor similar to that of Figure 9.13 if the plate area is 1 mm$^2$ and the instantaneous plate separation is 1 mm. What is the sensor sensitivity? Assume the plates are separated by air.

**SOLUTION**
With $A = 1$ mm$^2$, $t = 1$ mm, $\varepsilon = 8.85 \times 10^{-15}$ F/mm, $c = 1$:

$$C = c\varepsilon A/t = (1)(8.85 \times 10^{-15} F/\text{mm})(1 \text{ mm}^2)/(1 \text{ mm}) = 8.85 \times 10^{-15} F$$

or $C = 8.85 \times 10^{-3}$ pF (picofarads). For small pressure changes, the diaphragm curvature has a negligible effect on the area-averaged plate gap so that pressure can be treated as linear with the area-average gap value.

The sensor sensitivity is $K_s = \partial C/\partial t = -8.85 \times 10^{-3} \times (cA/t^2)$ pF/mm.

---

The capacitance pressure transducer has the attractive features of other diaphragm transducers, including small size and a wide operating range, and is suitable for both static and dynamic pressure measurements. When the technique is used to measure acoustical sound pressure, the device is known as a condenser microphone. Many inexpensive pressure transducers use this measuring principle and are suitable for most engineering measurement demands, including a niche as the on-board car tire pressure sensor. However, the principle is sensitive to temperature changes and has a relatively high impedance output. If attention is paid to these shortcomings, an accurate and stable transducer can be made.

## Piezoelectric Crystal Elements

Piezoelectric crystals form effective secondary elements for dynamic pressure measurements, including measuring acoustical sound pressure and dynamic mechanical surface-to-surface contact pressure. They are not intended for static pressure measurements.

Under the action of compression, tension, or shear, a piezoelectric crystal deforms and develops a surface charge $q$, which is proportional to the force acting to bring about the deformation. In a piezoelectric pressure transducer, a preloaded crystal or stack of crystals is mounted to the diaphragm sensor as indicated in Figure 9.14. Pressure acts normal to the crystal axis and changes

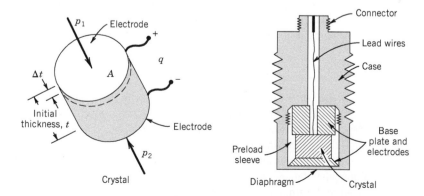

**FIGURE 9.14** Piezoelectric pressure transducer.

the crystal thickness $t$ by a small amount $\Delta t$. This sets up a charge, $q = K_q pA$, where $p$ is the pressure acting over the electrode area $A$ and $K_q$ is the crystal charge sensitivity, a material property. A charge amplifier (see Chapter 6) is used to convert charge to voltage so that the voltage developed across the electrodes is

$$E_o = q/C \tag{9.13}$$

where $C$ is the capacitance of the crystal-electrode combination, again given by Equation 9.12. The operating equation becomes

$$E_o = K_q tp/c\varepsilon = Kp \tag{9.14}$$

where $K$ is the overall transducer gain. The crystal sensitivity for quartz, the most common material used, is $K_q = 2.2 \times 10^{-9}$ coulombs/N.

# 9.5 Pressure Transducer Calibration

## Static Calibration

A static calibration of a pressure transducer is usually accomplished either by direct comparison against any of the pressure reference instruments discussed (Section 9.3) or against the output from a certified laboratory standard transducer. For the low-pressure range, the McLeod gauge or the appropriate manometric instruments along with the laboratory barometer serve as convenient working standards. The approach is to pressurize (or evacuate) a chamber and expose both the reference instrument, which serves as the standard, and the candidate transducer to the same pressure for a side-by-side measurement. For the high-pressure range, the deadweight tester is a commonly used pressure reference standard.

## Dynamic Calibration

The rise time and frequency response of a pressure transducer are found by dynamic calibration, and there are a number of clever ways to accomplish this [6]. As discussed in Chapter 3, the rise time of an instrument is found through a step change in input. For underdamped systems, natural frequency and damping ratio can be determined based on the ringing behavior (see Chapter 3) from either a step or an impulse test. The frequency response can also be found directly by applying a constant amplitude periodic input signal and varying its frequency. Although a flush-mounted transducer places the sensor in direct contact with the measurement site, transducers could also be attached by means of a pressure tap or by connecting tubing between the transducer and the tap. This length (called the transmission line) affects the overall response and should be included as part of the dynamic calibration.

An electrical switching valve or flow control valve can create a step change in pressure. But the mechanical lag of the valve limits its use to transducers having an expected rise time on the order of 50 ms or more. Faster applications might use a shock tube calibration or some equivalent diaphragm burst test.

As shown in Figure 9.15, the shock tube consists of a long pipe separated into two chambers by a thin diaphragm. The transducer is mounted into the pipe wall of one chamber, the expansion section, at pressure $p_1$. The pressure in the other chamber, the driver section, is raised from $p_1$ to $p_2$. Some mechanism, such as a mechanically controlled needle, is used to burst the diaphragm on command. Upon bursting, the pressure differential causes a pressure shock wave to move down the low-pressure chamber. A shock wave has a thickness on the order of 1 μm and moves at the speed of sound, $a$. So as the shock passes the transducer, the transducer experiences a change in

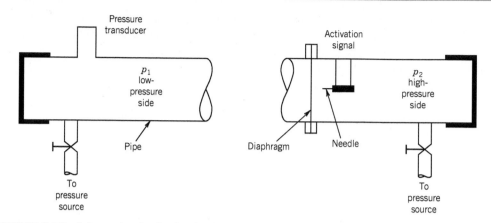

**FIGURE 9.15**    Schematic of a shock tube facility.

pressure from $p_1$ to $p_3$ over a time $t = d/a$, where $d$ is the diameter of the transducer pressure port, and pressure $p_3$ is

$$p_3 = p_1[1 + (2k/k + 1)(M_1^2 - 1)] \tag{9.15}$$

where $k$ is the gas-specific heat ratio and $M_1$ is the Mach number calculated using normal shock wave tables and absolute pressure $p_1$. The velocity of the shock wave can also be deduced from the output of fast-acting standard pressure sensors mounted in the shock tube wall. Typical values of $t$ are on the order of 1 to 10 μs, so this method is at least four orders of magnitude faster than a switching valve. The transducer rise time is calculated from the output record.

A common verification check for system response is the "pop test," which is well suited for liquids or gases and systems needing just moderate response times. In this situation, the transducer and connecting tubing are pressurized to a steady value, perhaps by using a syringe or hand pump. The system is suddenly vented to atmosphere. The recorded transducer response is indicative of the system rise time and ringing behavior. One variation of this approach attaches a balloon or similar flexible material to one end of the connecting tubing/transducer system. After pressurizing, the balloon is popped to suddenly vent the system.

Other techniques include an encased loudspeaker or an acoustically resonant enclosure, which serves as a frequency driver, a reciprocating piston and cylinder to develop a sinusoidal pressure variation, or using an oscillating flow control valve to vary system pressure with time [6].

---

**EXAMPLE 9.6**    **Pop Test**

A surgical pressure transducer is attached to a stiff-walled catheter and a small balloon is attached at its other end. The catheter is filled with saline using a syringe to pressurize the system to 80 mm Hg. At $t = 0$ s, the balloon is popped, forcing a step change in pressure from 80 to 0 mm Hg. The time-based signal is recorded (Figure 9.16). From the data, the ringing period is measured to be 45.5 ms at second peak amplitude, $y(0.0455) = 5.152$ mV. Would this system be suitable for measuring physiological pressures having frequency content up to 5 Hz? Static sensitivity $K = 1$ mV/mm Hg.

**KNOWN** $A = 80$ mm Hg, $K = 1$ mV/mm Hg.
       $y(0) = KA = 80$ mV; $y(0.0455) = 5.152$ mV.
       $T_d = 0.0455$ s.

**FIGURE 9.16**  Recorded output signal from the pop test of Example 9.6.

**FIND**  $M(f = 5\,\text{Hz})$.

**SOLUTION**

We use methods developed in Chapter 3. The step-function response has the form

$$y(t) = Ce^{-\omega_n \zeta t} \cos\left(\omega_n \sqrt{1 - \zeta^2} + \phi\right)$$

where for $y(0) = 80\,\text{mV}$, $C = 80\,\text{mV}$, which is consistent with the recorded signal. The steady-state value is $y(\infty) = 0$. The first peak amplitude is at $t = 0$, $y_1 = y(0) = 80\,\text{mV}$. The second peak amplitude is $y_2 = y(0.0455) = 5.152\,\text{mV}$. Using logarithmic decrement,

$$\zeta = \frac{1}{\sqrt{1 + \left(2\pi/\ln\left(y_1/y_2\right)\right)^2}} = \frac{1}{\sqrt{1 + (2\pi/\ln(80/5.152))^2}} = 0.40$$

$$f_n = 1/T_d \sqrt{1 - \zeta^2} = 23.98\,\text{Hz} \quad (\text{i.e., } \omega_n = 150\,\text{rad/s})$$

Then,

$$M(f = 5) = \frac{1}{\sqrt{\left[1 - (f/f_n)^2\right]^2 + (2\zeta f/f_n)^2}} = 1.03$$

With a dynamic error ($\delta(f) = M(f) - 1$) of only about 3%, this system should be suitable for the intended measurements.

## 9.6 Pressure Measurements in Moving Fluids

Pressure measurements in moving fluids warrant special consideration. Consider the flow over the bluff body shown in Figure 9.17. Assume that the upstream flow is uniform and steady with negligible losses. Along streamline $A$, the upstream flow moves with a velocity $U_1$, such as at point 1. As the flow approaches point 2, it must slow down and finally stop at the front end of the body. Above streamline $A$, flow moves over the top of the bluff body, and below streamline $A$, flow moves under the body. Point 2 is known as the stagnation point and streamline $A$ the stagnation streamline for this flow. Along streamline $B$, the velocity at point 3 is $U_3$ and because the upstream flow is considered to be uniform it follows that $U_1 = U_3$. As the flow along $B$ approaches the body,

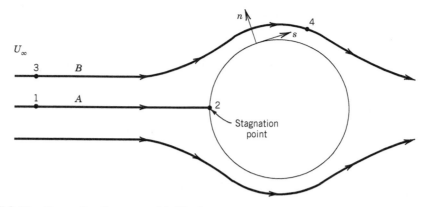

**FIGURE 9.17**   Streamline flow over a bluff body.

it is deflected around the body. From the conservation of mass principles, $U_4 > U_3$. Application of conservation of energy between points 1 and 2 and between 3 and 4 yields

$$p_1 + \rho U_1^2/2 = p_2 + \rho U_2^2$$
$$p_3 + \rho U_3^2/2 = p_4 + \rho U_4^2 \tag{9.16}$$

However, because point 2 is the stagnation point, $U_2 = 0$, and

$$p_2 = p_t = p_1 + \rho U_1^2/2 \tag{9.17}$$

it follows that $p_2 > p_1$ by an amount equal to $p_v = \rho U_1^2/2$, called the *dynamic pressure*, an amount equivalent to the kinetic energy per unit mass of the flow as it moves along the streamline. If no energy is lost through irreversible processes, such as through a transfer of heat,[1] this translational kinetic energy is transferred completely into $p_2$. The value of $p_2$ is known as the *stagnation* or the *total pressure* and is noted as $p_t$. The total pressure can be measured by bringing the flow to rest at a point in an isentropic manner.

The pressures at 1, 3, and 4 are known as static pressures[2] of the flow. The *static pressure* is that pressure sensed by a fluid particle as it moves with the local flow. The static pressure and velocity at points 1 and 3 are given the special names of the "freestream static pressure" and "freestream velocity." Because $U_4 > U_3$, Equation 9.16 shows that $p_4 < p_3$. It follows from Equation 9.17 that the total pressure is the sum of the static and dynamic pressures anywhere in the flow.

## Total Pressure Measurement

In practice, the total pressure is measured using an impact probe such as those depicted in Figure 9.18. A small hole in the impact probe is aligned with the flow so as to cause the flow to come to rest at the hole. The sensed pressure is transferred through the impact probe to a pressure transducer or other pressure-sensing device such as a manometer. Alignment with the flow is somewhat critical, although the probes in Figure 9.18a, b are relatively insensitive (within ~1% error in indicated reading) to misalignment within a ±7 degree angle [1]. A special type of impact probe shown in Figure 9.18c, known as a Kiel probe, uses a shroud around the impact port. The effect of the shroud is to force the local flow to align itself with the shroud axis so as

---

[1]This is a realistic assumption for subsonic flows. In supersonic flows, the assumption is not valid across a shock wave.
[2]The term "static pressure" conforms to common expression; the term "stream pressure" is also used.

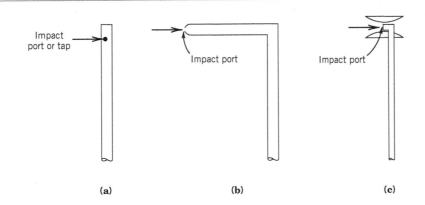

**FIGURE 9.18**    Total pressure measurement devices. (a) Impact cylinder. (b) Pitot tube. (c) Kiel probe.

to impact directly onto the impact port. This effectively eliminates total pressure sensitivity to misalignment up to ±40 degree [1]. A thermocouple temperature sensor is often built into the impact site so that the flow stagnation temperature can be measured simultaneously.

## Static Pressure Measurement

The local value of static pressure in a moving stream is measured by sensing the pressure in the direction that is normal to the flow streamline. Within ducted flows, static pressure is sensed by wall taps, which are small, burr-free, holes drilled into the duct wall perpendicular to the flow direction at the measurement point. The tap is fitted with a hose or tube, which is connected to a pressure gauge or transducer. A recommended design for a wall tap is shown in Figure 9.19. The tap hole diameter $d$ is typically between 1% and 10% of the pipe diameter, with the smaller size preferred [7]. The tap must be perpendicular to the local tangent to the wall and with no drilling burrs [8].

Alternatively, a static pressure probe can be inserted into the flow to measure local stream pressure. It should be a streamlined design to minimize the disturbance of the flow. It should be physically small so as not to cause more than a negligible increase in velocity in the vicinity of measurement. As a rule, the frontal area of the probe should not exceed 5% of the pipe flow area. The static pressure sensing port should be located well downstream of the leading edge of the probe so as to allow the streamlines to realign themselves parallel with the probe. Such a concept is built into the improved Prandtl tube design shown in Figure 9.20a.

A Prandtl tube probe consists of eight holes arranged about the probe circumference and positioned 8 to 16 probe tube diameters, $D$, downstream of the probe leading edge and 16 probe tube diameters upstream of its support stem. A pressure transducer or manometer is connected to the probe stem to measure the sensed pressure. The hole positions are chosen to minimize

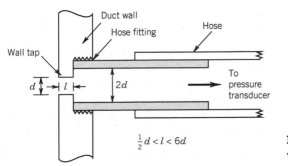

**FIGURE 9.19**    Anatomy of a static pressure wall tap.

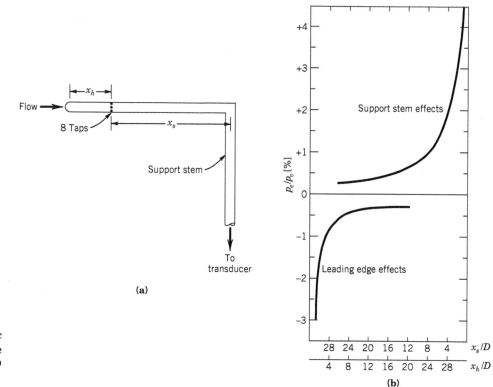

**FIGURE 9.20**
Improved Prandtl tube for static
pressure. (a) Design. (b) Relative
static error along tube length. $D$
is the probe's tube diameter.

static pressure error caused by the disturbance to the flow streamlines due to the probe's leading edge and stem. This is illustrated in Figure 9.20b, where the relative static error, $p_e/p_v = (p_i - p)/\left(\frac{1}{2}\rho U^2\right)$, is plotted as a function of tap location along the probe body. Real viscous effects around the static probe cause a slight discrepancy between the actual static pressure and the indicated static pressure. To account for this, a correction factor, $C_0$, is used with $p = C_0 p_i$ where $0.99 < C_0 < 0.995$ and $p_i$ is the indicated (measured) pressure.

## 9.7 Modeling Pressure–Fluid Systems

Fluid systems can be modeled using lumped parameter ideal elements just as common resistor-inductor-capacitor electrical loops and mass-damper-spring mechanical systems are used. The common elements are inertance, resistance, and compliance.

*Inertance* describes the inertial properties of a mass in motion, such as that of a mass of fluid moving within a vessel. For a fluid of density $\rho$ and a vessel of cross-sectional area $A$ and length $l$, the inertance is written

$$L_f = \rho l / A \tag{9.18}$$

When modeling inertial forces in laminar flows, this value should be increased by a factor of $\frac{4}{3}$. Inertance is the direct analog to electrical inductance.

Fluid *resistance* describes the opposition to motion. This is the pressure change required to move a volume of fluid per unit time, $Q$. It is written

$$R = \Delta p^n / Q = \Delta E / I \tag{9.19}$$

where $n = 1$ for laminar and $n = 0.5$ for turbulent flows. Hence, $Q = \frac{1}{R}\Delta p^n$.

The resistance of the laminar flow of a Newtonian fluid through a circular pipe is $R = \frac{128\mu l}{\pi d^4}$, where $\mu$ is the fluid viscosity. In electrical systems, it has the analogous meaning as the opposition to current flow for an imposed voltage potential, such as an electrical resistor.

*Compliance* describes a measure of the volume change associated with a corresponding pressure change, such as

$$C_{vp} = \Delta \forall / \Delta p \tag{9.20}$$

It is a measure of the flexibility in a structure, component, or substance, and so it is the inverse of the system *stiffness*. Compliance is the direct analog to electrical capacitance.

## 9.8 Design and Installation: Transmission Effects

Consider the configuration depicted in Figure 9.21 in which a tube of volume $\forall_t$ with length $l$ and diameter $d$ is used to connect a pressure tap to a pressure transducer of internal dead volume $\forall$ (e.g., Figure 9.12). Under static conditions, the pressure transducer indicates the static pressure at the tap. But if the pressure at the tap is a time-dependent pressure, $p_a(t)$, the response behavior of the tubing influences the time-indicated output from the transducer, $p(t)$.

By considering the one-dimensional pressure forces acting on a lumped mass of fluid within the connecting tube, balancing inertance, compliance, and resistance against forcing function, we can develop a model for the pressure system response. A network model is shown in Figure 9.22a in which the circuit is driven between two pressures, the applied pressure $p_a(t)$ at the tap and the

**FIGURE 9.21**    Wall tap to pressure transducer connection: the transmission line.

**(a)**

**(b)**

**FIGURE 9.22**    An equivalent lumped parameter network of the pressure transmission line model of Figure 9.21 using an electrical analogy.

measured pressure $p_m(t)$ at the transducer sensor. Using the electrical analog, inertance is modeled by the inductor, fluid resistance by a resistor, and compliance by a capacitor (Figure 9.22b). The circuit analysis of the two loops gives

$$L\ddot{I} + RI + \frac{1}{C}\int I dt = E_a \quad \text{and} \quad \frac{1}{C}\int I dt = E_m \tag{9.21}$$

Taking the derivative of the second loop to get $\dot{E}_m$ and $\ddot{E}_m$ in terms of $\dot{I}$ and $\ddot{I}$, then substituting these back into Equation 9.21 with $E_a = p_a$, $E_m = p_m$, $L = L_f$, and $C = C_{vp}$ gives

$$L_f C_{vp}\ddot{p}_m + RC_{vp}\dot{p}_m + p_m = p_a(t) \tag{9.22}$$

Substituting Equations 9.18 to 9.20 into Equation 9.22 gives the working system response equation for the applied and measured pressures,

$$\frac{16\ell\rho C_{vp}}{3\pi d^2}\ddot{p}_m + \frac{128\mu\ell C_{vp}}{\pi d^4}\dot{p}_m + p_m = p_a(t) \tag{9.23}$$

in which we have augmented the fluid inertial force by $\frac{4}{3}$ [9].

In this simple model, the system compliance lumps the compliance of the fluid, tube walls, and transducer into a single value, $C_{vp}$. Because these individual compliances could be modeled separately by using capacitors in parallel, the total capacitance is simply the sum of each. If one compliance dominates, the others can be neglected. Furthermore, the inertance and resistance of the connecting tube and the transducer cavity are lumped into single values. Improved models use distributed, lumped parameters such as are commonly used to model physiological vascular systems [10]. A longstanding approach to modeling the transmission line [11] examines the forces acting on a fluid element. In this model, small pressure changes act on the element, moving it back and forth by a distance $x$ within the tube. Summing the forces and substituting for $p_m$ again results in Equation 9.23 [12].[3]

We can study the transient and frequency response of the system represented by Equation 9.23 by extracting values for $\omega_n$ and $\zeta$,

$$\omega_n = \frac{d}{4}\sqrt{3\pi/\rho\ell C_{vp}} \tag{9.24}$$

$$\zeta = \frac{16\mu}{d^3}\sqrt{3\ell C_{vp}/\pi\rho} \tag{9.25}$$

The total system compliance could be measured using Equation 9.20 by closing the pressure tap, increasing the fluid volume in the tubing by a small known amount, such as by syringe, and measuring the corresponding pressure change.

## Liquids

Liquids are relatively incompressible, so the compression-restoring force in the transmission line results primarily from the compliance in the transducer, which is a stated transducer specification, for use in Equations 9.24 and 9.25. Connecting tubing can usually be considered rigid for the underlying assumptions of the above lumped parameter analysis. Thick-walled, flexible tubing is often used, but this is relatively rigid and its compliance can be ignored. If need be, the compliance can be measured.

---

[3]Be aware that models are based on simplifying assumptions and should be used only as a guide in the design of a system, not as a replacement for in situ calibration.

# Gases

For gases, we simplify by assuming that the system is rigid relative to the compressibility of the gas. Compliance is then modeled in terms of the fluid's adiabatic bulk modulus of elasticity, $E_m = \forall/C_{vp}$. This gives

$$\omega_n = \frac{d}{4}\sqrt{3\pi E_m/\rho\ell\forall} \tag{9.26}$$

$$\zeta = \frac{16\mu}{d^3}\sqrt{3\ell\forall/\pi\rho E_m} \tag{9.27}$$

Equations 9.26 and 9.27 can also be written in terms of the speed of sound for the gas, $a$, which is related to its compressibility by $a = \sqrt{E_m/\rho}$ and for a perfect gas by $a = \sqrt{kRT}$, where $T$ is the gas absolute temperature, giving

$$\omega_n = \frac{ad}{4}\sqrt{3\pi/\ell\forall} \tag{9.28}$$

$$\zeta = \frac{16\mu}{a\rho d^3}\sqrt{3\ell\forall/\pi} \tag{9.29}$$

When the tube volume, $\forall_t \gg \forall$, then a series of standing pressure waves develop and we can expect $\omega \sim O(a/l)$. Hougen et al. [13] discuss an improved prediction as

$$\omega_n = \frac{a}{\ell\sqrt{0.5 + \forall/\forall_t}} \tag{9.30}$$

$$\zeta = \frac{16\mu\ell}{\rho a d^2}\sqrt{0.5 + \forall/\forall_t} \tag{9.31}$$

Note that in all cases, larger diameter and shorter length tubes improve pressure system response.

---

**EXAMPLE 9.7**

The pressure in a water-filled pipe varies with time. A pressure transducer is connected to a wall tap using a length of nonrigid plastic tubing to measure this pressure. To estimate the pressure system compliance, the tubing is removed at the wall tap, filled with water, and purged of any residual air. The transducer output is noted. Using a syringe, 1 mL of water is then added to the system and the corresponding measured pressure increases by 100 mm Hg. Find the compliance.

**SOLUTION**

Using Equation 9.20, the compliance of the transducer-tubing system is

$$C_{vp} = \Delta\forall/\Delta p = 1\ \text{mL}/100\ \text{mm Hg} = 0.01\ \text{mL/mm Hg}$$

---

**EXAMPLE 9.8**

A pressure transducer with a natural frequency of 100 kHz is connected to a 0.10-in. static wall pressure tap using a 0.10-in. i.d. rigid tube that is 5 in. long. The transducer dead volume is 1 in.$^3$ Determine the system frequency response to fluctuating pressures of air at 72°F if the pressure fluctuates about an average value of 1 atm abs. Express the frequency response in terms of $M(\omega)$. $R_{\text{air}} = 53.3\ \text{ft-lb/lb}_m\text{-R}$; $\mu = 4 \times 10^{-7}\ \text{lb-s/ft}^2$; $E_m = 20.5\ \text{lb/in.}^2$

**KNOWN** $\ell = 5$ in. $k = 1.4$

$$d = 0.1\text{in.} \quad T = 72°\text{F} = 532\text{R}$$
$$\forall = 1\text{in.}^3 \quad \rho = p/RT = 0.075\text{Ib}_\text{m}/\text{ft}^3$$

**ASSUMPTION**  Air behaves as a perfect gas.

**FIND**  $M(\omega)$.

**SOLUTION**

The frequency-dependent magnitude ratio is given by Equation 3.22

$$M(\omega) = \frac{1}{\sqrt{\left[1 - (\omega/\omega_n)^2\right]^2 + \left[2\zeta(\omega/\omega_n)\right]^2}}$$

We need $\omega_n$ and $\zeta$ to solve for $M(\omega)$ for various frequencies.

With $\forall_t < \forall$, we use Equations 9.26 and 9.27, or alternately using Equations 9.28 and 9.29 with $a = 1,130$ ft/s, to find

$$\omega_n = \frac{d}{4}\sqrt{3\pi E_m/\rho\ell\forall} = 470\text{rad/s}$$

$$\zeta = \frac{16\mu}{d^3}\sqrt{3\ell\forall/\pi\rho E_m} = 0.06$$

This system is lightly damped. The magnitude frequency response, $M(\omega)$, is shown in Figure 9.23. The transmission line effects control the transducer response. In effect, the transmission line acts as a mechanical low-pass filter for the transducer.

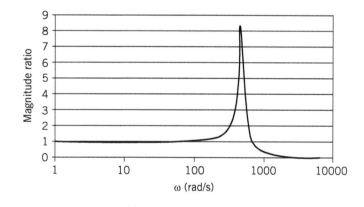

**FIGURE 9.23**    Magnitude response of the transmission line for Example 9.10.

## Heavily Damped Systems

In systems in which we estimate a damping ratio greater than 1.5, the frequency response model can be simplified further. The behavior of the pressure-measuring system closely follows that of a first-order system. Again, the typical pressure transducer–tubing system has a compliance $C_{vp}$, which is a measure of the transducer and tubing volume change relative to an applied pressure change. The response of the first-order system is indicated through its time constant, which is estimated by neglecting the second-order term in Equation 9.23 so that [12]

$$\tau = \frac{128\mu\ell C_{vp}}{\pi d^4} \tag{9.32}$$

Equation 9.32 shows that the time constant is proportional to $\ell/d^4$. Long- and small-diameter connecting tubes promote a more sluggish system response to changes in pressure.

## 9.9  Acoustical Measurements

Acoustical measurements involve the measurement of sound pressure. There are many other applications that require measuring sound pressure. For example, sound measurements are used to amplify or record speech and music and to tune large rooms and theater acoustics to achieve appropriate sound level reproduction [14]. A common application is to monitor for increased sound levels or changes in frequency in rotating equipment, which are indicators of failing bearings, rotors, or gears [15]. A significant engineering application involves measuring sound originating from appliances, machinery, or office equipment, and within vehicles in an effort to reduce noise levels that affect the comfort and safety of occupants and workers, as well as to ensure compliance with national noise standards.

Sound is quantified by sound pressure (1 Pa = 1 μbar), a dynamic variation of pressure from the prevailing ambient pressure. It is described by a spectrum of signal amplitudes at corresponding frequencies (e.g., sound frequency spectra are shown in Figures 2.20 and 2.21). Sound pressure is usually time-integrated, with the resulting value referenced against the value taken as the threshold of human hearing at 1,000 Hz. This value is $p_0 = 2 \times 10^{-5}$ Pa = 0.0002 μbar. Sound power is the energy per unit time from a sound source. Sound intensity is power per unit area with a reference value of $I_0 = 10^{-16}$ W/cm$^2$. Sound pressure and intensity levels are usually expressed in decibels:

$$\text{Pressure level (dB)} = 20 \, \log \, (p/p_0) \tag{9.33a}$$

$$\text{Intensity level (dB)} = 10 \, \log \, (I/I_0) \tag{9.33b}$$

Hence, doubling of the sound pressure and intensity equates to changes of 6 dB and 3 dB, respectively. Some typical sound pressure levels as detected by the human ear are listed in Table 9.1, a range that spans a factor of 1:10,000,000.

### Signal Weighting

There are applications that require reporting the sound level exactly as it is measured across the different frequencies, such as in reporting the frequency response of a transducer or for measuring the force of sound pressure acting against a wall or boundary. The frequency bandwidth of transducers varies with individual design; some extend up to 1 GHz. There are other applications that require reporting the sound levels in ways that better describe how the human ear would perceive them, such as in noise isolation, noise abatement, or audio playback. Normal human hearing is sensitive to frequencies ranging from 20 to 20,000 Hz, but the human ear does not respond to sound equally over this frequency range. The ear is most sensitive to sound at frequencies between 1 and 5 kHz and attenuates sound at all frequencies above and below these frequencies. Signal weighting designations were developed to simulate this frequency response. The A-weighting scale compensates sound measurements in a manner that mimics the human

**Table 9.1  Sound Pressure Level References**

| | | | |
|---|---|---|---|
| 0 dB = 0.00002 Pa | Threshold of hearing | 100 dB = 2 Pa | Jackhammer |
| 60 dB = 0.02 Pa | Conversation | 120 dB = 20 Pa | Airplane takeoff |
| 94 dB = 1 Pa | Large truck passing | 140 dB = 200 Pa | Threshold of pain |

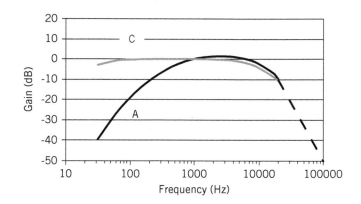

**FIGURE 9.24** Frequency weighting: A- and C-scales.

**Table 9.2  A-Weighting and C-Weighting Scales**

| Relative response (dB) | Octave-band center frequency f in Hz | | | | | | | | | |
|---|---|---|---|---|---|---|---|---|---|---|
| | **31.25** | **62.5** | **125** | **250** | **500** | **1,000** | **2,000** | **4,000** | **8,000** | **16,000** |
| dB(A) | −39.4 | −26.2 | −16.1 | −8.6 | −3.2 | 0 | +1.2 | +1.0 | −1.1 | −6.6 |
| dB(C) | −3.0 | −0.8 | −0.2 | 0 | 0 | 0 | −0.2 | −0.8 | −3.0 | −8.5 |

ear frequency response at common sound pressure levels. A-weighting is used in the reporting of industrial and environmental noise measurements, as well as in reporting audio equipment response and music reproduction. At higher sound levels (>100 dB), the human ear tends to have a flatter frequency response and the C-weighting scale is used.

Both scales are shown in the spectral plot of Figure 9.24 with the corresponding octave-band center frequencies listed in Table 9.2. An octave is just the interval between two points where the frequency at the second point is twice the frequency of the first. The A- and C-weighting scales are both equal to 0dB at 1 kHz. Measurements using A-weighting or C–weighting are designated dB(A) or dB(C), respectively. Exact measurements are scaled according to these weightings. For example, to apply A-weighting to a measured sound pressure level at a particular frequency or octave center-band frequency, add the A-weighting value, which may be positive or negative per Table 9.2:

$$\text{reported value in dB(A)} = \text{measured value in dB} + \text{A-weighting in dB} \qquad (9.34)$$

In this way, the weighting adjusts the measured value to the level that the human ear would hear it. When sound pressure is measured at several discrete octaves, an overall sound pressure can be calculated by logarithmic addition. The overall sound pressure level, SPL (dB), is formed from the corresponding sound pressures measured over each octave:

$$\overline{SPL} = 10\log_{10}\left(\sum_i 10^{(SPL_i/10)}\right) \qquad (9.35)$$

## Microphones

Acoustical sound pressure is measured using a microphone that measures pressure amplitude and frequency. There are different microphone designs available: dynamic, condenser, and piezoelectric.

Grid Cap

Diaphragm

Backplane

Body

Insulator

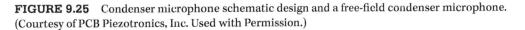

**FIGURE 9.25** Condenser microphone schematic design and a free-field condenser microphone. (Courtesy of PCB Piezotronics, Inc. Used with Permission.)

Dynamic microphones are passive devices that operate on electromagnetic principles (Chapter 6) similar to those of a loudspeaker. A diaphragm attached to a voice coil acts as the sensor. The wire-wound voice coil is wrapped about a permanent magnet. Dynamic sound pressure contacting the diaphragm moves the coil generating an AC voltage signal. The output signal follows the frequency and amplitude of the measured sound pressure.

Condenser microphones (Figure 9.25) contain a metal diaphragm sensor and a metal backplane with an air gap between them to form a capacitor (also known as a condenser), a design that mirrors a capacitance pressure transducer (Section 9.4). Sound pressure waves move the diaphragm so as to change this air gap width. This alters the capacitance value, which in turn affects the value of the output signal. The output signal follows the frequency and amplitude of the sound pressure. A DC excitation voltage powers the backplane of the capacitor. An electret condenser microphone places a permanently charged electret material either in the diaphragm, in the backplane, or in the interval between them to provide a polarizing voltage caused by diaphragm motion, eliminating the need for the excitation voltage. The low-level signal output needs amplification.

Piezoelectric microphones contain a quartz or ceramic crystal coupled to a diaphragm sensor. As described with piezoelectric pressure transducers (Figure 9.14), the pressure force deforms the preloaded crystal, which develops a surface charge $q$, proportional to the sound pressure. A charge amplifier is used to convert charge to voltage. Despite low sensitivities, these sensors are quite durable, have a wide frequency bandwidth (up to 1 GHz), and are able to measure very high sound levels. These are the microphone of choice in acoustic emissions testing, such as for detecting mechanical bearing failure or structural cracks.

Microphones are calibrated using a reference noise source, usually a constant pressure (94 dB) signal at a fixed frequency. Microphones can be made to have different sensitivities to sound levels coming from different angles about a 360° axis and to operate within different sound fields. Unidirectional microphones have a preferred single direction of sensitivity. Bidirectional microphones are sensitive to sounds coming from both front and rear. Omnidirectional microphones are nearly equally sensitive to sound from all directions.

Microphones are designed to meet certain applications. A free-field microphone is designed to measure sound pressure levels that radiate from a single source or direction. These might be used to measure the sound coming from an appliance or machine. These microphones work best in open areas or within anechoic chambers. A pressure field microphone is designed to measure the

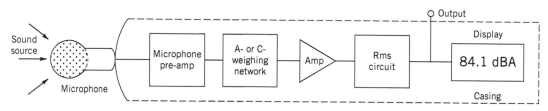

**FIGURE 9.26**    Internal block schematic of a sound level meter.

sound pressure at a position or wall boundary. These are often mounted with only the diaphragm exposed and flush with the wall. These work well to measure the pressure exerted on a wall or structure such as an airplane wing or within housings and cavities. A random incidence type of microphone is omnidirectional so as to measure sound pressure coming from multiple directions, multiple sources, and multiple reflections. Examples of their use include sound measurements within a car or a music hall.

Sound level meters (SLM) are self-contained, hand-held devices used in the sound measurement industry to measure sound pressure. An SLM consists of a microphone, signal conditioning including weighting scale algorithms and root-mean-square circuits, and a readout device (Figure 9.26). The standard IEC 61672-1 specifies sound level meter performance [16].

### EXAMPLE 9.9

Apply A-weighting to these octave-band measurements. Find the overall average sound pressure level.

| | Octave-band center frequency $f$ in Hz | | | | | | | | |
|---|---|---|---|---|---|---|---|---|---|
| Relative response (dB) | 31.25 | 62.5 | 125 | 250 | 500 | 1,000 | 2,000 | 4,000 | 8,000 |
| Measured SPL (dB) | 94 | 95 | 92 | 95 | 97 | 97 | 102 | 97 | 92 |
| A-weighting (dB) | −39.4 | −26.2 | −16.1 | −8.6 | −3.2 | 0 | +1.2 | +1.0 | −1.1 |
| Corrected values dB(A) | 55 | 69 | 76 | 86 | 94 | 97 | 103 | 98 | 91 |

The overall average sound pressure level is

$$\overline{SPL} = 10 \log_{10}\left(10^{SPL_i/10}\right)$$
$$= 10 \log\left(10^{5.5} + 10^{6.9} + 10^{7.6} + 10^{8.6} + 10^{9.4} + 10^{9.7} + 10^{10.3} + 10^{9.8} + 10^{9.1}\right) = 105 \text{ dB(A)}$$

### EXAMPLE 9.10    Case Study: Active Noise Reduction Headsets

Traditional passive headsets are very good at attenuating external ambient high-frequency sounds (>1 kHz), effectively reducing those background noise levels as heard by the user. They are less effective at attenuating low-frequency background sounds. Active noise reduction (ANR) is a method that improves a headset's ability to attenuate the low-frequency noise signals, including the subaural sounds that can fatigue a listener. They are particularly effective at reducing constant level background ambient noise, such as ventilation system or machinery noise. Developed in the 1950s and becoming popular with general aviation pilots in the 1990s to attenuate the constant drone of aircraft propeller noise, these types of headsets are now popular with the general public to filter out unwanted continuous sounds.

In practical design, a microphone measures the ambient low-frequency background noise present within the earpiece of the listener. The measured noise is passed through signal conditioning circuitry and broadcast through a speaker mounted within the earpiece at exactly 180° out of phase with itself. These measured and broadcast noise signals sum to zero, canceling each other out. Non-repetitive sounds and sounds at other frequencies are not canceled; these are transmitted through the earpiece and are unaffected by the out-of-phase signal. The net effect is that background noise levels that occur over a certain range of frequencies are reduced. The example spectrum in Figure 9.27 compares the sound reduction as heard using either a passive or ANR headset. The typical overall sound level reduction (using Equation 9.35) is −10 dB(A) beyond the reduction of a quality passive headset. Desirable sounds are clearer with ANR owing to the higher signal-to-noise ratio. A headset provides a controlled small-volume listening environment requiring very little power for implementing ANR methods. The future challenge is taking ANR to larger-volume environments. Expect to find ANR applied to car cabins, appliance cabinets, and high noise devices in the work environment.

**FIGURE 9.27**  Typical ambient noise reduction as experienced by a listener using a headset of either passive only or active noise reduction design.

## 9.10  Fluid Velocity Measuring Systems

Velocity measuring systems are used to measure the local velocity in a moving fluid. Desirable information can consist of the mean velocity as well as any of the dynamic components of the velocity. Dynamic components are found in pulsating, phasic, or oscillating flows or in turbulent flows. For most general engineering applications, information about the mean flow velocity is usually sufficient. The dynamic velocity information is often sought during applied and basic fluid mechanics research and development, such as in attempting to study airplane wing response to air turbulence, a complex periodic waveform as the wing sees it. In general, the instantaneous velocity can be written as

$$U(t) = \overline{U} + u \tag{9.36}$$

where $\overline{U}$ is the mean velocity and $u$ is the time-dependent dynamic (fluctuating) component of the velocity. The instantaneous velocity can also be expressed in terms of a Fourier series:

$$U(t) = \overline{U} + \sum_i C_i \sin(\omega_i t + \phi_i') \tag{9.37}$$

so that the mean velocity and the amplitude and frequency information concerning the dynamic velocity component can be resolved with a Fourier analysis of the time-dependent velocity signal.

## Pitot–Static Pressure Probe

For a steady, incompressible, isentropic flow, Equation 9.16 can be written at any arbitrary point $x$ in the flow field as

$$p_t = p_x + \frac{1}{2}\rho U_x^2 \tag{9.38}$$

or, rearranging,

$$p_v = p_t - p_x = \frac{1}{2}\rho U_x^2 \tag{9.39}$$

Again $p_v$, the difference between the total and static pressures at any point in the flow, is the *dynamic pressure*. Hence measuring the dynamic pressure of a moving fluid at a point provides a method for estimating the local velocity,

$$U_x = \sqrt{\frac{2p_v}{\rho}} = \sqrt{\frac{2(p_t - p_x)}{\rho}} \tag{9.40}$$

In practice, Equation 9.40 is used through a device known as a *pitot–static pressure probe*. Such an instrument has an outward appearance similar to that of an improved Prandtl static pressure probe (Figure 9.20a), except that the pitot–static probe contains an interior pressure tube attached to an impact port at the leading edge of the probe, as shown in Figure 9.28. This creates two coaxial internal cavities within the probe, one exposed to the total pressure and the second exposed to the static pressure. The two pressures are typically measured using a differential pressure transducer so as to indicate $p_v$ directly.

The pitot–static pressure probe is relatively insensitive to misalignment over the yaw angle range of ±15 degrees [1]. When possible, the probe can be rotated until a maximum signal is measured, a condition that indicates that it is aligned with the mean flow direction. However, the probes have a lower velocity limit of use that is brought about by strong viscous effects in the entry regions of the pressure ports. In general, viscous effects should not be a concern, provided that the Reynolds number based on the probe radius, $\mathrm{Re}_r = \overline{U}r/\nu$, is greater than 500, where $\nu$

**FIGURE 9.28**   Pitot–static pressure probe.

is the kinematic viscosity of the fluid. For $10 < \text{Re}_r < 500$, a correction to the dynamic pressure should be applied, $p_v = C_v p_i$, where

$$C_v = 1 + (4/\text{Re}_r) \tag{9.41}$$

and $p_i$ is the indicated dynamic pressure from the probe. However, even with this correction, the measured dynamic pressure has a systematic uncertainty on the order of 40% at $\text{Re}_r \approx 10$ but decreases to 1% for $\text{Re}_r \geq 500$.

In high-speed gas flows, compressibility effects near the probe leading edge require a closer inspection of the governing equation for a pitot–static pressure probe. An energy balance along a streamline for a perfect gas between any point $x$ and the stagnation point can be written as

$$\frac{U^2}{2} = c_p(T_t - T_\infty) \tag{8.34}$$

For an isentropic process, the relationship between temperature, pressure, and density can be stated as

$$\frac{T_t}{T_x} = \left(\frac{p_t}{p_x}\right)^{\frac{k-1}{k}} = \left(\frac{\rho_t}{\rho_x}\right)^{k-1} = \left(1 + \frac{k-1}{2}M^2\right) \tag{9.42}$$

where $k$ is the ratio of specific heats for the gas, $k = c_p/c_v$. The Mach number of a moving fluid relates its local velocity to the local speed of sound,

$$M = U/a \tag{9.43}$$

where the speed of sound, also called the acoustic wave speed, for a perfect gas is $a = \sqrt{kRT_x}$ where $T_x$ is the absolute temperature of the gas at the point of interest. Combining these equations and using a binomial expansion gives a relationship between total pressure and static pressure at any point $x$ in a moving compressible flow,

$$p_v = p_t - p_x = \frac{1}{2}\rho U_x^2 \left[1 + M^2/4 + (2-k)M^4/24 + ...\right] \tag{9.44}$$

Equation 9.44 reduces to Equation 9.39 when $M \ll 1$. The error in between the two equations relative to the true dynamic pressure becomes significant for $M > 0.3$, as shown in Figure 9.29. Thus $M \sim 0.3$ is used as the incompressible limit for perfect gas flows.

**FIGURE 9.29**  Relative error in the dynamic pressure between using a compressible and incompressible formulation at increasing flow speeds.

For $M > 1$, the local velocity is found through iteration using the Rayleigh relation:

$$p_t/p = \left( \frac{(k+1)^2 M^2}{4kM^2 - 2(k-1)} \right)^{k/k-1} \left( \frac{1 - k + 2kM^2}{k+1} \right) \tag{9.45}$$

where $p_t$ is the measured value at the impact port and $p$ refers to the static pressure upstream of the probe.

## Thermal Anemometry

The rate at which energy, $\dot{Q}$, is transferred between a warm body at $T_s$ and a cooler moving fluid at $T_f$ is proportional both to the temperature difference between them and to the thermal conductance of the heat transfer path, $hA$. This thermal conductance increases with fluid velocity, thereby increasing the rate of heat transfer at any given temperature difference. Hence a relationship between the rate of heat transfer and velocity exists, forming the working basis of a thermal anemometer.

A thermal anemometer uses a sensor, a metallic resistance temperature detector (RTD) element, which makes up one active leg of a Wheatstone bridge circuit, as indicated in Figure 9.30. The resistance–temperature relation for such a sensor was shown in Chapter 8 to be well represented by

$$R_s = R_0[1 + \alpha(T_s - T_0)] \tag{9.46}$$

so that sensor temperature $T_s$ can be inferred through a resistance measurement. A current is passed through the sensor to heat it to some desired temperature above that of the host fluid. The relationship between the rate of heat transfer from the sensor and the cooling fluid velocity is given by King's law [17] as

$$\dot{Q} = I^2 R = A + BU^n \tag{9.47}$$

where $A$ and $B$ are constants that depend on the fluid and sensor physical properties and operating temperatures and $n$ is a constant that depends on sensor dimensions [18]. Typically, $0.45 \leq n \leq 0.52$ [19]. $A$, $B$, and $n$ are found through calibration.

Two types of sensors are common: the hot wire and the hot film. As shown in Figure 9.31, the hot-wire sensor consists of a tungsten or platinum wire typically ranging 1 to 4 mm in length and 1.5 to 15 μm in diameter. The wire is supported between two rigid needles that protrude from a ceramic tube that houses the lead wires. A hot-film sensor usually consists of a thin (2 μm) platinum or gold film deposited onto a glass substrate and covered with a high thermal conductivity coating. The coating acts to electrically insulate the film and offers some mechanical protection. Hot wires are used in electrically nonconducting fluids, whereas hot films can be used in conducting fluids or wherever a more rugged sensor is needed.

Two anemometer bridge operating modes are possible: (1) constant current and (2) constant resistance. In constant-current operation, a fixed current is passed through the sensor to heat it. The sensor resistance and thus its temperature, based on Equation 9.46, are permitted to vary

**FIGURE 9.30**  Thermal anemometer circuit, shown in constant resistance mode.

**FIGURE 9.31** Schematic of a hot-wire probe.

with the rate of heat transfer between the sensor and its environment. Bridge-deflection voltage provides a measure of the cooling velocity. The more common mode of operation is constant resistance. In constant-resistance operation, the sensor resistance is maintained constant in measurement by using a differential feedback amplifier to sense small changes in bridge balance; that is, the circuit acts as a closed loop controller using the bridge balance as the error signal. The feedback amplifier rapidly readjusts the bridge applied voltage, thereby adjusting the sensor current to bring the sensor back to its set-point resistance and corresponding temperature. The instantaneous power ($I^2 R_s$) required to maintain constant temperature is equivalent to the instantaneous rate of heat transfer from the sensor ($\dot{Q}$), which varies with cooling velocity. In terms of the instantaneous applied bridge voltage, $E$, the velocity is found by the general correlation

$$E^2 = C + DU^n \tag{9.48}$$

where the values of $C$, $D$, and $n$ are found by calibration under a fixed sensor and fluid temperature condition. An electronic or digital linearizing scheme is usually employed to condition the signal by performing the transformation

$$E_1 = K\left(\frac{E^2 - C}{D}\right)^{1/n} \tag{9.49}$$

such that the measured output from the linearizer, $E_1$, is

$$E_1 = KU \tag{9.50}$$

where $K$ is found through a static calibration.

For mean velocity measurements, the thermal anemometer is a straightforward device to use. Multiple velocity components can be measured by using multiple sensors, each sensor aligned differently to the mean flow direction and operated by independent anemometer circuits [18, 20]. Because it has a high-frequency response, fluctuating (dynamic) velocities can be measured. In highly turbulent flows with root-mean-square (rms) fluctuations of $\sqrt{\overline{u^2}} \geq 0.1\overline{U}$, signal interpretation can become complicated, but it has been well investigated [20]. Low-frequency fluid temperature fluctuations can be compensated for by placing resistor $R_3$ directly adjacent to the sensor and exposed to the flow. An extensive bibliography of thermal anemometry theory and signal interpretation can be found elsewhere [21].

In constant temperature mode, a hot-wire system can attain a frequency response that is flat up to 100,000 Hz, which makes it particularly useful in fluid mechanics turbulence research. However, less expensive and more rugged systems are commonly used for industrial flow monitoring, where a fast dynamic response is desirable. There is an upper frequency limit on a cylindrical sensor of diameter $d$ that is brought about by the natural oscillation in the flow immediately downstream of a body. This vibrates the sensor. The frequency of this oscillation, known as the Strouhal frequency (and explained further in Section 10.6), occurs at approximately

$$f \approx 0.22[\overline{U}/d] \qquad (10^2 < \text{Re}_d < 10^7) \tag{9.51}$$

The heated sensor warms the fluid within its proximity. Under flowing conditions, this does not cause any measurable problems so long as the condition

$$\text{Re}_d \geq Gr^{1/3} \tag{9.52}$$

is met where $\text{Re}_d = \overline{U}d/v$, $Gr = d^3 g \beta (T_s - T_{\text{fluid}})/v^2$, and $\beta$ is the coefficient of thermal expansion of the fluid. Equation 9.52 ensures that the inertial forces of the moving fluid dominate over the buoyant forces brought on by the heated sensor. This forms a lower velocity limit on the order of 0.6 m/s for using hot-wire sensors in air.

## Doppler Anemometry

The Doppler effect describes the phenomenon experienced by an observer whereby the frequency of light or sound waves emitted from a source that is traveling away from or toward the observer is shifted from its original value and by an amount proportional to its speed. Most readers are familiar with the change in pitch of a train as heard by an observer as the train changes from approaching to receding. Any radiant energy source, such as a sound or light wave, demonstrates a Doppler effect, first recognized and modeled by Christian Johann Doppler (1803–1853). The observed shift in frequency, called the Doppler shift, is directly related to the speed of the emitter relative to the observer.

*Doppler anemometry* refers to a class of techniques that use the Doppler effect to measure the local velocity at a point in a moving fluid. The emission source is a stationary coherent narrow incident wave. Either acoustic waves or light waves are used. Small particles suspended in and moving with the fluid are used to generate the Doppler effect by scattering the incident waves. Here we discuss point or full-field velocity measurement methods. In Chapter 10, we discuss a flow rate measurement method using Doppler (ultrasonic) principles.

A laser Doppler anemometer (LDA) uses a laser beam as the emission source [22–25] to measure the time-dependent velocity at a point in the flow. As a moving particle suspended in the fluid passes through a laser beam, it scatters light in all directions. An observer viewing this encounter perceives the scattered light at a frequency, $f_s$:

$$f_s = f_i \pm f_D \tag{9.53}$$

where $f_i$ is the frequency of the incident laser beam and $f_D$ is the Doppler shift frequency. Using visible light, an incident laser beam frequency is on the order of $10^{14}$ Hz. For most engineering applications, the velocities are such that the Doppler shift frequency, $f_D$, is on the order of $10^3$ to $10^7$ Hz. An operating mode that distinctly isolates the Doppler frequency and focuses on a small volume of flow is shown in Figure 9.32. In this dual-beam mode, a single laser beam is divided into two coherent beams using an optical beam splitter, which are then focused through a lens at a point in the flow. This focal point forms the measuring volume (sensor) of the instrument. Particles suspended in and moving with the fluid scatter light as they pass through the beams. Within the measuring volume incident information from the two incident beams mix, a process known as optical heterodyne. The outcome of this mixing separates the incident frequency from the Doppler frequency into two distinct signal components.

**FIGURE 9.32** Laser Doppler anemometer, shown here in the dual-beam mode of operation.

Richard Figliola

**FIGURE 9.33** Oscilloscope trace of a photodiode output showing the Doppler frequency from a single particle moving through the measuring volume.

The related technique of phase Doppler anemometry [26] is used to simultaneously measure particle sizes. It shares a similar optical setup but uses two or more photodetectors. Phase differences between the detected signals are proportional to the light-scattering particle size. Engineering applications include emissions monitoring, combustion, and pneumatic transport.

For the setup shown in Figure 9.32, the velocity is related directly to the Doppler shift by

$$U = \frac{\lambda}{2 \sin \theta/2} f_D = d_f f_D \qquad (9.54)$$

The component of the velocity measured is that which is in the plane of and bisector to the crossing beams [24]. By using beams of different color or polarization, different velocity components can be measured simultaneously. The output from the photodiode transducer is a current of a magnitude proportional to the square of the amplitude of the scattered light seen and of a frequency equal to $f_D$. This effect is seen as a Doppler "burst" shown in the trace of Figure 9.33. In a time-varying flow field, the Doppler shift from successive scatters will vary with time. The most common way to measure this time-dependent frequency variation is through a burst analyzer using either fast Fourier transform or autocorrelation methods. At very low light levels, photon correlation techniques are successful [23, 24]. As a point measuring technique, the measuring volume needs to be translated about to map out the flow field.

**EXAMPLE 9.11**

An LDA made up of a He–Ne laser ($\lambda = 632.8$ nm) is used to measure the velocity of water at a point in a flow. A 150-mm lens, with $\theta = 11.0$ degrees, is used to operate the LDA in a dual-beam mode. If an average Doppler frequency of 1.41 MHz is measured, estimate the velocity of water.

**KNOWN** $\bar{f}_D = 1.411$ MHz.

$\lambda = 632.8$ nm    $\theta = 11.0$ degrees.

**ASSUMPTIONS**  Scattering particles follow the water exactly.

**FIND** $\overline{U}$.

**SOLUTION**

Applying Equation 9.54,

$$\overline{U} = \frac{1}{2\sin\theta/2}\overline{f}_D = \left(\frac{632.8 \times 10^{-9}\text{ m}}{2\sin(11°/2)}\right)(1.411 \times 10^6\text{ Hz}) = 4.655\text{ m/s}$$

## Particle Image Velocimetry

*Particle image velocimetry* (PIV) measures the full-field instantaneous velocities in a planar cross section of a flow. The technique tracks the time displacement of particles, which are assumed to follow the flow. Principle components for the technique are a coherent light source (laser beam), optics, a CCD camera, and dedicated signal interrogation software.

In a simple overview, the image of particles suspended in the flow are illuminated and recorded repetitively at very short intervals in time. Sequential image pairs are compared. The distance traveled by a particle during the period between images is a measure of its velocity. By repeatedly recording the particle field, the particle positions can be tracked and velocity as a function of space and time obtained.

In a typical layout, such as shown in Figure 9.34, a laser beam is passed through a cylindrical lens, which converts the beam into a two-dimensional (2D) sheet of light. This laser sheet is mechanically situated to illuminate an appropriate cross section of the flow field. The camera is positioned and focused to record the view of the illuminated field. The time interval between image pairs can be controlled either by using a pulsed-light laser synchronized with the camera shutter or by using a continuous light laser and a high speed motion camera. The acquired digital images are stored and processed by interrogation software, resulting in a full-field instantaneous velocity mapping of the flow.

The operating principle is based on particle displacement with time

$$\vec{U} = \Delta\vec{x}/\Delta t \tag{9.55}$$

where $\vec{U}$ is the instantaneous particle velocity vector based on its spatial position $\vec{x}(x, y, z, t)$. The camera records particle position into separate image frames. To obtain velocity data in a rapid manner, each image is divided into small areas, called interrogation areas. The corresponding interrogation areas between two images, $I_1$ and $I_2$, are cross-correlated with each other, on a

**FIGURE 9.34**   Basic layout of a digital particle image velocimeter.

pixel-by-pixel basis. A particular particle movement from position $\vec{x}_1$ to $\vec{x}_2$ shows up as a signal peak in the correlation $R_{12}(\Delta\vec{x})$, where

$$R_{12}(\Delta\vec{x}) = \int_A \int I_1(\Delta\vec{x})I_2(\vec{x} + \Delta\vec{x})d\vec{x} \tag{9.56}$$

identifying the common particle and allowing the estimate of the particle displacement $\Delta\vec{x}$. By repeating the cross-correlation between images for each interrogation area, a velocity vector map of the full area results.

A number of variations of this process have been developed but the concept remains the same. The technique works in gases or liquids. Three-dimensional information can be obtained by using two cameras. Particle size and properties should be chosen relative to the fluid and flow velocities expected so that the particles move with the fluid, but the particles must be large enough relative to camera pixel size to avoid peak-locking errors [27]. The maximum flow speed measurable is limited by the interrogation area size. The resolution depends on laser flash width and separation time, flow velocity, camera recording time, and image magnification. By using a high magnification lens, called microscopic PIV, measurements with a very small length scale are possible. Raffel et al. [27] discuss the technique, its variations, and error estimates.

## Selection of Velocity Measuring Methods

Selecting the best velocity measuring system for a particular application involves a number of factors that an engineer needs to weight accordingly:

1. Required spatial resolution

2. Required velocity range

3. Sensitivity to velocity changes only

4. Required need to quantify dynamic velocity

5. Acceptable probe blockage of flow

6. Ability to be used in hostile environments

7. Calibration requirements

8. Low cost and ease of use

When used under appropriate conditions, the uncertainty in velocity determined by any of the discussed methods can be as low as 1% of the measured velocity [26].

### Pitot–Static Pressure Methods

The pressure probe methods are best suited for finding the mean velocity in fluids of constant density. Relative to other methods, they are the simplest and cheapest method available to measure velocity at a point. Probe blockage of the flow is not a problem in large ducts and away from walls. They are subject to mean flow misalignment errors. They require no calibration and are frequently used in the field and laboratory alike.

### Thermal Anemometer

Thermal anemometers are best suited for use in very clean fluids of constant temperature and density. They are well suited for measuring dynamic velocities with very high resolution. However, signal interpretation in strongly dynamic flows can be complicated [20, 28]. Hot-film sensors are less fragile and less susceptible to contamination than hot-wire sensors. Probe blockage is not significant in large ducts and away from walls. The cylindrical sensors are 180 degrees directionally ambiguous (i.e., flows from the left or right directions give the same output signal), an important factor in flows that may contain flow reversal regions. An industrial-grade system can be built rather inexpensively. The thermal anemometer is usually calibrated against pressure probes.

## Doppler Anemometry

The LDA is a relatively expensive point velocity measuring technique for transparent fluids that is well suited to hostile, combusting, or dynamic (unsteady, pulsatile, or highly turbulent) flow environments. The method offers good frequency response, small spatial and temporal resolutions, no probe blockage, and simple signal interpretation, but requires optical access and the presence of scattering particles [29, 30].

## Particle Image Velocimetry

PIV is a relatively expensive and technically advanced full-field velocity measuring technique that can be used for most types of flows, including hostile and combusting flows. There is no probe blockage of the flow, but it requires optical access and the presence of scattering particles. The method provides an instantaneous snapshot of the flow, providing excellent views of flow structures. Time-dependent quantification of such dynamic flows is possible, but frequency bandwidth is limited by camera frame rate and spatial resolution. Careful planning is required in particle selection to ensure that the particle velocities represent the fluid velocity exactly and that particle size is properly matched to pixel resolution.

## SUMMARY

Several reference pressure instruments have been presented that form the working standards for pressure transducer calibration. Pressure transducers convert sensed pressure into an output form that is readily quantifiable. These transducers come in many forms but tend to operate on hydrostatic principles, expansion techniques, or force-displacement methods.

In moving fluids, special care must be taken when measuring pressure to delineate between static and total pressure. Methods for the separate measurement of static and total pressure or for the measurement of the dynamic pressure are readily available and well documented. But improper measuring technique causes errors, lowering the total pressure or increasing the static pressure.

Measuring the local velocity within a moving fluid can be accomplished in a number of ways. Selecting the proper tool requires assessment of the need: mean or fluctuating velocity, point or full-field measurement, optical access or opaque boundaries or fluid. Specifically, dynamic pressure, thermal anemometry, Doppler anemometry, and particle velocimetry methods have been presented. As discussed, each method offers advantages over the others, and the best technique must be carefully weighed against the needs and constraints of a particular application.

## REFERENCES

1. American Society of Mechanical Engineers (ASME), *PTC 19.2–1987: Pressure Measurement*, ASME International, New York, 1987.

2. Brombacher, W. G., D. P. Johnson, and J. L. Cross, Mercury Barometers and Manometers, *National Bureau Standards Monograph 8*, 1960.

3. McLeod, H. G., Vacuum Gauge, *Philosophical Magazine*, 48, 110, 1874. Reprinted in *History of Vacuum Science and Technology*, eds. T. E. Madey and W. C. Brown, American Vacuum Society, New York, pp. 102–105, 1984.

4. Hetenyi, M. ed., *Handbook of Experimental Stress Analysis*, Wiley, New York, 1950.

5. Way, S., Bending of Circular Plates with Large Deflection, *Transactions of the ASME* 56, 1934.

6. Instrument Society of America (ISA), *A Guide for the Dynamic Calibration of Pressure Transducers*, ISA-37.16.01-2002, Instrument Society of America, 2002.

7. Franklin, R. E., and J. M. Wallace, Absolute Measurements of Static-Hole Error Using Flush Transducers, *Journal of Fluid Mechanics* 42, 1970.

8. Rayle, R. E., Influence of Orifice Geometry on Static Pressure Measurements, ASME Paper No. 59-A-234, 1959.

9. Munson, B., D. Young, T. Okishi, *Fundamentals of Fluid Mechanics*, 6th ed., Wiley, New York, 2009.

10. Migliavacca, F. et al., Multiscale Modelling in Biofluidynamics: Application to Reconstructive Paediatric Cardiac Surgery, *Journal of Biomechanics*, 39(6), 2006.

11. Delio, G., G. Schwent, and R. Cesaro, Transient Behavior of Lumped-Constant Systems for Sensing Gas Pressures, National Advisory Council on Aeronautics (NACA) TN-1988, 1949.

12. Doebelin, E. O., *Measurement Systems: Application and Design*, 5th ed., McGraw-Hill Science/Engineering/Math, New York, 2003.

13. Hougen, J., O. Martin, and R. Walsh, Dynamics of Pneumatic Transmission Lines, *Control Engineering*, September 1963.

14. Acoustics—Measurement of Room Acoustic Parameters, ISO 3382, International Standards Organization, Geneva, Switzerland, 2012.

15. Mechanical Vibration—Evaluation of Machine Vibration by Measurements on Non-Rotating Parts, ISO 10816, International Organization for Standardization, Geneva, Switzerland, 2016.

16. Electroacoustics—Sound Level Meters—Part 1: Specifications, IEC61672, International Electrotechnical Commission (IEC). Geneva, Switzerland, September 2013.

17. King, L. V., On the Convection from Small Cylinders in a Stream of Fluid: Determination of the Convection Constants of Small Platinum Wires with Application to Hot-Wire Anemometry, *Proceedings of the Royal Society, London 90*, 1914.

18. Hinze, J. O., *Turbulence*, McGraw-Hill, New York, 1959.

19. Collis, D. C., and M. J. Williams, Two-Dimensional Convection from Heated Wires at Low Reynolds Numbers, *Journal of Fluid Mechanics* 6, 1959.

20. Rodi, W., A New Method for Analyzing Hot-Wire Signals in a Highly Turbulent Flow and its Evaluation in a Round Jet, *DISA Information*, Dantek Electronics, Denmark, 1975. See also Bruun, H. H., Interpretation of X-Wire Signals, *DISA Information*, Dantek Electronics, Denmark, 1975.

21. Freymuth, P., A Bibliography of Thermal Anemometry, *TSI Quarterly*, 4, 1978.

22. Yeh, Y., and H. Cummins, Localized Fluid Flow Measurement with a He–Ne Laser Spectrometer, *Applied Physics Letters*, 4, 1964.

23. Cummins H. Z., and E. R. Pike, eds., Photon Correlation and Light Beating Spectroscopy, *Proceedings of the NATO ASI*, Plenum, New York, 1973.

24. Durst, F., A. Melling, and J. H. Whitelaw, *Principles and Practice of Laser Doppler Anemometry*, Academic Press, New York, 1976.

25. Goldstein, R. J., ed., *Fluid Mechanics Measurements*, 2nd ed., CRC Press, New York, 1996.

26. Goldstein, R. J., and D. K. Kried, Measurement of Laminar Flow Development in a Square Duct Using a Laser Doppler Flowmeter, *Journal of Applied Mechanics*, 34, 1967.

27. Raffel, M., C. E. Willert, S. T. Wereiey, and J. Kompenhans, *Particle Image Velocimetry*, 2nd ed., Springer, Heidelberg, 2007.

28. Yavuzkurt, S., A Guide to Uncertainty Analysis of Hot-Wire Data, *Transactions of the ASME, Journal of Fluids Engineering*, 106, 1984.

29. Maxwell, B. R., and R. G. Seaholtz, Velocity Lag of Solid Particles in Oscillating Gases and in Gases Passing through Normal Shock Waves, National Aeronautics and Space Administration TN-D-7490, 1974.

30. Dring, R. P., and M. Suo, Particle Trajectories in Swirling Flows, *Transactions of the ASME, Journal of Fluids Engineering*, 104, 1982.

## NOMENCLATURE

| | | | | |
|---|---|---|---|---|
| $d$ | diameter ($l$) | | $p_m$ | measured pressure ($m^{-1}lt^{-2}$) |
| $e_i$ | elemental errors | | $p_t$ | total or stagnation pressure ($m^{-1}lt^{-2}$) |
| $d_f$ | fringe spacing ($l$) | | $p_v$ | dynamic pressure ($m^{-1}lt^{-2}$) |
| $f_D$ | Doppler frequency (Hz) | | $q$ | charge (C) |
| $h$ | depth ($l$) | | $r$ | radius ($l$) |
| $h_0$ | reference depth ($l$) | | $t$ | thickness ($l$) |
| $k$ | ratio of specific heats | | $y$ | displacement ($l$) |
| $p$ | pressure ($m^{-1}lt^{-2}$) | | $\rho$ | density ($ml^{-3}$) |
| $p_a$ | applied pressure ($m^{-1}lt^{-2}$) | | $\iota$ | time constant ($t$) |
| $p_{abs}$ | absolute pressure ($m^{-1}lt^{-2}$) | | $\phi$ | latitude |
| $p_e$ | relative static pressure error ($m^{-1}lt^{-2}$) | | $\omega$ | frequency ($t^{-1}$) |
| $p_i$ | indicated pressure ($m^{-1}lt^{-2}$) | | $\omega_n$ | natural frequency ($t^{-1}$) |

| | | | | |
|---|---|---|---|---|
| $z$ | altitude ($l$) | | $R$ | fluid resistance[4] |
| $C$ | capacitance (F); compliance[4] ($l^4t^2m^{-1}$) | | $\mathrm{Re}_d$ | Reynolds number, $\mathrm{Re} = Vd/v$ |
| $E$ | voltage (V) | | $S$ | specific gravity |
| $E_m$ | bulk modulus of elasticity ($m^{-1}lt^{-2}$) | | $U$ | velocity ($lt^{-l}$) |
| Gr | Grashof number | | $\forall$ | volume ($l^3$) |
| $H$ | manometer deflection height ($l$) | | $\gamma$ | specific weight ($ml^{-2}t^{-2}$) |
| $K$ | static sensitivity | | $\zeta$ | damping ratio |
| $K_q$ | charge sensitivity ($m^{-1}lt^{-2}$) | | $\varepsilon$ | dielectric constant |
| $K_E$ | voltage sensitivity ($\mathrm{V}m^{-1}t^{-2}$) | | $\lambda$ | wavelength ($l$) |
| L | inductance (H); inertance[4] ($ml^{-4}$) | | $\mu$ | absolute viscosity ($mt^{-1}l^{-1}$) |
| $M$ | Mach number | | $v$ | kinematic viscosity ($l^2/t$) |
| $R$ | resistance ($\Omega$) | | $\upsilon_p$ | Poisson ratio |

---

[4] See Section 9.7 for definition and dimensions.

# Flow Measurements

10

## 10.1 Introduction

The rate at which a fluid moves through a conduit is measured in terms of flow rate. This chapter discusses some of the most common and accepted methods for measuring flow rate. Flow rate can be expressed in terms of a volume flow per unit time, known as the *volume flow rate*, or as a mass flow per unit time, known as the *mass flow rate*. Flow rate devices, called flow meters, are used to quantify, totalize, or monitor flowing processes. Type, accuracy, size, pressure drop, pressure losses, capital and operating costs, and compatibility with the fluid are important engineering design considerations for choosing a flow metering device. Each method has both desirable and undesirable features that necessitate compromise in the selection of the best method for the particular application, and many such considerations are discussed in this chapter. Inherent uncertainties in fluid properties, such as density, viscosity, or specific heat, can affect the accuracy of a flow measurement. However, some techniques, such as those incorporated into Coriolis mass flow meters, do not require knowledge of exact fluid properties, allowing for highly accurate mass flow measurements in demanding engineering applications. The chapter objective is to present both an overview of basic flow metering techniques for proper meter selection, as well as those design considerations important in the integration of a flow rate device with the process system it will meter.

Upon completion of this chapter, the reader will be able to:

- Relate velocity distribution to flow rate within conduits

- Use engineering test standards to select and specify size, specify installation considerations, and use common obstruction meters

- Understand the differences between volume flow rate and mass flow rate and the means required to measure each

- Describe the physical principles employed in various commercially available types of flow meters and the engineering terms common to their use

- Describe calibration methods for flow meters

## 10.2 Historical Background

Their importance to engineered systems give flow measurement methods their rich history. The earliest available accounts of flow metering were recorded by Hero of Alexandria (ca. 150 B.C.) who proposed a scheme to regulate water flow using a siphon pipe attached to a constant head reservoir. The early Romans developed elaborate water systems to supply public baths and private homes. In fact, Sextus Frontinius (A.D. 40–103), commissioner of water works for Rome, authored a treatise on design methods for urban water distribution systems. Evidence suggests that Roman designers understood correlation between volume flow rate and pipe flow area. Weirs were used to regulate bulk flow through aqueducts, and the cross-sectional area of terracotta pipe was used to regulate fresh running water supplies to individual buildings.

After a number of experiments conducted using olive oil and water, Leonardo da Vinci (1452–1519) first formally proposed the modern continuity principle: that duct area and fluid velocity were related directly to flow rate. However, most of his writings were lost until centuries later, and Benedetto Castelli (ca. 1577–1644), a student of Galileo, has been credited in some texts with developing the same steady, incompressible continuity concepts in his day. Isaac Newton (1642–1727), Daniel Bernoulli (1700–1782), and Leonhard Euler (1707–1783) built the mathematical and physical bases on which modern flow meters would later be developed. By the nineteenth century, the concepts of continuity, energy, and momentum were sufficiently understood for practical exploitation. Relations between flow rate and pressure losses were developed that would permit the tabulation of the hydraulic coefficients necessary for the quantitative engineering design of many modern flow meters.

## 10.3 Flow Rate Concepts

The flow rate through a conduit, be it a pipeline, duct, or channel, depends on fluid density, average fluid velocity, and conduit cross-sectional area. Consider fluid flow through a circular pipe of radius $r_1$ and having a velocity profile at some axial pipe cross section given by $u(r,\theta)$. The mass flow rate depends on the average mass flux, $\rho u(r,\theta)$ flowing through a cross-sectional area. Then the mass flow rate is given by

$$\dot{m} = \int \int_A \rho u(r,\theta)dA = \overline{\rho U}A \tag{10.1}$$

where the overbar denotes an area-averaged quantity. The pipe cross-sectional area is $A = \pi r_1^2$. To directly measure mass flow rate, a device must be sensitive in some way to the area-averaged mass flux, $\overline{\rho U}$, or to the fluid mass passing through the device per unit time. Mass flow rate has the dimensions of mass per unit of time (e.g., units of kg/s, $lb_m$/s, etc.).

The volume flow rate depends only on the area-averaged velocity over a cross section of flow as given by

$$Q = \int \int_A u dA = \overline{U}A \tag{10.2}$$

So to directly measure volume flow rate requires a device that is in some way sensitive either to the average velocity, $\overline{U}$, or to the fluid volume passing through it per unit time. Volume flow rate has dimensions of volume per unit time (e.g., units of $m^3$/s, $ft^3$/s).

The difference between Equations 10.1 and 10.2 is quite significant in that either requires a very different approach to its measurement: one sensitive to the product of density and velocity or to mass rate and the other sensitive only to the average velocity or to volume rate. In the simplest case, where density is a constant, the mass flow rate can be inferred by multiplying the measured volume flow rate by the density. But in the metering of many fluids, this assumption may not be good enough to achieve necessary accuracy. This can be because either the density changes during the metered process, it may not be well known, such as in the transport of polymers or petrochemicals, or because small errors in the assumed density accumulate into large errors, such as in the transport of millions of cubic meters of product per day.

The flow character can affect the accuracy of a flow meter. The flow through a pipe or duct can be characterized as being laminar, turbulent, or a transition between the two. Flow character is determined through the engineering parameter known as the Reynolds number, defined by

$$\text{Re}_{d_1} = \frac{\overline{U}d_1}{v} = \frac{4Q}{\pi d_1 v} \tag{10.3}$$

where $v$ is the fluid kinematic viscosity and $d_1$ is the diameter for circular pipes. In pipes, the flow is laminar when $\text{Re}_{d_1} < 2000$ and turbulent at higher Reynolds number. The Reynolds number is a necessary parameter in estimating flow rate when using most of the types of flow meters discussed. In estimating the Reynolds number in noncircular conduits, the hydraulic diameter, $d_H = 4r_H$, is used in place of diameter $d_1$, where $r_H$ is the wetted conduit area divided by its wetted perimeter.

# 10.4 Volume Flow Rate through Velocity Determination

Volume flow rate can be determined with direct knowledge of the velocity profile as indicated by Equation 10.2. This requires measuring the velocity at multiple points along a cross section of a conduit to estimate the velocity profile or a method sensitive to measuring an area-averaged velocity. For highest accuracy when measuring at points across a conduit, several traverses should be made at differing circumferential locations to account for flow nonsymmetry. Methods for determining the velocity at a point include any of those previously discussed in Chapter 9. Because it is a tedious method, this procedure is most often used for the one-time verification or calibration of system flow rates. For example, the procedure is often used in ventilation system setup and problem diagnosis, where the installation of an inline flow meter is not necessary because operation does not require continuous monitoring. Or the method can be used to verify the calibration of a flow rate device.

When using this technique in circular pipes, a number of discrete measuring positions $n$ are chosen along each of $m$ flow cross sections spaced at $360/m$ degrees apart, as shown in Figure 10.1. A velocity probe is traversed along each flow cross section, with readings taken at each of several measurement positions. There are options in selecting the measuring positions for different-shaped ducts, and such details are specified in available engineering test

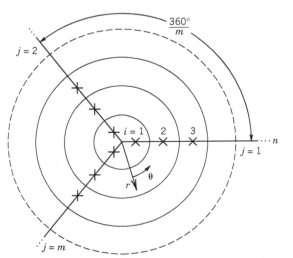

**FIGURE 10.1** Location of $n$ measurements along $m$ radial lines in a pipe.

standards [1–4]. The simplest method is to divide the flow area into smaller equal areas, making measurements at the centroid of each small area and assigning the measured velocity to that area. Regardless of the option selected, the average flow rate is estimated along each cross section traversed using Equation 10.2 and the pooled mean of the flow rates for the $m$ cross sections calculated to yield the best estimate of the duct flow rate. Uncertainty is assessed both by repeating the measurements and by analyzing for spatial variation effects. Example 10.1 illustrates this method for estimating volume flow rate.

### EXAMPLE 10.1

A steady flow of air at 20°C passes through a 25.4-cm-inside-diameter (i.d.) circular pipe. A velocity-measuring probe is traversed along three cross-sectional lines ($j = 1, 2, 3$) of the pipe, and measurements are made at four radial positions ($i = 1, 2, 3, 4$) along each traverse line, such that $m = 3$ and $n = 4$. The locations for each measurement are selected at the centroids of equally spaced areal increments as indicated below and based on existing engineering standards [1, 5]. Determine the volume flow rate in the pipe.

| Radial location, $i$ | $r/r_1$ | $U_{ij}$ (m/s) Line 1 ($j = 1$) | Line 2 ($j = 2$) | Line 3 ($j = 3$) |
|---|---|---|---|---|
| 1 | 0.3536 | 8.71 | 8.62 | 8.78 |
| 2 | 0.6124 | 6.26 | 6.31 | 6.20 |
| 3 | 0.7906 | 3.69 | 3.74 | 3.79 |
| 4 | 0.9354 | 1.24 | 1.20 | 1.28 |

**KNOWN**   $U_{ij}(r/r_1)$ for $i = 1,2,3,4$; $j = 1,2,3$

$$d_1 = 25.4\text{cm}(A = \pi d_1^2/4 = 0.05067 \text{ m}^2).$$

**ASSUMPTIONS**   Constant and steady pipe flow during all measurements; incompressible flow.

**FIND**   Volume flow rate, $Q$.

**SOLUTION**
The flow rate is found by integrating the velocity profile across the duct along each line and subsequently averaging the three values. For discrete velocity data, Equation 10.2 is written along each line, $j = 1, 2, 3 \dots m$ as

$$Q_j = 2\pi \int_0^{r_1} U r \, dr \approx 2\pi \sum_{i=1}^{4} U_{ij} r \Delta r$$

where $\Delta r$ is the radial distance separating each position of measurement. Here, this can be further simplified since the velocities are located at positions that make up the centroids of equal areas:

$$Q_j = \frac{A}{4} \sum_{i=1}^{4} U_{ij}$$

Then, the mean flow rate along each line of traverse is

$$Q_1 = 0.252 \text{ m}^3/\text{s} \quad Q_2 = 0.252 \text{ m}^3/\text{s} \quad Q_3 = 0.254 \text{ m}^3/\text{s}$$

The average pipe flow rate $\overline{Q}$ is equal to the pooled mean of the individual flow rates $\langle\overline{Q}\rangle$

$$\overline{Q} = \langle\overline{Q}\rangle = \frac{1}{3}\sum_{j=1}^{3} Q_j = 0.253 \text{ m}^3/\text{s}$$

---

**EXAMPLE 10.2**

Referring to Example 10.1, determine a value of the systematic standard uncertainty in mean flow rate due to the measured spatial variation.

**KNOWN** Data over three ($m = 3$) traverse sections.

**SOLUTION**
The systematic standard uncertainty in mean flow rate due to error introduced by the measured spatial variations is estimated by

$$b_{\overline{Q}} = \frac{s_{\langle Q\rangle}}{\sqrt{m}} = \frac{\sqrt{\sum_{j=1}^{3}(Q_j - \langle\overline{Q}\rangle)^2/2}}{\sqrt{3}} = \frac{0.0012}{\sqrt{3}} = 0.000693 \text{ m}^3/\text{s} \quad \text{with} \quad \nu = 2$$

**COMMENT** Here we used statistical information to estimate a systematic uncertainty; it estimates the possible offset in $\overline{Q}$ and is about 0.3% of the mean flow rate.

---

# 10.5 Pressure Differential Meters

The operating principle of a pressure differential flow rate meter is based on the relationship between volume flow rate and the pressure difference $\Delta p = p_1 - p_2$, between two defined locations along the flow path,

$$Q \propto (p_1 - p_2)^n \tag{10.4}$$

where $n = 1$ for laminar flow occurring between the pressure measurement locations and $n = \frac{1}{2}$ for fully turbulent flow. Under steady flow conditions, an intentional reduction in flow area between locations 1 and 2 causes a measurable local pressure drop across this flow path. This reduced flow area leads to a concurrent local increase in velocity due to flow continuity (conservation of mass) principles. The pressure change is in part due to the so-called Bernoulli effect, the inverse relationship between local velocity and pressure, but is also due to flow energy losses. A class of pressure differential flow rate meters that use area reduction methods is commonly called *obstruction meters*.

## Obstruction Meters

Three common obstruction meters are the *orifice plate*, the *venturi*, and the *flow nozzle*. Flow area profiles of each are shown in Figure 10.2 and detailed construction dimensions can be found in test standards [1, 3, 5]. These meters are usually inserted in-line with a pipe, such as between pipe flanges. This class of meters as a whole operates using similar physical reasoning to relate volume flow rate to pressure drop. Referring to Figure 10.3, consider an energy balance between two control surfaces for an incompressible fluid flow through the arbitrary control volume shown. Under the assumptions of incompressible, steady and one-dimensional flow with no external

**FIGURE 10.2**   Flow area profiles of common obstruction meters. **(a)** Square-edged orifice plate meter. **(b)** American Society of Mechanical Engineers (ASME) long-radius nozzle. **(c)** ASME Herschel venturi meter.

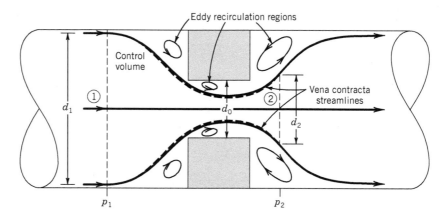

**FIGURE 10.3**   Control volume concept as applied between two streamlines for flow through an obstruction meter.

energy transfer, the energy equation is

$$\frac{p_1}{\gamma} + \frac{\overline{U}_1^2}{2g} = \frac{p_2}{\gamma} + \frac{\overline{U}_2^2}{2g} + h_{L_{1-2}} \tag{10.5}$$

where $h_{L_{1-2}}$ denotes the head losses occurring between control surfaces 1 and 2.

For incompressible flows, conservation of mass between cross-sectional areas 1 and 2 gives

$$\overline{U}_1 = \overline{U}_2 \frac{A_2}{A_1} \tag{10.6}$$

Then, substituting Equation 10.6 into Equation 10.5 and rearranging yields the incompressible volume flow rate

$$Q_I = \overline{U}_2 A_2 = \frac{A_2}{\sqrt{1 - (A_2/A_1)^2}} \sqrt{\frac{2(p_1 - p_2)}{\rho} + 2gh_{L_{1-2}}} \tag{10.7}$$

where the subscript $I$ emphasizes that Equation 10.7 gives an incompressible flow rate. Later we drop the subscript.

When the flow area changes abruptly, the effective flow area immediately downstream of the area reduction is not necessarily the same as the pipe flow area. This was originally investigated by Jean Borda (1733–1799) and illustrated in Figure 10.3. A central core flow called the vena contracta forms, which is bounded by regions of slower-moving recirculating eddies. The pressure sensed with pipe wall taps corresponds to the higher moving velocity within the vena contracta with its unknown flow area, $A_2$. To account for this unknown, we introduce a contraction coefficient $C_c$, where $C_c = A_2/A_0$, with $A_0$ based on the meter throat diameter, into Equation 10.7. This gives

$$Q_I = \frac{C_c A_0}{\sqrt{1-(C_c A_0/A_1)^2}} \sqrt{\frac{2(p_1 - p_2)}{\rho} + 2gh_{L_{1-2}}} \qquad (10.8)$$

Furthermore, the frictional head losses can be incorporated into a friction coefficient, $C_f$, such that Equation 10.8 becomes

$$Q_I = \frac{C_f C_c A_0}{\sqrt{1-(C_c A_0/A_1)^2}} \sqrt{\frac{2(p_1 - p_2)}{\rho}} \qquad (10.9)$$

For convenience, the coefficients are factored out of Equation 10.9 and replaced by a single coefficient known as the *discharge coefficient, C.* The discharge coefficient represents the ratio of the actual flow rate through a meter to the ideal flow rate for the pressure drop measured—that is, $C = Q_{I_{actual}}/Q_{I_{ideal}}$. Reworking Equation 10.9 leads to the incompressible form of the working equation used with obstruction meters

$$Q_I = CEA_0 \sqrt{\frac{2\Delta p}{\rho}} = K_0 A_0 \sqrt{\frac{2\Delta p}{\rho}} \qquad (10.10)$$

where $E$, known as the velocity of approach factor, is defined by

$$E = \frac{1}{\sqrt{1-(A_0/A_1)^2}} = \frac{1}{\sqrt{1-\beta^4}} \qquad (10.11)$$

with the beta ratio defined as $\beta = d_0/d_1$, and where $K_0 = CE$ is called the *flow coefficient.*

The discharge coefficient and the flow coefficient are tabulated quantities found in test standards [1, 3, 5]. Each is a function of the Reynolds number, $\beta$ ratio, and pressure tap locations for each particular obstruction flow meter design, $C = f(\text{Re}_{d_1}, \beta)$ and $K_0 = f(\text{Re}_{d_1}, \beta)$. Specific pressure tap locations for $p_1$ and $p_2$ are defined in the test standards.

## Compressibility Effects

In compressible gas flows, compressibility effects in obstruction meters can be accounted for by introducing the compressible adiabatic expansion factor, $Y$. Here $Y$ is defined as the ratio of the actual compressible volume flow rate, $Q$, divided by the assumed incompressible flow rate $Q_I$. Combining with Equation 10.10 yields

$$Q = YQ_I = CEA_0 Y\sqrt{2\Delta p/\rho_1} \qquad (10.12)$$

where $\rho_1$ is the upstream fluid density. When $Y = 1$, the flow is incompressible, and Equation 10.12 reduces to Equation 10.10. *Equation 10.12 represents the most general form of the working equation for volume flow rate determination when using an obstruction meter.*

The expansion factor, $Y$, depends on several values: the type of meter, the $\beta$ ratio, the gas specific heat ratio, $k$, and the relative pressure difference across the meter, $(p_1 - p_2)/p_1$, that is, $Y = f[\beta, k, (p_1 - p_2)/p_1]$. As a general rule, compressibility effects should be considered whenever $(p_1 - p_2)/p_1 \geq 0.1$.

## Standards for Obstruction Meters

The flow behaviors of the most common obstruction meters, namely, the orifice plate, venturi, and flow nozzle, have been studied to such an extent that these meters are used extensively without calibration but with strict adherence to the test standards. Values for the discharge coefficients, flow coefficients, and expansion factors are tabulated and available in standard U.S. and international flow handbooks along with standardized construction, installation, and operation techniques [1, 3, 5, 6]. Equations 10.10 and 10.12 are very sensitive to pressure tap location. For steam or gas flows, pressure taps should be oriented on the top or side of the pipe; for liquids, pressure taps should be oriented on the side. We discuss some recommended standard tap locations with each meter [1, 3]. A nonstandard installation or design requires an in situ calibration.

## Orifice Meter

An orifice meter consists of a circular plate having a central hole (orifice). The plate is inserted into a pipe so as to effect a flow area change. The orifice hole is smaller than the pipe diameter and arranged to be concentric with the pipe's i.d. The square-edged orifice plate is shown in Figure 10.4. Installation is simplified by housing the orifice plate between two pipe flanges.

**FIGURE 10.4**    Square-edged orifice meter installed in a pipeline with optional 1 $d$ and $\frac{1}{2} d$, and flange pressure tap locations shown. Relative flow pressure drop along pipe axis is shown.

With this installation technique, any particular orifice plate is interchangeable with others of different $\beta$ values. The simplicity of the design allows for a range of $\beta$ values to be maintained on hand at modest expense.

For an orifice meter, plate dimensions and use are specified by engineering standards [1, 3]. Equation 10.12 is used with values of $A_0$ and $\beta$ based on the orifice (hole) diameter, $d_0$. The plate thickness should be between $0.005\,d_1$ and $0.02\,d_1$; otherwise, a taper must be added to the downstream side [1, 3]. The exact placement of pressure taps is crucial when using standard coefficients. Standard pressure tap locations include (1) flange taps where pressure tap centers are located 25.4 mm (1 in.) upstream and 25.4 mm (1 in.) downstream of the nearest orifice face, (2) $d$ and $d/2$ taps located one pipe diameter upstream and one-half pipe diameter downstream of the upstream orifice face, and (3) vena contracta taps. Nonstandard tap locations will require in situ meter calibration.

Values for the flow coefficient, $K_0 = (\mathrm{Re}_{d_1}, \beta)$ and for the expansion factor, $Y = f[\beta, k, (p_1 - p_2)/p_1]$, for a square-edged orifice plate are given in Figures 10.5 and 10.6 based on the use of flange taps. The relative instrument systematic uncertainty in the discharge coefficient [5] is $\sim 0.6\%$ of $C$ for $0.2 \le \beta \le 0.6$ and $\beta\%$ of $C$ for all $\beta > 0.6$. The relative instrument systematic uncertainty for the expansion factor is about $[4(p_1 - p_2)/p_1]\%$ of $Y$. Realistic estimates of the overall systematic uncertainty in estimating $Q$ using an orifice meter are between 1% (high $\beta$) and 3% (low $\beta$) at high Reynolds numbers when using standard tables.[1] Although the orifice plate represents a relatively inexpensive flow meter solution with an easily measurable pressure drop, it also introduces a large permanent pressure loss, $(\Delta p)_{\text{loss}} = \rho g h_L$, into the flow system. Larger pressure losses must be able to be overcome by the system prime mover (e.g., pump, blower, compressor, fan, etc.). The pressure drop is illustrated in Figure 10.4 with the pressure loss estimated from the charts or equations given in Figure 10.7.

$$K_0 = \frac{1}{(1 - \beta^4)^{1/2}} \left( 0.5959 + 0.0312\beta^{2.1} - 0.184\beta^8 + 91.71\beta^{2.5}\mathrm{Re}_{d_1}^{-0.75} \right)$$

**FIGURE 10.5**  Flow coefficients for a square-edged orifice meter having flange pressure taps. (Compiled from data in reference 1.)

---

[1]These uncertainties assume proper placement of the flow meter, that is, sufficiently far away from out-of-plane elbows in the flow path.

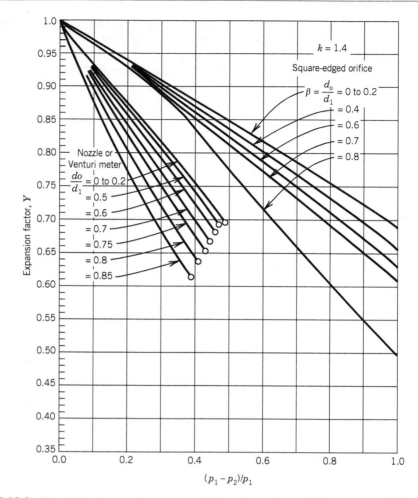

**FIGURE 10.6**  Expansion factors for common obstruction meters with $k = c_p/c_v = 1.4$. (Compiled from data in reference 1.)

Rudimentary versions of the orifice plate meter have existed for several centuries. Both Torricelli and Newton used orifice plates to study the relation between pressure head and efflux from reservoirs, although neither ever got the discharge coefficients quite right [7].

## Venturi Meter

A venturi meter consists of a smooth converging (21 degrees ± 1 degree) conical contraction to a narrow throat followed by a shallow diverging conical section, as shown in Figure 10.8. The engineering standard venturi meter design uses either a 15-degree or 7-degree divergent section [1, 3]. The meter is installed between two flanges intended for this purpose. Pressure taps are located just ahead of the upstream contraction and at the throat. Equation 10.12 is used with values for both $A$ and $\beta$ based on the throat diameter, $d_0$.

The quality of a venturi meter ranges from cast to precision-machined units. The discharge coefficient varies little for pipe diameters above 7.6 cm (3 in.). In the operating range $2 \times 10^5 \le \mathrm{Re}_{d_1} \le 2 \times 10^6$ and $0.4 \le \beta \le 0.75$, use a value of $C = 0.984$ with a systematic uncertainty of 0.7% (95%) for cast units and $C = 0.995$ with a systematic uncertainty of 1% (95%) for machined units [1, 3, 5]. Values for expansion factor are shown in Figure 10.6 and have an instrument

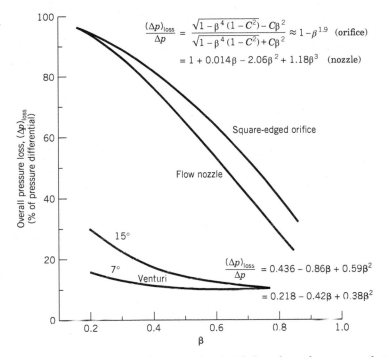

$$\frac{(\Delta p)_{loss}}{\Delta p} = \frac{\sqrt{1 - \beta^4 (1 - C^2)} - C\beta^2}{\sqrt{1 - \beta^4 (1 - C^2)} + C\beta^2} \approx 1 - \beta^{1.9} \quad \text{(orifice)}$$

$$= 1 + 0.014\beta - 2.06\beta^2 + 1.18\beta^3 \quad \text{(nozzle)}$$

Square-edged orifice

Flow nozzle

15°

7°

Venturi

$$\frac{(\Delta p)_{loss}}{\Delta p} = 0.436 - 0.86\beta + 0.59\beta^2$$

$$= 0.218 - 0.42\beta + 0.38\beta^2$$

**FIGURE 10.7** The permanent pressure loss associated with flow through common obstruction meters. (Compiled from data in reference 1.)

**FIGURE 10.8**
The Herschel venturi meter with the associated flow pressure drop along its axis.

systematic uncertainty of $[(4 + 100\beta^2)(p_1 - p_2)/p_1]\%$ of $Y$ [5]. Although a venturi meter presents a much higher initial cost over an orifice plate and requires a much longer section of pipe to install, Figure 10.7 demonstrates that the meter shows a much smaller permanent pressure loss for a given installation. This translates into lower system operating costs for the pump or blower used to move the flow.

The modern venturi meter was first proposed by Clemens Herschel (1842–1930). Herschel's design was based on his understanding of the principles developed by several men, most notably those of Daniel Bernoulli. However, he cited the studies of contraction/expansion angles and their corresponding resistance losses by Giovanni Venturi (1746–1822), and, later, those by James Francis (1815–1892), as being instrumental to his design of a practical flow meter.

## Flow Nozzles

A flow nozzle consists of a gradual contraction from the pipe's i.d. converging down to a narrow throat. It needs less installation space than a venturi meter at about 80% of the initial cost. Common forms are the ISO 1932 nozzle and the ASME long-radius nozzle [1, 3]. The long-radius nozzle contraction is that of the quadrant of an ellipse, with the major axis aligned with the flow axis, as shown in Figure 10.9. The nozzle is typically installed inline but can also be used at the inlet to or the outlet from a plenum or reservoir or at the exit of a pipe. Pressure taps are usually located (1) at one pipe diameter upstream of the nozzle inlet and at the nozzle throat using either wall or throat taps, or (2) $d$ and $d/2$ wall taps located one pipe diameter upstream and one-half diameter downstream of the upstream nozzle face. The flow rate is determined from

**FIGURE 10.9**    The ASME long-radius nozzle with the associated flow pressure drop along its axis.

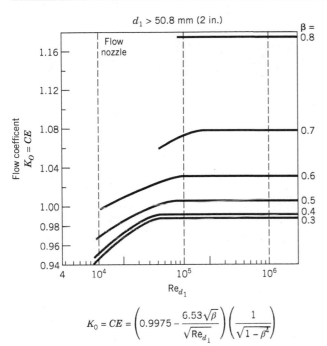

$d_1 > 50.8$ mm (2 in.)

Flow nozzle

$$K_0 = CE = \left(0.9975 - \frac{6.53\sqrt{\beta}}{\sqrt{Re_{d_1}}}\right)\left(\frac{1}{\sqrt{1-\beta^4}}\right)$$

**FIGURE 10.10** Flow coefficients for an ASME long-radius nozzle with a throat pressure tap. (Compiled from data in reference 1.)

Equation 10.12 with values for $A_o$ and $\beta$ based on the throat diameter. Typical values for the flow coefficient and expansion factor are given in Figures 10.10 and 10.6. The relative instrument systematic uncertainty at 95% confidence for the discharge coefficient is about 2% of $C$ and for the expansion factor is about $[2(p_1 - p_2)/p_1]\%$ of $Y$ [5]. The permanent loss associated with a flow nozzle is larger than for a comparable venturi but significantly smaller than for an orifice (Figure 10.7) for the same pressure drop.

The idea of using a nozzle as a flow meter was first proposed in 1891 by John Ripley Freeman (1855–1932), an inspector and engineer employed by a fire insurance firm. His work required tedious tests to quantify pressure losses in pipes, hoses, and fittings. He noted a consistent relationship between pressure drop and flow rate through fire nozzles at higher flow rates.

---

### EXAMPLE 10.3

The U-tube manometer filled with manometer fluid (of specific gravity $S_m$) of Figure 10.11 is used to measure the pressure drop across an obstruction meter. A fluid having specific gravity $S$ flows through the meter. Determine a relationship between the meter flow rate and the measured manometer deflection, $H$.

**KNOWN** Fluid ($S$ and $\gamma$).

Manometer fluid ($S_m$ and $\gamma_m$).

**FIND** $Q = f(H)$.

**SOLUTION**
Equation 10.12 provides a relationship between flow rate and flow pressure drop. From Figure 10.11 and hydrostatic principles (see Chapter 9), the pressure drop measured by the manometer is

$$\Delta p = p_1 - p_2 = \gamma_m H - \gamma H = (\gamma_m - \gamma)H = \gamma H\left[(S_m/S) - 1\right]$$

**FIGURE 10.11**    Manometer and flow meter of Example 10.3. A taper must be added to the downstream side of the orifice hole when the plate thickness exceeds $0.02d_1$ [1].

Substituting this relation into Equation 10.12 gives the working equation based on the equivalent pressure head:

$$Q = CEA_0 Y \sqrt{2gH[(S_m/S) - 1]} \qquad (10.13)$$

**EXAMPLE 10.4**

A 10-cm-diameter, square-edged orifice plate is used to meter the steady flow of 16°C water through a 20-cm pipe. Flange taps are used and the pressure drop measured is 50 cm Hg. Determine the pipe flow rate. The specific gravity of mercury is 13.5.

**KNOWN** $d_1 = 20$ cm  $H = 50$ cm Hg  $d_0 = 10$ cm
            Water properties (properties are from Appendix B).
              $\mu = 1.08 \times 10^{-3}$ N-s/m$^2$  $\rho = 999$ kg/m$^3$

**ASSUMPTIONS**  Incompressible flow of a liquid ($Y = 1$).

**FIND**  Volume flow rate, $Q$.

**SOLUTION**
Equation 10.12 is used with an orifice plate, and it requires knowledge of the product $CE$. The beta ratio is $\beta = d_0/d_1 = 0.5$, so the velocity of approach factor is calculated from Equation 10.11:

$$E = \frac{1}{\sqrt{1 - \beta^4}} = \frac{1}{\sqrt{1 - 0.5^4}} = 1.0328$$

We know that $C = f(\mathrm{Re}_{d_1}, \beta)$. The flow Reynolds number is estimated using Equation 10.3, which can be modified with the relation, $\nu = \mu/\rho$,

$$\mathrm{Re}_{d_1} = \frac{4Q}{\pi d_1 \nu} = \frac{4\rho Q}{\pi d_1 \mu}$$

We see that without information concerning $Q$, we cannot determine the Reynolds number, and so $C$ cannot be determined explicitly.

Instead, a trial-and-error iteration is undertaken: Guess a value for $K_0$ (or for $C$) and iterate. A good start is to guess a value at a high value of $\mathrm{Re}_{d_1}$. This is the flat region of the curve ($\beta = 0.5$) in Figure 10.5, so choose a value of $K_0 = CE = 0.625$.

Based on the manometer deflection and Equation 10.13 (see Example 10.2),

$$Q = CEA_0 Y \sqrt{2gH[(S_m/S) - 1]} = K_0 A_0 Y \sqrt{2gH[(S_m/S) - 1]}$$
$$= (0.625)(\pi/4)(0.10 \text{ m})^2 (1)\sqrt{2(9.8 \text{ m/s}^2)(0.50 \text{ m})[(13.5/1) - 1]}$$
$$= 0.054 \text{ m}^3/\text{s}$$

Next we have to test the guessed value for $K_0$ to determine if it was correct. For this value of $Q$,

$$\mathrm{Re}_{d_1} = \frac{4\rho Q}{\pi d_1 \mu} = \frac{4(999\ \mathrm{kg/m^3})(0.054\ \mathrm{m^3/s})}{\pi(0.20\ \mathrm{m}(1.08\times10^{-3}\ \mathrm{N\text{-}s/m^2}))} = 3.2\times10^5$$

From Figure 10.5, at this Reynolds number, $K_0 \approx 0.625$. So our guess was spot on. The solution is converged, so we conclude that $Q = 0.054\ \mathrm{m^3/s}$.

## EXAMPLE 10.5

Air flows at 20°C through a 6-cm pipe. A square-edged orifice plate with $\beta = 0.4$ is chosen to meter the flow rate. A pressure drop of 250 cm $H_2O$ is measured at the flange taps with an upstream pressure of 93.7 kPa abs. Find the flow rate.

**KNOWN**  $d_1 = 6$ cm      $p_1 = 93.7$ kPa abs
$\beta = 0.4$          $T_1 = 20°C = 293$ K
$H = 250$ cm $H_2O$

Properties (found in Appendix B):

Air: $v = 1.0\times10^{-5}\mathrm{m^2/s}$

Water: $\rho_{H_2O} = 999\ \mathrm{kg/m^3}$

**ASSUMPTIONS** Treat air behavior as an ideal gas ($p = \rho RT$).

**FIND** Volume flow rate, $Q$.

## SOLUTION
The orifice flow rate is found from Equation 10.12, which requires information about $E$, $C$, and $Y$. From the given information, we calculate both the orifice area, $A_{d_0} = 4.52\times10^{-4}\ \mathrm{m^2}$, and, from Equation 10.11, the velocity of approach factor, $E = 1.013$. The air density is found from the ideal gas equation of state:

$$\rho_1 = \frac{p_1}{RT_1} = \frac{93,700\ \mathrm{N/m^2}}{(287\ \mathrm{N\text{-}m/kg})(293\ \mathrm{K})} = 1.114\ \mathrm{kg/m^3}$$

or use air-property tables. The pressure difference is $p_1 - p_2 = \rho_{H_2O}gH = 24,500\ \mathrm{N/m^2}$.

The pressure ratio for this gas flow, $(p_1 - p_2)/p_1 = 0.26$. For pressure ratios greater than 0.1 the compressibility of the air should be considered. From Figure 10.6, $Y = 0.92$ for $k = 1.4$ (air) and a pressure ratio of 0.26.

As in the previous example, the discharge coefficient cannot be found explicitly unless the flow rate is known, since $C = f(\mathrm{Re}_{d_1}, \beta)$. So a trial-and-error iterative approach is used. From Figure 10.5, we start with a guess of $K_0 = CE \approx 0.61$ (or $C = 0.60$). Then,

$$Q = CEYA_0\sqrt{\frac{2\Delta p}{\rho_1}} = (0.60)(1.013)(0.92)(4.52\times10^{-4}\ \mathrm{m^2})\sqrt{\frac{(2)(24,500\ \mathrm{N/m^2})}{1.114\ \mathrm{kg/m^3}}}$$
$$= 0.053\ \mathrm{m^3/s}$$

**CHECK** For this flow rate, $\mathrm{Re}_{d_1} = 4Q/\pi d_1 v = 7.4\times10^4$ and from Figure 10.5, $K_0 \approx 0.61$, as assumed. The flow rate through the orifice is taken to be 0.053 $\mathrm{m^3/s}$.

## EXAMPLE 10.6

A pump can often impose pressure oscillations into a pipe flow that are sensed in the pressure measurements. Figure 10.12 plots separate pressure measurements taken upstream, $p_1$, and downstream, $p_2$, of an orifice meter, as well as pressure differential measurements, $\Delta p = p_1 - p_2$, taken across the meter. The systematic effect of the oscillations on the internal pipe pressure is seen in comparing $p_1$ and $p_2$, which rise and fall together with a correlation coefficient of $r_{p_1 p_2} = 0.998$. Estimate the contribution to random standard uncertainty in estimating pressure differential due to the data scatter. The following information is known:

$$\bar{p}_1 = 9.25 \text{ kPa} \quad s_{p_1} = 2.813 \text{ kPa} \quad N = 20$$
$$\bar{p}_2 = 7.80 \text{ kPa} \quad s_{p_2} = 2.870 \text{ kPa} \quad N = 20$$
$$\overline{\Delta p} = 1.45 \text{ kPa} \quad s_{\Delta p} = 0.188 \text{ kPa} \quad N = 20$$

**FIGURE 10.12**    Upstream and downstream pressures and differential pressures across an orifice meter for Example 10.6.

## SOLUTION

As seen in Figure 10.12, pump-induced oscillations impose a trend on the individual pressure measurements. This is confirmed by the high correlation coefficient (see Chapter 4) between $p_1$ and $p_2$. Notably, the pressure differential signal is unaffected by this periodicity, because although the pressure within the pipe may rise and fall with time, the pressure difference across the orifice plate remains largely unaffected. Hence, analyzing the individual pressure data directly without accounting for this systematic effect would overstate the random uncertainty, which should be due only to the data scatter; analyzing the pressure differential data eliminates the systematic effect. We address these issues here and show how to compensate for them.

The measured pressure differential is $\overline{\Delta p} = 1.45$ kPa, or, alternately, when calculated from the individual pressure measurements, $\bar{p}_1 - \bar{p}_2 = 1.45$ kPa. So it is independent of how we measure it.

We can estimate the random standard uncertainty in pressure differential directly from the data variation in $\Delta p$ as

$$s_{\overline{\Delta p}} = \frac{s_{\Delta p}}{\sqrt{20}} = 0.042 \text{ kPa} \quad \text{with } \nu = 19.$$

Alternatively, we could analyze the individual pressure data and use these to estimate the random standard uncertainty in $\overline{\Delta p}$. With $\Delta p = p_1 - p_2$, the sensitivities are

$$\frac{\partial \Delta p}{\partial p_1} = 1 \quad \frac{\partial \Delta p}{\partial p_2} = -1$$

The elemental random standard uncertainties are

$$s_{\bar{p}_1} = \frac{s_{p_1}}{\sqrt{20}} = 0.629 \text{ kPa} \quad s_{\bar{p}_2} = \frac{s_{p_2}}{\sqrt{20}} = 0.642 \text{ kPa}$$

Then, accounting for the correlated error influence on the pressures, we write

$$s_{\overline{\Delta p}} = \left[ \left( \frac{\partial \Delta p}{\partial p_1} s_{\bar{p}_1} \right)^2 + \left( \frac{\partial \Delta p}{\partial p_2} s_{\bar{p}_2} \right)^2 + 2 \frac{\partial \Delta p}{\partial p_1} \frac{\partial \Delta p}{\partial p_2} r_{p_1 p_2} s_{\bar{p}_1} s_{\bar{p}_2} \right]^{1/2} = 0.042 \text{ kPa} \qquad (10.14)$$

with $v = 19$. Here the last term under the radical corrects for the correlated error between $p_1$ and $p_2$.

Thus, either method gives the same result for $\overline{\Delta p}$ and for $s_{\overline{\Delta p}}$. However, the first method is preferred because it directly estimates the desired measured variable, $\Delta p$, and its random uncertainty is unaffected by the systematic effect.

If we neglected the correlated effect of the pump oscillation, we would instead be tempted to calculate

$$s_{\overline{\Delta p}} = \left[ \left( \frac{\partial \Delta p}{\partial p_1} s_{\bar{p}_1} \right)^2 + \left( \frac{\partial \Delta p}{\partial p_2} s_{\bar{p}_2} \right)^2 \right]^{1/2} = 0.899 \text{ kPa}$$

which overstates the random standard uncertainty significantly and is wrong.

**COMMENT** Equation 10.14 revisits a method to correct for imposed correlated effects on uncertainty estimates as discussed in Chapter 5. This example shows the importance of reviewing data and data trends in a measurement to uncover test procedure tendencies that affect test interpretation.

## Sonic Nozzles

Sonic nozzles are used both to measure and to regulate the flow rate of compressible gases [8]. They usually take the form of either a flow nozzle or venturi meter. The sonic condition refers to achieving the speed of sound of the fluid at the throat of the meter. With sonic operating conditions within the flow meter, flow rate will depend only on upstream pressure and temperature. Sonic meters are often used as a transfer standard with uncertainties as low as 0.5%. Both large pressure drops and system pressure losses must be tolerated with the technique [1].

The theoretical basis for such a meter stems from the early work of Bernoulli, Venturi, and Saint-Venant (1797–1886). In 1866, Julius Weisbach (1806–1871) developed a direct relation between pressure drop and a maximum mass flow rate. When the pressure drop across an obstruction meter becomes sufficiently high, the sonic condition will be reached at the meter throat as the fluid accelerates. At that point the throat is considered to be choked meaning that the mass flow rate through the meter is at the maximum for the given inlet conditions. Lowering the downstream pressure or raising the value of upstream pressure will not increase the mass flow rate further.

For a perfect gas undergoing an isentropic process, the pressure ratio corresponding to the onset of the choked flow condition at the meter's minimum area, the meter throat, is given by the *critical pressure ratio*:

$$\left( \frac{p_{th}}{p_t} \right)_{\text{critical}} = \left( \frac{2}{k+1} \right)^{k/(k-1)} \qquad (10.15)$$

where $p_{th}$ is the throat pressure and subscript "$t$" denotes stagnation properties. If $(p_{th}/p_t) \leq (p_{th}/p_t)_{\text{critical}}$ then the meter throat is choked. This critical condition gives the maximum mass flow rate through the nozzle for the prevailing conditions existing upstream of the throat. The mass flow rate can be found from

$$\dot{m} = CA_{th}\frac{p_t}{\sqrt{RT_t}}\sqrt{\frac{2k}{k+1}\left(\frac{2}{k+1}\right)^{2/(k-1)}} = CC^*A_{th}\frac{p_t}{\sqrt{RT_t}} \tag{10.16}$$

with

$$C^* = \sqrt{\frac{2k}{k+1}\left(\frac{2}{k+1}\right)^{2/(k-1)}} \tag{10.17}$$

where $k$ is the specific heat ratio of the gas, $C^*$ is the critical ratio, and $C$ is the discharge coefficient [1, 5]. For air and other diatomic gases, $k = 1.4$ and so $C^* = 0.6847$. Commercial units are supplied with a discharge coefficient, otherwise $C = 0.98 \pm 2\%$ (95%) is used [1]. Corrections (e.g., Equations 9.42–9.45) can be applied to use measured upstream static properties when $A_{th} \ll A_1$.

## Obstruction Meter Selection

Selecting between obstruction meter types depends on a number of factors and engineering compromises. Primary considerations include meter placement, overall pressure loss, overall (capital and operating) costs, accuracy, and operating range (turndown).

### Placement

It is important to provide sufficient straight pipe lengths to either side of a flow meter to allow for proper flow development. The flow development dissipates swirl, promotes a symmetric velocity distribution, and allows for proper pressure recovery downstream of the meter so as to reduce systematic error. The available standards provide detailed recommendations regarding installation placement for best practice [1, 3, 5]. Even with recommended lengths, an additional 0.5% systematic uncertainty should be added to the uncertainty in the discharge coefficient to account for swirl effects downstream of turns [5]. Unorthodox installations require in situ meter calibration to determine flow coefficients or the use of special correction procedures [3, 5, 9]. The physical size of a meter could become an important consideration in itself, since venturi and nozzle meter installations can take up a considerable length of pipe.

### Pressure Loss

The unrecoverable overall pressure loss, $\Delta p_{loss}$, associated with a flow meter depends on the β ratio and flow rate (Figure 10.7). These losses must be overcome by the system prime mover (e.g., a pump, fan, or compressor driven by a motor) in addition to any other pipe system losses. In established systems, flow meters must be chosen on the basis of the pressure loss that the prime mover can accommodate and still maintain a desired system flow rate. The power, $\dot{W}$, required to overcome any loss in a system for flow rate $Q$ is given by

$$\dot{W} = Q\frac{\Delta p_{loss}}{\eta} \tag{10.18}$$

where $\eta$ is the prime mover efficiency.

### Costs

The actual cost of a meter depends on its initial capital cost, the costs of installing it (including system down time), calibration costs, and the added capital and operating costs associated with flow meter pressure losses to the system. Indirect costs may include product losses resulting from the meter flow rate uncertainty.

## Accuracy

The ability of an obstruction flow meter to accurately estimate the volume flow rate in a pipeline depends on both the method used to calculate flow rate and factors inherent to the meter. If standard tabulated values for meter coefficients are used, factors that contribute to the overall uncertainty of the measurement enter through the following data-acquisition elemental errors: (1) actual $\beta$ ratio error and pipe eccentricity, (2) pressure and tap position errors, (3) temperature effects leading to relative thermal expansion of components, and (4) actual upstream flow profile. These errors are in addition to the errors inherent in the coefficients themselves and errors associated with the estimation of the upstream fluid density [1]. Direct in situ calibration of any flow meter installation can assess these effects and their contributions to overall uncertainty. For example, the actual $\beta$ ratio value, pressure tap locations, and other installation effects would be included in a system calibration and accommodated in the computed flow coefficient.

## Turndown

The relative range, $(Q_{max} - Q_{min})/Q_{min}$, over which a meter is designed to be used is known as the meter *turndown*. The selection of a meter should consider whether the system into which the meter is to be installed will be used at more than one flow rate, and its effect on system performance over the entire anticipated flow rate range should be considered.

### EXAMPLE 10.7

For the orifice meter in Example 10.3 with $\beta = 0.5$ and a pressure drop of 50 cm Hg, calculate the permanent pressure loss due to the meter that must be overcome by a pump.

**KNOWN** $H = 50$cm Hg

$\qquad \beta = 0.5$

**ASSUMPTION** Steady flow.

**FIND** $\Delta p_{loss}$.

**SOLUTION**

For a properly designed and installed orifice meter ($\beta = 0.5$), the permanent pressure loss of the meter is about 75% of the pressure drop across the meter (Figure 10.7). Hence

$$\Delta p_{loss} = 0.75 \times \gamma_{Hg}H = 49.9 \text{ kPa} \quad \text{(or } H_{loss} = 38 \text{ cm Hg)}.$$

**COMMENT** In comparison, for $Q = 0.053$ m$^3$/s a typical venturi (say a 15-degree outlet) with a $\beta = 0.5$ would provide a pressure drop equivalent to 19.6 cm Hg with a permanent loss of only 3.1 cm Hg. Since the pump needs to supply enough power to create the flow rate and to overcome losses, a smaller pump could be used with the venturi.

### EXAMPLE 10.8

For the orifice of Example 10.7, estimate the operating costs required to overcome these permanent losses if commercial electricity is available at \$0.08/kW-h, the pump is used 6,000 h/year, and the pump motor efficiency is 60%.

**KNOWN**    Water at 16°C      $Q = 0.053 \text{ m}^3/\text{s}$
                 $\Delta p_{\text{loss}} = 49.9 \text{ kPa}$    $\eta = 0.60$

**ASSUMPTIONS**   Meter installed according to standards [1].

**FIND**   Annual cost due to $\Delta p_{\text{loss}}$ alone.

**SOLUTION**

The pump power required to overcome the orifice meter permanent pressure loss is estimated from Equation 10.18,

$$\dot{W} = Q \frac{\Delta p_{\text{loss}}}{\eta}$$

$$= \frac{(0.55 \text{ m}^3/\text{s})(49{,}900 \text{ Pa})(3600 \text{ s/h})(1 \text{ N/m}^2/\text{Pa})(1 \text{ W/N-m/s})}{0.6} = 4530 \text{ W} = 4.53 \text{ kW}$$

The additional annual pump operating cost required to overcome this is

$$\text{Cost} = (4.53 \text{ kW})(6000 \text{ h/year})(\$0.08/\text{kW-h}) = \$2175/\text{year}.$$

**COMMENT** In comparison, the venturi discussed in the comment to Example 10.7 would cost about \$180/year to operate. If installation space allows, the venturi meter may be the less expensive long-term solution.

## Laminar Flow Elements

Laminar flow meters take advantage of the linear relationship between volume flow rate and the pressure drop over a length for laminar pipe flow ($\text{Re}_d < 2000$). For a pipe of diameter $d$ with pressure drop over a length $L$, the governing equations of motion lead to

$$Q = \frac{\pi d^4}{128\mu} \frac{p_1 - p_2}{L} \quad \text{where Re}_d < 2000 \tag{10.19}$$

This relation was first demonstrated by Jean Poiseulle (1799–1869), who conducted meticulous tests documenting the resistance of flow through capillary tubes.

The simplest type of laminar flow meter consists of two pressure taps separated by a length of piping.[2] However, because the Reynolds number must remain low, the flow rate range is limited. This limitation is overcome in commercial units by using bundles of small-diameter tubes or passages placed in parallel and called laminar flow elements (Figure 10.13). The strategy of a laminar flow element is to divide the total flow by passing it through the smaller tubes within the tube bundle. The flow rate within each small tube is much less than the total flow in a way that keeps Re < 2000. The total flow is then the sum of the flows through all of the tubes in the bundle. Pressure drop is measured between the entrance and the exit of the laminar flow element, and a flow coefficient is used to account for inefficiency, such that

$$Q = K_1 \Delta p \tag{10.20}$$

where coefficient $K_1$ is the meter static sensitivity and is called its *K-factor*. Commercial units come calibrated with $K_1$ a constant over the useful meter range.

---

[2]Alternately, if flow rate can be measured or is known, Equation 10.19 provides the basis for a capillary tube viscometer. The method also achieves a linear resistance with flow rate.

**FIGURE 10.13** Laminar flow element flow meter concept using pipe flanges for installation. For noncircular tubes, Reynolds number is based on $d = 4r_H$.

A laminar flow element has an upper limit on usable flow rate (i.e., its laminar flow limit). Various meter sizes and designs are available to accommodate user needs. Turndowns of up to 100:1 are available.

Laminar flow elements offer some distinct advantages over other pressure differential meters. These include (1) a high sensitivity even at low flow rates, (2) an ability to measure flow from either meter direction, (3) a wide usable flow range, and (4) the ability to indicate an average flow rate in pulsating flows. The instrument systematic uncertainty in flow rate determination is as low as 0.25% (95%) of the flow rate. However, these meters are limited to clean fluids for considerations of clogging potential. The measured pressure drop remains a system pressure loss.

## 10.6 Insertion Volume Flow Meters

Dozens of volume flow meter types based on a number of different principles have been proposed, developed, and sold commercially. A large group of meters is based on some phenomenon that is actually sensitive to the average velocity across a control surface of known area, that is, $Q = f(\overline{U}, A) = \overline{U}A$. Several of these designs are included in the discussion below. Another common group, called *positive displacement meters*, actually measure parcels of a volume of fluid per unit time—that is, $Q = f(\forall, t) = \forall/t$.

### Electromagnetic Flow Meters

The operating principle of an *electromagnetic flow meter* [10] is based on the fundamental principle that an electromotive force (emf) of electric potential $E$ is induced in a conductor of length $L$ that moves with velocity $U$ through a magnetic field of magnetic flux, $B$. Simply, when an electrically conductive liquid moves through a magnetic field, a voltage is induced in the liquid at a right angle to the field. The voltage is detected by metal electrode sensors with the voltage magnitude and polarity directly proportional to the volume flow rate and the flow direction, respectively. This physical behavior was first recorded by Michael Faraday (1791–1867). In principle,

$$E = U \times B \cdot L \tag{10.21}$$

where $U$, $B$, and $L$ are vector quantities.

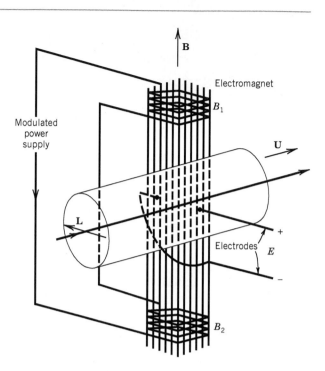

**FIGURE 10.14** Electromagnetic principle as applied to a working flow meter.

A practical use of the principle is shown in Figure 10.14. From Equation 10.21 the magnitude of $E$ is affected by the average velocity, $\overline{U}$, as

$$E = \overline{U}BL \sin \alpha = f(\overline{U})$$

where $\alpha$ is the angle between the mean velocity vector and the magnetic flux vector, usually at 90 degrees. In general, electrodes are located either in or mounted on the pipe wall in a diametrical plane that is normal to the known magnetic field. The electrodes are separated by a scalar length, $L$. The average magnitude of the velocity, $\overline{U}$, across the pipe is thus inferred through the measured emf. The flow rate is found by

$$Q = \overline{U}\frac{\pi d_1^2}{4} = \frac{E}{BL}\frac{\pi d_1^2}{4} = K_1 E \tag{10.22}$$

The value of $L$ is on the order of the pipe diameter, the exact value depending on the meter construction and magnetic flux lines. The static sensitivity (or meter $K$-factor) $K_1$ is found by calibration and supplied by a manufacturer such that the relationship between the flow rate and the measured potential is linear.

The electromagnetic flow meter comes commercially as a packaged flow device, which is installed directly inline and connected to an external electronic output unit. Special designs include an independent flow sensor unit that can clamp over a nonmagnetic pipe, a design favored to monitor blood flow rate through major arteries during surgery. The sensor is connected to a control unit by wiring. External sensors are also available as an inline union that can measure flow rate in either direction with instantaneous changes (Figure 10.15).

The electromagnetic flow meter has a very low pressure loss associated with its use due to its open tube, no obstruction design, and is suitable for installations that can tolerate only a small pressure drop. It can be used for either steady or unsteady flow measurements, providing either time-averaged or instantaneous data in the latter. This absence of internal parts is very attractive for metering corrosive and "dirty" fluids. The operating principle is independent of fluid density

**FIGURE 10.15**  Example of an in-line electromagnetic flow probe. The black housing contains the electromagnet and two electrodes. The connecting plug has wires for power and electrode signals to/from an external control box.

and viscosity, responding only to average velocity, and there is no difficulty with measurements in either laminar or turbulent flows, provided that the velocity profile is reasonably symmetrical. Uncertainty down to 0.25% (95%) of the measured flow rate can be attained, although values from 1% to 5% (95%) are more common for these meters in industrial settings. The fluid must be conductive, but the minimum conductivity required depends on a particular meter's design. Fluids with values as low as 0.1 microsieman (µsieman)/cm have been metered. Adding salts to a fluid increases its conductivity. Stray electronic noise is perhaps the most significant barrier in applying this type of meter. Grounding close to the electrodes and increasing fluid conductivity reduce noise.

## Vortex Shedding Meters

An oscillating street sign and the "singing" of power lines on a windy day are examples of the effects induced by vortex shedding from bluff-shaped bodies, a natural phenomenon wherein alternating vortices are shed in the wake of the body. The vortices formed on opposite sides of the body are carried downstream in the body's wake, forming a "vortex street," with each vortex having an opposite sign of rotation. This behavior is seen in Figure 10.16, a photograph that captures the vortex shedding downstream of a section of an aircraft wing. The aerodynamicist Theodore von Karman (1881–1963) first deduced the existence of a vortex street, although Leonardo da Vinci appears to have been the first to actually record the phenomenon [11].

A *vortex flow meter* operates on the principle that the frequency of vortex shedding depends on the average velocity of the flow past the body and the body shape. The basic relationship between shedding frequency, $f$, where $f$ [$Hz$] $= \omega/2\pi$, and average velocity, $\overline{U}$, for a given shape is given by the Strouhal number,

$$\text{St} = \frac{fd}{\overline{U}} \tag{10.23}$$

where $d$ is a characteristic length for the body.

**FIGURE 10.16** Smoke lines in this photograph reveal the vortex shedding behind a streamlined wing-shaped body in a moving flow.

Richard Figliola

**FIGURE 10.17** Vortex shedding flow meter. Different shedder shapes are available.

A typical design is shown in Figure 10.17. The shedder spans the pipe, so its length $l \approx d_1$, and $d/l \approx 0.3$, so as to provide for strong, stable vortex strength. Although the Strouhal number is a function of the Reynolds number, various geometrical shapes, known as shedders, can produce a stable vortex flow that has a constant Strouhal number over a broad range of flow Reynolds numbers (for $\text{Re}_d > 10^4$). The oscillation stability, quality, and strength of the shedding is improved over common circular cylinders by using abrupt edges on the rear end of the shape and by providing a slightly concave upstream body face that traps the stagnation streamline at a point. Examples are given in Table 10.1.

For a fixed body and constant Strouhal number, the flow rate for a pipe of i.d. $d_1$ is

$$Q = \overline{U}A = c\frac{\pi d_1^2}{4St}fd = K_1 f \qquad (10.24)$$

where the constant $c$ accounts for shedder blockage effects that tend to increase the average velocity sensed. The value of $K_1$, known as the $K$-factor, is the meter static sensitivity and remains essentially constant for $10^4 < \text{Re}_d < 10^7$. Shedding frequency can be measured in many ways. The shedder strut can be instrumented to detect the force oscillation by using strut-mounted strain gauges or capacitance sensor, for example, or a piezoelectric crystal wall sensor can be used to detect the pressure oscillations in the flow.

The lower flow rate limit on vortex meters is at Reynolds numbers (for Re based on $d$ in Figure 10.17) near 10,000, below which the Strouhal number can vary nonlinearly with flow rate and shedding becomes unstable. This can be a limitation in metering high-viscosity pipe flows ($\mu > 20$ centipoise). The upper flow bound is limited either by the onset of cavitation in liquids or by the onset of compressibility effects in gases at Mach numbers exceeding 0.2.

**Table 10.1   Shedder Shape and Strouhal Number**

| Cross section | Strouhal number[a] |
|---|---|
| | 0.16 |
| | 0.19 |
| | 0.16 |
| | 0.15 |
| | 0.12 |

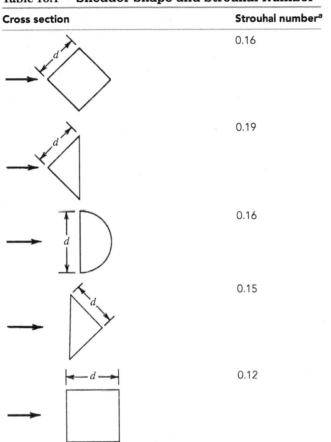

[a]For Reynolds number $Re_d \geq 10^4$. Strouhal number $St = fd/\bar{u}$.

Property variations affect meter performance only indirectly. Density variations affect the strength of the shed vortex, and this places a lower limit on fluid density, which is based on the sensitivity of the vortex-shedding detection equipment. Viscosity affects the operating Reynolds number. Otherwise, within bounds, the meter is insensitive to property variations.

The meter has no moving parts and relatively low pressure losses compared to obstruction meters. A single meter can operate over a flow range of up to 20:1 above its minimum with a linearity in $K_1$ of 0.5%. Because $d$ can be replaced by a constant times $d_1$ in Equation 10.24, we see that the strength of the shedding frequency sensed over the pipe area decreases as the pipe diameter cubed. This indicates that a meter's accuracy decreases as pipe diameter increases, placing an upper limit on the meter size.

## Rotameters

The rotameter is a widely used insertion meter for volume flow rate indication. As depicted in Figure 10.18, the meter consists of a float within a vertical tube, tapered to an increasing cross-sectional area at its outlet. Flow entering through the bottom passes over the float, which is free to move. The equilibrium height of the float indicates the flow rate.

The operating principle of a rotameter is based on the balance between the drag force, $F_D$, and the weight, $W$, and buoyancy forces, $F_B$, acting on the float in the moving fluid. It is the drag force that varies with the average velocity over the float.

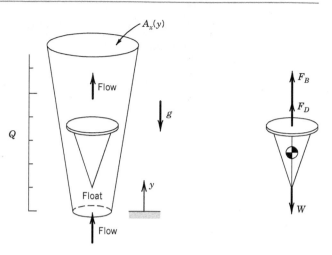

**FIGURE 10.18**    Concept of a rotameter.

The force balance in the vertical direction $y$ yields

$$\sum F_y = 0 = +F_D - W + F_B$$

with $F_D = \frac{1}{2} C_D \rho \overline{U}^2 A_x$, $W = \rho_b g \forall_b$, and $F_B = \rho g \forall_b$. The average velocity sensed by the float depends on its height in the tube and is given by

$$\overline{U} = \overline{U}(y) = \sqrt{2(\rho_b - \rho)g \forall_b / C_D \rho A_x} \qquad (10.25)$$

where

$\rho_b$ = density of float (body)

$\rho$ = density of fluid

$C_D$ = drag coefficient of the float

$A_x$ = tube cross-sectional area

$\overline{U}$ = average velocity past the float

$\forall_b$ = volume of float (body)

In operation, the float rises to an equilibrium position. The height of this position increases with flow velocity and, hence, flow rate. This flow rate is found by

$$Q = \overline{U} A_a(y) = K_1 A_a(y) \qquad (10.26)$$

where $A_a(y)$ is the annular area between the float and the tube, and $K_1$ is a meter constant. Both the average velocity and the annular area depend on the height of the float in the tube. So the float's vertical position is a direct measure of flow rate, which can be read from a graduated scale, electronically sensed with an optical cell, or detected magnetically. Floats with sharp edges are less sensitive to fluid viscosity changes with temperature. A typical meter turndown is 10:1 with an instrument systematic uncertainty of ~2% (95%) of flow rate.

## Turbine Meters

Turbine meters make use of angular momentum principles to meter flow rate. In a typical design (Figure 10.19), a rotor is encased within a bored housing through which the fluid to be metered is

Signal

Pickup coil

Magnet

Rotor    Sleeve

Retaining nut

**FIGURE 10.19**   Cutaway view of a turbine
flow meter.

passed. Its housing contains flanges or threads for direct insertion into a pipeline. In principle, the
exchange of momentum within the flow turns the rotor at a rotational speed that is proportional
to the flow rate. Rotor rotation can be measured in a number of ways. For example, a reluctance
pickup coil can sense the passage of magnetic rotor blades, producing a pulse train signal at a
frequency that is directly related to rotational speed. This can be directly output as a TTL pulse
train, or the frequency can be converted to an analog voltage.

The rotor angular velocity, $\omega$, depends on the average flow velocity, $\overline{U}$, and the fluid kinematic
viscosity, $\nu$, through the meter bore of diameter, $d_1$. Dimensionless analysis of these parameters
[12] yields

$$Q = K_1 \omega \tag{10.27}$$

where $K_1$ is the meter's $K$-factor but the relation is a function of Reynolds number. In practice,
there is a region in which the rotor angular velocity varies linearly with flow rate, and this region
becomes the meter operating range.

Turbine meters offer a low-pressure drop and very good accuracy. A typical instrument's systematic uncertainty in flow rate is $\sim 0.25\%$ (95%) with a turndown of 20:1. The measurements
are exceptionally repeatable, making these meters good candidates for local flow rate standards.
However, their use must be restricted to clean fluids to avoid fouling their rotating parts. The turbine meter rotational speed is sensitive to temperature changes, which affect fluid viscosity and
density and thus $K_1$. Some compensation for viscosity variations can be made electronically [13].
The turbine meter is very susceptible to installation errors caused by pipe flow swirl [9], and a
careful selection of installation position must be made.

## Transit Time and Doppler (Ultrasonic) Flow Meters

Ultrasonic meters use sound waves to determine flow rate. *Transit time flow meters* use the travel
time of ultrasonic waves to estimate average flow velocity. Figure 10.20 shows a pair of transducers, separated by some distance, fixed to the outside of a pipe wall. A reflector applied to
the opposite outside wall of the pipe increases the signal-to-noise ratio. Each transducer acts as
a transmitter and a receiver for ultrasonic waves. An ultrasonic wave emitted by one transducer
passes through the fluid, reflects off the pipe wall, and is received by the other transducer. The
difference in transit time for a wave to travel from transducer 1 to transducer 2 and from transducer 2 to transducer 1 is directly related to the average velocity of flow in the pipe. For a fluid

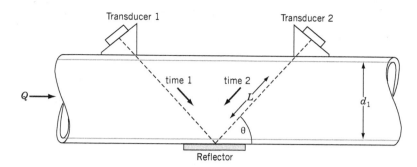

**FIGURE 10.20**   Principle of a transit time (ultrasonic) flow meter.

with speed of sound $a$, for flow with average velocity $\overline{U}$ based on a flow rate $Q$, and for a beam oriented at angle $\theta$ relative to the pipe flow axis,

$$t_1 = \frac{2L}{a + \overline{U}\cos\theta} \qquad t_2 = \frac{2L}{a - \overline{U}\cos\theta}$$

with $L = d_1/\sin\theta$. When $\overline{U} \ll a$, $\overline{U} = (a^2/4d_1\cot\theta)(t_2 - t_1)$. Or, eliminating $a$, we obtain $\overline{U} = (L/\cos\theta)(t_2 - t_1)/t_2 t_1$. The flow rate is

$$Q = \overline{U}A = K_1 A(t_2 - t_1) \tag{10.28}$$

where $K_1$ is a constant called the meter $K$-factor. The method requires accurate measurements of $t_1$ and $t_2$.

Transit time meters are noninvasive and thus offer no pressure drop. They can function with clean fluids. Portable models are strapped to the outside of the pipe, making them useful for field diagnostics. They can be set up to measure time-dependent velocity, including flow direction with time. Instrument relative systematic uncertainty ranges from ~1% to 5% (95%) of the flow rate.

*Doppler flow meters* use the Doppler effect (as introduced in Chapter 9) to measure the average velocity of fluid particles (contaminants, particulate, or small bubbles) suspended in the pipe flow. In one approach for a volume flow meter, a single ultrasonic transducer is mounted to the outside of a pipe wall. A wave of frequency $f$ on the order of 100 kHz to 1 MHz is emitted. The emitted wave is bounced back by particles suspended in the moving fluid, and this scattered wave is detected by the transducer. The scattered waves have a slightly different frequency, the difference being the Doppler frequency $f_D$. The flow rate is

$$Q = \overline{U}A = \frac{\pi d_1^2 a f_D}{8f\cos\theta} = K_1 f_D \tag{10.29}$$

These devices can measure time-dependent flow rates. The value $K_1$ is sensitive to pipe wall material and thickness and fluid properties. Instrument systematic uncertainty is ~2% (95%) of the flow rate.

## Positive Displacement Meters

Positive displacement meters contain mechanical elements that define a known volume filled with the fluid being metered. The free-moving elements are displaced or rotated by the action of the moving fluid. A geared counting mechanism counts the number of element displacements to provide a direct reading of the volume of fluid passed through the meter, ∀. This metering method is common to water, gasoline, and natural gas meters.

To be used as volume flow rate meters, volume measurement can be used in conjunction with a timer so that

$$Q = \forall / t \tag{10.30}$$

These units serve applications needing ruggedness and accuracy in steady flow metering. Common applications are metering related to domestic water and natural gas consumption.

*Diaphragm meters* contain two opposing flexible bellows coupled by linkages to exhaust valves that alternately displace a known volume of fluid through the meter. Through the coupled linkage, the flow of fluid through one bellows drives the motion of the opposite bellows. Thus, as entering gas causes one bellows to expand within its chamber sealed by a closed valve, the gas in the opposite bellows exhausts through an open valve to allow its bellows to collapse; this alternating motion cycles the gears of its counter register (output display). Such "dry gas" meters are common to natural gas or propane lines for volume measurement. Thomas Glover invented the first diaphragm meter in 1843, and modern meters retain the major elements of his design.

*Wobble meters* contain a disk that is seated in a chamber. The flow of liquid through the chamber causes the disk to oscillate so that a known volume of fluid moves through the chamber with each oscillation. The disk is connected to a counter that records each oscillation as volume is displaced. Wobble meters are used for domestic water applications, where they must have an uncertainty of no greater than 1% of actual delivery.

Frequently found on oil trucks and at the gasoline pump, *rotating vane meters* use rotating cups or vanes that move about an annular opening, displacing a known volume with each rotation.

With any of these meters, uncertainty can be as low as 0.2% (95%) of actual delivery. Because of the good accuracy of these meters, they are often used as local working standards to calibrate other types of volume flow meters.

### EXAMPLE 10.9    Focus on the Vortex Flow Meter

A modern vortex flow meter finds applications in liquids, gases, and steam measurements. Most meters use a proprietary shedder mounted on an instrumented strut on which a piezoelectric or capacitance sensor responds to the strut vibrations induced by vortex shedding. The sensor signal is passed through a charge amplifier, then into a low-pass filter and amplifier having enough gain to boost the signal into the mV to V range.

An example of a conditioned and strong vortex shedding signal from the sensor is shown in Figure 10.21(b). A high/low trigger level within the circuitry detects when the sensor signal crosses the threshold. As the signal passes through the high and then low trigger levels, this generates a TTL (true transistor logic, see Section 7.4) signal of corresponding duration. With regular vortex shedding, the TTL pulses per unit time will equal the shedding frequency. The flow meter applies the preprogrammed $K$-factor to the counted frequency. The output signal (e.g., LCD display, 4–20 mA, voltage, or pulse) will correspond to the measured flow rate.

If the meter is used outside of its range, the vortex signal will be erratic, noisy, at too high of a frequency, or of insufficient magnitude to even be detected by the sensor. This is illustrated in Figure 10.21(a), in which the TTL signal is erratic or missing due to too low a flow rate. To avoid such mishaps, the manufacturer will specify the flow rate range of the meter for both liquids and gases, the pipe size range that matches the meter size, and the permanent pressure loss for the meter.

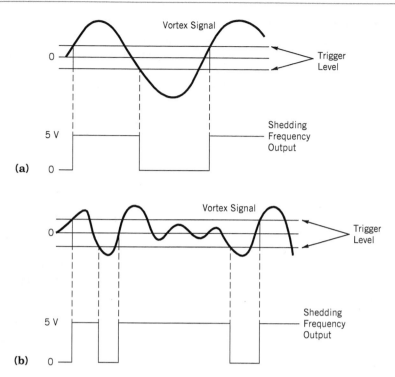

**FIGURE 10.21**    Triggering of vortex shedding signal creates a TTL pulse sequence. (a) Good signal. (b) Improperly sized meter signal.

## 10.7  Mass Flow Meters

There are many situations in which the mass flow rate is the quantity of interest. Direct mass flow rate determination demands measuring the mass flux, $\overline{\rho U}$, to yield $\dot{m} = \overline{\rho U}A$.

If the density of the metered fluid is known under the exact conditions of measurement, then a direct estimate can be made based on volume flow rate measurements. But not all fluids have a constant, known value of density (e.g., petroleum products, polymers, and cocoa butter have densities that are very sensitive to temperature and shear rate), and many processes are subject to significant changes in density. The direct measurement of mass flow rate is desirable because it eliminates the uncertainties associated with estimating or measuring actual density.

The difference between a meter that is sensitive to $Q$ as opposed to $\dot{m}$ is not trivial. Before the 1970s, reliable commercial mass flow meters with sufficient accuracy to circumvent volume flow rate corrections were generally not available, even though the basic principles and implementation schemes for such meters had been understood in theory for several decades. U.S. patents dating back to the 1940s record schemes for using heat transfer, Coriolis forces, and momentum methods to infer mass flow rate directly.

### Thermal Flow Meter

The rate at which energy, $\dot{E}$, must be added to a flowing fluid to raise its temperature between two control surfaces is directly related to the mass flow rate by

$$\dot{E} = \dot{m}c_p \Delta T \tag{10.31}$$

where $c_p$ is the fluid specific heat. Methods to utilize this effect to directly measure mass flow rate incorporate an in-line meter having some means to input energy to the fluid over the meter length. The passing of a current through an immersed filament is a common method. Fluid temperatures are measured at the upstream and downstream locations of the meter. This type of meter is quite easy to use and appears to be reliable. It is widely used for gas flow applications. In fact, in the 1980s, the technique was adapted for use in automobile fuel injection systems to provide an exact air–fuel mixture to the engine cylinders despite short-term altitude, barometric, and seasonal environmental temperature changes.

The operating principle of this meter assumes that $c_p$ is known and remains constant over the length of the meter. For common gases, such as air, this assumption is quite good. Flow rate turndown of up to 100:1 is possible with uncertainties of 0.5% (95%) of flow rate with very little pressure drop. But the assumptions become restrictive for liquids and for gases for which $c_p$ may be a strong function of temperature.

A second type of thermal mass flow meter is a velocity-sensing meter and thermal sensor together in one direct insertion unit. The meter uses both hot-film anemometry methods (see Chapter 9) to sense fluid velocity through a conduit of known diameter and an adjacent resistance temperature detector (RTD) sensor for temperature measurement. For sensor and fluid temperatures, $T_s$ and $T_f$, respectively, mass flow rate is inferred from the correlation

$$E = \left[ C + B(\rho \overline{U})^{1/n} \right] (T_s - T_f) \tag{10.32}$$

where $C$, $B$, and $n$ are constants that depend on fluid properties [14] and are determined through calibration. In a scheme to reduce the fluid property sensitivity of the meter, the RTD may be used as an adjacent resistor leg of the anemometer Wheatstone bridge circuit to provide a temperature-compensated velocity output over a wide range of fluid temperatures with excellent repeatability (0.25%). Gas velocities of up to 12,000 ft/min and flow rate turndown of 50:1 are possible with uncertainties down to 2% of the flow rate and very little pressure drop.

## Coriolis Flow Meter

The term "Coriolis flow meter" refers to the family of insertion meters that meter mass flow rate by inducing a Coriolis acceleration on the flowing fluid and measuring the resulting developed Coriolis force [15]. The developed force is directly related to the mass flow rate independent of the fluid properties. The Coriolis effect was proposed by Gaspard de Coriolis (1792–1843) after his studies of accelerations in rotating systems. Coriolis meters pass a fluid through a rotating or vibrating pipe system to develop the Coriolis force. A number of methods using this effect have been proposed since the first U.S. patent for a Coriolis effect meter was issued in 1947. This family of meters has seen steady growth in market share since the mid-1980s.

In the most common scheme for commercially available units, the pipe flow is diverted from the main pipe and divided between two bent, parallel, adjacent tubes of equal diameter, such as shown for the device in Figure 10.22, a device developed in part by the first author of this text.[3] The tubes themselves are mechanically vibrated in a relative out-of-phase sinusoidal oscillation by an electromagnetic driver. In general, a fluid particle passing through the meter tube, which is rotating (due to the oscillating tube) relative to the fixed pipe, experiences an acceleration at any arbitrary position $S$. The total acceleration at $S$, $\ddot{\mathbf{r}}$, is composed of several components (Figure 10.23),

$$\ddot{\mathbf{r}} = \ddot{\mathbf{R}}o' + \dot{\boldsymbol{\omega}} \times \mathbf{r}_{S/O'} + \boldsymbol{\omega} \times \boldsymbol{\omega} \times \mathbf{r}_{S/O'} + \ddot{\mathbf{r}}_{S/O'} + 2\boldsymbol{\omega} \times \dot{\mathbf{r}}_{S/O'} \tag{10.33}$$

---

[3]In its initial beta test, the unit shown in Figure 10.22 was installed in a chocolate factory to meter the mass flow of cocoa butter.

**FIGURE 10.22**   Cutaway view of a Coriolis mass flow meter.

Courtesy of Red Seal Measurement, Greenwood, SC. Used with Permission.

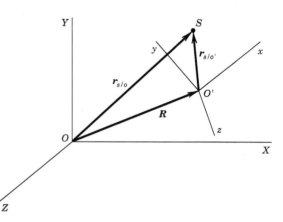

**FIGURE 10.23**   Fixed and rotating reference frames.

where boldface refers to vector quantities and

$\ddot{\mathbf{R}}_{O'}$ = translation acceleration of rotating origin $O'$ relative to fixed origin $O$

$\boldsymbol{\omega} \times \boldsymbol{\omega} \times \mathbf{r}_{S/O'}$ = centripetal acceleration of $S$ relative to $O'$

$\dot{\boldsymbol{\omega}} \times \mathbf{r}_{S/O'}$ = tangential acceleration of $S$ relative to $O'$

$\ddot{\mathbf{r}}_{S/O'}$ = translational acceleration of $S$ relative to $O'$

$2\boldsymbol{\omega} \times \dot{\mathbf{r}}_{S/O'}$ = Coriolis acceleration at $S$ relative to $O'$

with $\omega$ the angular velocity of point $S$ relative to $O'$. In these meters, the tubes are rotated but not translated, so that the translational accelerations are zero. A fluid particle experiences forces because of the remaining accelerations that cause equal and opposite reactions on the meter tube walls. The Coriolis acceleration distinguishes itself by acting in a plane perpendicular to the tube axes and develops a force gradient that creates a twisting motion or oscillating rotation about the tube plane.

The use of the Coriolis force depends on the shape of the meter. However, the basic principle is illustrated in Figure 10.24. Rather than rotating the tubes a complete 360 degrees about the pipe axis, the meter tubes are vibrated continuously at a drive frequency, $\omega$, with amplitude displacement, $z$, about the pipe axis. This eliminates rotational seal problems. The driving frequency is selected at the tube resonant frequency that places the driven tube into what is called a limit cycle,

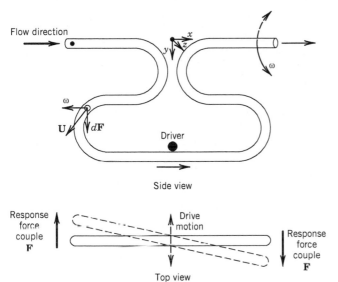

Side view

Top view

**FIGURE 10.24**   Concept of the operating principle of a Coriolis mass flow meter. Use the right-hand rule to relate the vectors of ω, U, and $d\mathbf{F}$.

a continuous, single-frequency oscillation. This configuration is essentially a self-sustaining tuning fork, because the meter naturally responds to any disturbance at this frequency with a minimum in input energy. By driving the tube up and down, the mass flow within the tube causes the tube to wobble. The magnitude of wobble is directly related to the mass flow rate.

As a fluid particle of elemental mass, $dm$, flows through a section of the flow meter, it experiences the Coriolis acceleration, $2\boldsymbol{\omega} \times \dot{\mathbf{r}}_{S/O'}$, and an inertial Coriolis force

$$d\mathbf{F} = (2\boldsymbol{\omega} \times \dot{\mathbf{r}}_{S/O'})dm \tag{10.34}$$

acting in the $z$ direction. For the flow meter design of Figure 10.22 and as depicted in Figure 10.24, as the particle travels along the meter the direction of the velocity vector changes. This results, using the right-hand rule, in a change in direction of vector force, $d\mathbf{F}$, between the left and right sides. The resultant forces experienced by the tube are of equal magnitude but the opposite sign of those experienced by the particle. Each tube segment senses a corresponding differential torque, $d\mathbf{T}$, a rotation about the $y$ axis at a frequency $\boldsymbol{\omega}_c$,

$$d\mathbf{T} = x' \times d\mathbf{F} = x' \times (2\boldsymbol{\omega} \times \dot{\mathbf{r}}_{S/O'})dm \tag{10.35}$$

where $x'$ refers to the $x$ distance between the elemental mass and the $y$ axis, $\mathbf{x}' = \mathbf{x}'\widehat{e}_x$. The total magnitude of torque experienced by each tube is found by integration along the total path length of the tube, $L$,

$$\mathbf{T} = \int_0^L d\mathbf{T} \tag{10.36}$$

A differential element of fluid has mass $dm = \rho A dl$, for an elemental cross section of fluid $A$, of differential length $dl$, and of density $\rho$. If this mass moves with an average velocity, $\overline{U}$, then the differential mass can be written as

$$dm = \rho A dl = \dot{m}(dl/\overline{U}) \tag{10.37}$$

The Coriolis cross-product can be expressed as

$$2\boldsymbol{\omega} \times \dot{\mathbf{r}}_{S/O'} = (2\boldsymbol{\omega} \times \dot{\mathbf{r}}_{S/O'} \sin\theta)\widehat{e}_z = (2\omega_c \overline{U} \sin\theta)\widehat{e}_z \tag{10.38}$$

where $\theta$ is the angle between the Coriolis rotation and the velocity vector. Then,

$$\dot{m} = \frac{T}{2\displaystyle\int_0^L (\rho x' \omega_c \sin \theta)dl} \, \widehat{e}_y \tag{10.39}$$

Because the velocity direction changes by 180 degrees, the Coriolis-developed torque acts in opposite directions on each side of the tube.[4] The meter tubes twist about the $y$ axis of the tube (i.e., wobbles) at an angle $\delta$. For small angles of rotation the twist angle is related to torque by

$$\delta = k_s T = \text{constant} \times \dot{m} \tag{10.40}$$

where $k_s$ is related to the stiffness of the tube. The objective becomes to measure the twist angle, which can be accomplished in many ways. For example, by splitting the flow into two parallel tubes, as discussed, and driving the tubes 180 degrees out of phase, the relative phase at any time between the two tubes is directly related to the mass flow rate. The exact relationship is linear over a wide flow range and determined by calibration with any fluid [16].

The tangential and centripetal accelerations remain of minor consequence due to the tube stiffness in the directions in which they act. However, at high mass flow rates they can excite modes of vibration in addition to the driving mode, $\omega$, and response mode, $\omega_c$. In doing so, they affect the meter linearity and zero error (drift), which affects mass flow uncertainty and meter turndown. Essentially, the magnitude of these undesirable effects is inherent to the particular meter shape and is controlled, if necessary, through tube-stiffening members.

Another interesting problem occurs mostly in flow meters of small tube diameter where the tube mass may approach the mass of the fluid in the tube. At flow rates that correspond to flow transition from a laminar to turbulent regime, the driving frequency can excite the flow instabilities responsible for the flow transition. The fluid and tube can go out of phase, reducing the response amplitude and its corresponding torque. This affects the meter's calibration linearity, but a good design can contain this effect to within 0.5% of the meter reading.

The meter principle is unaffected by changing fluid properties, but temperature changes affect the overall meter stiffness, an effect that can be compensated for electronically. A very desirable feature is an apparent insensitivity to installation position. Commercially available Coriolis flow meters can measure mass flow rate with an instrument systematic uncertainty in liquids to 0.10% (95%), gases to 0.5% (95%). Turndown is about 20:1. The meter is also used as an effective densitometer that can measure to within 0.0005 g/cm$^3$.

---

**EXAMPLE 10.10**    **The Automobile Mass Air Flow Sensor**

The mass air flow sensor (MAF) in a modern automobile engine measures the mass flow rate (grams/s) of air entering through the engine intake and delivered to the engine cylinders. This sensor sends a signal to the electronic control unit (ECU) that makes corresponding adjustments to maintain a desirable combustion air–fuel ratio for the applied throttle (power requirement) setting.

A common scheme for an MAF unit is depicted in Figure 10.25. The device uses a heated element, which can take on various forms such as a metallic hot-film probe (Chapter 9) or a platinum hot-wire, and an air temperature sensor, such as a thermistor. These form two active arms of a resistance bridge used to monitor the sensor temperature in relation to the incoming

---

[4]We can envision a design in which the velocity does not reverse direction along the flow path but $\omega$ does.

air temperature (IAT). With the engine operating, the element is heated to a preset temperature above the ambient temperature. As air flow moves through the intake duct, it will cool the heated sensor down. As a conductor's resistance is a nearly linear function of temperature (Chapter 8), the sensor resistance changes with this cooling. The resulting resistance change unbalances the bridge and so the bridge circuitry acts to increase the current to the element to bring its temperature (and resistance) back to the preset value. The resulting bridge deflection voltage is directly proportional to the mass flow rate, and this information is sent to the controlling ECU for appropriate action.

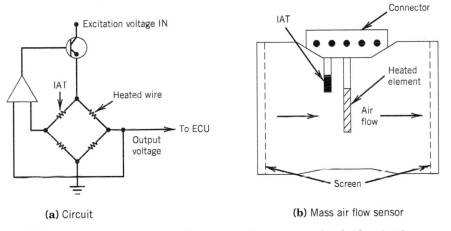

(a) Circuit                                  (b) Mass air flow sensor

**FIGURE 10.25**   Schematic of an automobile mass air flow sensor and its bridge circuit.

## 10.8 Flow Meter Calibration and Standards

While a fundamental primary standard for flow rate does not exist per se, there are a number of calibration test code procedures in place, and government and private bureaus that perform calibrations. The general procedure for the calibration of inline flow meters requires establishing a steady flow in a calibration flow loop and then determining the volume or mass of flowing fluid passing through the flow meter in an accurately determined time interval. Such flow loop calibration systems are known as *provers*. Often calibration is by comparison to a meter of proven accuracy. Several methods to establish the flow rate are discussed.

In liquids, variations of a "catch-and-weigh" technique are often employed in flow provers. One variation of the technique consists of a calibration loop with a catch tank as depicted in Figure 10.26. Tank A is a large tank from which fluid is pumped back to a constant head reservoir, which supplies the loop with a steady flow. Tank B is the catch-and-weigh tank into which liquid can be diverted for an accurately determined period of time. The liquid volume is measured, either directly using a positive displacement meter, or indirectly through its weight, and the flow rate deduced through time. The ability to determine the volume and the uncertainty in the initial and final time of the event are fundamental limitations to the accuracy of this technique. Neglecting installation effects, the ultimate limits of uncertainty (at 95%) in the flow rate of liquids are on the order of 0.03%, a number based on $u_{meter} \sim 0.02\%$, $u_{\forall} \sim 0.02\%$, and $u_t \sim 0.01\%$.

Flow meter calibration by measured pipe velocity profile is particularly effective for in situ calibration in both liquids and gases, provided that the gas velocity does not exceed about 70% of the sonic velocity. Velocity traverses at any cross-sectional location some 20 to 40 pipe diameters downstream of any pipe fitting in a long section of straight pipe are preferred.

Comparison calibration against a local standard flow meter is another common means of establishing the flow rate. Flow meters are installed in tandem with the standard and directly calibrated

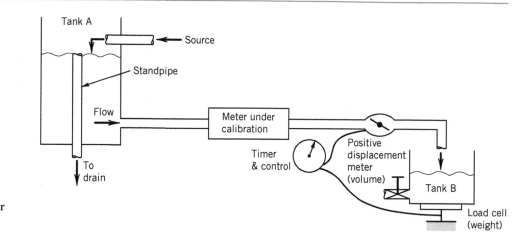

**FIGURE 10.26**
Flow diagram of a flow meter
prover for liquids.

against it. Turbine and vortex meters and Coriolis mass flow meters have consistent, highly accurate calibration curves and are often used as local standards. Other provers use an accurate positive displacement meter to determine flow volume over time. Of course, these standards must be periodically recalibrated. In the United States, the National Institute of Standards and Technology (NIST) maintains flow meter calibration facilities for this purpose. However, installation effects in the end-user facility are not accounted for in an NIST calibration. This last point accounts for much of the uncertainty in a flow meter calibration and use.

In any of these methods, the calibration uncertainty is limited by the standard used, installation effects, end-use variations, and the inherent limitations of the flow meter calibrated.

## 10.9  Estimating Standard Flow Rate

When the range of pressures and temperatures for flow processes vary in use, the measured "actual" flow rates can be adjusted to a standard temperature and pressure for comparison. This adjusted flow rate is called the *standard flow rate*. Standard flow rates are often reported in units such as standard cubic meters per minute (SCMM) or standard cubic feet per minute (SCFM). They are found by converting from the actual conditions at the measurement point, with the actual flow rate noted with corresponding units, such as ACMM or ACFM, in which A stands for "actual." The mass flow rate remains the same regardless of conditions, so using subscript $s$ for the standard conditions and $a$ for the actual measured conditions, we have $\dot{m}_s = \dot{m}_a$, so that

$$Q_s = Q_a \frac{\rho_a}{\rho_s} \tag{10.41}$$

For example, the actual flow rates of common gases are commonly "standardized" to 760 mm Hg absolute and 20°C. Assuming ideal gas behavior, $\rho = p/RT$, the standard flow rate is found from Equation 10.41 as

$$Q_s = Q_a \frac{T_s}{T_a} \frac{p_a}{p_s} = Q_a \left( \frac{293}{273 + T_a} \right) \left( \frac{760 + p_a}{760} \right) \tag{10.42}$$

Check the cited code or manufacturer specifications for the definition of the reference standard condition values. For example, many manufacturers use 20°C (often called the "normal" temperature). However, for air or natural gas, 15°C (atmospheric standard temperature) is also often used. However, in U.S. Customary units, a temperature of 60°F (15.6°C) is considered standard temperature and often used as the reference standard value. Standard pressure is almost always

101.325 kPa abs or 760 mm Hg abs or 14.69 psia. If you know the reference standard used, it is straightforward to convert back and forth.

## SUMMARY

Flow quantification has been an important engineering task for well over two millennia. This chapter presents methods to determine the volume rate of flow and the mass rate of flow. The engineering decision involving the selection of a particular meter depends on a number of constraining factors. In general, flow rate can be determined to within about 0.25% of actual flow rate with the best of present technology but practical values for industrial installations are more nearly 3% to 6% for obstruction meters and 1% to 3% for insertion meters. However, new methods may push these lower limits even further, provided that calibration standards can be developed to document method uncertainties and the effects of installation.

## REFERENCES

1. American Society of Mechanical Engineers, *PTC 19.5 –Flow Measurements*, ASME International, New York, 2004 (Rev. 2013).

2. American Society of Heating, Refrigeration and Air Conditioning Engineers (ASHRAE), *ASHRAE Fundamentals*, rev. ed., ASHRAE, Atlanta, 2009.

3. International Organization for Standardization (ISO), *Measurement of Fluid Flow by Means of Pressure Differential Devices Inserted in Circular Cross-Section Conduits: Parts 1 through 4*, ISO 5167, ISO, Geneva, 2003.

4. Jorgensen, R. (ed.), *Fan Engineering*, 9th ed., Buffalo Forge Co., Buffalo, New York, 1999.

5. American Society of Mechanical Engineers, *Measurement of Fluid Flow in Pipes Using Orifice, Nozzle and Venturi*, ASME Standard MFC-3M-2004, ASME International, New York, 2004.

6. Crane Company, *Flow of Fluids through Valves and Fittings*, Technical Paper No. 410, Spiral Edition, Crane Co., Chicago, IL, 2009.

7. Rouse, H., and S. Ince, *History of Hydraulics*, Dover, New York, 1957.

8. Amberg, B. T., A Review of Critical Flowmeters for Gas Flow Measurements, *Transactions of the ASME*, 84: 447–460, 1962.

9. Mattingly, G., *Fluid Measurements: Standards, Calibrations and Traceabilities*, Proceedings of the ASME/AIChE National Heat Transfer Conference, Philadelphia, PA, 1989.

10. Shercliff, J. A., *Theory of Electromagnetic Flow Measurement*, Cambridge University Press, New York, 1962.

11. da Vinci, L., *Del Moto e Misura Dell'Aqua* (English translation), E. Carusi and A. Favaro, eds., Zanichelli, Bologna, 1923.

12. Hochreiter, H. M., Dimensionless Correlation of Coefficients of Turbine-Type Flow-Meters, *Transactions of the ASME*, 80: 1363–1368, 1958.

13. Lee, W. F., and H. Karlby, A Study of Viscosity Effects and Its Compensation On Turbine Flow Meters, *Journal of Basic Engineering*, 82: 717–728, 1960.

14. Hinze, J. O., *Turbulence*, McGraw-Hill, New York, 1953.

15. American Society of Mechanical Engineers, *Measurement of Fluid Flow by Means of Coriolis Mass Flow Meters*, MFC-11M, ASME International, 2003.

16. Corwon, M., and R. Oliver, Omega-shaped Coriolis-type Mass Flow Meter System, U.S. Patent 4,852,410, 1989.

## NOMENCLATURE

| | |
|---|---|
| $c_p, c_v$ | specific heats ($l^2/t^2$-°) |
| $d$ | diameter ($l$) |
| $d_0$ | flow meter throat or minimum diameter ($l$) |
| $d_1$ | pipe diameter ($l$) |
| $d_2$ | vena contracta diameter (see Figure 10.3) ($l$) |
| $f$ | cyclical frequency ($\omega/2\pi$) ($t^{-1}$) |
| $g$ | gravitational acceleration constant ($lt^{-2}$) |
| $h_{L_{1-2}}$ | energy head loss between points 1 and 2 ($l$) |
| $k$ | ratio of specific heats, $c_p/c_v$ |
| $C$ | discharge coefficient |
| $C_D$ | drag coefficient |
| $E$ | velocity of approach factor; voltage |
| Gr | Grashof number |
| $H$ | manometer deflection ($l$) |
| $K_0$ | flow coefficient ($= CE$) |
| $K_1$ | flow meter constant; $K$-factor; static sensitivity |
| $L$ | length ($l$) |
| $Q$ | volume flow rate ($l^3 t^{-1}$) |
| $R$ | gas constant |
| $\mathrm{Re}_{d_1}$ | Reynolds number (based on $d_1$) |
| $\dot{m}$ | mass flow rate ($mt^{-1}$) |
| $p$ | pressure ($ml^{-1}t^{-2}$) |
| $p_1-p_2, \Delta p$ | pressure differential ($ml^{-1}t^{-2}$) |
| $\Delta p_{\mathrm{loss}}$ | permanent pressure loss ($ml^{-1}t^{-2}$) |
| $r$ | radial coordinate ($l$) |
| $r_1$ | pipe radius ($l$) |

| | | | |
|---|---|---|---|
| $r_H$ | hydraulic radius ($l$) | $\forall$ | volume ($l^3$) |
| $A$ | area ($l^2$) | $Y$ | expansion factor |
| $A_0$ | area based on $d_0$ ($l^2$) | $\beta$ | diameter ratio |
| $A_1$ | area based on pipe diameter $d_1$ ($l^2$) | $\rho$ | density ($ml^{-3}$) |
| $A_2$ | vena contracta area ($l^2$) | $\delta$ | twist angle |
| $\boldsymbol{B}$ | magnetic field flux vector | $\omega$ | frequency ($t^{-1}$); angular velocity ($t^{-1}$) |
| $S$ | specific gravity | $\mu$ | absolute viscosity ($ml^{-1}t^{-1}$) |
| $T$ | temperature (°); torque ($ml^2t^{-2}$) | $v$ | kinematic viscosity ($=\mu/\rho$) ($l^2t^{-1}$) |
| $U$ | velocity ($lt^{-1}$) | | |

# Strain Measurement

## 11.1 Introduction

The design of safe load-carrying components for machines and structures requires information concerning the distribution of forces within the particular component. Proper design of devices such as shafts, pressure vessels, and support structures must consider load-carrying capacity and allowable deflections. Mechanics of materials provides a basis for predicting these essential characteristics of a mechanical design; however, theoretical analysis is often not sufficient, and experimental measurements are required to achieve a final design.

Design criteria are based on stress levels within a part. In most cases, stress cannot be measured directly. But the length of the rod in Figure 11.1 changes when the load is applied, and such changes in length or shape of a material can be measured. This chapter discusses the measurement of physical displacements in engineering components. The stress is calculated from these measured deflections.

Upon completion of this chapter, the reader will be able to:

- Define strain and delineate the difficulty in measuring stress
- State the physical principles underlying mechanical strain gauges
- Analyze strain gauge bridge circuits
- Describe methods for optical strain measurement

## 11.2 Stress and Strain

Before we proceed to develop techniques for strain measurements, we briefly review the relationship between deflections and stress. The experimental analysis of stress is accomplished by measuring the deformation of a part under load and inferring the existing state of stress from the measured deflections. Again, consider the rod in Figure 11.1. If the rod has a cross-sectional area of $A_c$, and the load is applied only along the axis of the rod, the *normal stress* is defined as

$$\sigma_a = F_N/A_c \tag{11.1}$$

where $A_c$ is the cross-sectional area, and $F_N$ is the tension force applied to the rod normal to the area $A_c$. The ratio of the change in length of the rod (which results from applying the load) to the original length is the *axial strain*, defined as

$$\varepsilon_a = \delta L/L \tag{11.2}$$

where $\varepsilon_a$ is the average strain over the length $L$, $\delta L$ is the change in length, and $L$ is the original unloaded length. For most engineering materials, strain is a small quantity; strain is usually

**FIGURE 11.1** Free-body diagram illustrating internal forces for a rod in uniaxial tension.

**FIGURE 11.2** A typical stress–strain curve for mild steel.

reported in units of $10^{-6}$ in./in. or $10^{-6}$ m/m. These units are equivalent to a dimensionless unit called a microstrain ($\mu\varepsilon$).

Stress–strain diagrams are very important in understanding the behavior of a material under load. Figure 11.2 is such a diagram for mild steel (a ductile material). For loads less than that required to permanently deform the material, most engineering materials display a linear relationship between stress and strain. The range of stress over which this linear relationship holds is called the *elastic region*. The relationship between uniaxial stress and strain for this elastic behavior is expressed as

$$\sigma_a = E_m \varepsilon_a \tag{11.3}$$

where $E_m$ is the modulus of elasticity, or Young's modulus, and the relationship is called Hooke's law. Hooke's law applies only over the range of applied stress where the relationship between stress and strain is linear. Different materials respond in a variety of ways to loads beyond the linear range, largely depending on whether the material is ductile or brittle. For almost all engineering components, stress levels are designed to remain well below the elastic limit of the material; thus, a direct linear relationship may be established between stress and strain.

## Lateral Strains

Consider the elongation of the rod shown in Figure 11.1 that occurs as a result of the load $F_N$. As the rod is stretched in the axial direction, the cross-sectional area will, for most materials, decrease. In the elastic range, there is a constant rate of change in the lateral strain as the axial strain increases. In the same sense that the modulus of elasticity is a property of a given material, the ratio of lateral strain to axial strain is also a material property. This property is called Poisson's ratio, defined as

$$v_p = -\frac{\text{Lateral strain}}{\text{Axial strain}} = -\frac{\varepsilon_L}{\varepsilon_a} \tag{11.4}$$

For the case of axial tension and for most materials, the lateral strain will be negative and Poisson's ratio a positive number. Also for the case of uniaxial tension as in Figure 11.1, the volume of the rod will change. The dilatation, $e$, is defined as

$$e = \frac{\Delta V}{V_0} = \varepsilon_L(1 - 2v_p) \tag{11.5}$$

The total volume is only conserved for the case of $v_p = 0.5$.

Engineering components are seldom subject to one-dimensional axial loading. The relationship between stress and strain must be generalized to a multidimensional case. Consider a two-dimensional geometry, as shown in Figure 11.3, subject to tensile loads in both the $x$ and $y$ directions, resulting in normal stresses $\sigma_x$ and $\sigma_y$. In this case, for a biaxial state of stress, the stresses and strains are

$$\varepsilon_y = \frac{\sigma_y}{E_m} - v_p\frac{\sigma_x}{E_m} \qquad \varepsilon_x = \frac{\sigma_x}{E_m} - v_p\frac{\sigma_y}{E_m}$$

$$\sigma_x = \frac{E_m(\varepsilon_x + v_p\varepsilon_y)}{1 - v_p^2} \qquad \sigma_y = \frac{E_m(\varepsilon_y + v_p\varepsilon_x)}{1 - v_p^2} \tag{11.6}$$

$$\tau_{xy} = G\gamma_{xy}$$

In this case, all the stress and strain components lie in the same plane. The state of stress in the elastic condition for a material is similarly related to the strains in a complete three-dimensional situation [1, 2]. Since stress and strain are related, it is possible to determine stress from measured strains under appropriate conditions. However, strain measurements are made at the surface of an engineering component. The measurement yields information about the state of stress on

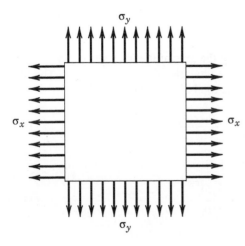

**FIGURE 11.3**  Biaxial state of stress.

the surface of the part. The analysis of measured strains requires application of the relationship between stress and strain at a surface. Such analysis of strain data is described elsewhere [3], and an example is provided in this chapter.

## 11.3 Resistance Strain Gauges

The measurement of the small displacements that occur in a material or object under mechanical load can be accomplished by methods as simple as observing the change in the distance between two scribe marks on the surface of a load-carrying member or as advanced as optical holography. In any case, the ideal sensor for the measurement of strain would (1) have good spatial resolution, implying that the sensor would measure strain at a point, (2) be unaffected by changes in ambient conditions, and (3) have a high-frequency response for dynamic (time-resolved) strain measurements. A sensor that closely meets these characteristics is the bonded resistance strain gauge.

In practical application, the bonded resistance strain gauge is secured to the surface of the test object by an adhesive so that it deforms as the test object deforms. The resistance of a strain gauge changes when it is deformed, and this is easily related to the local strain. Both metallic and semiconductor materials experience a change in electrical resistance when they are subjected to a strain. The amount that the resistance changes depends on how the gauge is deformed, the material from which it is made, and the design of the gauge. Gauges can be made quite small for good resolution and with a low mass to provide a high-frequency response. With some ingenuity, ambient effects can be minimized or eliminated.

In an 1856 publication in the *Philosophical Transactions of the Royal Society* in England, Lord Kelvin (William Thomson) [4] laid the foundations for understanding the changes in electrical resistance that metals undergo when subjected to loads, which eventually led to the strain gauge concept. Two individuals began the modern development of strain measurement in the late 1930s—Edward Simmons at the California Institute of Technology and Arthur Ruge at the Massachusetts Institute of Technology. Their development of the bonded metallic wire strain gauge led to commercially available strain gauges. The resistance strain gauge also forms the basis for a variety of other transducers, such as load cells, pressure transducers, and torque meters.

### Metallic Gauges

To understand how metallic strain gauges work, consider a conductor having a uniform cross-sectional area $A_c$ and length $L$ made of a material having an electrical resistivity, $\rho_e$. For this electrical conductor, the resistance, $R$, is given by

$$R = \rho_e L / A_c \tag{11.7}$$

If the conductor is subjected to a normal stress along the axis of the wire, the cross-sectional area and the length change resulting in a change in the total electrical resistance, $R$. The total change in $R$ is due to several effects, as illustrated in the total differential:

$$dR = \frac{A_c(\rho_e dL + L d\rho_e) - \rho_e L dA_c}{A_c^2} \tag{11.8}$$

which may be expressed in terms of Poisson's ratio as

$$\frac{dR}{R} = \frac{dL}{L}(1 + 2\upsilon_p) + \frac{d\rho_e}{\rho_e} \tag{11.9}$$

Hence, the changes in resistance are caused by two basic effects: the change in geometry as the length and cross-sectional area change and the change in the value of the resistivity, $\rho_e$. The dependence of resistivity on mechanical strain is called piezoresistance and may be expressed in terms of a piezoresistance coefficient, $\pi_1$, defined by

$$\pi_1 = \frac{1}{E_m}\frac{d\rho_e/\rho_e}{dL/L} \tag{11.10}$$

With this definition, the change in resistance may be expressed

$$dR/R = dL/L(1 + 2v_p + \pi_1 E_m) \tag{11.11}$$

## EXAMPLE 11.1

Determine the total resistance of a copper wire having a diameter of 1 mm and a length of 5 cm. The resistivity of copper is $1.7 \times 10^{-8}$ $\Omega$-m.

**KNOWN** $D = 1$ mm

$L = 5$ cm

$\rho_e = 1.7 \times 10^{-8}$ $\Omega$-m

**FIND** The total electrical resistance.

**SOLUTION**
The resistance may be calculated from Equation 11.7 as

$$R = \rho_e L/A_c$$

where

$$A_c = \frac{\pi}{4}D^2 = \frac{\pi}{4}(1 \times 10^{-3})^2 = 7.85 \times 10^{-7}\,\text{m}^2$$

The resistance is then

$$R = \frac{(1.7 \times 10^{-8}\,\Omega\text{m})(5 \times 10^{-2}\,\text{m})}{7.85 \times 10^{-7}\,\text{m}^2} = 1.08 \times 10^{-3}\,\Omega$$

**COMMENT** If the material were nickel ($\rho_e = 7.8 \times 10^{-8}$ $\Omega$-m) instead of copper, the resistance would be $5 \times 10^{-3}$ $\Omega$ for the same diameter and length of wire.

## EXAMPLE 11.2

A very common material for the construction of strain gauges is the alloy constantan (55% copper with 45% nickel), which has a resistivity of $49 \times 10^{-8}$ $\Omega$-m. A typical strain gauge might have a resistance of 120 $\Omega$. What length of constantan wire of diameter 0.025 mm would yield a resistance of 120 $\Omega$?

**KNOWN** The resistivity of constantan is $49 \times 10^{-8}$ $\Omega$-m.

**FIND** The length of constantan wire needed to produce a total resistance of 120 $\Omega$.

**SOLUTION**

From Equation 11.7, we may solve for the length, which yields in this case

$$L = \frac{RA_c}{\rho_e} = \frac{(120\,\Omega)(4.91 \times 10^{-10}\,\mathrm{m}^2)}{49 \times 10^{-8}\,\Omega\text{-m}} = 0.12\,\mathrm{m}$$

The wire would then be 12 cm long to achieve a resistance of 120 Ω.

**COMMENT** As shown by this example, a single straight conductor is normally not practical for a local strain measurement with meaningful resolution. Instead, a simple solution is to bend the wire conductor so that several lengths of wire are oriented along the axis of the strain gauge, as shown in Figure 11.4.

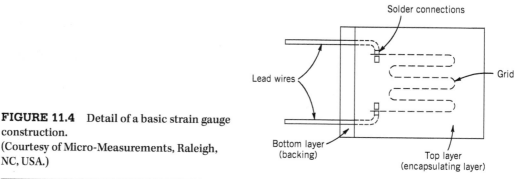

**FIGURE 11.4**   Detail of a basic strain gauge construction.
(Courtesy of Micro-Measurements, Raleigh, NC, USA.)

## Strain Gauge Construction and Bonding

Figure 11.5 illustrates the construction of a typical metallic-foil-bonded strain gauge. Such a strain gauge consists of a metallic foil pattern that is formed in a manner similar to the process used to produce printed circuits. This photoetched metal foil pattern is mounted on a plastic backing material. The gauge length, as illustrated in Figure 11.5, is an important specification for a particular application. Because strain is usually measured at the location on a component where the stress is a maximum and the stress gradients are high, the strain gauge averages the measured strain over

**FIGURE 11.5**   Construction of a typical metallic foil strain gauge.
(Courtesy of Micro-Measurements, Raleigh, NC, USA.)

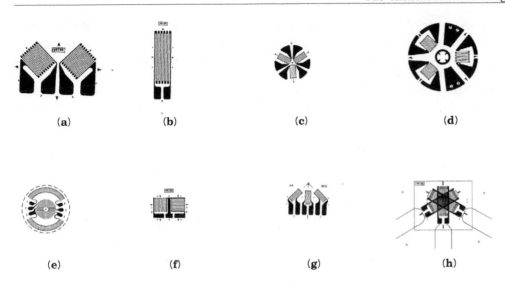

**FIGURE 11.6**
Strain gauge configurations.
(a) Torque Rosette; (b) Linear
Pattern; (c) Delta Rosette;
(d) Residual Stress Pattern;
(e) Diaphragm Pattern; (f) Tee
Pattern; (g) Rectangular
Rosette; (h) Stacked Rosette.
(Courtesy of Micro-
Measurements, Raleigh, NC,
USA.)

the gauge length. Because the maximum strain is the quantity of interest and the gauge length is the resolution, errors from averaging can result from improper choice of a gauge length [5].

The variety of conditions encountered in particular applications require special construction and mounting techniques, including various backing materials, grid configurations, bonding techniques, and total gauge electrical resistance. Figure 11.6 shows a variety of strain gauge configurations. The adhesives used in the bonding process and the mounting techniques for a particular gauge and manufacturer vary according to the specific application. However, some fundamental aspects are common to all bonded resistance gauges.

The strain gauge backing serves several important functions. It electrically isolates the metallic gauge from the test specimen and transmits the applied strain to the sensor. A bonded resistance strain gauge must be appropriately mounted to the specimen for which strain is to be measured. The backing provides the surface used for bonding with an appropriate adhesive. Backing materials are available that are useful over temperatures that range from −270 to 290°C.

The adhesive bond serves as a mechanical and thermal coupling between the metallic gauge and the test specimen. As such, the strength of the adhesive should be sufficient to accurately transmit the strain experienced by the test specimen, and should have thermal conduction and expansion characteristics suitable for the application. If the adhesive shrinks or expands during the curing process, apparent strain can be created in the gauge. A wide array of adhesives are available for bonding strain gauges to a test specimen. Among these are epoxies, cellulose nitrate cement, and ceramic-based cements.

## Gauge Factor

The change in resistance of a strain gauge is normally expressed in terms of an empirically determined parameter called the gauge factor (*GF*). For a particular strain gauge, the gauge factor is supplied by the manufacturer. The gauge factor is defined as

$$GF \equiv \frac{\delta R/R}{\delta L/L} = \frac{\delta R/R}{\varepsilon_a} \tag{11.12}$$

Relating this definition to Equation 11.11, we see that the gauge factor is dependent on Poisson's ratio for the gauge material and its piezoresistivity. For metallic strain gauges, Poisson's ratio is approximately 0.3, and the resulting gauge factor is ~2.

The gauge factor represents the total change in resistance for a strain gauge, under a calibration loading condition. The calibration loading condition generally creates a biaxial strain field, and the lateral sensitivity of the gauge influences the measured result. Strictly speaking then, the sensitivity to normal strain of the material used in the gauge and the gauge factor are not the same. Generally, gauge factors are measured in a biaxial strain field that results from the deflection of a beam having a value of Poisson's ratio of 0.285. Thus, for any other strain field, there is an error in strain indication due to the transverse sensitivity of the strain gauge. The percentage error due to transverse sensitivity for a strain gauge mounted on any material at any orientation in the strain field is

$$e_L = \frac{K_t(\varepsilon_L/\varepsilon_a + \upsilon_{po})}{1 - \upsilon_{po}K_t} \tag{11.13}$$

where

$\varepsilon_a$, $\varepsilon_L$ = axial and lateral strains, respectively (with regard to the axis of the gauge)

$\upsilon_{po}$ = Poisson's ratio of the material on which the manufacturer measured $GF$ (usually 0.285 for steel)

$e_L$ = error as a percentage of axial strain (with respect to the axis of the gauge)

$K_t$ = lateral (transverse) sensitivity of the strain gauge

The uncorrected estimate can be used as the uncertainty.

Typical values of the transverse sensitivity for commercial strain gauges range from $-0.19$ to 0.05. Figure 11.7 shows a plot of the percentage error for a strain gauge as a function of the

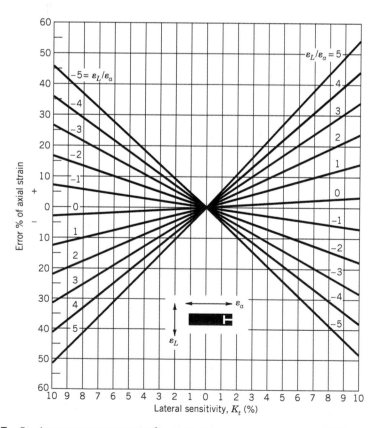

**FIGURE 11.7** Strain measurement error due to strain gauge transverse sensitivity. (Courtesy of Micro-Measurements, Raleigh, NC, USA.)

ratio of lateral loading to axial loading and the lateral sensitivity. It is possible to correct for the lateral sensitivity effects [6].

## Semiconductor Strain Gauges

When subjected to a load, a semiconductor material exhibits a change in resistance, and therefore can be used for the measurement of strain. Silicon crystals are the basic material for semiconductor strain gauges; the crystals are sliced into very thin sections to form strain gauges. Mounting such gauges in a transducer, such as a pressure transducer, or on a test specimen requires backing and adhesive techniques similar to those used for metallic gauges. Because of the large piezoresistance coefficient, the semiconductor gauge exhibits a very large gauge factor, as large as 200 for some gauges. These gauges also exhibit higher resistance, longer fatigue life, and lower hysteresis under some conditions than metallic gauges. However, the output of the semiconductor strain gauge is nonlinear with strain, and the strain sensitivity or gauge factor may be markedly dependent on temperature.

Semiconductor materials for strain gauge applications have resistivities ranging from $10^{-6}$ to $10^{-2}$ $\Omega$-m. Semiconductor strain gauges may have a relatively high or low density of charge carriers [3, 7]. Semiconductor strain gauges made of materials having a relatively high density of charge carriers ($\sim$1,020 carriers/cm$^3$) exhibit little variation of their gauge factor with strain or temperature. On the other hand, for the case where the crystal contains a low number of charge carriers ($<$1,017 carriers/cm$^3$), the gauge factor may be approximated as

$$GF = \frac{T_0}{T}GF_0 + C_1\left(\frac{T_0}{T}\right)^2 \varepsilon \qquad (11.14)$$

where $GF_0$ is the gauge factor at the reference temperature $T_0$, under conditions of zero strain [8], and $C_1$ is a constant for a particular gauge. The behavior with temperature of a high-resistivity P-type semiconductor is shown in Figure 11.8.

**FIGURE 11.8** Temperature effect on resistance for various impurity concentrations for P-type semiconductors (reference resistance at 81°F).
(Courtesy of Kulite Semiconductor Products, Inc. Used with permission.)

Semiconductor strain gauges find their primary application in the construction of transducers, such as load cells and pressure transducers. Because of the capability for producing small gauge lengths, silicon semiconductor strain gauge technology provides for the construction of very small transducers. For example, flush-mount pressure transducers having diameters of less than 8 mm provide pressure measurements up to 15,000 psi, with excellent frequency response characteristics. However, silicone diaphragm pressure transducers require special construction and procedures for measuring in liquid environments. Semiconductor strain gauges are somewhat limited in the maximum strain that they can measure, approximately 5,000 $\mu\varepsilon$ for tension, but larger in compression. Because of the possibility of an inherent sensitivity to temperature, careful consideration must be given to each application to provide appropriate temperature compensation or correction. Temperature effects can result, for a particular measurement, in zero drift for the duration of a measurement.

## 11.4   Strain Gauge Electrical Circuits

A bridge circuit is generally used to detect the small changes in resistance that are the output of a strain gauge measurement circuit. A typical strain gauge measuring installation on a steel specimen has a sensitivity of $10^{-6}$ $\Omega$/(kPa). Thus, a high-sensitivity device such as a Wheatstone bridge is desirable for measuring resistance changes for strain gauges. The fundamental relationships for the analysis of such bridge circuits are discussed in Chapter 6. Equipment is commercially available that can measure changes in gauge resistance of less than 0.0005 $\Omega$ (0.000001 $\mu\varepsilon$).

A simple strain gauge Wheatstone bridge circuit is shown in Figure 11.9. The bridge output under these conditions is given by Equation 6.12:

$$E_0 + \delta E_0 = E_i \frac{(R_1 + \delta R)R_4 - R_3 R_2}{(R_1 + \delta R + R_2)(R_3 + R_4)} \tag{6.12}$$

where $E_0$ is the bridge output at initial conditions, $\delta E_0$ is the bridge deflection associated with the change in the strain gauge resistance $\delta R$. Consider the case where all the fixed resistors and the strain gauge resistance are initially equal, and the bridge is balanced such that $E_0 = 0$. If the strain gauge is then subjected to a state of strain, the change in the output voltage, $\delta E_0$, from Equation 6.12 reduces to (as shown in Equation 6.13)

$$\frac{\delta E_o}{E_i} = \frac{\delta R/R}{4 + 2(\delta R/R)} \approx \frac{\delta R/R}{4} \tag{11.15}$$

**FIGURE 11.9**   Basic strain gauge Wheatstone bridge circuit.

Circuit arrangement for shunt balance    Differential shunt balance arrangement

**FIGURE 11.10**    Balancing schemes for bridge circuits.

under the assumption that $\delta R/R \ll 1$. This simplified form of Equation 6.12 is suitable for all but those measurements that demand the highest accuracy. Using the relationship from Equation 11.12 that $\delta R/R = GF\varepsilon$,

$$\frac{\delta E_o}{E_i} = \frac{GF\varepsilon}{4 + 2GF\varepsilon} \approx \frac{GF\varepsilon}{4} \tag{11.16}$$

Equations 11.15 and 11.16 yield two practical equations for strain gauge measurements using a single gauge in a Wheatstone bridge.

The Wheatstone bridge has several distinct advantages for use with electrical resistance strain gauges. The bridge may be balanced by changing the resistance of one arm of the bridge. Therefore, once the gauge is mounted in place on the test specimen under a condition of zero loading, the output from the bridge may be zeroed. Two schemes for circuits to accomplish this balancing are shown in Figure 11.10. Shunt balancing provides the best arrangement for strain gauge applications, since the changes in resistance for a strain gauge are small. Also, the strategic placement of multiple gauges in a Wheatstone bridge can both increase the bridge output and cancel out certain ambient effects and unwanted components of strain as discussed in the next two sections.

### EXAMPLE 11.3

A strain gauge, having a gauge factor of 2, is mounted on a rectangular steel bar ($E_m = 200 \times 10^6$ kN/m$^2$), as shown in Figure 11.11. The bar is 3 cm wide and 1 cm high and is subjected to a tensile force of 30 kN. Determine the resistance change of the strain gauge if the resistance of the gauge was 120 $\Omega$ in the absence of the axial load.

**KNOWN**    GF $= 2$    $E_m = 200 \times 10^6$ kN/m$^2$    $F_N = 30$ kN

$$R = 120\Omega \quad A_c = 0.03\,\text{m} \times 0.01\,\text{m}$$

**FIND**  The resistance change of the strain gauge for a tensile force of 30 kN.

**FIGURE 11.11** Strain gauge circuit subject to uniaxial tension.

**SOLUTION**

The stress in the bar under this loading condition is

$$\sigma_a = \frac{F_N}{A_c} = \frac{30\,\text{kN}}{(0.03\,\text{m})(0.01\,\text{m})} = 1 \times 10^5\,\text{kN/m}^2$$

and the resulting strain is

$$\varepsilon_a = \frac{\sigma_a}{E_m} = \frac{1 \times 10^5\,\text{kN/m}^2}{200 \times 10^6\,\text{kN/m}^2} = 5 \times 10^{-4}\,\text{m/m}$$

For strain along the axis of the strain gauge, the change in resistance from Equation 11.12 is

$$\delta R/R = \varepsilon GF$$

or

$$\delta R = R\varepsilon GF = (120\,\Omega)(5 \times 10^{-4})(2) = 0.12\,\Omega.$$

---

**EXAMPLE 11.4**

Suppose the strain gauge described in Example 11.3 is to be connected to a measurement device capable of determining a change in resistance with a stated uncertainty of $\pm 0.005\,\Omega$ (95%). This stated uncertainty includes a resolution of $0.001\,\Omega$. What uncertainty in stress would result when using this resistance measurement device?

**KNOWN** A stress is to be inferred from a strain measurement using a strain gauge having a gauge factor of 2 and a zero load resistance of $120\,\Omega$. The measurement of resistance has a stated uncertainty of $\pm 0.005\,\Omega$ (95%).

**FIND** The design-stage uncertainty in stress.

**SOLUTION**

The design-stage uncertainty in stress, $(u_d)_\sigma$, is given by

$$(u_d)_\sigma = \frac{\partial \sigma}{\partial(\delta R)}(u_d)_{\delta R}$$

with

$$\sigma = \varepsilon E_m = \frac{\delta R/R}{GF}E_m$$

Then with $(u_d)_{\delta R} = 0.005 \; \Omega$ and

$$\frac{\partial \sigma}{\partial(\delta R)} = \frac{E_m}{R(GF)}$$

we can express the uncertainty as

$$(u_d)_\sigma = \frac{E_m}{R(GF)}(u_d)_{\delta R} = \frac{200 \times 10^6 \; \text{kN/m}^2}{120 \; \Omega(2)}(0.005 \; \Omega)$$

This results in a design-stage uncertainty in stress of $(u_d)_\sigma = \pm 4.17 \times 10^3 \; \text{kN/m}^2 \; (95\%)$ or ~2.4% of the expected stress.

## 11.5 Practical Considerations for Strain Measurement

This section describes some characteristics of strain gauge applications that allow practical implementation of strain measurement.

### The Multiple Gauge Bridge

The output from a bridge circuit can be increased by the appropriate use of more than one strain gauge. This increase can be quantified by employing a bridge constant as illustrated in the following discussion. In addition, multiple gauges can be used to compensate for unwanted effects, such as temperature or specific strain components. Consider the case when all four resistances in the bridge circuit of Figure 11.9 represent strain gauges. In general, the bridge output is given by

$$E_0 = E_i \left[ \frac{R_1}{R_1 + R_2} - \frac{R_3}{R_3 + R_4} \right] \tag{11.17}$$

The strain gauges $R_1$, $R_2$, $R_3$, and $R_4$ are assumed initially to be in a state of zero strain. If these gauges are now subjected to strains such that the resistances change by $dR_i$, where $i = 1, 2, 3$, and 4, then the change in the bridge output voltage can be expressed as

$$dE_0 = \sum_{i=1}^{4} \frac{\partial E_0}{\partial R_i} dR_i \tag{11.18}$$

Evaluating the appropriate partial derivatives from Equation 11.17 yields

$$dE_0 = E_i \left[ \frac{R_2 dR_1 - R_1 dR_2}{(R_1 + R_2)^2} + \frac{R_3 dR_4 - R_4 dR_3}{(R_3 + R_4)^2} \right] \tag{11.19}$$

Then, from Equations 11.2 and 11.12, $dR_i = R_i \varepsilon_i GF_i$, and the value of $dE_0$ can be determined. Assuming that $dR_i \ll R_i$, the resulting change in the output voltage, $\delta E_0$, may now be expressed as

$$\delta E_0 = E_i \left[ \frac{R_1 R_2}{(R_1 + R_2)^2}(\varepsilon_1 GF_1 - \varepsilon_2 GF_2) + \frac{R_3 R_4}{(R_3 + R_4)^2}(\varepsilon_4 GF_4 - \varepsilon_3 GF_3) \right] \tag{11.20}$$

If $R_1 = R_2 = R_3 = R_4$, then

$$\frac{\delta E_0}{E_i} = \frac{1}{4}(\varepsilon_1 GF_1 - \varepsilon_2 GF_2 + \varepsilon_4 GF_4 - \varepsilon_3 GF_3) \tag{11.21}$$

It is possible and desirable to purchase matched sets of strain gauges for a particular application, so that $GF_1 = GF_2 = GF_3 = GF_4$, and

$$\frac{\delta E_0}{E_i} = \frac{GF}{4}(\varepsilon_1 - \varepsilon_2 + \varepsilon_4 - \varepsilon_3) \tag{11.22}$$

Equation 11.22 is important and forms the basic working equation for a strain gauge bridge circuit using multiple gauges (compare this equation with Eq. 11.16).

Equation 11.22 shows that for a bridge containing two or four strain gauges, equal strains on opposite bridge arms sum, whereas equal strains on adjacent arms of the bridge cancel. These characteristics can be used to increase the output of the bridge, to provide temperature compensation, or to cancel unwanted components of strain. Practical means of achieving these desirable characteristics will be explored further after the concept of the bridge constant is developed.

## Bridge Constant

Commonly used strain gauge bridge arrangements may be characterized by a bridge constant, $\kappa$, defined as the ratio of the actual bridge output to the output that would result from a single gauge sensing the maximum strain, $\varepsilon_{max}$. The output for a single gauge experiencing the maximum strain may be expressed as,

$$\frac{\delta R}{R} = \varepsilon_{max} GF \tag{11.23}$$

so that, again for a single gauge,

$$\frac{\delta E_0}{E_i} \cong \frac{\varepsilon_{max} GF}{4} \tag{11.24}$$

The bridge constant, $\kappa$, is found from the ratio of the actual bridge output given by Equation 11.22 to the output for a single gauge given by Equation 11.24. When more than one gauge is used in the bridge circuit, Equation 11.16 becomes

$$\frac{\delta E_0}{E_i} = \frac{\kappa \delta R/R}{4 + 2\delta R/R} = \frac{\kappa GF\varepsilon}{4 + 2GF\varepsilon} \approx \frac{\kappa GF\varepsilon}{4} \tag{11.25}$$

Application of the bridge constant is illustrated in Example 11.5.

**EXAMPLE 11.5**

Determine the bridge constant for two strain gauges mounted on a structural member, as shown in Figure 11.12. The member is subject to uniaxial tension, which produces an axial strain $\varepsilon_a$ and a lateral strain $\varepsilon_L = -\upsilon_p\varepsilon_a$. Assume that all the resistances in Figure 11.12 are initially equal so that the bridge is initially balanced. Let $GF_1 = GF_2$.

**KNOWN** Strain gauge installation shown in Figure 11.12.

**FIND** The bridge constant for this installation.

**SOLUTION**
The gauges are mounted so that gauge 1 is aligned with the axial tension and gauge 2 is mounted transversely on the member. The changes in resistance for the gauges may be expressed

$$\frac{\delta R_1}{R_1} = \varepsilon_a(GF)_1 \tag{11.26}$$

**FIGURE 11.12**   Bridge circuit with two arms active; strain gauge installation for increased sensitivity.

and

$$\frac{\delta R_2}{R_2} = -\upsilon_p \varepsilon_a (GF)_2 = -\upsilon_p \frac{\delta R_1}{R_1} \qquad (11.27)$$

With only one gauge active, Equation 11.15 would be applicable and yields

$$\frac{\delta E_0}{E_i} = \frac{\delta R_1 / R_1}{4 + 2(\delta R_1 / R_1)} \approx \frac{\delta R_1 / R_1}{4} \qquad (11.15)$$

But with both gauges installed and active, as shown in Figure 11.12, the output of the bridge is determined from an analysis of the bridge response, which results in

$$\frac{\delta E_0}{E_i} = \frac{(\delta R_1 / R_1)(1 + \upsilon_p)}{4 + 2(\delta R_1 / R_1)(1 + \upsilon_p)}$$

In most practical applications, changes in resistance are small in comparison to the resistance values, that is, $\delta R / R \ll 1$; thus,

$$\frac{\delta E_0}{E_i} = \frac{(\delta R_1 / R_1)(1 + \upsilon_p)}{4} \qquad (11.28)$$

And the bridge constant is the ratio of Equation 11.28 to Equation 11.15:

$$\frac{(\delta R_1 / R_1)(1 + \upsilon_p)/4}{(\delta R_1 / R_1)/4} \qquad (11.29)$$

And the bridge constant is

$$\kappa = 1 + \upsilon_p$$

Comparing Equation 11.28 to Equation 11.15 shows that the use of two gauges oriented as described has increased the output of the bridge by a factor of $1 + \upsilon_p$ over that of using a single gauge.

# 11.6 Apparent Strain and Temperature Compensation

Apparent strain is caused by any change in gauge resistance that is not due to the component of strain being measured. Techniques for accomplishing temperature compensation, eliminating certain components of strain, and increasing the value of the bridge constant can be devised by

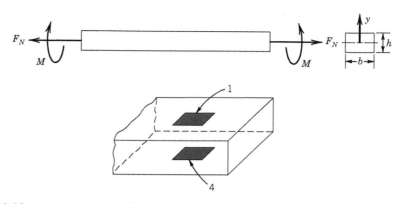

**FIGURE 11.13**   Strain gauge installation for bending compensation.

examining more closely Equation 11.22. The bridge constant is influenced by (1) the location of strain gauges on the test specimen and (2) the gauge connection positions in the bridge circuit.

Let us examine how a component of strain can be removed (compensation) from the measured signal. Consider a beam having a rectangular cross section and subject to the loading condition shown in Figure 11.13, where the beam is subject to an axial load $F_N$ and a bending moment $M$. The stress distribution in this cross section is given by

$$\sigma_x = \frac{-12My}{bh^3} + \frac{F_N}{bh} \tag{11.30}$$

To remove the effects of bending strain, identical strain gauges are mounted to the top and bottom of the beam as shown in Figure 11.13, and they are connected to bridge locations 1 and 4 (opposite bridge arms). The gauges experience equal but opposite bending strains (Eq. 11.30), and both strain gauges are subject to the same axial strain caused by $F_N$. The bridge output under these conditions is

$$\frac{\delta E_0}{E_i} = \frac{GF}{4}(\varepsilon_1 + \varepsilon_4) \tag{11.31}$$

where $\varepsilon_1 = \varepsilon_{a1} + \varepsilon_{b1}$ and $\varepsilon_4 = \varepsilon_{a4} - \varepsilon_{b4}$, with subscripts $a$ and $b$ referring to axial and bending strain, respectively. Hence, because $|\varepsilon_{a1}| = |\varepsilon_{a4}|$ and $|\varepsilon_{b1}| = |\varepsilon_{b4}|$, the bending strains cancel, but the axial strains sum, giving

$$\frac{\delta E_0}{E_i} = \frac{GF}{2}\varepsilon_a \tag{11.32}$$

For a single gauge experiencing the maximum strain,

$$\frac{\delta E_0}{E_i} = \frac{GF}{4}\varepsilon_a \tag{11.33}$$

The ratio of the output represented in Equation 11.32 to that represented in Equation 11.33 has a value of 2. This is the bridge constant ($\kappa = 2$) for this strain gauge arrangement. The bending strain cancels from Equation 11.32, indicating that this arrangement compensates for the bending strain.

A guide for some practical bridge-gauge configurations is provided in Table 11.1.

## Temperature Compensation

Differential thermal expansion between the gauge and the specimen on which it is mounted creates an apparent strain in the strain gauge. So temperature sensitivity of strain gauges is caused

**Table 11.1  Common Gauge Mountings**

| | Arrangement | Compensation Provided | Bridge Constant $\kappa$ |
|---|---|---|---|
| | Single gauge in uniaxial stress | None | $\kappa = 1$ |
| | Two gauges sensing equal and opposite strain—typical bending arrangement | Temperature | $\kappa = 2$ |
| | Two gauges in uniaxial stress | Bending only | $\kappa = 2$ |
| | Four gauges with pairs sensing equal and opposite strains | Temperature and bending | $\kappa = 4$ |
| | One axial gauge and one Poisson gauge | | $\kappa = 1 + \nu$ |
| | Four gauges with pairs sensing equal and opposite strains—sensitive to torsion only; typical shaft arrangement | Temperature and axial | $\kappa = 4$ |

by temperature changes in the gauge itself and the strain experienced by the gauge as a result of differential thermal expansion between the gauge and the material on which it is mounted. Using gauges of identical alloy composition as the specimen minimizes this latter effect. However, even keeping the specimen at a constant temperature may not be enough to eliminate the effect of gauge thermal expansion. Heating of the strain gauge as a result of current flow from the measuring device may be a source of significant error since the gauge is also a temperature-sensitive element. The temperature sensitivity of a strain gauge is an obstacle to accurate mechanical strain measurement that must be overcome. Fortunately, there are effective ways to deal with it.

Figure 11.14 shows two circuit arrangements that provide temperature compensation for a strain measurement. The strain gauge mounted on the test specimen experiences changes in resistance caused by temperature changes and by applied strain, whereas the compensating gauge experiences resistance changes caused only by temperature changes. As long as the compensating gauge, shown in Figure 11.14, experiences an identical thermal environment to that experienced by the measuring gauge, temperature effects will be eliminated from the circuit. To show this, consider the case in which all bridge resistances are initially equal and the bridge is thus balanced.

**FIGURE 11.14**   Bridge arrangements for temperature compensation.

If the temperature of the strain gauges now changes, their resistance changes as a result of thermal expansion, creating an apparent thermal strain. Under an applied axial load, the output of the bridge is derived from Equation 11.22,

$$\delta E_0 = E_i \frac{GF(\varepsilon_1 - \varepsilon_2)}{4} = E_i \frac{GF\varepsilon_a}{4} \tag{11.34}$$

where $\varepsilon_1 = \varepsilon_a + \varepsilon_T$ and $\varepsilon_2 = \varepsilon_T$, where $\varepsilon_T$ refers to the temperature-induced apparent strain. Under this condition, the output value is not affected by changes in temperature. The result is the same if the compensating gauge is mounted in arm 3 instead, as shown in Figure 11.14b. Furthermore, any two active gauges mounted on adjacent bridge arms compensate for temperature changes.

Looking back, the arrangement in Figure 11.13 does not provide temperature compensation; gauges 1 and 4 are on opposite bridge arms. However, temperature compensation could be provided for that installation by having two additional strain gauges that are at the same temperature as gauges 1 and 4 and that occupy arms 2 and 3 of the bridge.

## Bridge Static Sensitivity

The static sensitivity of the bridge arrangement in Figure 11.14a (method 1) is

$$K_B = \frac{E_0}{\varepsilon} = E_i \frac{R_1 R_2}{(R_1 + R_2)^2} GF \tag{11.35}$$

and with $E_i = (2R_g)I_g$ and $R_g = R_1 = R_2$, the static sensitivity may be expressed in terms of the current flowing through the gauge, ($I_g$), as

$$K_B = \frac{1}{2} GF \sqrt{(I_g^2 R_1)R_1} \tag{11.36}$$

Note that $I_g^2 R_g$ is the power dissipated in the strain gauge as a result of the bridge current. Excessive power dissipation in the gauge would cause temperature changes and introduce uncertainty into a strain measurement. These effects can be minimized by good thermal coupling between the strain gauge and the object to which it is bonded to allow effective dissipation of thermal energy.

Consider the static sensitivity of the bridge arrangement in Figure 11.14b. With identical gauges at positions $R_1$ and $R_3$ and equal resistance changes for the two gauges, no change in bridge output would occur. However, the static sensitivity for this arrangement is not the same as for method 1 but rather is given by

$$K_B = \frac{R_1/R_2}{1 + R_1/R_2} GF \sqrt{(I_g^2 R_1) R_1} \qquad (11.37)$$

Here the sensitivity is the same as for a bridge having a single active gauge and without temperature compensation. However, the sensitivity depends on the choice of the fixed resistor $R_2$. If $R_1 = R_2$, the resulting sensitivity is the same as for method 1. However, resistor $R_2$ can be chosen to provide the desired static sensitivity for the circuit, within the limitations of measurement capability and allowable bridge current.

## Practical Considerations

An assumption in the definition of the gauge factor is that the change in resistance of the gauge is linear with applied strain for a particular gauge. However, a strain gauge can exhibit some nonlinearity. Also, in cycling between a loaded and unloaded condition, there is some degree of hysteresis and a shift in the resistance for a state of zero strain. A typical cycle of loading and unloading is shown in Figure 11.15. The strain gauge typically indicates lower values of strain during unloading than are measured as the load is increased. The extent of these behaviors is determined not only by the strain gauge characteristics but also by the characteristics of the adhesive and by the previous strains that the gauge has experienced. For properly installed gauges, the deviation from linearity should be on the order of 0.1% [3]. On the other hand, first-cycle hysteresis and zero shift are difficult to predict. The effects of first-cycle hysteresis and zero shift can be minimized by cycling the strain gauge between zero strain and a value of strain above the maximum value to be measured before taking measurements.

In dynamic measurements of strain, the dynamic response of the strain gauge itself is generally not the limiting factor for such dynamic measurements. The rise time (90%) of a bonded resistance strain gauge may be approximated [9] as

$$t_{90} \approx 0.8(L/a) + 0.5 \ \mu s \qquad (11.38)$$

where $L$ is the gauge length and $a$ is the speed of sound in the material on which the gauge is mounted. Typical response times for gauges mounted on steel specimens are on the order of 1 μs.

## Analysis of Strain Gauge Data

Strain gauges mounted on the surface of a test specimen respond only to the strains that occur at the surface of the test specimen. As such, the results from strain gauge measurements must be analyzed to determine the state of stress occurring at the strain gauge locations. The complete determination of the stress at a point on the surface of a particular test specimen generally requires the measurement of three strains at the point under consideration. The result of these measurements yields the principal strains and allows determination of the maximum stress [3].

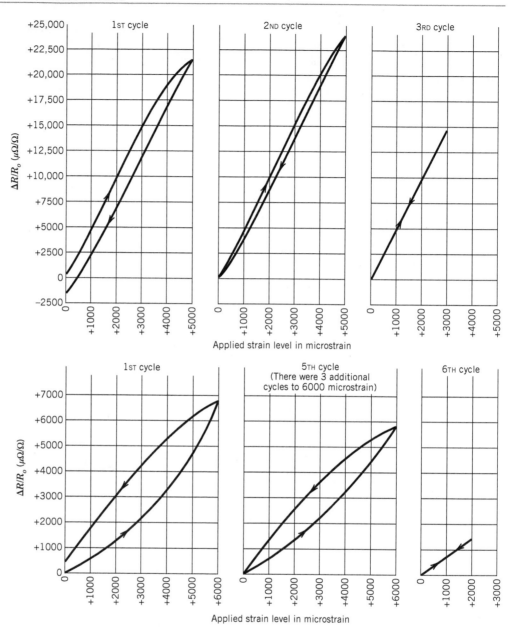

**FIGURE 11.15**
Hysteresis in initial loading cycles for two strain gauge materials.
(Courtesy of Micro-Measurements, Raleigh, NC, USA.)

The multiple-element strain gauges used to measure more than one strain at a point are called strain gauge rosettes. An example of a two-element rosette is shown in Figure 11.16. For general measurements of strain and stress, commercially available strain rosettes can be chosen that have a pattern of multiple-direction gauges that is compatible with the specific nature of the particular application [5]. In practice, the applicability of the single-axis strain gauge is extremely limited, and improper use can result in large errors in the measured stress.

Strain gauge rosettes are commercially available in a variety of configurations, including gauges placed at angles of 0, 45, 90 degrees and 60, 60, 60 degrees. Consider the rectangular rosettes with angles of 0, 45, 90 degrees as shown in Figure 11.17. As we have emphasized, measuring surface strain allows calculation of the principal stresses. Assume that the measured strains from the stacked gauge in Figure 11.17b are $\varepsilon_1$ (from the 0-degree gauge, oriented with

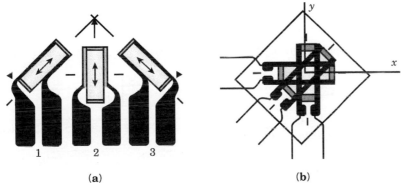

**(a)**

**(b)**

**FIGURE 11.16** Biaxial strain gauge rosettes.
(a) Single-plane type. (b) Stacked type.
(Courtesy of Micro-Measurements, Raleigh, NC, USA.)

**(a)**　　　　　　　　　**(b)**

**FIGURE 11.17** Rectangular (0, 45, 90 degrees) strain gauge rosettes. (a) Planar configuration.
(b) Stacked configuration.

the $x$-axis), $\varepsilon_2$ (from the 45-degree gauge), and $\varepsilon_3$ (from the 90-degree gauge, oriented with the $y$-axis). From these measured strains, the principal stresses, $\sigma_{max}$ and $\sigma_{min}$, and the maximum shear stress, $\tau_{max}$, may be calculated as

$$\sigma_{max} = \frac{E_m}{2}\left[\frac{\varepsilon_1 + \varepsilon_3}{1 - \upsilon_p} + \frac{1}{1 + \upsilon_p}\sqrt{(\varepsilon_1 - \varepsilon_3)^2 + [2\varepsilon_2 - (\varepsilon_1 + \varepsilon_3)]^2}\right]$$

$$\sigma_{min} = \frac{E_m}{2}\left[\frac{\varepsilon_1 + \varepsilon_3}{1 - \upsilon_p} - \frac{1}{1 + \upsilon_p}\sqrt{(\varepsilon_1 - \varepsilon_3)^2 + [2\varepsilon_2 - (\varepsilon_1 + \varepsilon_3)]^2}\right]$$

$$\tau_{max} = \frac{E_m}{2(1 + \upsilon_p)}\sqrt{(\varepsilon_1 - \varepsilon_3)^2 + [2\varepsilon_2 - (\varepsilon_1 + \varepsilon_3)]^2} \tag{11.39}$$

The angle between the $x$-axis and the maximum principal stress is given by

$$\phi = \frac{1}{2}\tan^{-1}\frac{2\varepsilon_2 - (\varepsilon_1 + \varepsilon_3)}{\varepsilon_1 - \varepsilon_3} \tag{11.40}$$

**EXAMPLE 11.6**

A rectangular strain gauge rosette is composed of strain gauges oriented at relative angles of 0, 45, and 90 degrees, as shown in Figure 11.17b. The rosette is used to measure strain on an aluminum structural member ($E_m = 69$ MPa, $v_p = 0.334$). The measured values of strain are

$$\varepsilon_1 = 20,000 \ \mu\varepsilon$$
$$\varepsilon_2 = 5000 \ \mu\varepsilon$$
$$\varepsilon_3 = 10,000 \ \mu\varepsilon$$

Determine the principal stresses and the angle of the maximum principle stress relative to the $x$-axis.

**KNOWN**  $\varepsilon_1 = 20,000 \ \mu\varepsilon$
$\varepsilon_2 = 5000 \ \mu\varepsilon$
$\varepsilon_3 = 10,000 \ \mu\varepsilon$

**ASSUMPTIONS** Strain gauge 1 is oriented along the $x$-axis. The material behaves in a homogeneous, isotropic manner and the loads are below the elastic limit.

**FIND** The values and direction of the principal stresses.

**SOLUTION**
Equation 11.39 can be applied directly to compute the stresses. For the maximum principal stress,

$$\sigma_{max} = \frac{E_m}{2} \left[ \frac{\varepsilon_1 + \varepsilon_3}{1 - v_p} + \frac{1}{1 + v_p} \sqrt{(\varepsilon_1 - \varepsilon_3)^2 + [2\varepsilon_2 - (\varepsilon_1 + \varepsilon_3)]^2} \right]$$

$$\sigma_{max} = \frac{69 \text{ MPa}}{2} \left[ \frac{(0.02 + 0.01)}{1 - 0.334} + \frac{1}{1 + 0.334} \sqrt{(0.02 - 0.01)^2 + [2 \times 0.005 - (0.02 + 0.01)]^2} \right]$$

$$= 2.13 \text{ MPa}$$

Equations 11.39 and 11.40 yield the remaining values

$$\sigma_{min} = -0.976 \text{ MPa}$$
$$\tau_{max} = 0.578 \text{ MPa}$$
$$\phi = -63.4°$$

**COMMENT** Other orientations of strain gauge rosettes result in different equations relating stress and measured strain.

## Signal Conditioning

The most common form of signal conditioning in strain gauge bridge circuits is amplification. For an amplifier of gain $G_A$, Equation 11.25 becomes

$$\frac{\delta E_0}{E_i} = \frac{G_A \kappa (\delta R/R)}{4 + 2(\delta R/R)} = \frac{G_A \kappa GF\varepsilon}{4 + 2(\delta R/R)} \approx \frac{G_A \kappa GF\varepsilon}{4} \tag{11.41}$$

A common means of recording strain gauge bridge circuit signals is the automated data acquisition system. A schematic diagram of such a setup is shown in Figure 11.18.

**FIGURE 11.18** Data acquisition and reduction system for Example 11.7.

## EXAMPLE 11.7

The strain developed at $M$ different locations about a large test specimen is to be measured at each location using a setup similar to that shown in Figure 11.18. At each location, similar strain gauges are appropriately mounted and connected to a Wheatstone bridge that is powered by an external supply. The bridge deflection voltage is amplified and measured. A similar setup is used at all $M$ locations. The output from each amplifier is applied as an input through an $M$-channel multiplexer to an automated data acquisition system. If $N$ (say 30) readings for each of the $M$ setups are taken while the test specimen is maintained in a condition of uniform zero strain, what information is obtained?

**KNOWN** Strain setup of Figure 11.18
$M$ (j = 1, 2, ..., $M$) setups
$N$ (i = 1, 2, ..., $N$) readings per setup.

**ASSUMPTIONS** Each setup is operated in a similar manner.

**SOLUTION**
Each setup is exposed to the same strain and should indicate the same strain. We can calculate the mean strain at each of the $M$ locations. Any difference between the measured mean strain and the applied strain at each position is an offset error for that channel. For zero applied strain, we can use this information to correct each channel to zero. We can estimate the uncertainty in the correction from the standard deviation of the means for the offset deviations (see Example 5.13). The variation in measured strain at each of the $M$ locations is a measure of the noise.

**COMMENT** The test data do not include uncertainties resulting from temporal variation of the measured strain or procedural variations under loading, instrument calibration errors, temperature variation effects, and dynamic effects on the gauges, including differences in creep and fatigue or reduction curve fit errors.

| Example 11.8 | Case Study—Strain Measurement at the Nanoscale: Carbon Nanotubes |
|---|---|

Carbon nanotubes are hollow cylinders whose walls are formed by regular arrays of carbon atoms; single-walled carbon nanotubes (SWNTs) typically have a diameter of about 1 nanometer. The structure of the tube walls is a member of the fullerene family of structures. These nanotubes have unique mechanical and electronic properties; the carbon–carbon chemical bond can be among the strongest in nature; carbon nanotubes are one of the strongest and stiffest materials ever found, with SWNT having a measured modulus of elasticity up to 5 TPa. There are several motivations for wanting to measure strain in carbon nanotubes, including determining the strength of the material and quantifying the change in electronic properties with strain. Measuring strain at the nanoscale requires unique approaches. A direct measurement of bending strain can be accomplished using the atomic force microscope, described in Chapter 1, to displace the center of a nanotube that is fixed at each end, creating a bending stress. However, this raises the question, "How can the ends of the SWNT be held?" when we are working at the nanoscale.

A method is described in reference [10]. In this method, long carbon nanotubes are grown, using chemical vapor deposition, over H-shaped slits in a silicone wafer. The width of the slits was approximately 100 μm. The process yields numerous SWNT, and, again at the nanoscale, one clean separate fiber was identified. Then a layer of gold was deposited on the end of the SWNT to "clamp" it to the silicone wafer. This assembly was glued to a steel substrate, with appropriate cutouts and the steel heated to induce thermal expansion and apply a strain.

**COMMENT** Future applications of nanoscale materials and devices will increase and affect diverse areas, including sports equipment and delivery of medicine in the bloodstream. There is a critical need for measurement technology to keep up with the challenges to measure the properties of the new materials created at this small scale.

# 11.7 Optical Strain Measuring Techniques

Optical methods for experimental stress analysis can provide fundamental information concerning directions and magnitudes of the stresses in parts under design loading conditions. Optical techniques have been developed for the measurement of stress and strain fields, either in models made of materials having appropriate optical properties or through coating techniques for existing specimens. Photoelasticity takes advantage of the changes in optical properties of certain materials that occur when these materials are strained. A second optical method of stress analysis is based on the development of a moiré pattern, which is an optical effect resulting from the transmission or reflection of light from two overlaid grid patterns. The fringes that result from relative displacement of the two grid patterns can be used to measure strain; each fringe corresponds to the locus of points of equal displacement.

Recent developments in strain measurement include the use of lasers and holography to very accurately determine whole field displacements for complex geometries.

## Basic Characteristics of Light

To use optical strain measurement techniques, we must first examine some basic characteristics of light. Electromagnetic radiation, such as light, may be thought of as a transverse wave with sinusoidally oscillating electric and magnetic field vectors that are at right angles to the direction of propagation. In general, a light source emits a series of waves containing vibrations in all perpendicular planes, as illustrated in Figure 11.19. A light wave is said to be plane-polarized

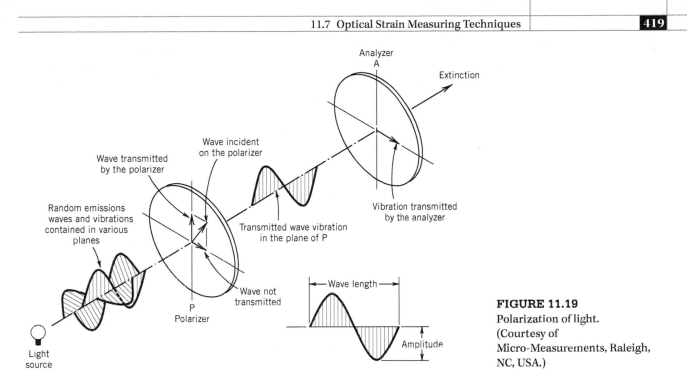

**FIGURE 11.19**
Polarization of light.
(Courtesy of
Micro-Measurements, Raleigh,
NC, USA.)

if the transverse oscillations of the electric field are parallel to each other at all points along the direction of propagation of the wave.

Figure 11.19 illustrates the effect of a polarizing filter on an incident light wave; the transmitted light is plane polarized, with a known direction of polarization. Complete extinction of the light beam could be achieved by introduction of a second polarizing filter, with the axis of polarization at 90 degrees to the first filter (labeled as Analyzer in Figure 11.19). These behaviors of light are employed to measure direction and magnitude of strain in photoelastic materials.

## Photoelastic Measurement

Photoelastic methods of stress analysis take advantage of the anisotropic optical characteristics of some materials, notably plastics, when subject to an applied load to determine the strain field. Stress analysis may be accomplished either by constructing a model of the part to be analyzed from a material selected for its optical properties or by coating the actual part or prototype with a photoelastic coating. If a model is constructed from a suitable plastic, the required loads for the model are significantly less than the service loads of the actual part, reducing effort and expense in testing.

The anisotropy that occurs in photoelastic materials results in two refracted beams of light and one reflected beam, produced for a single incident beam of appropriately polarized light. The two refracted beams propagate at different velocities through the material because of an anisotropy in the index of refraction. In an appropriately designed photoelastic (two-dimensional) model, these two refracted components of the incident light travel in the same direction and can be examined in a polariscope. The degree to which the two light waves are out of phase is related to the stress by the stress–optic law:

$$\delta \propto n_x - n_y \tag{11.42}$$

where

$\delta$ = relative retardation between the two light beams

$n_x, n_y$ = indices of refraction in the directions of the principal strains

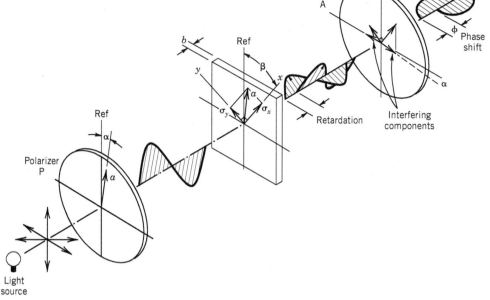

**FIGURE 11.20**
Construction of a plane polariscope. (Courtesy of Micro-Measurements, Raleigh, NC, USA.)

The index of refraction changes in direct proportion to the amount the material is strained, such that

$$n_x - n_y = K(\varepsilon_x - \varepsilon_y) \qquad (11.43)$$

The strain optical coefficient, $K$, is generally assumed to be a material property independent of the wavelength of the incident light. Figure 11.20 shows the use of a plane polariscope to examine the strain in a photoelastic model. Plane polarized light enters the specimen and emerges with two planes of polarization along the principal strain axes. This light beam is then passed through a polarizing filter, called the analyzer, which transmits only the component of each of the light waves that is parallel to the plane of polarization. The transmitted waves interfere, since they are out of phase, and the resulting light intensity is a function of the angle between the analyzer and the principal strain direction and the phase shift between the beams. The variations in strain in the specimen produce a pattern of fringes that can be related to the strain field through the strain optic relation.

When a photoelastic model is observed in a plane polariscope, a series of fringes is observed. The complete extinction of light occurs at locations where the principal strain directions coincide with the axes of the analyzer or where either the strain is zero or $\varepsilon_x - \varepsilon_y = 0$. These fringes, termed isoclinics, are used to determine the principal strain directions at all points in the photoelastic model. Figure 11.21 shows the isoclinics in a ring subject to a compression load (as shown in the figure). A reference direction is selected along the horizontal compression load and labeled 0 degrees. For each measurement angle, one of the principal strains at a point on an isoclinic is parallel to the specified angle and the other is perpendicular. For the 0-degree isoclinics, the principal strains are oriented at 0 and 90 degrees.

Using the facts that the direction of the principal axes is known at a free surface and that the shear stress is zero on a free surface, the magnitude of the stress on the boundary can be determined. The primary applications for photoelasticity, especially in a historical sense, have been in the study of stress concentrations around holes or reentrant corners. In these cases, the maximum stress is at the boundary and corresponds to one of the principal stresses. This maximum stress can be obtained directly by the optical method, because the shear stress is zero on the boundary.

0 degree          15 degrees          30 degrees

45 degrees          60 degrees          75 degrees

**FIGURE 11.21**   Construction of a plane polariscope.
(Courtesy of Micro-Measurements, Raleigh, NC, USA.)

## Moiré Methods

A moiré pattern results from two overlaid, relatively dense patterns that are displaced relative to each other. This observable optical effect occurs, for example, in color printing, where patterns of dots form an image. If the printing is slightly out of register, a moiré pattern results. Another common example is the striking shimmering effect that occurs with some patterned clothing on television. This effect results when the size of the pattern in the fabric is essentially the same as the resolution of the television image.

In experimental mechanics, moiré patterns are used to measure surface displacements, typically in a model constructed specifically for this purpose. The technique uses two gratings, or patterns of parallel lines spaced equally apart. Figure 11.22 shows two line gratings. There are

$\alpha = \dfrac{t}{p} = 0.5$
for this grating

$\alpha = \dfrac{t}{p} = 0.75$

**FIGURE 11.22**   Moiré gratings.

two important properties of line gratings for moiré techniques. The pitch is defined as the distance between the centers of adjacent lines in the grating, and for typical gratings has a value of from 1 to 40 lines/mm. The second characteristic of gratings is the ratio of the open, transparent area of the grating to the total area, or, for a line grating, the ratio of the distance between adjacent lines to the center-to-center distance, as illustrated in Figure 11.22. Clearly, a greater density of lines per unit width allows a greater sensitivity of strain measurement, but as line densities increase, coherent light is required for practical measurement.

To determine strain using the moiré technique, a grating is fixed directly to the surface to be studied. This can be accomplished through photoengraving, cementing film copies of a grating to the surface, or interferometric techniques. The master or reference grating is next placed in contact with the surface, forming a reference for determining the relative displacements under loaded conditions. A series of fringes result when the gratings are displaced relative to each other; the bright fringes are the loci of points where the projected displacements of the surface are integer multiples of the pitch. The technique then is two-dimensional, providing information concerning the projection of the displacements into the plane of the master grating. Once a fringe pattern is recorded, data reduction techniques are employed to determine the stress and strain field. Graphical techniques exist that allow the strain components in two orthogonal directions to be determined. Further information on moiré techniques, along with additional references, may be found in the review article by Sciammarella [11].

Recently, techniques such as moiré-fringe multiplication have greatly increased the sensitivity of moiré techniques, with possible grating density of 1200 lines/mm. Moiré interferometry is an extension of moiré-fringe multiplication that uses coherent light and has sensitivities on the order of 0.5 mm/fringe [12]. A reflective grating is applied to the specimen, which experiences deformation under load conditions. The technique provides whole field readings of in-plane strain, with a four-beam optical arrangement currently in use [13].

## Fiber Bragg Strain Measurement

Associated with recent progress in optical fiber technology is an emerging area of sensor technology based on fiber-optic Bragg gratings. A fiber-optic Bragg grating is illustrated in Figure 11.23.

The Bragg grating itself is an area of the fiber-optic core having a periodic variation of the refractive index of the material. As shown in Figure 11.23, a broadband light input interacts with

**FIGURE 11.23**   Fiber optic Bragg grating.

the Bragg grating in two ways. Some of the incoming light is transmitted, and some of reflected by the Bragg grating. The Bragg wavelength corresponds to the largest intensity of the reflected light. If the Bragg condition is met [14] and the Bragg grating is of infinite length, there will be one Bragg frequency with maximum reflection given by

$$\lambda_B = 2n_{eff}L \tag{11.44}$$

where

$\lambda_B$ = Bragg wavelength

$n_{eff}$ = effective refractive index of the Bragg grating

$L$ = spacing of the periodic variations in refractive index in the Bragg grating

A representative value of $n_{eff}$ is 1.45 [15]. Fiber-optic Bragg grating sensors are sensitive to temperature because thermal expansion changes the spacing, $L$.

## EXAMPLE 11.9

Determine the required spacing of the periodic variations in a fiber Bragg grating if the wavelength of the peak in reflected light is to be 1545 nm.

Assuming that the refractive index is 1.45 [15], the resulting spacing is given by

$$L = \frac{\lambda_B}{2n_{eff}} = \frac{1545 \text{ nm}}{2(1.45)} = 533 \text{ nm}$$

**COMMENT** During manufacture, fiber-optic Bragg gratings are created by exposing the fiber to an ultraviolet interference pattern that creates the periodic variation in refractive index. Spacings down to several hundred nanometers are easily achievable.

## EXAMPLE 11.10

A steel bridge is a structure that requires continuous maintenance and periodic inspection. However, increasingly bridges and other structures such as high-rise buildings are being monitored continuously using a variety of sensors that indicate the structural health of the structure. Structural health monitoring is an important application where strain measurement plays a vital role. For a steel bridge, there are a variety of issues that can negatively affect the structural health, including corrosion, increased traffic loads, and increased vehicle weight.

Some aspects of bridge structural health can be monitored using accelerometers to characterize bridge vibration during active use, looking for changes in modal frequencies and damping ratios as compared to a healthy baseline. Static displacements and resulting strain can also be used to monitor bridge structural health. A large number of point strain measurements using standard strain gauges yield data that can infer something about the state of the structure. However, if damage exists in a region without a sensor, it may not be detected.

Sensors having the ability to sense strain over a very large gauge length can overcome the limitations of point measurements. Fiber-optic Bragg grating sensors can be constructed with gauge lengths of several meters [16]. Using such sensors and the fact that the ratio of the strain experienced by two different locations in a structure is constant under various loading conditions, a method for structural health monitoring results [16]. These sensors and method have been successfully employed in monitoring a bridge in Japan.

## SUMMARY

Experimental stress analysis can be accomplished through several practical techniques, including electrical resistance, photoelastic, and moiré strain measurement techniques. Each of these methods yields information concerning the surface strains for a test specimen. The design and selection of an appropriate strain measurement system begins with the choice of a measurement technique.

The bonded electrical resistance strain gauge provides a versatile means of measuring strain at a specific location on a test specimen. Strain gauge selection involves the specification of strain gauge material, the backing or carrier material, and the adhesive used to bond the strain gauge to the test specimen, as well as the total electrical resistance of the gauge. Other considerations include the orientation and pattern for a strain gauge rosette and the temperature limit and maximum allowable elongation. In addition, for electrical resistance strain gauges, appropriate arrangement of the gauges in a bridge circuit can provide temperature compensation and elimination of specific components of strain.

Optical methods are useful in the initial determination of a stress field for complex geometries and the determination of whole-field information in model studies. Such whole-field methods provide the basis for design and establish information necessary to make detailed local strain measurements.

The techniques for strain measurement described in this chapter provide the basis for determining surface strains for a test specimen. Although the focus here has been the measurement techniques, the proper placement of strain gauges and the interpretation of the measured results require further analysis. The inference of load-carrying capability and safety for a particular component is a result of the overall experimental program.

## REFERENCES

1. Hibbeler, R. C., *Mechanics of Materials*, 9th ed., Prentice-Hall, Upper Saddle River, NJ, 2013.

2. Timoshenko, S. P., and J. M. Goodier, *Theory of Elasticity*, 3rd ed., Engineering Society Monographs, McGraw-Hill, New York, 1970.

3. Dally, J. W., and W. F. Riley, *Experimental Stress Analysis*, 3rd ed., McGraw-Hill, New York, 1991.

4. Thomson, W. (Lord Kelvin), On the electrodynamic qualities of metals, *Philosophical Transactions of the Royal Society* (London) 146: 649–751, 1856.

5. Micro-Measurements Division, Measurements Group, Inc., Strain Gauge Selection: Criteria, Procedures, Recommendations, Technical Note 505-1, Raleigh, NC, 1989.

6. Micro-Measurements Division, Measurements Group, Inc., Strain Gauge Technical Data, Catalog 500, Part B, and TN-509: Errors Due to Transverse Sensitivity in Strain Gauges, Raleigh, NC, 2011.

7. Kulite Semiconductor Products, Inc, Bulletin KSG-5E: *Semiconductor Strain Gauges*, Leonia, NJ.

8. Weymouth, L. J., J. E. Starr, and J. Dorsey, Bonded resistance strain gauges, *Experimental Mechanics* 6(4): 19A, 1966.

9. Oi, K., Transient response of bonded strain gauges, *Experimental Mechanics* 6(9): 463, 1966.

10. Huang, M., Wu, Y, Chandra, B, Yan, H., Shan, Y., Heinz, T, and J. Hone, Direct measurement of strain-induced changes in the band structure of carbon nanotubes, *Physical Review Letters* 100, 136803, April 2008.

11. Sciammarella, C. A., The moiré method—a review, *Experimental Mechanics* 22(11): 418, 1982.

12. Post, D., Moiré interferometry at VPI & SU, *Experimental Mechanics* 23(2): 203, 1983.

13. Post, D., Moiré interferometry for deformation and strain studies, *Optical Engineering* 24(4): 663, 1985.

14. Yao, J., A tutorial on microwave photonics, *IEEE Photonics Society News*, 26, 4 2012.

15. Julich, F. and J. Roths, Determination of the effective refractive index of various single mode fibres for fibre Bragg grating sensor applications, Sensor & Test Conference 2009, OPTO2009 Proceedings.

16. Serker, K. and Z. Wu, Structural health monitoring using distributed macro-strain response, *Journal of Applied Sciences* 9(7): 1276–1284, 2009.

## NOMENCLATURE

| | | | | |
|---|---|---|---|---|
| $b$ | width ($l$) | | $E_m$ | modulus of elasticity ($ml^{-1}t^{-2}$) |
| $c$ | speed of sound ($l\,t^{-1}$) | | $E_i$ | input voltage (V) |
| $h$ | height ($l$) | | $E_o$ | output voltage (V) |
| $n_i$ | index of refraction in the direction of the principal strain in the $i$ direction | | $e_i$ | strain gauge lateral sensitivity error as a percentage of axial strain |
| $(u_d)_x$ | design-stage uncertainty in the variable $x$ | | $F_N$ | force normal to $A_c$ ($mlt^{-2}$) |
| $A_c$ | cross-sectional area ($l^2$) | | $G$ | shear modulus ($mt^{-2}$) |
| $D$ | diameter ($l$) | | $G_A$ | amplifier gain |

GF     gauge factor

$K$     strain optical coefficient

$K_B$     bridge sensitivity (V)

$K_I$     strain gauge lateral sensitivity

$L$     length ($l$)

$\delta E_o$     voltage change (V)

$\delta L$     change in length ($l$)

$\delta R$     resistance change ($\Omega$)

$M$     bending moment ($ml^2t^{-2}$)

$R$     electrical resistance ($\Omega$)

$T$     temperature (°)

$T_0$     reference temperature (°)

$\alpha$     moiré grating width to spacing parameter

$\gamma_{xy}$     shear strain in the $xy$ plane ($ml^{-1}t^{-2}$)

$\delta$     relative retardation between two light beams in a photoelastic material

$\varepsilon_a$     axial strain

$\varepsilon_b$     bending strain

$\varepsilon_r$     strain in the $i$ coordinate direction (i.e., $x$ direction)

$\varepsilon_l$     lateral or transverse strain

$\varepsilon_t$     apparent strain due to temperature

$\kappa$     bridge constant

$\nu_p$     Poisson ratio

$\nu_{po}$     Poisson ratio for gauge factor calibration test

$\pi_1$     piezoresistance coefficient ($t^2lm^{-1}$)

$\rho_e$     electrical resistivity ($\Omega l$)

$\sigma$     stress ($ml^{-1}t^{-2}$)

$\sigma_a$     axial stress ($ml^{-1}t^{-2}$)

$\tau_{xy}$     shear stress in the $xy$ plane ($ml^{-1}t^{-2}$)

# 12 Mechatronics[1]: Sensors, Actuators, and Controls

## 12.1 Introduction

Rapid advances in microprocessors have led to a dramatic increase in electronically controlled devices and systems, and many are connected to each other (as in the Internet of things). Many of these devices and systems require sensors and actuators to be interfaced with the electronics. Understanding the operating principles and limitations of sensors is essential to selecting and interfacing sensors for linear motion, rotary motion, and engineering variables such as force, torque, and power. *Actuators* are required to produce motion, such as to move an electric car seat, for a fly-by-wire throttle for automotive applications, or for positioning a precision laser welder. This chapter discusses sensors and actuators and provides a brief introduction to linear control theory.

Upon completion of this chapter, the reader will be able to

- Describe and analyze methods for displacement measurement
- State the physical principles underlying velocity and acceleration measurements
- Describe various load cells and their appropriate applications
- Describe various methods for measuring torque and power
- Describe various actuators and their role in mechatronic systems
- Analyze and design proportional–integral–derivative (PID) control schemes as part of a mechatronic system

## 12.2 Sensors

The previous chapters described a wide variety of measurement sensors and the fundamentals of their operation. Thermocouples, strain gauges, flow meters, and pressure sensors represent the means to measure the very important engineering process variables of temperature, strain, flow rate, and pressure. In this chapter, we add to this base by introducing methods and sensors for the measurement of linear and rotary displacement, acceleration and vibration, velocity, force or load, torque, and mechanical power.

### Displacement Sensors

Methods to measure position or displacement are a key element in many mechatronic systems. Often potentiometric or linear variable differential transformer (LVDT) transducers are employed in these applications.

---

[1]The term "mechatronics" is derived from the terms "mechanical" and "electronic" and refers to the integration of mechanical and electronic devices.

**FIGURE 12.1** Potentiometer construction.

## Potentiometers

A potentiometer is a device employed to measure linear or rotary displacement. The principle of operation relies on a change in electrical resistance with displacement.

A wire-wound potentiometer[2] or variable electrical resistance transducer is depicted in Figure 12.1. This transducer is composed of a sliding contact and a winding. The winding is made of many turns of wire wrapped around a nonconducting substrate. Output signals from such a device can be realized by imposing a known voltage across the total resistance of the winding and measuring the output voltage, which is proportional to the fraction of the distance the contact point has moved along the winding. Potentiometers can also be configured in a rotary form, with numerous total revolutions of the contact possible in a helical arrangement. The output from the sliding contact as it moves along the winding is actually discrete, as illustrated in Figure 12.1; the resolution is limited by the number of turns per unit distance. The loading errors associated with voltage-dividing circuits, discussed in Chapter 6, should be considered in choosing a measuring device for the output voltage from the potentiometer.

Wire-wound potentiometers display a stepwise output as the wiper contacts successive turns of the wire winding. Conductive plastic potentiometers were developed to eliminate this step-wise output and are now widely employed in mechatronic systems. Two such potentiometers are shown in Figure 12.2. The key components are the structural support and mechanical interface, the conductive plastic resistor, and the wiper where electrical contact occurs. A linear output with displacement is most often the design objective for these sensors. Typical linearity errors introduce instrument uncertainty from 0.2% to 0.02% of the reading.

---

[2]The potentiometer–transducer should not be confused with the potentiometer–instrument. Although both are based on voltage-divider principles, the latter measures low-level voltages ($10^{-6}$ to $10^{-3}$ V) as described in Chapter 6.

**FIGURE 12.2**   Conductive plastic potentiometer.

Courtesy of Novotechnik U.S., Inc. Used with permission

**FIGURE 12.3**   Construction of a linear variable differential transformer (LVDT).

## Linear Variable Differential Transformers

The LVDT, as shown in Figure 12.3, produces an AC output with an amplitude that is proportional to the displacement of a movable core. The movement of the core causes a mutual inductance in the secondary coils for an AC voltage applied to the primary coil. The waveform of the output from the LVDT is sinusoidal; the amplitude of the sine wave is proportional to the displacement of the core for a limited range of core motion.

In 1831, the English physicist Michael Faraday (1791–1867) demonstrated that a current could be induced in a conductor by a changing magnetic field. An interesting account of the development of the transformer may be found elsewhere [1]. Recall that for two coils in close proximity, a change in the current in one coil induces a voltage in the second coil according to Faraday's law. The application of this inductance principle to the measurement of distance begins by applying an AC voltage to the primary coil of the LVDT. The two secondary coils are connected in a series circuit, such that when the iron core is centered between the two secondary coils, the output voltage is zero (Figure 12.3). Motion of the magnetic core changes the mutual inductance of the

**FIGURE 12.4** LVDT output as a function of core position.

coils, which causes a different voltage to be induced in each of the two secondary coils. Over a limited range of operation, the output voltage is essentially linear with core displacement, as first noted in a U.S. patent by G. B. Hoadley in 1940 [2]. The output of a differential transformer is illustrated in Figure 12.4, in which both the linear range and nonlinear behavior are observed. Beyond the linear range, the output amplitude rises in a nonlinear manner to a maximum, and eventually falls to zero. The output voltages on either side of the zero displacement position are 180 degrees out of phase. Thus, with appropriate phase measurement, it is possible to determine positive or negative displacement of the core. However, note that owing to harmonic distortion in the supply voltage and to the two secondary coils' not being identical, the output voltage with the coil centered is not zero but instead reaches a minimum. Resolution of an LVDT strongly depends on the resolution of the measurement system used to determine its output. Resolutions down to the nanometer range are possible.

## Dynamic Response

The dynamic response of an LVDT is directly related to the frequency of the applied AC voltage, since the output voltage of the secondary coil is induced by the variation of the magnetic field induced by the primary coil. For this reason, the excitation voltage should have a frequency at least 10 times the maximum frequency in the measured input. An LVDT can be designed to operate with input frequencies ranging from 60 Hz up to 25 kHz (for specialized applications, frequencies in the megahertz range can be used).

The maximum allowable applied voltage for an LVDT is determined by the current-carrying capacity of the primary coil, typically in the 1- to 10-V range. A constant current source is preferable for an LVDT to limit temperature effects. For other than a sine wave input voltage form, harmonics in the input signal increase the voltage output at the null position of the core. The appropriate means of measuring and recording the output signal from an LVDT and the AC frequency applied to the primary coil should be chosen based on the highest frequencies present in the input signal to the LVDT. For example, for static measurements and signals having frequency content much lower than the excitation frequency of the primary coil, an AC voltmeter may be an appropriate choice for measuring the output signal. In this case, it is likely that the frequency response of the measuring system would be limited by the averaging effects of the AC voltmeter. For higher frequency signals, it is possible to create a DC voltage output that follows the input motion to the LVDT through demodulation and amplification of the resulting signal using a dedicated electronic circuit. Alternatively, the output signal can be sampled at a sufficiently high frequency using a computer data acquisition system to allow signal processing for a variety of purposes.

The measurement of distance using an LVDT is accomplished using an assembly known as an LVDT gauge head. Such devices are widely used in machine tools and various types of gauging equipment. Control applications have similar transducer designs. The basic construction is shown in Figure 12.5. The gauge head can reduce instrument errors to as low as 0.05% with repeatability of 0.0001 mm.

Courtesy of Measurement Specialties, Inc.; from reference 2. Used with permission.

**FIGURE 12.5**    LVDT gauge head. Top: Cross section of a typical LVDT gauge head.

Angular displacement can also be measured using inductance techniques employing a rotary variable differential transformer (RVDT). The output curve of an RVDT and a typical construction are shown in Figure 12.6, in which the linear output range is approximately ±40 degrees.

## Measurement of Acceleration and Vibration

The measurement of acceleration is required for a variety of purposes, ranging from machine design to guidance systems to structural health. Because of the range of applications for acceleration and vibration measurements, there exists a wide variety of transducers and measurement techniques, each associated with a particular application. Displacement, velocity, and acceleration measurements are also referred to as shock or vibration measurements depending on the waveform of the forcing function that causes the acceleration. A forcing function that is periodic in nature generally results in accelerations that are analyzed as vibrations. On the other hand, a force input having a short duration and large amplitude would be classified as a shock load.

The fundamental aspects of acceleration, velocity, and displacement measurements can be discerned through examination of the most basic device for measuring acceleration and velocity, a seismic transducer.

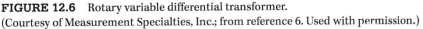

**FIGURE 12.6** Rotary variable differential transformer.
(Courtesy of Measurement Specialties, Inc.; from reference 6. Used with permission.)

## Seismic Transducer

A seismic transducer consists of three basic elements, as shown in Figure 12.7: a spring–mass–damper system, a protective housing, and an appropriate output transducer. Through the appropriate design of the characteristics of this spring–mass–damper system, the output is a direct indication of either displacement or acceleration. To accomplish a specific measurement, this basic seismic transducer is rigidly attached to the object experiencing the motion that is to be measured.

Consider the case where the output transducer senses the position of the seismic mass; a variety of transducers could serve this function. Under some conditions, the displacement of the seismic mass serves as a direct measure of the acceleration of the housing and the object to which it is attached. To illustrate the relation between the relative displacement of the seismic mass and acceleration, consider the case in which the input to the seismic instrument is a constant acceleration. The response of the instrument is illustrated in Figure 12.8. At steady-state conditions, under this constant acceleration, the mass is at rest with respect to the housing. The spring deflects an amount proportional to the force required to accelerate the seismic mass, and since the mass is known, Newton's second law yields the corresponding acceleration. The relationship between

**FIGURE 12.7**    Seismic transducer.

**FIGURE 12.8**    Response of a seismic transducer to a constant acceleration.

a constant acceleration and the displacement of the seismic mass is linear for a linear spring (where $F = kx$).

We might want to measure not only constant accelerations, but also complex acceleration waveforms. Consider an input to our seismic instrument such that the displacement of the housing is a sine wave, $y_{hs} = A\sin \omega t$, and the absolute value of the resulting acceleration of the housing is $A\omega^2 \sin \omega t$. If we consider a free-body diagram of the seismic mass, the spring force and the damping force must balance the inertial force for the mass. (Notice that gravitational effects could play an important role in the analysis of this instrument—for instance, if the instrument were installed in an aircraft.) The spring force and damping force are proportional to the relative displacement and velocity between the housing and the mass, whereas the inertial force is dependent only on the absolute acceleration of the seismic mass.

We define a relative displacement, $y_r$, as the difference between the housing displacement, $y_{hs}$, and the displacement of the seismic mass, $y_m$, given by

$$y_r = y_m - y_{hs} \qquad (12.1)$$

Newton's second law may then be expressed

$$m\frac{d^2 y_m}{dt^2} + c\frac{dy_r}{dt} + ky_r = 0 \tag{12.2}$$

Substituting Equation 12.1 into Equation 12.2 yields

$$m\left(\frac{d^2 y_{hs}}{dt^2} + \frac{d^2 y_r}{dt^2}\right) + c\frac{dy_r}{dt} + ky_r = 0 \tag{12.3}$$

But we know that $y_{hs} = A\sin\omega t$ and

$$\frac{d^2 y_{hs}}{dt^2} = -A\omega^2 \sin\omega t \tag{12.4}$$

Thus,

$$m\frac{d^2 y_r}{dt^2} + c\frac{dy_r}{dt} + ky_r = mA\omega^2 \sin\omega t \tag{12.5}$$

This equation is identical in form to Equation 3.12. As in the development for a second-order system response, we examine the steady-state solution to this governing equation. The transducer senses the relative motion between the seismic mass and the instrument housing. Thus, for the instrument to be effective, the value of $y_r$ must provide indication of the desired output.

For the input function $y_{hs} = A\sin\omega t$, the steady-state solution for $y_r$ is

$$(y_r)_{steady} = \frac{(\omega/\omega_n)^2 A \cos(\omega t - \varphi)}{\left\{\left[1 - (\omega/\omega_n)^2\right]^2 + \left[2\zeta(\omega/\omega_n)\right]^2\right\}^{1/2}} \tag{12.6}$$

where

$$\omega_n = \sqrt{\frac{k}{m}} \quad \zeta = \frac{c}{2\sqrt{km}} \quad \varphi = \tan^{-1}\left(\frac{2\zeta(\omega/\omega_n)}{1 - (\omega/\omega_n)^2}\right) \tag{12.7}$$

The characteristics of this seismic instrument can be discerned by examining Equations 12.6 and 12.7. The natural frequency and damping are fixed for a particular design. We wish to examine the motion of the seismic mass and the resulting output for a range of input frequencies.

For vibration measurements, it is desirable to measure the amplitude of the displacements associated with the vibrations; thus, the desired behavior of the seismic instrument would be to have an output that gives a direct indication of $y_{hs}$. For this to occur, the seismic mass should remain stationary in an absolute frame of reference, and the housing and output transducer should move with the vibrating object. To determine the conditions under which this behavior would occur, the amplitude of $y_r$ at steady state can be examined. The ratio of the maximum amplitude of the output divided by the equivalent static output is $(y_r)_{max}/A$. For vibration measurements, this ratio should have a value of 1. Figure 12.9 shows $(y_r)_{max}/A$ as a function of the ratio of the input frequency to the natural frequency as determined from Equation 12.6. Clearly, as the input frequency increases, the output amplitude, $y_r$, approaches the input amplitude, $A$, as desired. Thus, a seismic instrument that is used to measure vibration displacements should have a natural frequency smaller than the expected input frequency. Damping ratios near 0.7 are common for such an instrument. The seismic instrument designed for this application is called a *vibrometer*.

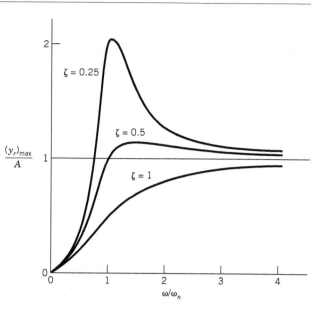

**FIGURE 12.9** Displacement amplitude at steady state as a function of input frequency for a seismic transducer.

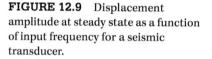

## EXAMPLE 12.1

A seismic instrument such as the one shown in Figure 12.7 is used to measure a periodic vibration having an amplitude of 0.05 in. and a frequency of 15 Hz.

1. Specify an appropriate combination of natural frequency and damping ratio such that the dynamic error in the output is less than 5%.

2. What spring constant and damping coefficient would yield these values of natural frequency and damping ratio?

3. Determine the phase lag for the output signal. Would the phase lag change if the input frequency were changed?

**KNOWN** Input function $y_{hs} = 0.05 \sin 30\pi t$.

**FIND** Values of $\omega_n$, $\zeta$, $k$, $m$, and $c$ to yield a measurement with less than 5% dynamic error. Examine the phase response of the system.

**SOLUTION**
Numerous combinations of the mass, spring constant, and damping coefficient would yield a workable design. Let's choose $m = 0.05 \; \text{lb}_m$ (22.7 g) and $\zeta = 0.7$. We know that for a spring–mass–damper system

$$\omega = \sqrt{\frac{k}{m}} \qquad c_c = 2\sqrt{km}$$

where $c_c$ is the critical damping coefficient. With $\zeta = c/c_c = 0.7$, the damping coefficient, $c$, is found as

$$c = 2(0.7)\sqrt{km}$$

We can now examine the values of $(y_r)_{max}/A$ and the phase angle, $\varphi$, for $\zeta = 0.7$. The largest magnitude in Equation 12.6 occurs when $\cos(\omega t - \varphi) = 1$, so

$$(y_r)_{max}/A = \frac{(\omega/\omega_n)^2}{\left\{ \left[1 - (\omega/\omega_n)^2\right]^2 + \left[2\zeta(\omega/\omega_n)\right]^2 \right\}^{1/2}}$$

and φ is found from Equation 12.7. Here is a sample of the results:

| $\dfrac{\omega}{\omega_n}$ | $\dfrac{(y_r)_{max}}{A}$ | $\varphi$(degrees) |
|---|---|---|
| 10 | 1.000 | 172 |
| 8 | 1.000 | 169.9 |
| 6 | 1.000 | 166.5 |
| 4 | 0.999 | 159.5 |
| 3 | 0.996 | 152.3 |
| 2 | 0.975 | 137.0 |
| 1.7 | 0.951 | 128.5 |

Using the dynamic error as $\left| \dfrac{(y_r)_{max}}{A} - 1 \right|$, acceptable behavior is achieved for values of $\omega/\omega_n \geq$ 1.7 for $\zeta = 0.7$, and an acceptable maximum value of the natural frequency is

$$\omega_n = \omega/1.7 = 30\pi/1.7 = 55.4\,\text{rad/s}$$

with[3]

$$\omega_n = \sqrt{k/m}$$

for $m = 0.05\ \text{lb}_m$ (22.7 g), then the value of $k$ is 4.8 lb/ft (70 N/m). So the value of the damping coefficient is

$$c = 2(0.7)\sqrt{km} = 0.12\,\text{lb-s/ft} = 1.76\,\text{N-s/m}$$

**COMMENTS** We should consider other constraints, including size, cost, and operating environment. The phase shift–frequency behavior for this combination of design parameters could allow distortion of complex waveform signals so if that is important, another set of parameters may be preferable. This and other issues can be explored using the program file *seismic_transducer.vi*.

### Acceleration Measurement with a Seismic Instrument

If it is desired to measure acceleration, the behavior of the seismic mass must be quite different from that in Example 12.1. The amplitude of the acceleration input signal is $A\omega^2$. To have the output value $y_r$ represent the acceleration, it is clear from Equation 12.6 that the value of $y_r/A(\omega/\omega_n)^2$ must be a constant over the design range of input frequencies. If this is true, the output will be proportional to the acceleration. The amplitude of $y_r/A(\omega/\omega_n)^2$ is

$$\frac{(y_r)_{steady}}{A(\omega/\omega_n)^2} = \frac{\cos(\omega t - \varphi)}{\left\{ \left[1 - (\omega/\omega_n)^2\right]^2 + \left[2\zeta(\omega/\omega_n)\right]^2 \right\}^{1/2}} \tag{12.8}$$

and

$$M(\omega) = \frac{1}{\left\{ \left[1 - (\omega/\omega_n)^2\right]^2 + \left[2\zeta(\omega/\omega_n)\right]^2 \right\}^{1/2}} \tag{12.9}$$

where $M(\omega)$ is the magnitude ratio as given by Equation 3.22 in Chapter 3. The magnitude ratio is plotted as a function of input frequency and damping ratio in Figure 3.16.

---

[3]Note: $1\text{lb} = 32.2\text{lb}_m\text{-ft}/s^2$, whereas $1\text{N} = 1\text{kg-m/s}^2$.

From Figure 3.16, we find that the desired flat magnitude response is achieved over a range of input frequency ratios from $0 \leq \omega/\omega_n \leq 0.4$. Typically, in an accelerometer, the damping ratio is designed to be near 0.7 so that the phase shift is linear with frequency such that distortion is minimized.

In summary, the seismic instrument can be designed so that the output can be interpreted in terms of either the input displacement or the input acceleration. Acceleration measurements may be integrated to yield velocity information; the differentiation of displacement data to determine velocity or acceleration introduces significantly more difficulties than the integration process.

## Transducers for Shock and Vibration Measurement

In general, the destructive forces generated by vibration and shock are best quantified through the measurement of acceleration. Although a variety of accelerometers are available, strain gauge and piezoelectric transducers are widely employed for the measurement of shock and vibration [3].

A piezoelectric accelerometer employs the principles of a seismic transducer through the use of a piezoelectric element to provide a portion of the spring force. Figure 12.10 illustrates one basic construction of a piezoelectric accelerometer. A preload is applied to the piezoelectric element simply by tightening the nut that holds the mass and piezoelectric element in place. Upward or downward motion of the housing changes the compressive forces in the piezoelectric element, resulting in an appropriate output signal. Instruments are available with a range of frequency response from 0.03 to 10,000 Hz. Depending on the piezoelectric material used in the transducer construction, the static sensitivity can range from 1 to 100 mV/g. Steady accelerations cannot be effectively measured using such a piezoelectric transducer.

Strain gauge accelerometers are generally constructed using a mass supported by a flexure member, with the strain gauge sensing the deflection that results from an acceleration of the mass. The frequency response and range of acceleration of these instruments are related such that instruments designed for higher accelerations have a wider bandwidth but significantly lower static sensitivity. Table 12.1 provides typical performance characteristics for two strain gauge accelerometers that employ semiconductor strain gauges.

**FIGURE 12.10**   Basic piezoelectric accelerometer.

Table 12.1   **Representative Performance Characteristics for Piezoresistive Accelerometers**[a]

| Characteristic | 25-g range | 2,500-g range |
| --- | --- | --- |
| Sensitivity [mV/g] | 50 | 0.1 |
| Resonance frequency [Hz] | 2,700 | 30,000 |
| Damping ratio | 0.4–0.7 | 0.03 |
| Resistance [Ω] | 1,500 | 500 |

[a]Adapted from reference 7.

Many other accelerometer designs are available, including potentiometric or reluctive or accelerometers that use closed-loop servo systems to provide a high output level. Piezoelectric transducers have the highest frequency response and range of acceleration but have relatively lower sensitivity; amplification can overcome this drawback to some degree. Semiconductor strain gauge transducers have a lower frequency response limit than do piezoelectric transducers, but they can be used to measure steady accelerations.

### Laser Vibrometer

Laser vibrometers employ the same basic principles used in Laser Doppler Anemometry as described in Chapter 9. Interferometry allows vibration measurements to nanometer scales. Because the method is noncontact, the range of applications is enormous. Biological measurements can be made on living and moving creatures. Testing in automotive and aerospace applications can be done in real time.

## Velocity Measurements

Linear and angular velocity measurements use a variety of approaches ranging from radar and laser systems for speed measurement to mechanical counters to provide an indication of a shaft rotational speed. For many applications, the sensors employed provide a scalar output of speed. However, sensors and methods exist that can provide indication of both speed and direction when properly employed. Here, we consider techniques for the measurement of linear and angular speed.

Displacement, velocity, and acceleration measurements are made with respect to some frame of reference. Consider the case of a game of billiards in a moving railway car. Observers on the platform and on the train would assign different velocity vectors to the balls during play. The velocity vectors would differ by the relative velocity of the two observers. Simply differentiating the vector velocity equation, however, shows that the accelerations of the balls are the same in all reference frames moving relative to one another with constant relative velocity.

### Linear Velocity Measurements

As previously discussed, measurements of velocity require a frame of reference. For example, the velocity of a conveyor belt might be measured relative to the floor of the building where it is housed. On the other hand, advantages may be realized in a control system if the velocity of a robotic arm, which "picks" parts from this same conveyor belt, is measured relative to the moving conveyor. In the present discussion, it is assumed that velocity is measured relative to a ground state, which is generally defined by the mounting point of a transducer.

Consider the measurement of velocity relative to a fixed frame of reference. Typically, if the measurement of linear velocity is to be made on a continuous basis, an equivalent angular rotational speed is measured, and the data is analyzed in such a way so as to produce a measured linear velocity. For example, a speedometer on an automobile provides a continuous record of the speed of the car, but the output is derived from measuring the rotational speed of the drive shaft or transmission.

### Velocity from Displacement or Acceleration

Velocity, in general, can be directly measured by mechanical means only over very short times or small displacements, due to limitations in transducers. However, if the displacement of a rigid body is measured at identifiable time intervals, the velocity can be determined through differentiation of the time-dependent displacement. Alternatively, if acceleration is measured, the velocity may be determined from integration of the acceleration signal. The following example demonstrates the effect of integration and differentiation on the uncertainty of velocities computed from acceleration or displacement.

**EXAMPLE 12.2**

Suppose our goal is to assess the merits of measuring velocity through the integration of an acceleration signal as compared to differentiating a displacement signal. The following conditions are assumed to apply.

For both $y(t)$ and $y''(t)$, the data acquisition system and transducers have the characteristics described in Table 12.2. Note that the uncertainties are derived directly from the analog-to-digital (A/D) resolution error and accuracy statements. Assume that the signals for both acceleration and displacement are sampled at 10 Hz and numerical techniques are used to differentiate or integrate the resulting signals.

**SOLUTION**

Consider first determining the velocity through differentiation of the displacement signal. Displacement is measured digitally by the data acquisition system, with a digitized value of displacement recorded at time intervals $\delta t$. The velocity at any time $n\delta t$ can be approximated as

$$v(t) = y'(t) = \frac{y_{n+1} - y_n}{\delta t} \tag{12.10}$$

where

$y_{n+1}$ = the $(n + 1)$ measurement of displacement at time $(n + 1)\delta t$

$y_n$ = the $n$th measurement of displacement at time $n\delta t$

$v(t)$ = velocity at time $t$

If the signal is sampled at 10 Hz, $\delta t$ is 0.1 s. For the present, we assume that the uncertainty in time is negligible, so that the uncertainty in $v$, $u_v$ can be expressed as

$$u_v = \left\{ \left[ \frac{\partial v}{\partial y_{n+1}} u_{y_{n+1}} \right]^2 + \left[ \frac{\partial v}{\partial y_n} u_{y_n} \right]^2 \right\}^{1/2} \tag{12.11}$$

The uncertainties in the measured displacements, $y_n$ and $y_{n+1}$, will be equal and are listed in Table 12.2. Substituting these values for uncertainty in Equation 12.11 yields an uncertainty of $\pm 3$ cm/s in the velocity measurement. Notice that this uncertainty magnitude is not a function of

**Table 12.2    Specification and Uncertainty Analyses for Displacement and Acceleration**

| Measured variable | Functional form | Full-scale output range |
|---|---|---|
| Displacement (cm) | $y(t) = 20 \sin 2t$ | 0 to 10 V |
| Acceleration (cm/s$^2$) | $y''(t) = -80 \sin 2t$ | −5 to 5 V |

| Uncertainty values for displacement and acceleration | |
|---|---|
| **Measured variable** | **Uncertainty** |
| Displacement | Accuracy: 1% full scale = 0.2 cm |
| | A/D 8 bit (0.04 V) = 0.08 cm |
| | Total uncertainty = $\pm 0.22$ cm |
| Acceleration | Accuracy: 1% full scale = 0.05 cm/s$^2$ |
| | A/D 8 bit (0.04 V) = 0.04 cm/s$^2$ |
| | Total uncertainty = $\pm 0.064$ cm/s$^2$ |

time or the measured velocity. This corresponds to a minimum uncertainty of 30% in the velocity measurement.

To determine velocity from acceleration, the measured values of acceleration must be integrated. Because we have a digital signal, the integration can be accomplished numerically as

$$v(t) = y'(t) = \sum_i y_i'' \, \delta t \tag{12.12}$$

Assuming that the uncertainty in time is negligible, the uncertainty in velocity at any time $t$ is simply

$$u_v = u_y'' t \tag{12.13}$$

Clearly, the integration process tends to accumulate error as the calculation of velocity proceeds in time, as illustrated in Figure 12.11, but can produce acceptable results for very short time periods.

**COMMENT** This example can be used to illustrate several useful principles for data analysis. Consider the effects of adding noise, or a degree of random error, to measured data; such a signal is shown in Figure 12.12. In general, noise tends to be minimized through a process of integration

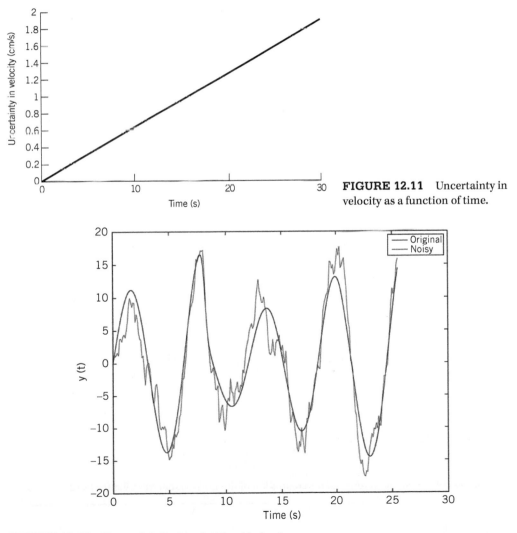

**FIGURE 12.11**  Uncertainty in velocity as a function of time.

**FIGURE 12.12**  Deterministic signal with added noise.

and amplified through differentiation. Figure 12.13 shows the result of integrating the noisy data from Figure 12.12, with the result that the effects of noise are reduced. However, because of the accumulation of error, the technique must balance the integration period against error accumulation. Differentiation tends to be extremely sensitive to low-amplitude, high-frequency noise, which can create very large errors in derivatives, especially at high sampling frequencies. This is apparent in Figure 12.14, in which the signal in Figure 12.12 has been differentiated numerically. Appropriate filtering or smoothing techniques are generally effective at reducing errors associated with noise in derivatives in cases in which the noise has low amplitude.

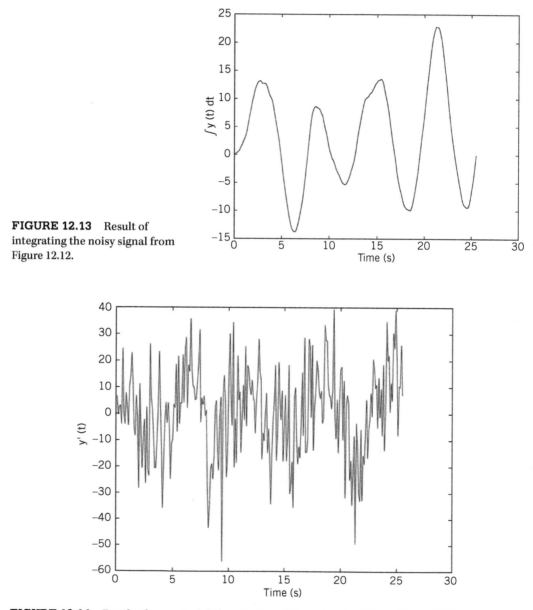

**FIGURE 12.13** Result of integrating the noisy signal from Figure 12.12.

**FIGURE 12.14** Result of numerical differentiation of the noisy signal from Figure 12.12.

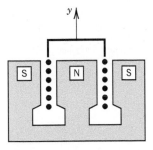

**FIGURE 12.15** Moving coil transducer.

## Moving Coil Transducers

Moving coil transducers take advantage of the voltage generated when a conductor experiences a displacement in a magnetic field. An illustration of a moving coil velocity pickup is provided in Figure 12.15. Recall that a current-carrying conductor experiences a force in a magnetic field and that a force is required to move a conductor in a magnetic field. For the latter case, a voltage is induced in the conductor. Consider the case in which the magnetic field strength is at right angles to the conductor. The induced voltage is given by

$$E = \pi B D_c l N \frac{dy}{dt} \qquad (12.14)$$

where

$B$ = magnetic field strength

$D_c$ = coil diameter

$E$ = induced voltage

$N$ = number of turns in coil

$dy/dt$ = velocity of coil linear motion

$l$ = coil length

A moving coil transducer is appropriate for vibration applications in which the velocities of small amplitude motions are measured. The output voltage is proportional to the coil velocity, and the output polarity indicates the velocity direction. Static sensitivities on the order of 2 V-s/m are typical. Moving coil transducers find application in seismic measurements as well as in vibration applications.

## Angular Velocity Measurements

The measurement of angular velocity finds a wide range of applications, including such familiar examples as speedometers on automobiles. We consider a variety of applications and measurement techniques.

### Mechanical Measurement Techniques

Mechanical means of measuring angular velocity or rotational speed were developed primarily to provide feedback for control of engines and steam turbines. Mechanical governors and centrifugal tachometers [4] operate on the principle illustrated in Figure 12.16. Here, the centripetal acceleration of the flyball masses results in a steady-state displacement of the spring, which provides a control signal or is a direct indication of rotational speed. For this arrangement, the spring force is proportional to the square of the angular velocity.

Flyball mass

A  Vertical motion only
B  Vertical motion only
C  Vertical and rotational motion
D  Rotational motion only

Output signal

Output

**FIGURE 12.16**  Mechanical angular velocity sensor.

## Stroboscopic Angular Velocity Measurements

A stroboscopic light source provides high-intensity flashes of light, which can be caused to occur at a precise frequency. A stroboscope is illustrated in Figure 12.17. Stroboscopes permit the intermittent observation of a periodic motion in a manner that appears to stop or slow the motion. Figure 12.18 illustrates the use of a strobe to measure rotational speed. A timing mark on the rotating object is illuminated with the strobe, and the strobe frequency is adjusted such that the

**FIGURE 12.17**  Stroboscope.

| 3600 rpm | 3600 rpm | 3600 rpm |
|---|---|---|
| 1800  fpm | 7200  fpm | 10,800  fpm |
| 1200  fpm | 2400  fpm | 5400  fpm |
| 900  fpm | 1440  fpm | 2700  fpm |
| 720  fpm | | 2160  fpm |
| 600  fpm | **(b)** | |
| 3600/n  fpm | | **(c)** |
| **(a)** | | |

**FIGURE 12.18**  Images resulting from harmonic and subharmonic flashing rates for stroboscopic angular speed measurement; rotation rate being measured is 3,600 rpm. Note: fpm indicates *flashes per minute.*

mark appears to remain motionless, as shown in Figure 12.18a. Thus, the highest synchronous speed is the actual rotational speed. At this speed, the output value is available from a calibration of the stroboscopic lamp flash frequency, with uncertainties to less than 0.1%. Clearly, a timing mark would appear motionless for integer multiples of the actual rotational speed and for integral submultiples, as illustrated in Figure 12.18.

The synchronization of images at flashing rates other than the actual speed requires some practical approaches to ensuring the accurate determination of speed. Spurious images can easily result for symmetrical objects, and some asymmetric marking is necessary to prevent misinterpretation of stroboscopic data. To distinguish the actual speed from a submultiple, the flashing rate can be decreased until another single synchronous image appears. If this flashing rate corresponds to half the original rate, then the original rate is the actual speed. If it does not occur at half the original value, then the original value is a submultiple.

The upper limit of the flash rate of the strobe does not limit the ability of the stroboscope to measure rotational speed. For high speeds, synchronization can be achieved $N$ times, with $\omega_1$ representing the maximum achievable synchronization speed, and $\omega_2, \omega_3, \ldots$ representing successively lower synchronization speeds. The measured rotational speed is then calculated as

$$\omega = \frac{\omega_1 \omega_N (N-1)}{\omega_1 - \omega_N} \tag{12.15}$$

The program file *stroboscope.vi* on the companion software disk illustrates the basic operation and interesting stroboscopic effects.

### Electromagnetic Techniques

Several measurement techniques for rotational velocity use transducers that generate electrical signals, which are indicative of angular velocity. One of the most basic is illustrated in Figure 12.19. This transducer consists of a toothed wheel and a magnetic pickup; the pickup consists of a magnet and a coil. As the toothed wheel rotates, a voltage is induced in the coil as a result of changes in the magnetic field. As each ferromagnetic tooth passes the pickup, the reluctance of the magnetic circuit changes in time, yielding a voltage in the coil given by

$$E = C_B N_t \omega \sin N_t \omega t \tag{12.16}$$

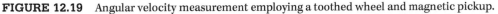

**FIGURE 12.19** Angular velocity measurement employing a toothed wheel and magnetic pickup.

where

$E$ = output voltage

$C_B$ = proportionality constant

$N_t$ = number of teeth

$\omega$ = angular velocity of the wheel

The angular velocity can be found either from the amplitude or the frequency of the output signal. The voltage amplitude signal is susceptible to noise and loading errors. Thus, less error is introduced if the frequency is used to determine the angular velocity; typically, some means of counting the pulses electronically is employed. This frequency information can be transmitted digitally for recording, which eliminates the noise and loading error problems associated with voltage signals.

## Force Measurement

The measurement of force is most familiar as the process of weighing, ranging from weighing micrograms of a medicine to weighing trucks on the highway. Force is a quantity derived from the fundamental dimensions of mass, length, and time. Standards and units of measure for these quantities are defined in Chapter 1. The most common techniques for force measurement are described in this section.

### Load Cells

"Load cell" is a term used to describe a transducer that generates a voltage signal as a result of an applied force, usually along a particular direction. Such force transducers often consist of an elastic member and a deflection sensor. These deflection sensors may employ changes in capacitance, resistance, or the piezoelectric effect to sense deflection. A technology overview for such devices is provided elsewhere [5]. Consider first load cells that are designed using a linearly elastic member instrumented with strain gauges.

**Strain Gauge Load Cells**  Strain gauge load cells are most often constructed of a metal, and have a shape such that the range of forces to be measured results in a measurable output voltage over the desired operating range. The shape of the linearly elastic member is designed to meet the following goals: (1) provide an appropriate range of force-measuring capability with necessary accuracy, (2) provide sensitivity to forces in a particular direction, and (3) have low sensitivity to force components in other directions.

A variety of designs of linearly elastic load cells are shown in Figure 12.20. In general, load cells may be characterized as beam-type load cells, proving rings, or columnar-type designs. Beam-type load cells may be characterized as bending beam load cells or shear beam load cells.

A bending beam load cell as shown in Figure 12.21 is configured such that the sensing element of the load cell functions as a cantilever beam. Strain gauges are mounted on the top and bottom of the beam to measure normal or bending stresses. Figure 12.21 provides qualitative indication of the shear and normal stress distributions in a cantilever beam. In the linear elastic range of the load cell, the bending stresses are linearly related to the applied load.

In a shear beam load cell, the beam cross section is that of an I-beam. The resulting shear stress in the web is nearly constant, allowing placement of a strain gauge essentially anywhere in the web with reasonable accuracy. Such a load cell is illustrated schematically in Figure 12.22, along with the shear stress distribution in the beam. In general, bending beam load cells are less costly because of their construction, but the shear beam load cells have several advantages, including lower creep and faster response times.

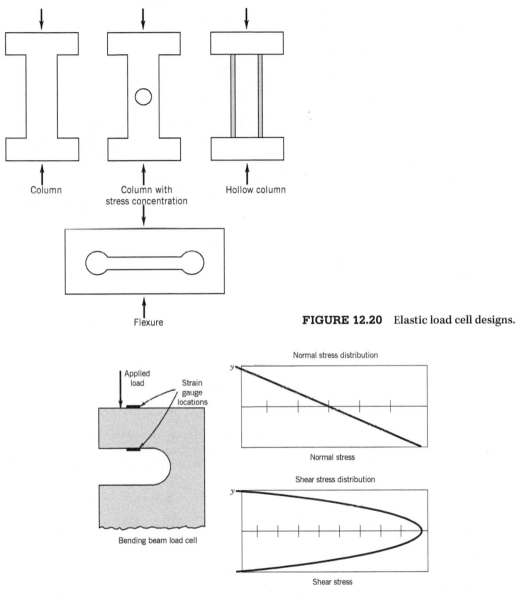

**FIGURE 12.20**   Elastic load cell designs.

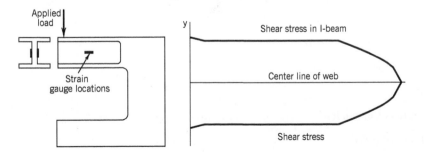

**FIGURE 12.21**   Bending beam load cell and stress distributions.

**FIGURE 12.22**   Shear beam load cell and shear stress distribution.

**FIGURE 12.23**    Piezoelectric load cell design.
(Courtesy of the Kistler Instrument Co. Used with permission.)

**Piezoelectric Load Cells**    Piezoelectric materials are characterized by their ability to develop a charge when subject to a mechanical strain. The most common piezoelectric material is single-crystal quartz. The basic principle of transduction that occurs in a piezoelectric element may best be thought of as a charge generator and a capacitor. The frequency response of piezoelectric transducers is very high, because the frequency response is determined primarily by the size and material properties of the quartz crystal. The modulus of elasticity of quartz is approximately 85 GPa, yielding load cells with typical static sensitivities ranging from 0.05 to 10 mV/N, and a frequency response up to 15,000 Hz. A typical piezoelectric load cell construction is shown in Figure 12.23.

**Proving Ring**    A ring-type load cell can be employed as a local force standard. Such a ring-type load cell, as shown in Figure 12.24, is often employed in the calibration of materials testing machines because of the high degree of precision and accuracy possible with this arrangement of transducer and sensor. If the sensor is approximated as a circular right cylinder, the relationship between applied force and deflection is given by

$$\delta y = \left( \frac{\pi}{2} - \frac{4}{\pi} \right) \frac{F_n D^3}{16EI} \tag{12.17}$$

where

$\delta y$ = deflection along the applied force

$F_n$ = applied force

$D$ = diameter

$E$ = modulus of elasticity

$I$ = moment of inertia

The application of the proving ring involves measuring the deflection of the proving ring in the direction of the applied force. Typical methods for this displacement measurement include

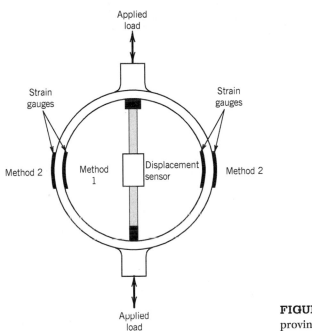

Applied
load

Strain
gauges

Strain
gauges

Method 2

Method
1

Displacement
sensor

Method 2

Applied
load

**FIGURE 12.24**   Ring type load cell, or
proving ring.

displacement transducers, which measure overall displacement, and strain gauges. These methods
are illustrated in Figure 12.24.

## Torque Measurements

Torque and mechanical power measurements are often associated with the energy conversion processes that serve to provide mechanical and electrical power to our industrial world. Such energy
conversion processes are largely characterized by the mechanical transmission of power produced
by prime movers such as internal combustion engines. From automobiles to turbine-generator
sets, mechanical power transmission occurs through torque acting through a rotating shaft.

The measurement of torque is important in a variety of applications, including sizing of load-
carrying shafts. This measurement is also a crucial aspect of the measurement of shaft power,
such as in an engine dynamometer. Strain gauge–based torque cells are constructed in a manner
similar to load cells, in which a torsional strain in an elastic element is sensed by strain gauges
appropriately placed on the elastic element. Figure 12.25 shows a circular shaft instrumented with
strain gauges for the purpose of measuring torque.

Consider the stresses created in a shaft of radius $R_0$ subject to a torque $T$. The maximum
shearing stress in a circular shaft occurs on the surface and may be calculated from the torsion
formula[4]

$$\tau_{\max} = TR_0/J \tag{12.18}$$

where

$\tau_{\max}$ = maximum shearing stress

$T$ = applied torque

$J$ = polar moment of inertia ($\pi R_0^2/2$ for a solid circular shaft)

---

[4]Coulomb developed the torsion formula in 1775 in connection with electrical instruments.

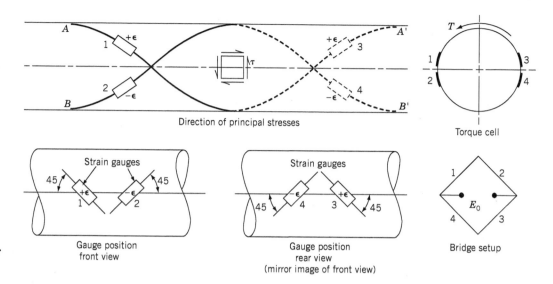

**FIGURE 12.25**
Shaft instrumented for
torque measurement.

For a shaft in pure torsion, there are no normal stresses, $\sigma_x$, $\sigma_y$, or $\sigma_z$. The principal stresses lie along a line that makes a 45-degree angle with the axis of the shaft, as illustrated in Figure 12.25, and have a value equal to $\tau_{max}$. Strains that occur along the curve labeled A–A′ are opposite in sign from those that occur along B–B′. These locations allow placement of four active strain gauges in a Wheatstone bridge arrangement and the direct measurement of torque in terms of bridge output voltage.

## Mechanical Power Measurements

Almost universally, prime movers such as internal combustion (IC) engines and gas turbines convert chemical energy in a fuel to thermodynamic work transmitted by a shaft to the end use. The resulting power is transmitted through a mechanical coupling. This section discusses the measurement of such mechanical power transmission.

### Rotational Speed, Torque, and Shaft Power

Shaft power is related to rotational speed and torque as

$$\vec{P}_s = \vec{\omega} \times \vec{T} \tag{12.19}$$

where $\vec{P}_s$ is the shaft power, $\vec{\omega}$ is the rotational velocity vector, and $\vec{T}$ is the torque vector. In general, the orientation of the torque and rotational velocity vectors are such that the equation may be written in scalar form as

$$P_s = \omega T \tag{12.20}$$

Table 12.3 provides a summary of useful equations related to shaft power, torque, and speed as employed in mechanical measurements. Historically, a device called a Prony brake was used to measure shaft power. A typical Prony brake arrangement is shown in Figure 12.26. Consider using the Prony brake to measure power output for an IC engine. The Prony brake serves to provide a well-defined load for the engine, with the power output of the engine dissipated as thermal energy in the braking material. By adjusting the load, the power output over a range of speeds and throttle settings can be realized. The power is measured by recording the torque acting on the torque arm and the rotational speed of the engine. Clearly, this device is limited in speed and power, but it

## Table 12.3    Shaft Power, Torque, and Speed Relationships

|  | SI | U.S. customary |
|---|---|---|
| Shaft power, $P$ | $P = \omega T$ | $P = \dfrac{2\pi n T}{550}$ |
| Power | $P$ (W) | $P$ (hp) |
| Rotational speed | $\omega$ (rad/s) | $n$ (rev/s) |
| Torque | $T$ (N m) | $T$ (ft lb) |

**FIGURE 12.26**    Prony brake.
(Courtesy of the American Society of Mechanical Engineers, New York, NY. Reprinted from PTC 19.7-1980 6.)

does demonstrate the operating principles of power measurement, and is historically significant as the first technique for measuring power.

## Cradled Dynamometers

A Prony brake is an example of an absorbing dynamometer. The term "dynamometer" refers to a device that absorbs and measures the power output of a prime mover. Prime movers are large mechanical power-producing devices such as gasoline or diesel engines or gas turbines. Several methods of energy dissipation are used in various ranges of power, but the measurement techniques are governed by the same underlying principles. Thus, we consider first the measurement of power and then discuss means for dissipating the sometimes large amounts of power generated by prime movers.

The cradled dynamometer measures mechanical power by measuring the rotational speed of the shaft, which transmits the power, and the reaction torque required to prevent movement of the stationary part of the prime mover. This reaction torque is impressively illustrated by the so-called wheelies performed by motorcycle riders. A cradled dynamometer is supported in bearings, called trunnion bearings, such that the reaction torque is transmitted to a torque or force-measuring device.

In principle, the operation of the dynamometer involves the steady-state measurement of the load $F_r$ created by the reaction torque and the measurement of shaft speed. Using Equation 12.19, the transmitted shaft power can be calculated directly.

The American Society of Mechanical Engineers (ASME) Performance Test Code (PTC) 19.7 [6] provides guidelines for the measurement of shaft power. According to PTC 19.7, overall uncertainty in the measurement of shaft power by a cradled dynamometer results from (1) trunnion-bearing friction error uncertainty, (2) force measurement uncertainty ($F_r$), (3) moment arm length uncertainty ($L_r$), (4) static unbalance of dynamometer error uncertainty, and (5) uncertainty in rotational speed measurement.

A means of supplying a controllable load to the prime mover and dissipating the energy absorbed in the dynamometer are integral parts of the design of any dynamometer. Several techniques are described for providing an appropriate load.

**Eddy Current Dynamometers:**  A direct current field coil and a rotor allow shaft power to be dissipated by eddy currents in the stator winding. The resulting conversion to thermal energy by Joulian heating of the eddy currents necessitates that some cooling be supplied, typically from cooling water.

**Alternating Current and DC Generators:**  Cradled AC and DC machines are employed as power-absorbing elements in dynamometers. The AC applications require variable frequency capabilities to allow a wide range of power and speed measurements. The power produced in such dynamometers may be dissipated as thermal energy using resistive loads.

**Waterbrake Dynamometers:**  A waterbrake dynamometer employs fluid friction and momentum transport to create a means of energy dissipation. Two representative designs are provided in Figure 12.27. The viscous shear-type brake is useful for high rotational speeds, and the agitator type unit is used over a range of speeds and loads. Waterbrakes may be employed for applications up to 10,000 HP (7,450 kW). The load absorbed by waterbrakes can be adjusted using water level and flow rates in the brake. Figure 12.28 shows a state-of-the-art waterbrake engine dynamometer. Figure 12.28a shows the dynamometer without an engine installed, and Figure 12.28b illustrates the installed engine and test setup.

## 12.3  Actuators

### Linear Actuators

The task of a linear actuator is to provide motion in a straight line. We discuss three ways to achieve linear motion:

1. Conversion of rotary motion into linear motion. This can be accomplished using a linkage, as in the slider-crank mechanism, or using screw threads coupled to a rotary motion source.

2. Use of a fluid pressure to move a piston in a cylinder. When air or another gas is used as the working fluid, the system is called a pneumatic system. When a fluid such as oil is used as the working fluid, the system is termed hydraulic.

3. Electromagnetic force applied to a moveable component.

### Slider–Crank Mechanism

A common means of generating a reciprocating linear motion, or converting linear motion to rotary motion, is the slider–crank mechanism, as illustrated in Figure 12.29. Such a mechanism is the basis of transforming the reciprocating motion of the piston in an internal combustion engine to the rotary motion of a crankshaft. This or similar linkages could also be applied in pick-and-place operations or in a variety of automation applications.

### Screw-Drive Linear Motion

A common means for translating rotary motion into linear motion is a lead screw. A lead screw has helical threads that are designed for minimum backlash to allow precise positioning. Numerous designs exist for such actuating threads. The basic principle is illustrated in Figure 12.30. The rotary motion of the lead screw is translated into linear motion of the nut, with the torque required to drive the lead screw directly related to the thrust the particular application requires.

**FIGURE 12.27** Waterbrake dynamometers.
(Courtesy of the American Society of Mechanical Engineers, New York. Reprinted from PTC 19.7-1980 6.)

(a)

(b)

Courtesy of SuperFlow Dynamometers & Flowbenches.

**FIGURE 12.28**   Waterbrake engine dynamometer. (a) Dynamometer without test engine installed. (b) Dynamometer with test engine in place.

**FIGURE 12.29**   Slider–crank mechanism.

**FIGURE 12.30**   Linear actuation using a lead screw.

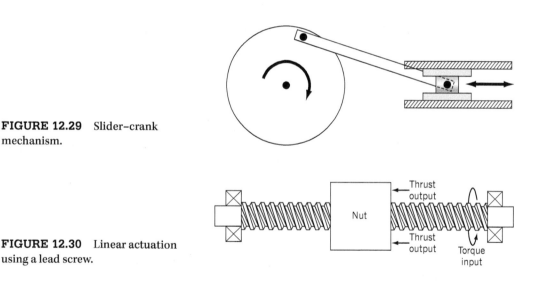

Common applications that employ a lead screw include the worktable for a mill, as well as a variety of other precision positioning translation tables, such as the one shown in Figure 12.31.

## Pneumatic and Hydraulic Actuators

The term "pneumatic" implies a component or system that uses compressed air as the energy source. On the other hand, a hydraulic system or component uses incompressible oil as the working fluid. An example of a hydraulic system is the power steering on an automobile; such a system is illustrated in Figure 12.32. Hydraulic fluid is supplied at an elevated pressure from the power steering pump. When a steering input is made from the driver, the rotary valve allows high-pressure fluid to enter the appropriate side of the piston, and aid in turning the wheels. By maintaining a direct connection between the steering column and the rack and pinion, the car can be steered even if the hydraulic system fails.

UniSlide® from Velmex, Inc. Used with permission.

**FIGURE 12.31** Precision translation table.

**FIGURE 12.32** Schematic diagram of a power steering system.

## Pneumatic Actuators

When compressed air is the energy source of choice, a pneumatic cylinder can create linear motion. In general, the purpose of a pneumatic cylinder is to provide linear motion between two fixed locations. Figure 12.33 shows a pneumatic cylinder and a cutaway of such a cylinder. By applying high-pressure compressed air to either side of the piston, linear actuation between two defined positions can easily be accomplished.

## Solenoids

"Solenoid" is a term used to describe an electromagnetic device that is employed to create linear motion of a plunger, as shown in Figure 12.34. The initial force available from a solenoid can be determined from

$$F = \frac{1}{2}(NI)^2 \frac{\mu A}{\delta^2} \tag{12.21}$$

Courtesy of Parker Hannifin Corp. Used with permission.

**FIGURE 12.33** Construction of a pneumatic cylinder.

**FIGURE 12.34** Construction of a solenoid linear actuator.

where

$F$ = force on plunger

$N$ = number of turns of wire in the electromagnet

$I$ = current

$\mu$ = magnetic permeability of air ($4\pi \times 10^{-7}$ H/m)

$\delta$ = size of the air gap

$A$ = plunger cross-sectional area

When the electromagnet is actuated, the resulting magnetic force pulls the plunger into the C-frame. Because the air gap is largest when the electromagnet is actuated, the minimum force occurs at actuation and the force increases as the air gap decreases.

## Rotary Actuators

### Stepper Motors

There is a class of electric motors that has the primary purpose of providing power to a process. An example would be the electric motor that drives an elevator, an escalator, or a centrifugal blower. In these applications, the electric motor serves as a prime mover, with clear and specific requirements for rotational speed, torque, and power. However, some applications have stringent requirements for positioning.

Rotary positioning presents a significant engineering challenge, but one that is so ubiquitous that it has been addressed through a variety of design strategies. One design strategy is to employ a free-rotating DC motor to supply the motive power and impose precise control on the resulting motion through gearing and some control scheme. DC motors that are subject to feedback control are generally described as servo-motors. Although this may be appropriate and necessary for some applications, the stepper motor has found wide-ranging applications in precision rotary motion control and is a better choice for many applications.

Stepper or stepping motors are capable, as their name implies, of moving a fraction of a rotation with a great degree of precision. This is accomplished by the design of a rotor that aligns with the magnetic field generated by energized coils. The step size can range from 90 degrees to as little as 0.5 degrees or less. Two common types of stepper motors are variable reluctance and unipolar designs. The design of a variable reluctance stepping motor is illustrated in Figure 12.35. Let's consider the operation of this motor. There are three sets of windings, labeled 1, 2, and 3 in the figure, and there are two sets of teeth on the rotor, labeled X and Y. When the windings labeled 1 are energized, the rotor snaps to a position where one set of the teeth are aligned with the windings. This motion is a result of the magnetic field generated by the windings. Suppose that winding 1 is de-energized and winding 2 is energized. The rotor will turn until the teeth marked Y are aligned with winding 2. This produces a 30-degree step.

A useful characteristic of stepper motors is holding torque. As long as one of the windings is energized, the rotor resists motion, until the torque produced by the winding to rotor interaction is overcome.

The motor shown in Figure 12.36 is a variable reluctance design. Unipolar motors incorporate permanent magnets as the rotor. Figure 12.36 shows a rotor having six magnetic poles and two sets of windings. The motor moves in 30-degree increments as the windings are alternately energized.

## Flow-Control Valves

Valves are mechanical devices intended to allow, restrict, throttle, or meter fluid flow through pipes or conduits. Flow-control valves are used to regulate either the flow or the pressure of a

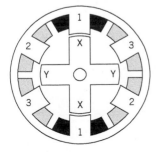

**FIGURE 12.35**   Variable reluctance stepper motor design.

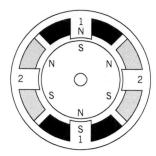

**FIGURE 12.36** Variable reluctance stepper motor design having six poles and two windings.

fluid by their actuation. They generally function by allowing flow while in their open position, stopping flow when closed, and metering flow or fluid pressure to a desired value with a position that is somewhere between these settings, which is called proportional control. These valves contain a valve positioning element that is driven by an actuator, such as a solenoid. Any valve type can be controlled. The common control valve design offers either a single chamber body containing a poppet with valve seat or a multichamber body containing a sliding spool with multiple poppets. Flow-control valves are used to transfer gases, liquids, and hydraulic fluids. The application ratings are as follows: general service, for working with common liquids and gases; cryogenic service, for fluids such as liquid oxygen; vacuum service, for low-pressure applications; and oxygen service, for contamination-free flow of oxygen.

The control valve can respond to signals from any type of process variable transducer. The signal determines the position of the actuating solenoid. A specific characteristic of any control valve refers to whether its nonenergized operating state is open or closed. This is referred to as its "fail position." The fail position of a control valve is determined by the nonenergized solenoid plunger position. This position is an important consideration for process safety.

These valves come in various configurations reflecting their number of ports. A two-way valve has two ports. Two-way position control takes on one of two values: open or closed. A two-way valve has two connections: supply port (P) and service port (A). Most common household valves fall into this category. A three-way valve has three port connections: supply (P), exhaust (T), and service (A). The service port may be switched between the supply and the exhaust. A four-way has four connections: supply (P), exhaust (T), and two service ports (A and B). The valve connects either P to A and B to T, or P to B and A to T. In general, an $N$-way valve has $N$ ports, with $N$ number of flow directions available. An example of a three-port sliding spool control valve is shown in Figure 12.37. The solenoid drives the spool, which contains two valve seats. In the fully activated position, port P is open to service port A. When the solenoid is deactivated, port T is open to service port A (shown). For example, in one application this valve can be used to pressurize a system (open the system to port P) for a period of time and then adjust the system pressure to another value (open the system to port A) for a period of time.

**FIGURE 12.37** Three-way flow-control valve (deactivated position shown).

All valve ports offer some level of flow resistance, and this is specified through a flow coefficient, $C_v$. Flow resistance can be adjusted in design by varying the internal dimensions of the valve chamber and can be set operationally by varying the element position within the chamber. The flow coefficient is found based on the formulation detailed in Chapter 10 or simply as

$$Q = C_v \sqrt{\Delta p} \tag{12.22}$$

where $Q$ is the steady flow rate through the valve and $\Delta p$ is the corresponding pressure drop. This loss is also expressed in terms of a $K$-factor based on the average velocity through the ports,

$$\Delta p = K \rho \overline{U}^2 / 2 \tag{12.23}$$

Flow-control valves are classified in a number of ways: the type of control, the number of ports in the valve housing, the specific function of the valve, and the type of valve element used in the construction of the valve. Directional control valves allow or prevent the flow of fluid through designated ports. Flow can move in either direction. Check valves are a special class of directional valve that allow flow in only one direction. Proportional valves can be positioned to control the amount, pressure, and direction of fluid flow. In a proportional valve, the valve is opened by an amount proportional to the applied current. These valves provide a way to control pressure or flow rate with a fast response rate.

In the simplest application, a solenoid is used to turn a valve either on or off. In a more demanding application, the solenoid is expected to cycle rapidly to open and close the valve. The time between each signal cycle coupled with the internal flow loss character of the valve determines the average flow. Valve response time can be defined in several ways but all are consistent with the methods used in Chapter 3. The 90% response time, $t_{90}$, is the time required to either fill or exhaust a target device chamber through a valve port, in effect a step-function response. There is a separate response time for filling or exhausting. Either way,

$$t_{90} = m + F\forall \tag{12.24}$$

where $m$ is the valve lag time between when the signal is applied and steady flow is established at the designated port, $F$ is the reciprocal of the average flow rate through the port, and $\forall$ is the volume of the target device chamber. For example, a valve having an $F$ of 0.54 ms/cc and a lag time of 20 ms requires $t_{90} = 155$ ms to fill a 250-cc chamber. Alternatively, the valve frequency response can be found by cycling the valve with a sine wave electrical signal and measuring the flow rate through the valve. The valve frequency bandwidth is thus established by its $-3$ dB point.

## 12.4 Controls

Control of a process or system can be exerted in a wide variety of ways. Suppose our goal is to create a healthy lawn by appropriate watering. Each day we could monitor the weather forecast, take into account the probability of precipitation, and choose whether to water and for how long. We could choose to water all of the lawn or just those parts most subject to stress from heat and lack of moisture. If we choose to water, we could place the sprinklers and turn on the faucet (remembering to shut off the flow at an appropriate later time).

All the functions described above for lawn care are completely reasonable for a person to accomplish, and they represent the functioning of an intelligent controller. Suppose we wish to introduce some automation into the process.

At the simplest level, a timer-based control system could be implemented, as shown in Figure 12.38. The functioning of this system would be to open and close the faucet at predetermined times of the day. At the simplest level, this could be a mechanical timer that watered

**FIGURE 12.38**    Open-loop control of a sprinkler system.

the lawn once each 24-hour period for a predetermined length of time. This type of control is called an *open-loop control*. For this control system, there are no sensors to monitor the amount of water applied to the lawn; in fact, all that the control system is accomplishing is to open and, later, close the faucet.

More advanced automatic control systems implement a *closed-loop control*. For the present example, it might be desired to apply 100 gallons (379 liters) of water to the lawn at a particular time of day. A flow meter that sensed the total water flow that had occurred could be used to provide feedback to the control system to allow the faucet to be closed when the flow totaled 100 gallons. The term *closed-loop* or *feedback control* simply means that the variable that is to be controlled is being measured, and that the control system in some way uses this measurement to exert the control.

Figure 12.39 illustrates a control system designed to apply 100 gallons of water to the lawn. There are two inputs to the controller: the time of day and the output of the totalizing flow meter. At the appropriate time of day, the controller opens the valve. The totalizing flow meter output is used by the controller to close the valve after the total flow reaches 100 gallons. This type of binary control scheme is termed *on–off control*. The valve controlling the water flow is either fully open or fully closed.

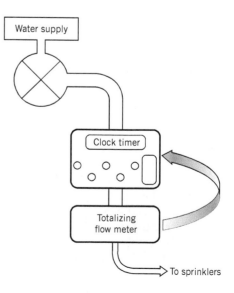

**FIGURE 12.39**    Feedback control of a sprinkler system.

**FIGURE 12.40** Operation of on–off controller with a dead band.

**FIGURE 12.41** Components of a thermostatic control for a home furnace.

Probably, the most familiar form of a binary on–off control system is the thermostat for a home furnace or air conditioner. Figure 12.40 shows the status of a home furnace and a time trace of the inside temperature during a winter day. A schematic representation of this control system is shown in Figure 12.41. A key element here is that there is the possibility of a *disturbance* that would influence the rate of change of the inside temperature. Suppose a delivery arrives and the door remains open for a period of time. The thermostat must then respond to this disturbance and attempt to maintain the inside temperature at the set point.

Essentially all practical implementations of on–off control systems require a dead band that creates a hysteresis loop in the control action. This is illustrated in Figure 12.42. The dead band is centered around zero error, and the action of the controller depends on the magnitude of the error. Here the error is defined as

$$e = T_{\text{setpoint}} - T_{\text{room}}$$

Recall that we are considering a furnace thermostat under winter conditions. As the room temperature falls relative to the set point, the error becomes a larger positive number. When the error reaches the value corresponding to "Furnace ON" in Figure 12.42, the furnace begins to add heat to the conditioned space. Room temperature begins to rise and the error decreases toward zero. The furnace remains on until the room temperature reaches a predetermined value that is greater than the set point. Here the error is negative. At this temperature, the furnace is turned off and room temperature begins once again to decrease. Because of the dead band in the controller, no further control action occurs until the error reaches the "Furnace ON" error magnitude.

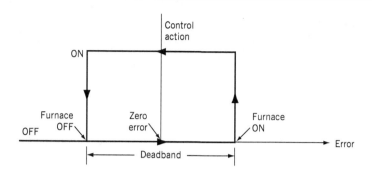

**FIGURE 12.42**   Control action and hysteresis loop of a binary on–off controller with a dead band.

**FIGURE 12.43**   Flow rate control system.

Many control systems are designed to maintain a specified set point without a dead band; clearly this is not possible using on–off control. Consider again watering a lawn. Suppose that as a result of varying water pressure, the lawn was being watered nonuniformly. We might choose to control both the water flow rate and the total water flow applied to the lawn. A scheme for accomplishing this is shown in Figure 12.43. The key components of the control system are a flow meter, the controller, and an actuator that can control the position of a gate valve. Here, a desired flow rate is set to, say, 2 gallons per minute. The task of the controller is to vary the position of the gate valve in order to maintain the flow rate at the set point.

## Dynamic Response

As another example of closed-loop control, consider the cruise-control system on an automobile. After a desired speed is set, the system varies the throttle position to ensure that the set point is maintained. Another issue in the analysis and design of control systems lies in the dynamic response of the physical process, the controller, and the actuators. An automobile does not respond instantaneously to changes in the throttle position, and time is required for a stepper motor to change the position of a valve. The complexities of the combined responses of the physical and control systems, especially in the presence of disturbances, are the subject of the remainder of this section.

## Laplace Transforms

In Chapter 3, we used Laplace transforms to model the response of simple systems. Now, let us apply Laplace transforms to understand how control systems function. The fundamental basis of the application of Laplace transforms to control systems is the solution of a mathematically well-posed initial value problem.

Consider an initial-value problem (here time is the independent variable) that is described by an ordinary differential equation. If we apply the Laplace transform to a differential equation, we convert the differential equation to an algebraic equation. For partial differential equations in time and one spatial variable, the Laplace transform converts the partial differential equation into an ordinary differential equation in the space variable. Appendix B reviews the application of Laplace transforms and provides a table of Laplace transform pairs.

We illustrate the application of the Laplace transform through solution of first-order and second-order differential equations, which are important for control systems.

## EXAMPLE 12.3

Consider the response of a first-order system to a unit step input forcing function, which we have previously modeled by Equation 3.4. The governing equation is

$$\tau \dot{y} + y = U_s(t) \tag{12.25}$$

Applying the Laplace transform (Table B.1 can be used as a reference) yields

$$\tau s Y(s) - y(0) + Y(s) = \frac{1}{s} \tag{12.26}$$

where we have employed the property of the Laplace transform that

$$\mathcal{L}\left[\tau \frac{dy(t)}{dt}\right] = \tau \mathcal{L}\left[\frac{dy(t)}{dt}\right]$$

The process of solving the differential equation involves the following steps:

1. Apply the Laplace transform to the governing differential equation; in this case, Equation 12.25.

2. Solve the resulting equation for $Y(s)$, Equation 12.26.

3. Employ a table of Laplace transform pairs to determine $y(t)$.

We assume that $y(0) = 0$. Solving Equation 12.26 for $Y(s)$ yields

$$Y(s) = \frac{1}{s(\tau s + 1)} \tag{12.27}$$

An important tool for the inversion of Laplace transforms is partial fractions. To accomplish the inversion of Equation 12.27, we start from the assumption

$$\frac{1}{s(\tau s + 1)} = \frac{A}{s} + \frac{B}{(\tau s + 1)} \tag{12.28}$$

Next, we find a common denominator for the right-hand side of Equation 12.28:

$$\frac{1}{s(\tau s + 1)} = \frac{A(\tau s + 1)}{s(\tau s + 1)} + \frac{Bs}{s(\tau s + 1)}$$

In this form, we see that

$$1 = A(\tau s + 1) + Bs$$

Equating powers of $s$ yields

$$1 = A$$
$$0 = A\tau + B \Rightarrow B = -\tau$$

We can now express $Y(s)$ as

$$Y(s) = \frac{1}{s} + \frac{-\tau}{(\tau s + 1)} = \frac{1}{s} + \frac{-1}{(s + 1/\tau)} \tag{12.29}$$

We find that (e.g., from transform pairs 1 and 5 in Table B.1)

$$y(t) = 1 - e^{-t/\tau} \tag{12.30}$$

With the appropriate initial condition and steady-state output value, this can be written as Equation 3.6:

$$\Gamma(t) = \frac{y(t) - y_\infty}{y_0 - y_\infty} = e^{-t/\tau}$$

where $\tau$ is the time constant for a first-order system and $\Gamma(t)$ is the error fraction.

---

### EXAMPLE 12.4

Employing Laplace transforms, determine the solution of the second-order, ordinary linear differential equation,

$$y'' + 4'_y - 5y = 0 \tag{12.31}$$

with the initial conditions

$$y(0) = -1$$
$$y'(0) = 4$$

### SOLUTION
We recall that

$$\mathcal{L}[y^{(n)}] = s^n \mathcal{L}[y] - s^{n-1}y(0) - s^{n-2}y'(0) - \ldots - y^{(n-1)}(0).$$

If we apply this relationship to Equation 12.31, we find that the Laplace transform is

$$s^2 Y(s) - sy(0) - y'(0) + 4[sY(s) - y(0)] - 5Y(s) = 0$$

Substituting the initial conditions yields

$$s^2 Y(s) - s - 4 + 4[sY(s) + 1] - 5Y(s) = 0$$

Solving for $Y(s)$ yields

$$Y(s) = \frac{-s}{s^2 + 4s - 5} \tag{12.32}$$

Now, we are faced with the task of determining the inverse Laplace transform of this expression. Factoring the denominator yields

$$Y(s) = \frac{-s}{(s + 5)(s - 1)} \tag{12.33}$$

Applying partial fractions allows Equation 12.33 to be expressed

$$\frac{-s}{(s + 5)(s - 1)} = \frac{-5/6}{s + 5} + \frac{-1/6}{s - 1} \tag{12.34}$$

Finding the inverse Laplace transform (e.g., from Table B.1) provides the solution for Equation 12.31 as

$$y(t) = -\frac{5}{6}e^{-5t} - \frac{1}{6}e^t \tag{12.35}$$

## Block Diagrams

A very useful representation of feedback control systems is accomplished using block diagrams. We first describe the basic elements of a block diagram.

### Operational Blocks

An operational block performs a defined operation on a signal. Consider a public-address system consisting of a microphone, an amplifier, and speakers. Figure 12.44 provides a single-input, single-output block representing the amplifier. An ideal amplifier would follow exactly the waveform of the input signal from the microphone, and simply multiply the voltage signal by a constant value, $\kappa$. This constant value represents a *gain*, so it is called a gain block.

In Figure 12.45, a gain block is shown that supplies a reference voltage signal based on a temperature set point. In most practical applications, signals are transmitted in control systems as voltage or current. The gain represented in Figure 12.45 would have units of volts/°C. We note that a pure linear gain is equivalent to the static sensitivity of a zero-order system.

An important type of block used to represent a control system is a comparator. Figure 12.46 illustrates the operation of a comparator. Two voltage signals are either added or subtracted by the comparator. Figure 12.46 uses a furnace thermostat as an example. The signals represent the desired temperature set point and the measured temperature, in terms of voltages. The difference in these two values represents the error in the temperature value, the difference between the measured and set-point values.

We can now construct a block diagram of the thermostat for the home furnace described in Figures 12.41 and 12.42. Figure 12.47 shows the block diagram of the controller, the furnace, and the house. Together, the furnace and the house are usually referred to as the plant or the process.

The detailed design of a control system requires that we consider the time-dependent behavior of both the controller and the process. We propose a process and derive the governing equation for the process. Then, a controller is implemented and the dynamic response of the system to a step change in the set point is derived. A key point in this model is the use of a stationary operating set point.

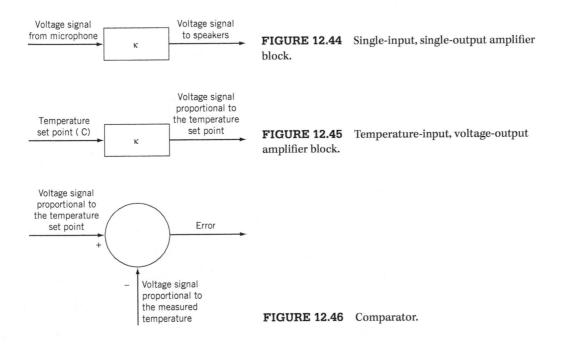

**FIGURE 12.44**   Single-input, single-output amplifier block.

**FIGURE 12.45**   Temperature-input, voltage-output amplifier block.

**FIGURE 12.46**   Comparator.

**FIGURE 12.47**   Block diagram representation of a furnace thermostat.

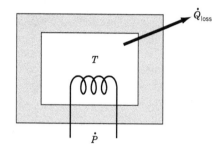

**FIGURE 12.48**   Energy flows for an oven.

## Model for Oven Control

### Plant Model

The system we wish to control is illustrated in Figure 12.48. An oven is maintained at a temperature above the ambient temperature by power input $P$ from an electric heater. The controller is implemented to maintain a desired oven temperature.

A first law analysis of the oven at steady-state conditions yields

$$\dot{P} = \dot{Q}_{loss}$$

The energy loss from the oven $\dot{Q}_{loss}$ may be expressed in terms of the oven temperature, $T$, the ambient temperature, $T_\infty$, the surface area, $A_s$, and an overall loss coefficient, $U$, as

$$\dot{Q}_{loss} = UA_s(T - T_\infty) \tag{12.36}$$

Consider a steady-state operating point for the oven, designated by the subscript $o$. At steady state, the power supplied by the heater is exactly balanced by the energy lost to the ambient,

$$\dot{P}_o = UA_s(T_o - T_\infty) \tag{12.37}$$

To aid in the mathematical analysis, we define a new temperature variable as

$$\theta = T - T_\infty \tag{12.38}$$

so that Equation 12.37 can be expressed as

$$\dot{P}_o = UA_s\theta_o \tag{12.39}$$

The first law for the oven for a transient condition, one where the power input is changed, is

$$\dot{P} - UA_s\theta = mc\frac{d\theta}{dt} \tag{12.40}$$

where $m$ is the mass of the oven and $c$ is the average specific heat. This product represents the total heat capacity of the oven. Equation 12.40 now provides the governing equation for the plant in this example.

## Controller Model

We now wish to implement a controller. The first controller we implement is termed a proportional controller and is illustrated in Figure 12.49. The error between the set-point temperature and the actual temperature of the oven is multiplied by a constant value to determine the power input to the heater. A block diagram of the controller and plant combined system is shown in Figure 12.50; two proportional gains are shown in this figure to emphasize that the error signal will most likely be a voltage.

The governing equation for the temperature of the oven may be expressed

$$\frac{mc}{[UA_s + \kappa_p\kappa_m]}\frac{d\theta}{dt} + \theta = \frac{\kappa_p\kappa_m}{[UA_s + \kappa_p\kappa_m]}\theta_{set} \tag{12.41}$$

By comparison with Equation 3.4,

$$\tau\dot{y} + y = KF(t)$$

it is clear that this represents the response of a first-order system having a time constant of

$$\tau = \frac{mc}{[UA_s + \kappa_p\kappa_m]} \tag{12.42}$$

The analysis for this step change is best accomplished by using Laplace transforms.

**FIGURE 12.49** Proportional control.

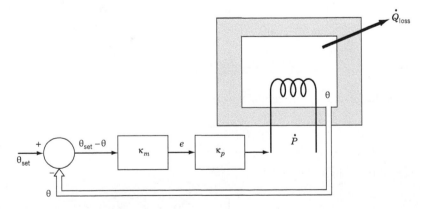

**FIGURE 12.50** Proportional control of an oven.

## Laplace Transform Analysis

Taking the Laplace transform of Equation 12.41 yields

$$\frac{mc}{[UA_s + \kappa_p \kappa_m]} s\Theta(s) + \Theta(s) = \frac{\kappa_p \kappa_m \Theta_{set}(s)}{[UA_s + \kappa_p \kappa_m]} \tag{12.43}$$

which is of the form

$$(\tau s + 1)\Theta(s) = KF(s)$$

Solving Equation 12.43 for the ratio $\Theta(s)/\Theta_{set}(s)$ yields

$$\frac{\Theta(s)}{\Theta_{set}(s)} = \frac{\dfrac{\kappa_p \kappa_m}{[UA_s + \kappa_p \kappa_m]}}{\left[\dfrac{mc}{UA_s + \kappa_p \kappa_m} s + 1\right]} = KG(s) \tag{12.44}$$

For convenience, define

$$C_1 = \frac{mc}{UA_s + \kappa_p \kappa_m}$$

$$C_2 = \frac{\kappa_p \kappa_m}{UA_s + \kappa_p \kappa_m} \tag{12.45}$$

With these definitions, Equation 12.44 becomes

$$\frac{\Theta(s)}{\Theta_{set}(s)} = \frac{C_2}{[C_1 s + 1]} = G(s) \tag{12.46}$$

Equation 12.46 represents the transfer function, $G(s)$, for the system consisting of the oven and the proportional controller. For the linear system consisting of the oven and controller, the transfer function represents the ratio of the Laplace transforms of the output to the input assuming a zero initial condition.

## Step Response

Suppose we consider the startup of the furnace, with an initial temperature equal to the ambient, or $\theta = 0$. From this condition, we impose a value of $\theta_{set}$ that is larger than the ambient temperature, say $\theta_1$; this represents a step-change input that is expressed in Laplace transform space as

$$\Theta_{set}(s) = \frac{\theta_1}{s} \tag{12.47}$$

The transform of the governing differential equation becomes

$$\Theta(s) = \frac{\theta_1 C_2}{s[C_1 s + 1]} \tag{12.48}$$

Employing partial fractions yields

$$\Theta(s) = \theta_1 C_2 \left[\frac{1}{s} - \frac{1}{s + 1/C_1}\right] \tag{12.49}$$

From Table B.1, we can determine the time-domain solution corresponding to this Laplace transform as

$$\theta(t) = \theta_1 C_2 \left[1 - e^{-t/C_1}\right] \tag{12.50}$$

Recall that we have assumed that $\theta(0) = 0$. In the limit as $t \to \infty$

$$\theta(t) = \theta_1 C_2 = \frac{\kappa_p \kappa_m \theta_1}{UA_s + \kappa_p \kappa_m}$$

For a step-change input, we have found that $\theta(t) \neq \theta_1$ unless $\frac{\kappa_p \kappa_m}{UA_s + \kappa_p \kappa_m} = 1$. This will not in general be the case. Thus, we find that a proportional controller, in general, is characterized by a nonzero steady-state error.

## EXAMPLE 12.5

Consider the proportional control of an oven that has a total mass of 20 kg, with an average specific heat of 800 J/kg-K. The oven is initially at room temperature. A step input is supplied to the controller, changing the set-point temperature to 100°C above the ambient. Plot the temperature of the oven and the power supplied to the oven as a function of time, and determine the steady-state error if the values of the gains in the system are $\kappa_m = 20$ and $\kappa_p = 12$.

### SOLUTION
Equation 12.50 provides the solution for oven temperature as a function of time, for the case in which there is a step change in the set-point temperature. By examining Figure 12.50, we see that our controller provides a power to the oven given by

$$\kappa_p \kappa_m (\theta_1 - \theta)$$

The temperature response and the supplied power are provided in Figure 12.51.

**COMMENT** The values of the gains in this problem are directly tied to the power required for the oven. The physical arrangement of heaters and the required electrical service are a strong consideration in a practical implementation of this conceptual design. The effect of the controller gains and the system parameters can be further explored using the program file *First Order PID.vi*.

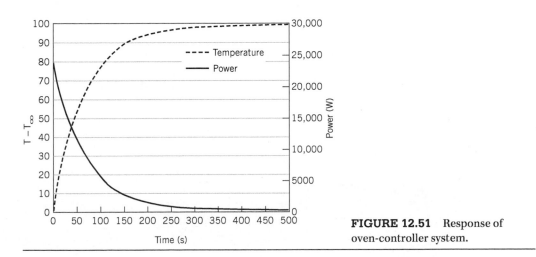

**FIGURE 12.51**   Response of oven-controller system.

**FIGURE 12.52**  Block diagram for an integral controller.

## Proportional–Integral (PI) Control

The steady-state error that exists for a proportional controller is not acceptable in many applications. Adding a control signal that is proportional to the time integral of the error improves the performance of the proportional controller. Let's first examine the behavior of a pure integral controller.

### Integral Control

Figure 12.52 provides a block diagram in the Laplace transform domain for an integral controller applied to control the temperature of our oven. In the time domain, this results in

$$\dot{P}(t) = \kappa_I \int_0^t e(t)dt + \dot{P}(0) \tag{12.51}$$

We immediately see that as long as the error remains finite and positive, the power will continue to increase. Clearly, a pure integral controller would have very limited application.

### Proportional–Integral (PI) Control

Suppose that the actions of the proportional and integral controllers are combined, as shown in Figure 12.53. In Laplace transform space, this may be expressed as

$$P(s) = \kappa_p \kappa_m E(s) + \frac{\kappa_I \kappa_m}{s} E(s) = \left[ \kappa_p \kappa_m + \frac{\kappa_I \kappa_m}{s} \right] E(s) \tag{12.52}$$

Expressed in terms of the set point, the closed-loop transfer function is

$$\frac{\Theta(s)}{\Theta_{set}(s)} = \frac{\kappa_m \left( \kappa_p + \dfrac{\kappa_I}{s} \right) \left( \dfrac{C_2}{C_1 s + 1} \right)}{1 + \kappa_m \left( \kappa_p + \dfrac{\kappa_I}{s} \right) \left( \dfrac{C_2}{C_1 s + 1} \right)} \tag{12.53}$$

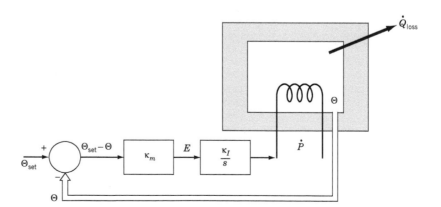

**FIGURE 12.53**  Block diagram for PI control of an oven.

Clearing fractions yields a form of the transfer function that allows the inverse transform to be determined:

$$\frac{\Theta(s)}{\Theta_{set}(s)} = \frac{\kappa_m(\kappa_p s + \kappa_I)C_2}{C_1 s^2 + s(1 + C_2\kappa_m\kappa_p) + C_2\kappa_m\kappa_I} \tag{12.54}$$

Once again, we impose a step change in the set-point temperature so that $\Theta_{set}(s) = \dfrac{\theta_1}{s}$, and

$$\Theta(s) = \frac{\theta_1[\kappa_m(\kappa_p s + \kappa_I)C_2]}{s[C_1 s^2 + s(1 + C_2\kappa_m\kappa_p) + C_2\kappa_m\kappa_I]} \tag{12.55}$$

Employing the final value theorem, we multiply by $s$ and take the limit as $s \to 0$:

$$\lim_{s \to 0} \Theta(s) = \lim_{s \to 0} \frac{s[\kappa_m(\kappa_p s + \kappa_I)C_2]\theta_1}{s[C_1 s^2 + s(1 + C_2\kappa_m\kappa_p) + C_2\kappa_m\kappa_I]} = \frac{C_2\kappa_m\kappa_I}{C_2\kappa_m\kappa_I}\theta_1 = \theta_1 \tag{12.56}$$

Clearly, the steady-state error for this PI controller is zero. This is a general result for PI control.

## Time Response

The time response of this system to a step change input could be determined by finding the inverse Laplace transform of Equation 12.57:

$$\frac{\Theta(s)}{\Theta_{set}(s)} = \frac{\kappa_m(\kappa_p s + \kappa_I)C_2}{C_1 s^2 + s(1 + C_2\kappa_m\kappa_p) + C_2\kappa_m\kappa_I} \tag{12.57}$$

However, by comparison with the Laplace transform of the governing differential equation for a second-order system, as provided in Equation 3.13, Palm [7] shows that

$$\varsigma = \frac{1 + C_2\kappa_m\kappa_p}{2\sqrt{C_1 C_2\kappa_m\kappa_I}} \tag{12.58}$$

and that for $\varsigma < 1$ the equivalent time constant is

$$\tau = \frac{2C_1}{1 + C_2\kappa_m\kappa_P} \tag{12.59}$$

The behavior of the oven-controller system can be further explored using the program file *First Order PID.vi*.

# Proportional–Integral–Derivative Control of a Second-Order System

Consider a spring–mass–damper system such as the one described in Figure 3.3. The governing equation describing the position of the mass as a function of time for a given forcing function $f(t)$ is

$$\frac{1}{\omega_n^2}\frac{d^2y}{dt^2} + \frac{2\varsigma}{\omega_n}\frac{dy}{dt} + y = f(t) \tag{12.60}$$

This system and the governing equation serve as a model to demonstrate the properties of PID control when applied to a second-order system.

The goal is to apply a step change in input to the system and have the mass move to the new equilibrium position; in other words, the ideal response of the system would be a step change.

Our first objective is to review the response of the second-order system without a controller in place. The behavior of the system is examined in two examples.

### EXAMPLE 12.6

A spring–mass–damper system has mass of 2 kg, a spring constant of 7,900 N/m, and a damping coefficient of 176 kg/s. Plot the open-loop time response of this system to a step-change input in force.

**ASSUMPTIONS** The initial conditions for velocity, $\dfrac{dy}{dt}$, and position, $y$, are zero.

**SOLUTION**

Although the solution to this example could be determined from the results in Chapter 3, we will determine the response using Laplace transforms. Taking the Laplace transform of Equation 12.60 yields

$$\frac{1}{\omega_n^2}s^2 Y(s) + \frac{2\zeta}{\omega_n}sY(s) + Y(s) = KF(s) \tag{12.61}$$

For the mass, spring constant, and damping ratio given in the problem statement, we compute the natural frequency and damping ratio as

$$\omega_n = \sqrt{\frac{k}{m}} = \sqrt{\frac{7900}{2}} = 62.85\,\text{rad/s} \quad \text{and} \quad \zeta = \frac{c}{2\sqrt{km}} = \frac{176}{2\sqrt{7900 \times 2}} = 0.7$$

The open-loop transfer function defines the dynamic response of the system and is defined as

$$\frac{Y(s)}{KF(s)} = \left[\frac{1}{\dfrac{1}{\omega_n^2}s^2 + \dfrac{2\zeta}{\omega_n}s + 1}\right] = \left[\frac{1}{\dfrac{1}{3950}s^2 + 0.02235s + 1}\right] \tag{12.62}$$

where $K$ is 1 for the present case. Options exist for calculating and plotting the open-loop response of this system. The program file *Second order PID.vi* can be used. Figure 12.54 shows the response of this system to a step input; the output has been normalized so that the displacement, as characterized by the amplitude, is from 0 to 1.

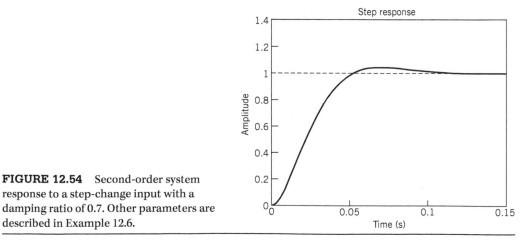

**FIGURE 12.54** Second-order system response to a step-change input with a damping ratio of 0.7. Other parameters are described in Example 12.6.

## EXAMPLE 12.7

Consider the second-order system described in Example 12.6. Plot the open-loop response of the system to a step change in input for damping ratios of 0.3 and 1.3.

### SOLUTION

Figures 12.55 and 12.56 show the response of the system to a step-change input for values of the damping coefficient of 0.3 and 1.3, respectively. By comparing Figures 12.54 through 12.56, we can learn some important characteristics of second-order systems. Suppose our goal is to have the system respond quickly to the input, but without oscillation and without exceeding the equilibrium value excessively. Let's characterize the response of the system having a damping ratio of 1.3 as being comparatively slow. This system requires 0.2 s to approach the equilibrium value of 1. The system having a damping coefficient of 0.3 first reaches 1 in a time less than 0.05 s, but there is both overshoot and oscillations. The system having a damping ratio of 0.7 has no oscillations, and minimal overshoot. The system with $\zeta = 0.3$ has a shorter rise time but longer settling time than that with $\zeta = 0.7$.

In the previous example, the response for a damping ratio of 0.7 appears to be the most likely candidate to meet our needs. But can we improve the situation by implementing a controller?

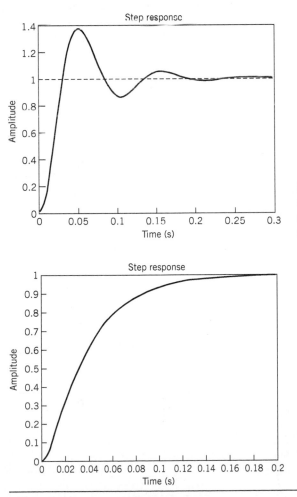

**FIGURE 12.55**  Second-order system response to a step-change input with a damping ratio of 0.3. Other parameters are described in Example 12.7.

**FIGURE 12.56**  Second-order system response to a step-change input with a damping ratio of 1.3. Other parameters are described in Example 12.7.

**FIGURE 12.57** Block diagram for proportional control of a second-order plant.

## Proportional Control

Consider the application of a proportional controller to the second-order system described in Example 12.7 for the case in which the damping coefficient is 0.3. The block diagram for the control system is shown in Figure 12.57.

The transfer function for this system when the damping ratio is 0.3 is

$$\frac{Y(s)}{KF(s)} = \left[ \frac{\kappa_p}{\frac{1}{\omega_n^2}s^2 + \frac{2\gamma}{\omega_n}s + 1} \right] = \left[ \frac{\kappa_p}{\frac{1}{3950}s^2 + 0.02235s + 1} \right] \qquad (12.63)$$

We implement a proportional controller having a proportional gain of 10, and examine the step response of the system. Figure 12.58 shows the resulting response of the system. Clearly, we have not eliminated the overshoot and oscillations in this system. On the other hand, if we implement the same controller for the system having a damping ratio of 1.3, the response is shown in Figure 12.59. Here, the proportional controller is helpful for the performance of the system.

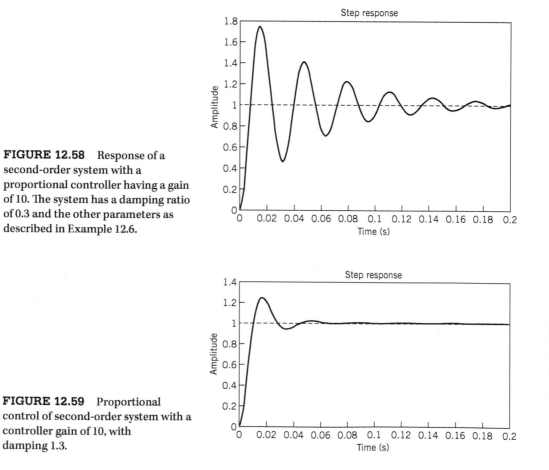

**FIGURE 12.58** Response of a second-order system with a proportional controller having a gain of 10. The system has a damping ratio of 0.3 and the other parameters as described in Example 12.6.

**FIGURE 12.59** Proportional control of second-order system with a controller gain of 10, with damping 1.3.

Let's explore whether a more sophisticated control scheme improves the response of the system having a damping ratio of 0.3.

## Proportional–Integral–Derivative Control

A block diagram representing a PID control for our spring–mass–damper system is shown in Figure 12.60. The transfer function for this system is

$$\frac{Y(s)}{KF(s)} = \left[ \frac{\kappa_D s^2 + \kappa_p s + \kappa_I}{\frac{1}{\omega_n^2} s^3 + \left( \frac{2\zeta}{\omega_n} + \kappa_D \right) s^2 + \kappa_p s + \kappa_I} \right] \tag{12.64}$$

Recall that taking a derivative in time space is multiplying by $s$ in Laplace transform space, and integrating in time space is dividing by $s$ in Laplace transform space. The choice of gains for the control actions, $\kappa_p$, $\kappa_I$, and $\kappa_D$, is a challenging task and one for which much theory has been established [7]. However, for our purposes, we only wish to demonstrate the possible improvement in the system performance.

We select gains for the controllers of unity and just test the performance. For our second-order system having a damping ratio of 0.3, the result is shown in Figure 12.61. We compare this figure with Figures 12.54 and 12.56 and clearly see that this control scheme has improved the performance of the system. The system behavior with a variety of parameters and control gains can be explored using the program file *Second order PID.vi*. Although there are interactions among the proportional, integral, and derivative actions in a PID controller, the basic trends with increasing gain are provided in Table 12.4.

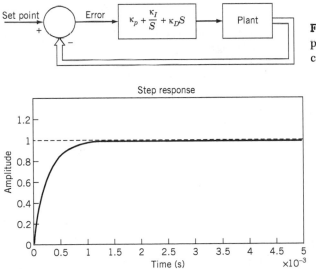

**FIGURE 12.60** Block diagram for proportional–integral–derivative (PID) control of a second-order system.

**FIGURE 12.61** Response of a second-order system having a damping ratio of 0.3 when subject to PID control.

Table 12.4  **Effect of Increasing Gain on Rise Time, Overshoot, and Steady-State Error for a PID Controller**

| Control action | Rise time | Overshoot | Steady-state error |
| --- | --- | --- | --- |
| Proportional | ⇓ | ⇑ | ⇓ |
| Integral | ⇓ | ⇑ | Eliminate |
| Derivative | | ⇑ | |

## SUMMARY

The integration of mechanical and electrical systems with advanced electronics is increasingly required for the design of systems as complex as an aircraft or as simple as a coffee maker that grinds beans and makes coffee at a set time each morning. Mechatronic systems require sensors, actuators, and control schemes to function as an integrated system. This chapter presented the operating principles and designs of those sensors used most often for linear and rotary displacement in mechatronic systems. "Actuators" is a term that represents a broad range of mechanical, electrical, pneumatic, and hydraulic devices designed to create motion.

Selection of an actuator should be done through a system integration approach, so that the sensors and control scheme are appropriate for the actuator chosen.

We have provided an introduction to block-diagram representation of dynamic systems based on Laplace transforms, and have provided a review of the application of Laplace transforms to the solution of ordinary, linear differential equations. A discussion of open- and closed-loop control methods provided examples of proportional, proportional–integral, and proportional–integral–derivative control.

## REFERENCES

1. Coltman, J. W., The Transformer, *Scientific American*, 86, January 1988.

2. Herceg, E. E., *Schaevitz Handbook of Measurement and Control*, Schaevitz Engineering, Pennsauken, NJ, 1976.

3. Bredin, H., Measuring Shock and Vibration, *Mechanical Engineering*, 30, February 1983.

4. Measurement of Rotary Speed, ASME Performance Test Codes, ANSI/ASME PTC 19.13-1961, American Society of Mechanical Engineers, New York, 1961.

5. Gindy, S. S., Force and Torque Measurement, A Technology Overview, *Experimental Techniques*, 9: 28, 1985.

6. *Measurement of Shaft Power*, ASME Performance Test Codes, ASME PTC 19.7-1980, (Reaffirmed Date: 1988), American Society of Mechanical Engineers, New York, 1980.

7. Nise, N., *Control Systems Engineering*, 6th ed., John Wiley and Sons, New York, 2011.

## NOMENCLATURE

| | | | | |
|---|---|---|---|---|
| $a$ | acceleration ($l\,t^{-2}$) | | $D_c$ | coil diameter for magnetic pickup ($l$) |
| $c$ | damping coefficient ($m\,t^{-1}$) | | $E_i$ | input voltage ($V$) |
| $c_c$ | critical damping coefficient ($m\,t^{-1}$) | | $E_o$ | output voltage ($V$) |
| $g$ | acceleration of gravity ($l\,t^{-2}$) | | $F$ | force, vector ($m\,l\,t^{-2}$) |
| $k$ | spring constant ($m\,t^{-2}$) | | $J$ | polar moment of inertia ($l^4$) |
| $m$ | mass ($m$) | | $L$ | length ($l$) |
| $p$ | pressure ($m\,l^{-1}t^{-2}$) | | $M$ | magnitude ratio |
| $\mathbf{r}$ | radius, vector ($l$) | | $M$ | moment, vector ($m\,l^2t^{-2}$) |
| $t$ | time ($t$) | | $P$ | power ($m\,l^2t$) |
| $\delta t$ | time interval for data sampling ($t$) | | $Q$ | volumetric flow rate ($l^3t^{-1}$) |
| $u$ | uncertainty | | $R_o$ | outer radius ($l$) |
| $v$ | linear velocity ($l\,t^{-1}$) | | $T$ | temperature (°) |
| $y$ | displacement ($l$) | | $T$ | torque ($m\,l^2t^{-2}$) |
| $y_{hs}$ | displacement of housing, seismic instrument ($l$) | | $\forall$ | volume ($l^3$) |
| $y_m$ | displacement of seismic mass ($l$) | | $\zeta$ | damping ratio |
| $y_r$ | relative displacement between seismic mass and housing in a seismic instrument ($l$) | | $\tau$ | time constant ($t$) |
| | | | $\tau_{max}$ | maximum shearing stress ($m\,t^2l^{-1}$) |
| $\delta y$ | deflection ($l$) | | $\varphi$ | phase angle |
| $A$ | amplitude | | $\kappa$ | controller gain |
| $B$ | magnetic field strength ($m\,C^{-1}t^{-1}$) | | $\omega$ | rotational speed ($t^{-1}$) |
| $C_v$ | discharge coefficient | | $\omega_n$ | natural frequency ($t^{-1}$) |

# PROPERTY DATA AND CONVERSION FACTORS    A

## Table A.1    Properties of Pure Metals and Selected Alloys

| | Density (kg/m³) | Modulus of Elasticity (GPa) | Coefficient of Thermal Expansion ($10^{-6}$ m/m-K) | Thermal Conductivity (W/m-K) | Electrical Resistivity ($10^{-6}$ Ω-cm) |
|---|---|---|---|---|---|
| *Pure Metals* | | | | | |
| Aluminum | 2 698.9 | 62 | 23.6 | 247 | 2.655 |
| Beryllium | 1 848 | 275 | 11.6 | 190 | 4.0 |
| Chromium | 7 190 | 248 | 6.2 | 67 | 13.0 |
| Copper | 8 930 | 125 | 16.5 | 398 | 1.673 |
| Gold | 19 302 | 78 | 14.2 | 317.9 | 2.01 |
| Iron | 7 870 | 208.2 | 15.0 | 80 | 9.7 |
| Lead | 11 350 | 12.4 | 26.5 | 33.6 | 20.6 |
| Magnesium | 1 738 | 40 | 25.2 | 418 | 4.45 |
| Molybdenum | 10 220 | 312 | 5.0 | 142 | 8.0 |
| Nickel | 8 902 | 207 | 13.3 | 82.9 | 6.84 |
| Palladium | 12 020 | — | 11.76 | 70 | 10.8 |
| Platinum | 21 450 | 130.2 | 9.1 | 71.1 | 10.6 |
| Rhodium | 12 410 | 293 | 8.3 | 150.0 | 4.51 |
| Silicon | 2 330 | 112.7 | 5.0 | 83.68 | $1 \times 10^5$ |
| Silver | 10 490 | 71 | 19.0 | 428 | 1.47 |
| Tin | 5 765 | 41.6 | 20.0 | 60 | 11.0 |
| Titanium | 4 507 | 99.2 | 8.41 | 11.4 | 42.0 |
| Zinc | 7 133 | 74.4 | 15.0 | 113 | 5.9 |
| *Alloys* | | | | | |
| Aluminum (2024, T6) | 2 770 | 72.4 | 22.9 | 151 | 4.5 |
| Brass (C36000) | 8 500 | 97 | 20.5 | 115 | 6.6 |
| Brass (C86500) | 8 300 | 105 | 21.6 | 87 | 8.3 |
| Bronze (C90700) | 8 770 | 105 | 18 | 71 | 1.5 |
| Constantan annealed (55% Cu 45% Ni) | 8 920 | — | — | 19 | 44.1 |
| Steel (AISI 1010) | 7 832 | 200 | 12.6 | 60.2 | 20 |
| Stainless Steel (Type 316) | 8 238 | 190 | — | 14.7 | — |

*Source:* Compiled from *Metals Handbook*, 9th ed., American Society for Metals, Metals Park, OH, 1978, and from other sources.

# Table A.2  Thermophysical Properties of Selected Metallic Solids

| Composition | Melting point (K) | Properties at 300 K | | | | Properties at various temperatures (K) | | | | | | | |
|---|---|---|---|---|---|---|---|---|---|---|---|---|---|
| | | $\rho$ (kg/m³) | $c_p$ (J/kg·K) | $k$ (W/m·K) | $\alpha \times 10^4$ (m²/s) | $k$(W/m·K) | | | | $c_p$ (J/kg·K) | | | |
| | | | | | | 100 | 200 | 400 | 600 | 100 | 200 | 400 | 600 |
| **Aluminum** | | | | | | | | | | | | | |
| Pure | 933 | 2702 | 903 | 237 | 97.1 | 302 | 237 | 240 | 231 | 482 | 796 | 949 | 1033 |
| Alloy 2024-T6 (4.5% Cu, 1.5% Mg, 0.6% Mn) | 775 | 2770 | 875 | 177 | 73.0 | 65 | 163 | 186 | 186 | 473 | 787 | 925 | 1042 |
| Alloy 195, cast (4.5% Cu) | — | 2790 | 883 | 168 | 68.2 | — | — | 174 | 185 | — | — | — | — |
| Chromium | 2118 | 7160 | 449 | 93.7 | 29.1 | 159 | 111 | 90.9 | 80.7 | 192 | 384 | 484 | 542 |
| **Copper** | | | | | | | | | | | | | |
| Pure | 1358 | 8933 | 385 | 401 | 117 | 482 | 413 | 393 | 379 | 252 | 356 | 397 | 417 |
| Commercial bronze (90% Cu, 10% Al) | 1293 | 8800 | 420 | 52 | 14 | — | 42 | 52 | 59 | — | 785 | 460 | 545 |
| Phosphor gear bronze (89% Cu, 11% Sn) | 1104 | 8780 | 355 | 54 | 17 | — | 41 | 65 | 74 | — | — | — | — |
| Cartridge brass (70% Cu, 30% Zn) | 1188 | 8530 | 380 | 110 | 33.9 | 75 | 95 | 137 | 149 | — | 360 | 395 | 425 |
| Constantan (55% Cu, 45% Ni) | 1493 | 8920 | 384 | 23 | 6.71 | 17 | 19 | — | — | 237 | 362 | — | — |
| **Iron** | | | | | | | | | | | | | |
| Pure | 1810 | 7870 | 447 | 80.2 | 23.1 | 134 | 94.0 | 69.5 | 54.7 | 216 | 384 | 490 | 574 |
| Armco (99.75%) | — | 7870 | 447 | 72.7 | 20.7 | 95.6 | 80.6 | 65.7 | 53.1 | 215 | 384 | 490 | 574 |
| **Carbon steels** | | | | | | | | | | | | | |
| Plain carbon (Mn ≤ 1%, Si ≤ 0.1%) | — | 7854 | 434 | 60.5 | 17.7 | — | — | 56.7 | 48.0 | — | — | 487 | 559 |
| AISI 1010 | — | 7832 | 434 | 63.9 | 18.8 | — | — | 58.7 | 48.8 | — | — | 487 | 559 |
| Carbon-silicon (Mn ≤ 1%, 0.1% < Si ≤ 0.6%) | — | 7817 | 446 | 51.9 | 14.9 | — | — | 49.8 | 44.0 | — | — | 501 | 582 |

| Composition | Melting Point (K) | ρ (kg/m³) | cp (J/kg·K) | k (W/m·K) | α·10⁶ (m²/s) | k (100 K) | k (200 K) | k (400 K) | k (600 K) | cp (100 K) | cp (200 K) | cp (400 K) | cp (600 K) |
|---|---|---|---|---|---|---|---|---|---|---|---|---|---|
| Carbon–manganese– silicon (1% < Mn ≤ 1.65%, 0.1% < Si ≤ 0.6%) | — | 8131 | 434 | 41.0 | 11.6 | — | — | 42.2 | 39.7 | — | — | 487 | 559 |
| Chromium (low) steels 1/2 Cr-1/4 Mo-Si (0.18% C, 0.65% Cr, 0.23% Mo, 0.6% Si) | — | 7822 | 444 | 37.7 | 10.9 | — | — | 38.2 | 36.7 | — | — | 492 | 575 |
| 1 Cr-1/2 Mo (0.16% C, 1% Cr, 0.54% Mo, 0.39% Si) | — | 7858 | 442 | 42.3 | 12.2 | — | — | 42.0 | 39.1 | — | — | 492 | 575 |
| 1 Cr–V (0.2% C, 1.02% Cr, 0.15% V) | — | 7836 | 443 | 48.9 | 14.1 | — | — | 46.8 | 42.1 | — | — | 492 | 575 |
| Stainless steels | | | | | | | | | | | | | |
| AISI 302 | 1670 | 8055 | 480 | 15.1 | 3.91 | — | — | 17.3 | 20.0 | — | — | 512 | 559 |
| AISI 304 | — | 7900 | 477 | 14.9 | 3.95 | 9.2 | 12.6 | 16.6 | 19.8 | 272 | 402 | 515 | 559 |
| AISI 316 | — | 8238 | 468 | 13.4 | 3.48 | — | — | 15.2 | 18.3 | — | — | 504 | 550 |
| AISI 347 | — | 7979 | 480 | 14.2 | 3.71 | — | — | 15.8 | 18.9 | — | — | 513 | 559 |
| Lead | 601 | 11 340 | 129 | 35.3 | 24.1 | 39.7 | 36.7 | 34.0 | 31.4 | 118 | 125 | 132 | 142 |
| Magnesium | 923 | 1740 | 1024 | 156 | 87.6 | 169 | 159 | 153 | 149 | 649 | 934 | 1074 | 1170 |
| Molybdenum | 2894 | 10 240 | 251 | 138 | 53.7 | 179 | 143 | 134 | 126 | 141 | 224 | 261 | 275 |
| Nickel | | | | | | | | | | | | | |
| Pure | 1728 | 8900 | 444 | 90.7 | 23.0 | 164 | 107 | 80.2 | 65.6 | 232 | 383 | 485 | 592 |
| Nichrome (80% Ni; 20% Cr) | 1672 | 8400 | 420 | 12 | 3.4 | — | — | 14 | 16 | — | — | 480 | 525 |
| Inconel X–750 (73% Ni; 15% Cr, 6.7% Fe) | 1665 | 8510 | 439 | 11.7 | 3.1 | 8.7 | 10.3 | 13.5 | 17.0 | — | 372 | 473 | 510 |

## Table A.3    Thermophysical Properties of Saturated Water (Liquid)

| T (K) | $\rho$ (kg/m³) | $c_p$ (kJ/kg · K) | $\mu \times 10^6$ (N · s/m²) | $k$ (W/m · K) | Pr | $\beta \times 10^6$ (K⁻¹) |
|---|---|---|---|---|---|---|
| 273.15 | 1000 | 4.217 | 1750 | 0.569 | 12.97 | −68.05 |
| 275.0 | 1000 | 4.211 | 1652 | 0.574 | 12.12 | −32.74 |
| 280 | 1000 | 4.198 | 1422 | 0.582 | 10.26 | 46.04 |
| 285 | 1000 | 4.189 | 1225 | 0.590 | 8.70 | 114.1 |
| 290 | 999 | 4.184 | 1080 | 0.598 | 7.56 | 174.0 |
| 295 | 998 | 4.181 | 959 | 0.606 | 6.62 | 227.5 |
| 300 | 997 | 4.179 | 855 | 0.613 | 5.83 | 276.1 |
| 305 | 995 | 4.178 | 769 | 0.620 | 5.18 | 320.6 |
| 310 | 993 | 4.178 | 695 | 0.628 | 4.62 | 361.9 |
| 315 | 991 | 4.179 | 631 | 0.634 | 4.16 | 400.4 |
| 320 | 989 | 4.180 | 577 | 0.640 | 3.77 | 436.7 |
| 325 | 987 | 4.182 | 528 | 0.645 | 3.42 | 471.2 |
| 330 | 984 | 4.184 | 489 | 0.650 | 3.15 | 504.0 |
| 335 | 982 | 4.186 | 453 | 0.656 | 2.89 | 535.5 |
| 340 | 979 | 4.188 | 420 | 0.660 | 2.66 | 566.0 |
| 345 | 977 | 4.191 | 389 | 0.664 | 2.46 | 595.4 |
| 350 | 974 | 4.195 | 365 | 0.668 | 2.29 | 624.2 |
| 355 | 971 | 4.199 | 343 | 0.671 | 2.15 | 652.3 |
| 360 | 967 | 4.203 | 324 | 0.674 | 2.02 | 679.9 |
| 365 | 963 | 4.209 | 306 | 0.677 | 1.90 | 707.1 |
| 370 | 961 | 4.214 | 289 | 0.679 | 1.79 | 728.7 |
| 373.15 | 958 | 4.217 | 279 | 0.680 | 1.73 | 750.1 |
| 400 | 937 | 4.256 | 217 | 0.688 | 1.34 | 896 |
| 450 | 890 | 4.40 | 152 | 0.678 | 0.99 | |
| 500 | 831 | 4.66 | 118 | 0.642 | 0.86 | |
| 550 | 756 | 5.24 | 97 | 0.580 | 0.88 | |
| 600 | 649 | 7.00 | 81 | 0.497 | 1.14 | |
| 647.3 | 315 | 0 | 45 | 0.238 | ∞ | |

Formulas for interpolation (T = absolute temperature)
$$f(T) = A + BT + CT^2 + DT^3$$

| $f(T)$ | A | B | C | D | Standard Deviation, $\sigma$ |
|---|---|---|---|---|---|
| | | | 273.15 < T < 373.15 K | | |
| $\rho$ | 766.17 | 1.80396 | $-3.4589 \times 10^{-3}$ | | 0.5868 |
| $c_p$ | 5.6158 | $-9.0277 \times 10^{-3}$ | $14.177 \times 10^{-6}$ | | $4.142 \times 10^{-3}$ |
| $k$ | −0.4806 | $5.84704 \times 10^{-3}$ | $-0.733188 \times 10^{-5}$ | | $0.481 \times 10^{-3}$ |
| | | | 273.15 < T < 320 K | | |
| $\mu \times 10^6$ | $0.239179 \times 10^6$ | $-2.23748 \times 10^3$ | 7.03318 | $-7.40993 \times 10^{-3}$ | $4.0534 \times 10^{-6}$ |
| $\beta \times 10^6$ | $-57.2544 \times 10^3$ | 530.421 | −1.64882 | $1.73329 \times 10^{-3}$ | $1.1498 \times 10^{-6}$ |
| | | | 320 < T < 373.15 K | | |
| $\mu \times 10^6$ | $35.6602 \times 10^3$ | −272.757 | 0.707777 | $-0.618833 \times 10^{-3}$ | $1.0194 \times 10^{-6}$ |
| $\beta \times 10^6$ | $-11.1377 \times 10^3$ | 84.0903 | −0.208544 | $0.183714 \times 10^{-3}$ | $1.2651 \times 10^{-6}$ |

Source: From Incropera, F. P., and D. P. DeWitt, *Fundamentals of Heat and Mass Transfer*, Wiley, New York, 1985.

## Table A.4  Thermophysical Properties of Air

| $T$ [K] | $\rho$ [kg/m³] | $c_p$ [kJ/kg · K] | $\mu \times 10^7$ [N · s/m²] | $v \times 10^6$ [m²/s] | $k \times 10^3$ [W/m · K] | $\alpha \times 10^5$ [m²/s] | Pr |
|---|---|---|---|---|---|---|---|
| 200 | 1.7458 | 1.007 | 132.5 | 7.590 | 18.1 | 10.3 | 0.737 |
| 250 | 1.3947 | 1.006 | 159.6 | 11.44 | 22.3 | 15.9 | 0.720 |
| 300 | 1.1614 | 1.007 | 184.6 | 15.89 | 26.3 | 22.5 | 0.707 |
| 350 | 0.9950 | 1.009 | 208.2 | 20.92 | 30.0 | 29.9 | 0.700 |
| 400 | 0.8711 | 1.014 | 230.1 | 26.41 | 33.8 | 38.3 | 0.690 |
| 450 | 0.7740 | 1.021 | 250.7 | 32.39 | 37.3 | 47.2 | 0.686 |
| 500 | 0.6964 | 1.030 | 270.1 | 38.79 | 40.7 | 56.7 | 0.684 |
| 550 | 0.6329 | 1.040 | 288.4 | 45.57 | 43.9 | 66.7 | 0.683 |
| 600 | 0.5804 | 1.051 | 305.8 | 52.69 | 46.9 | 76.9 | 0.685 |
| 650 | 0.5356 | 1.063 | 322.5 | 60.21 | 49.7 | 87.3 | 0.690 |
| 700 | 0.4975 | 1.075 | 338.8 | 68.10 | 52.4 | 98.0 | 0.695 |
| 750 | 0.4643 | 1.087 | 354.6 | 76.37 | 54.9 | 109.0 | 0.702 |
| 800 | 0.4354 | 1.099 | 369.8 | 84.93 | 57.3 | 120.0 | 0.709 |
| 850 | 0.4097 | 1.110 | 384.3 | 93.80 | 59.6 | 131.0 | 0.716 |
| 900 | 0.3868 | 1.121 | 398.1 | 102.9 | 62.0 | 143.0 | 0.720 |
| 950 | 0.3666 | 1.131 | 411.3 | 112.2 | 64.3 | 155.0 | 0.723 |
| 1000 | 0.3482 | 1.141 | 424.4 | 121.9 | 66.7 | 168.0 | 0.726 |

*Formulas for Interpolation (T = absolute temperature)*

$$\rho = \frac{348.59}{T} \quad (\sigma = 4 \times 10^{-4})$$
$$f(T) = A + BT + CT^2 + DT^3$$

| $f(T)$ | $A$ | $B$ | $C$ | $D$ | Standard Deviation, $\sigma$ |
|---|---|---|---|---|---|
| $c_p$ | 1.0507 | $-3.645 \times 10^{-4}$ | $8.388 \times 10^{-7}$ | $-3.848 \times 10^{-10}$ | $4 \times 10^{-4}$ |
| $\mu \times 10^7$ | 13.554 | 0.6738 | $-3.808 \times 10^{-4}$ | $1.183 \times 10^{-7}$ | $0.4192 \times 10^{-7}$ |
| $k \times 10^3$ | $-2.450$ | 0.1130 | $-6.287 \times 10^{-5}$ | $1.891 \times 10^{-8}$ | $0.1198 \times 10^{-3}$ |
| $\alpha \times 10^8$ | $-11.064$ | $7.04 \times 10^{-2}$ | $1.528 \times 10^{-4}$ | $-4.476 \times 10^{-8}$ | $0.4417 \times 10^{-8}$ |
| Pr | 0.8650 | $-8.488 \times 10^{-4}$ | $1.234 \times 10^{-6}$ | $-5.232 \times 10^{-10}$ | $1.623 \times 10^{-3}$ |

*Source*: From F. P. Incropera and D. P. DeWitt, *Fundamentals of Heat and Mass Transfer*, Wiley, New York, 1985.

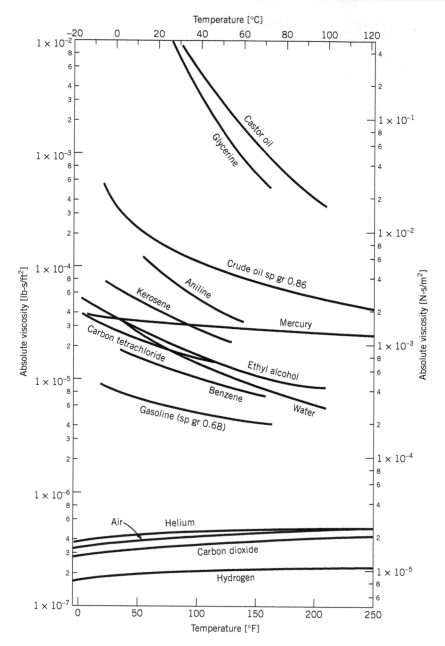

**FIGURE A.1**    Absolute viscosities of certain gases and liquids.
(Compiled from Pritchard, P.J., *Fox and McDonald's Introduction to Fluid Mechanics*, 8th ed., Wiley, New York, 2011 and other sources.)

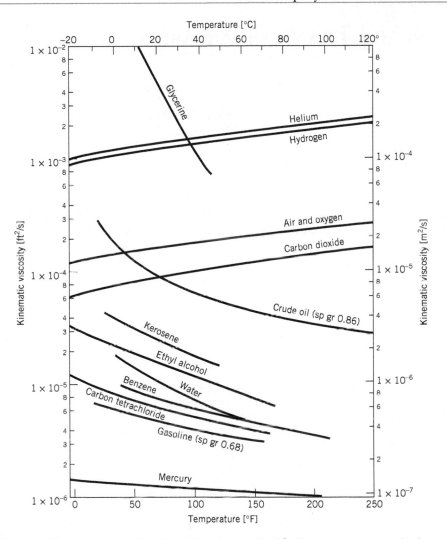

**FIGURE A.2** Kinematic viscosities of certain gases and liquids. The gases are at standard pressure. (Compiled from Pritchard, P.J., *Fox and McDonald's Introduction to Fluid Mechanics*, 8th ed., Wiley, New York, 2011, and other sources.)

# B    LAPLACE TRANSFORM BASICS

We offer the following primer of Laplace transforms for review purposes to meet the needs of this text. Extensive treatment of this subject is available elsewhere [1]. The Laplace transform of a function $y(t)$ is defined as

$$Y(s) = \mathscr{L}\,[Y(t)] = \int_0^\infty y(t)e^{-st}dt \tag{B.1}$$

There are many functions that are useful in modeling linear systems that have analytical forms for their Laplace transforms. Consider the unit step function, defined by

$$U(t) = \begin{Bmatrix} 0 & t < 0 \\ 1 & t > 0 \end{Bmatrix} \tag{B.2}$$

Applying the definition of the Laplace transform yields

$$\mathscr{L}\,[U(t)] = \int_0^\infty e^{-st}dt = \frac{1}{s} \tag{B.3}$$

The unit step function and its Laplace transform form a Laplace transform pair. Table B.1 provides a list of common Laplace transform pairs.

One of the most important properties of the Laplace transform results from the application of the Laplace transform to a derivative, as

$$\mathscr{L}\,\left[\frac{dy(t)}{dt}\right] = sY(s) - y(0) \tag{B.4}$$

Differentiation in the time domain is the same as multiplication in the Laplace transform domain. This relationship is derived through application of integration by parts of the definition of the Laplace transform, and can be generalized as

$$\mathscr{L}\,\left[\frac{d^n y(t)}{dt^n}\right] = s^n Y(s) - s^{n-1}y(0) - s^{n-2}\frac{dy(0)}{dt} - \cdots - \frac{d^{n-1}y(0)}{dt^{n-1}} \tag{B.5}$$

## Table B.1 Laplace Transform Pairs

| | $f(t)$ | $\mathscr{L}(f)$ |
|---|---|---|
| 1 | 1 | $\dfrac{1}{s}$ |
| 2 | $t$ | $\dfrac{1}{s^2}$ |
| 3 | $t^2$ | $\dfrac{2!}{s^3}$ |
| 4 | $t^n$ | $\dfrac{n!}{s^{n+1}}$ |

where $n$ is integer

| | | |
|---|---|---|
| 5 | $e^{-at}$ | $\dfrac{1}{s+a}$ |
| 6 | $te^{-at}$ | $\dfrac{1}{(s+a)^2}$ |
| 6 | $\dfrac{e^{-at}-e^{-bt}}{b-a}$ | $\dfrac{1}{(s+a)(s+b)}$ |
| 7 | $\sin at$ | $\dfrac{a}{s^2+a^2}$ |
| 8 | $\cos at$ | $\dfrac{s}{s^2+a^2}$ |

# B.1 Final Value Theorem

A useful property of the Laplace transform allows us to determine the value of the function $y(t)$ knowing only the Laplace transform of the function. The final value theorem states

$$\lim_{t \to \infty} y(t) = \lim_{s \to \infty} [sY(s)] \tag{B.6}$$

Let's apply this to the unit step function

$$\lim_{s \to 0} sY(s) = \lim_{s \to 0} s\frac{1}{s} = 1 \tag{B.7}$$

The unit step function has a limit of unity. Clearly this is the correct limit!

# B.2 Laplace Transform Pairs

Table B.1 provides Laplace transform pairs for a variety of important engineering functions. More extensive lists and methods for more complex transformations may be found in most advanced engineering mathematics texts.

## REFERENCE

Schiff, J. L., *The Laplace Transform: Theory and Applications*, Springer, New York, 1999.

# GLOSSARY

**absolute temperature scale** A temperature scale referenced to the absolute zero of temperature; equivalent to thermodynamic temperature.

**accelerometer** An instrument for measuring acceleration.

**accuracy** The closeness of agreement between the measured value and the true value.

**active filter** A filter circuit that includes one or more active electrical elements, such as an operational amplifier.

**actual flow rate** Flow rate specified under the actual temperature and pressure conditions of the measurement. *See* **standard flow rate**.

**actuator** Device that creates a desired linear or rotary motion; classes generally include electrical, mechanical, hydraulic, or pneumatic.

**A/D converter** A device that converts analog voltages into digital numbers.

**advanced-stage uncertainty** The uncertainty estimate beyond the design stage that includes the uncertainty due to measurement system, measurement procedure, and measured variable variation.

**aleatory uncertainty** Uncertainty due to the natural variability of the measurand.

**alias frequency** A false frequency appearing in a discrete data set that is not in the original signal.

**amplitude** Characteristic of a signal or component frequency of a signal that expresses a range of time-dependent magnitude variations.

**amplitude spectrum** A plot associating the individual components of frequency comprising a signal with their corresponding amplitudes.

**apparent strain** A false output from a strain gauge as a result of input from extraneous variables such as temperature.

**astable multivibrator** Switching circuit that toggles on or off continuously in response to an applied voltage command.

**band-pass filter** A device that allows a range of signal frequencies to pass unaltered while frequencies outside this range, both higher and lower, are attenuated.

**Bessel filter** Filter design having a maximally linear phase response over its passband.

**bias** An offset value.

**bias error** *See* **systematic error**. Archaic term no longer used in measurements.

**bimetallic thermometer** A temperature measuring device that uses differential thermal expansion of two materials.

**bistable multivibrator** *See* **flip-flop**.

**bit**  The smallest unit in binary representation.

**bit number**  The number of bits used to represent a word in a digital device.

**bonded resistance strain gauge**  A strain sensor that exhibits an electrical resistance change indicative of strain. The sensor is bonded to the test specimen in such a way that the gauge accurately measures the strain at the surface of the test specimen.

**Bragg grating**  Fiber optic device that has a periodic variation of the refraction of index.

**bridge constant**  The ratio of the actual bridge voltage to the output of a single strain gauge sensing the maximum strain.

**buffer**  Memory set aside to be accessed by a specific input/output (I/O) device, such as to temporarily store data.

**bus**  A major computer line used to transmit information.

**Butterworth filter**  Filter design having a maximally flat magnitude ratio over its passband.

**byte**  A collection of 8 bits.

**calibration**  Act of applying a known input value to a system so as to observe the system output value.

**calibration curve**  Plot of the input values versus the output values resulting from a calibration.

**cascading filters**  Multiple filters arranged in series.

**central tendency**  Tendency of data scatter to group about a central (mean or most expected) value.

**charge amplifier**  Amplifier used to convert a high impedance charge into a voltage.

**circular frequency**  Frequency specified in radians per second; also called angular frequency.

**closed-loop control**  Type of control action that is based on the output value from a sensor relative to a set-point.

**combined standard uncertainty**  Uncertainty estimate that includes both systematic and random standard uncertainties.

**common-mode voltage**  Voltage difference between two different ground points. *See* **ground loop**.

**concomitant method**  An alternate method for assessing the value of a variable.

**conduction error**  Error in temperature measurement caused by the diffusion of heat through a temperature sensor either toward or away from the measuring environment so as to change the sensor temperature.

**confidence interval**  Interval that defines the limits about a mean value within which the true value is expected to fall.

**confidence level**  Probability that the true value falls within the stated confidence interval.

**control**  A means of setting and maintaining a value during a test.

**control parameter**  Parameter whose value is set during a test and that has a direct effect on the value of the measured variable.

**conversion error**  The errors associated with converting an analog value into a digital number.

**conversion time** The duration associated with converting an analog value into a digital number.

**coupled system** The combined effect of interconnected instruments used to form a measurement system.

**critical pressure ratio** The pressure ratio necessary to accelerate a fluid to a velocity equal to its speed of sound.

**cutoff frequency** The 3 dB design point frequency of a filter.

**D/A converter** A device that converts a digital number into an analog voltage.

**damping ratio** A measure of system damping—a measure of a system's ability to absorb or dissipate energy.

**d'Arsonval movement** A device that senses the flow of current through torque on a current-carrying loop; the basis for most analog electrical meters.

**data acquisition module** A device that contains a minimum of components required for data acquisition; often connected to a computer by a link, wired or wireless.

**data acquisition system** A system that quantifies and stores data.

**deflection mode** Measuring method that uses signal energy to cause a deflection in the output indicator. *See* **loading error**.

**dependent variable** A variable whose value depends on the value of one or more other variables.

**design-stage uncertainty** The uncertainty due only to instrument error and instrument resolution; an uncertainty value based on the minimum amount of information that is known at the early stage of test design.

**deterministic signal** A signal that is predictable in time or in space, such as a sine wave or a ramp function.

**deviation plot** Graphical representation of the difference between a reference and expected value.

**diaphragm meter** A type of positive displacement flow meter that measures gas volume flow rate by alternately passing isolated but known volumes of the gas between two chambers.

**differential-ended connection** Dual-wire connection scheme that measures the difference between two signals with no regard to ground.

**digital signal** Representation of an analog signal that is quantized in both magnitude and time.

**digital signal processor** Microprocessor used to manage the transfer of digital data information and perform prescribed mathematical manipulation of the data.

**dilatation** The ratio of the change in volume to the original volume.

**direct memory access** Transfer protocol that allows a direct transmission path between a device and computer memory.

**discharge coefficient** Coefficient equating the ratio of actual flow rate across a flow meter to the ideal flow rate if no losses or flow separation occurred.

**discrete random variable** Random variable whose value is not continuous.

**discrete time signal** Signal in which information about the magnitude of the represented variable is available only at discrete points in time.

**dispersion** The scattering of values of a variable around a mean value.

**distortion** Any undesirable change in the shape of a waveform that occurs during signal conditioning or measurement.

**disturbance** An effect that changes the operating condition of a control system away from its set-point. Requires controller action to maintain or re-establish the set point.

**Doppler anemometry** Refers to various techniques that use the Doppler shift to measure velocity.

**Doppler flowmeter** Flowmeter that uses the Doppler shift in the frequency of ultrasonic waves passing through the fluid flow to indicate volume flow rate.

**dynamic calibration** Process of calibration such that the input variable is time-varying.

**dynamic pressure** Difference between the total and static pressures at any point.

**dynamic response** Response of a system to a well-defined sine wave input.

**dynamometer** Device for measuring shaft power.

**earth ground** A ground path whose voltage level is zero.

**elastic region** An operating range of strain versus load wherein a material returns to its initial length when the load is removed.

**electromagnetic flow meter** Device that bases fluid flow rate on the magnitude of the current induced in a magnetic field caused by the velocity of the fluid passing through that magnetic field.

**end standard** Length standard described the distance between two flat square ends of a bar.

**epistemic uncertainty** Uncertainty arising from unknowns about the measurand, process, or model.

**error** Difference between a measured value and its true (actual) value. In uncertainty analysis, it is an effect that causes a difference between a measured value and its true value. In controls, it is the difference between a sensor value and a set-point value.

**error fraction** A measure of the time-dependent error in the time response of a first-order system.

**expanded uncertainty** Uncertainty interval that includes the combined effects of random and systematic uncertainties at a stated confidence level (typically 95%) other than one standard deviation. *See* **standard uncertainty**.

**expansion factor** Coefficient used to account for compressibility effects in obstruction flow meters.

**extraneous variable** A variable that is not or that cannot be controlled during a measurement, but that affects the measured value.

**feedback-control stage** Stage of a measurement system that interprets the measured signal and exerts control over the process.

**filter** Circuit or algorithm used to reduce or remove the amplitude (or the effect) of unwanted frequencies from a signal.

**firewire** A serial method for data transmission between a computer and peripherals using a thin cable and based on the IEEE 1394 standard.

**first-order system** A system characterized as having time-dependent storage or dissipative ability but having no inertia.

**fixed point temperature** Temperatures specified through a reproducible physical behavior, such as a melting or boiling point.

**flip-flop** A switching circuit that toggles either on or off on command. Also called a bistable multivibrator.

**flow coefficient** Product of the discharge coefficient and the velocity of approach factor; used in a flow meter calculation.

**flow nozzle** Flow meter consisting of a gradual concentric contraction from a larger diameter to a smaller throat diameter that is inserted into a pipe to create a measurable pressure drop proportional to flow rate.

**Fourier analysis** A method for expressing a signal as a series of sines and cosine functions.

**Fourier transform** Mathematical function that allows a complex signal to be decomposed into its frequency components. Method to represent a signal in terms of amplitude as a function of frequency.

**frequency** Value indicating the rate of change of a signal with time.

**frequency bandwidth** The range of frequencies in which the magnitude ratio remains within 3 dB of unity.

**frequency response** The magnitude ratio and phase shift of a system as a function of applied input frequency.

**frequency spectrum** A representation, such as a plot, associating the individual frequency components comprising a signal in terms of the associated amplitude and/or phase angle.

**full-scale output (FSO)** The arithmetic difference between the end points (range) of a calibration expressed in units of the output variable.

**fundamental frequency** Smallest frequency component of a signal.

**gain** Result of amplification or some multiplier of a signal magnitude; also used to refer to a constant multiplier for a transfer function block.

**galvanometer** Instrument that is sensitive to current flow through torque on a current loop; typically used to detect a state of zero current flow.

**gauge blocks** Working length standard for machining and calibration.

**gauge factor** Property of a strain gauge that relates changes in electrical resistance to strain.

**gauge length** Effective length over which a strain gauge averages strain during a measurement.

**ground** Signal return path to earth.

**ground loop** A signal circuit formed by grounding a signal at more than one point with the ground points having different voltage potentials.

**Hall effect** Voltage induced across a current-carrying conductor situated within a magnetic field. The magnetic field exerts a transverse force pushing the charged electrons to one side of the conductor.

**handshake** An interface procedure to control data flow between different devices that includes start and stop (hello and goodbye) signals.

**harmonics** Frequencies present in a signal that are usually integer multiples of a fundamental frequency.

**heat flux sensor** A device that uses physical phenomena to indicate local heat flux.

**high-pass filter** A device that allows signal information at frequencies above the cutoff frequency to pass unaltered while information at lower frequencies are attenuated.

**Hooke's law** The fundamental linear relation between stress and strain. The proportionality constant is the modulus of elasticity.

**hypothesis test** Statistical test based on inferential statistics to accept or reject a statistical hypothesis that two measured parameters are equal, or one is greater than or less than the other.

**hysteresis** A difference in the indicated value for any particular input value when approached going upscale as opposed to going downscale.

**immersion errors** Temperature measurement errors due to the measuring environment leading to differences between sensor temperature and the temperature being measured.

**impedance matching** Process of setting either the input impedance or the output impedance in a way to keep signal loading errors to an acceptable level.

**independent variable** A variable whose value is changed directly to cause a change in the value of a dependent variable.

**indicated value** The value displayed or indicated by the measurement system.

**input value** Process information sensed by the measurement system.

**insertion error** Any error brought on by the interaction between a sensor and the immediate environment in which it is installed.

**in situ** In measurements, installed in the system as intended for use.

**interference** Extraneous effect that imposes a deterministic trend on the measured signal.

**interferometer** Device for measuring length that uses interference patterns of light sources; provides very high resolution.

**interrupt** A signal to initiate a different procedure; analogous to raising one's hand.

**junction** For thermocouples, an electrical connection between two dissimilar materials.

**laminar flow element** Device that measures flow rate by the linear pressure drop versus flow rate relationship occurring when fluid travels through a tube or cluster of tubes under laminar flow conditions.

**least squares regression** Analysis tool used to generate a curve fit equation that minimizes the sum of the squares of the deviations between the predicted values of the curve fit and the data set on which the curve fit is based.

**linearity** The closeness of a calibration curve to a straight line.

**linear variable differential transformer (LVDT)** A sensor/transducer that provides a voltage output as a function of core displacement.

**line standard** Standard for length made by scribing two marks on a dimensionally stable material.

**load cell** Sensor for measuring force or load.

**loading error** Error due to energy being extracted from the variable being measured by the very act of measurement so as to change the value measured.

**low-pass filter** A device that allows signal frequencies below the cutoff frequency to pass unaltered while higher frequencies are attenuated.

**Mach number** Ratio of the local speed to the speed of sound in the medium.

**magnitude** Indication of the size or level of a quantity.

**magnitude ratio** The ratio of the values of output amplitude to the input amplitude of a dynamic signal.

**mass flow rate** Mass flow per unit time.

**measurand** Another term for the measured variable.

**measured value** A number assigned to the measured variable.

**measured variable** A variable whose value is measured; also known as the measurand.

**measurement** The act of assigning a specific value to the variable being measured.

**mechatronics** Subject area focused on the interaction between electrical and mechanical components of devices and the control of such devices.

**metrology** The science of weights and measures.

**moiré patterns** An optical effect resulting from overlaying two dense grids that are displaced slightly in space from one another.

**monostable** One-shot device that toggles on and then off on a single voltage command.

**moving coil transducer** A sensor that uses a displacement of a conductor in a magnetic field to sense velocity.

**multiplexer** An input/output (I/O) switch having multiple input signal lines but only one output signal line.

**natural frequency** The frequency of the free oscillations of an undamped system.

**noise** An extraneous effect that imposes random variations on the measured signal.

**notch filter** A filter that only removes or reduces signal amplitude within a narrow frequency band termed as the notch.

**null mode** Measuring method that balances a known amount against an unknown amount to determine the measured value.

**Nyquist frequency** Half the sample rate or sampling frequency.

**obstruction flow meter** Device that measures flow rate by the magnitude of the pressure drop arising by purposefully creating an abrupt area change in the flow.

**ohmmeter** An instrument for measuring resistance.

**open-loop control** Type of control action that does not use a sensor to measure the variable being controlled, such as by controlling room lights using a light switch timer.

**operating conditions** Conditions of the test, including all equipment settings, sensor locations, and environmental conditions.

**optical pyrometer** Nonintrusive temperature measurement device. Measurement is made through an optical comparison.

**orifice meter** Flow meter consisting of a flat plate with a hole in its center that fits between a pipe flange to create a pressure drop proportional to flow rate; also known as an orifice plate.

**oscilloscope** An instrument for measuring a time-varying voltage signal and displaying a trace; very high frequency signals can be examined.

**outlier** A measured data point whose value falls outside the range of reasonable probability.

**output stage** Stage of a measurement system that indicates or records the measured value.

**output value** The value indicated by the measurement system.

**overall error** Instrument uncertainty estimate based on the square root of the sum of the squares (RSS) of the uncertainties of all known instrument errors.

**parallel communication** Technique that transmits information in groups of bits simultaneously.

**parameter** The value defined by some functional relationship between variables related to the measurement.

**particle image velocimetry** Optical method that measures flow velocity by observing the distance traveled by fluid particles per unit time suspended within the fluid.

**passband** Range of frequencies over which the signal amplitude is attenuated by no more than 3 dB. *See* **stopband**.

**passive filter** A filter circuit that only uses passive electrical elements such as resistors, inductors, and capacitors.

**Peltier effect** Describes the reversible conversion of energy from electrical to thermal when a current flows through a junction of dissimilar materials.

**phase shift** A shift or offset in value between the input signal and the output signal representing the measured signal; may be expressed in terms of an angle.

**phase spectrum** A plot associating the individual components of frequency comprising a signal with their corresponding phase shift.

**photoelastic** Describes a change in optical properties with strain.

**piezoresistance coefficient** Constant that describes the change in electrical resistance with applied strain.

**pitot–static pressure probe** Device used to measure dynamic pressure.

**Poisson's ratio** Ratio of lateral strain to axial strain.

**pooled value** Statistical value determined from separate data sets, such as replications, that are subsequently grouped together.

**population** The set representing all possible values for a variable.

**potentiometer** Refers either to a variable resistor or to an instrument for measuring small voltages with high accuracy.

**potentiometer transducer** Variable electrical resistance transducer for measurement of length or rotation angle.

**power spectrum** A plot associating the individual components of frequency comprising a signal with their corresponding power magnitudes.

**precision** The amount of dispersion expected with repeated measurements; archaic term used to describe the random uncertainty.

**precision** (instrument specification) Ultimate performance claim of instrument repeatability based on multiple tests conducted at multiple labs and on multiple units.

**precision error** Archaic term no longer used in measurements. *See* **random error**.

**precision indicator** Statistical measure that quantifies the dispersion in a reported value.

**precision interval** Interval evaluated from data scatter that defines the probable range of a statistical value and ignoring the effects of systematic errors.

**primary standard** The defining value of a unit.

**Prony brake** A historically significant dynamometer design that uses mechanical friction to dissipate and measure power.

**prover** Flow meter calibration device using some method placed in series with the candidate flow meter that accurately measures the volume of fluid passing per unit time.

**proving ring** An elastic load cell, which may be used as a local calibration standard.

**pyranometer** Optical instrument used to measure irradiation on a plane.

**quantization** Process of converting an analog value into a digital number.

**quantization error** An error brought on by the resolution of an A/D converter.

**radiation error** Temperature measurement insertion error caused by a radiative heat exchange between a temperature sensor and portions of its immediate environment not at the same temperature as the variable being measured.

**radiation shield** A device that reduces radiative heat transfer to a temperature sensor to reduce insertion errors associated with radiation.

**radiometer** Device for measuring radiative source temperature using a thermopile detector.

**random error** The portion of error that varies randomly on repeated measurements of the same variable.

**randomization methods** Test methods used to break up the interference effects from extraneous variables by strategies that randomize the effects of those variables on the data set.

**random standard uncertainty** The random uncertainty estimated at the one standard deviation level or at the 68% probability level for a normal distribution.

**random test** Measurement strategy that sets a random order to the changes in the independent variable.

**random uncertainty** Uncertainty assigned to a random error; contribution to combined uncertainty due to random errors.

**range** The lower to upper limits of an instrument or test.

**recovery errors** Temperature measurement error in a high-speed flowing gas stream related to the conversion of kinetic energy into a temperature rise at the sensor.

**reference junction** In reference to a thermocouple, the junction at which the temperature is known.

**relative uncertainty** Uncertainty value expressed as a fraction or percent of the measured value.

**repeatability** Instrument specification estimating the expected random uncertainty based on multiple tests within a given lab on a single unit.

**repetition** Repeating measurements during the same test.

**replication** Duplicating a test under similar operating conditions using similar instruments.

**reproducibility** Random uncertainty estimate based on multiple tests performed in different labs on a single unit.

**resistivity** Material property describing the basic electrical resistance of a material.

**resolution** The smallest detectable change in a measured value as indicated by the measuring system.

**resonance frequency** The frequency at which the magnitude ratio reaches a maximum value greater than unity.

**result** The final value determined from a set or sets of measurements.

**resultant** A variable calculated from the values of other variables.

**ringing frequency** The frequency of the free oscillations of a damped system; a function of the natural frequency and damping ratio.

**rise time** The time required for a first-order system to respond to 90% of a step change in input signal.

**rms value** Root-mean-square value.

**roll-off slope** Decay rate of a filter stopband.

**root-sum-square value (RSS)** The value obtained by taking the square root of the sum of a series of squared values.

**rotating cusp meter** A type of positive displacement flow meter that measures volume flow rate by the rate at which fluid is displaced within alternating rotating cusps.

**RTD** Resistance temperature detector; senses temperature through changes in electrical resistance of a conductor.

**sample** A data point or single value of a population.

**sample rate** The rate or frequency at which data points are acquired; reciprocal of sample time increment.

**sample statistics** Statistics of a variable determined from a set of data points whose set size is less than the population of that variable.

**sample time increment** The time interval between two successive data measurements. Reciprocal of sample rate.

**saturation error** Error due to an input value exceeding the device maximum.

**second-order system** A dynamic system model whose behavior includes inertia.

**Seebeck effect** Source of open circuit electromotive force in thermocouple circuits.

**seismic transducer** Instrument for measuring displacement in time, velocity, or acceleration; based on a spring-mass-damper system.

**sensitivity** The rate of change of a variable ($y$) relative to a change in some other variable ($x$), for example, $dy/dx$.

**sensitivity index** The sensitivity of a result relative to a variable evaluated at some operating point and used to weight individual uncertainties in uncertainty propagation. *See* **sensitivity**.

**sensor** The portion of the measurement system that senses or responds directly to the process variable being measured.

**serial communication** Technique that transmits information 1 bit at a time.

**settling time** The time required for a second-order system to settle to within $\pm 10\%$ of the final value of a step change in input value.

**shield** A metal foil or wire braid wrapped around signal-carrying wires that intercepts unwanted stray external electrical fields to reduce noise and interference effects.

**shielded twisted pair** Describes electrical wiring that reduces mutual induction between conductors through twisting the conductors around each other and that reduces noise and interference through shielding.

**signal** Information in transit through a measurement system; often exists in electrical or mechanical form.

**signal conditioning stage** Stage of a measurement system that modifies the signal to a desired magnitude and form.

**signal-to-noise ratio** Ratio of signal power to noise power.

**single-ended connection** Two-wire connection scheme that measures the signal relative to ground.

**single-sample uncertainty analysis** Method of uncertainty analysis used to evaluate uncertainty when very few repeated measurements are made.

**sound pressure level meter (SPL)** Stand-alone device with integral microphone and digital signal processor used to measure acoustical sound pressure levels.

**span** The difference between the maximum and minimum values of operating range of an instrument.

**stage** Refers to a general component in a measurement signal flow that has a specific purpose.

**stagnation pressure** Pressure at the point where a flow is brought to rest. *See* **total pressure**.

**standard** The known or reference value used as the basis of a calibration; also, a test method set by international agreement for quantifying some outcome or device performance.

**standard uncertainty** An uncertainty stated at one standard deviation level of confidence.

**static calibration** Calibration method whereby the input value is held constant while the output value is measured.

**static pressure** Pressure sensed by a fluid particle as it moves with the flow.

**static sensitivity** The rate of change of the output signal relative to a change in a static input signal. The slope of the static calibration curve at a point. Also called the static gain. *See* **sensitivity**.

**steady response** A portion of the time-dependent system response that either remains constant with time or repeats its waveform with time.

**step response** System time response to a step change in input value.

**stopband** Range of frequencies over which the signal amplitude is attenuated by 3 dB or more. *See* **passband**.

**strain** Elongation per unit length of a member subject to an applied force.

**strain gauge** Sensor used to measure the strain on the surface of a part when under load.

**strain gauge rosette** Two or more strain gauges arranged at precise angles and locations relative to each other so as to measure differing components of strain.

**stress** Internal force per unit area.

**stroboscope** High-intensity source of light that can be made to flash at a precise rate; used to measure the frequency of rotation.

**Student's *t*-distribution** Sampling distribution that arises when estimating the mean of a population when the sample size is small.

**systematic error** The portion of error that remains constant in repeated measurements of the same variable; a constant offset between the indicated value and the actual value measured.

**systematic standard uncertainty** The systematic uncertainty estimated at the one standard deviation level or at the 68% probability level for a normal distribution. When a correction for known systematic error is applied, it is the uncertainty in the correction. This value is often taken as being half of a systematic uncertainty assigned by judgement.

**systematic uncertainty** Uncertainty assigned to a systematic error. The uncertainty due to all the systematic errors affecting the result. (see **systematic standard uncertainty**)

**temperature scale** Establishes a universal means of assigning a quantitative value to temperatures.

**thermistor** Temperature-sensitive semiconductor resistor.

**thermocouple** Wire sensor consisting of at least two junctions of two dissimilar conductors; used to measure temperature.

**thermoelectric** Refers to a type of device that indicates a thermally induced voltage.

**thermopile** A multiple junction thermocouple circuit designed to measure a particular temperature, an average temperature, or a temperature difference.

**Thomson effect** Describes the presence of a voltage arising through a temperature difference in a homogeneous conductor.

**time constant** System property defining the time required for a first-order system to respond to 63.2% of a step input.

**time delay** The lag time between an applied input signal and the measured output signal.

**time response** The complete time-dependent system response to an input signal.

**total pressure** Pressure sensed at a point where the flow is brought to rest without losses; the sum of the static and dynamic pressures; the isentropic value of stagnation pressure.

**total sample period** Duration of the measured signal as represented by the data set.

**transducer** The portion of measurement system that transforms the sensed information into a different form; also, loosely, a device that houses the sensor, transducer, and, often, signal conditioning stages of a measurement system.

**transient response** Portion of the time-dependent system response that decays to zero with time.

**transit time flow meter** Device that determines flow rate by measuring the time it takes for an ultrasonic wave to travel through the moving flow along a well-defined path.

**trigonometric series**  Infinite series of trigonometric functions; applications include representing complex waveforms in terms of frequency content.

**true rms**  Data-reduction technique that can correctly provide the rms value for a nonsinusoidal signal; a signal conditioning method that integrates the signal.

**true value**  The actual or exact value of the measured variable.

**TTL (true-transistor-logic)**  Switched signal that toggles between a high (e.g., 5 V) and a low (e.g., 0 V) state.

**turndown**  Ratio of the highest flow rate to the lowest flow rate that a particular flow meter can measure within specification.

**type A uncertainty**  An uncertainty value estimated from the statistics of a variable.

**type B uncertainty**  An uncertainty value estimated by other than statistical methods.

**uncertainty**  An estimate of the probable error in a reported value.

**uncertainty analysis**  A process of identifying the errors in a measurement and quantifying their effects on the value of a stated result.

**uncertainty interval**  An interval about a variable or result that is expected to contain the true value.

**USB (universal serial bus)**  A serial method of data transmission between a computer and peripherals using a four-wire cable.

**validation**  Process or test to assess whether test model data behave in the same way as the physical process the test simulates; can be used to assign uncertainty in the model.

**velocity of approach factor**  A coefficient used in obstruction flow meter calculations that depends on the ratio of flow meter throat diameter to pipe diameter.

**venturi meter**  In-line flow meter whose internal diameter narrows from the pipe diameter to a smaller diameter throat and then expands back to the pipe diameter so as to create a pressure drop proportional to flow rate.

**verification**  Test designed to ensure that a measurement system as a whole or some portion of the system is working correctly; can be used to assign uncertainty values to aspects of a measurement system.

**vernier calipers**  Tool for measuring both inside and outside dimensions.

**vernier scale**  Technique that allows increased resolution in reading a length scale.

**voltage divider circuit**  An electrical circuit that proportions the input to output voltages using a variable resistance.

**volume flow rate**  Flow volume per unit time.

**VOM**  Acronym for volt-ohm meter.

**vortex flow meter**  Device that measures flow rate by measuring the frequency of vortex shedding from a body placed within the flow.

**waveform**  The shape and form of a signal as plotted with time or space.

**Wheatstone bridge**  Electrical circuit for measuring resistance with a high measure of accuracy and sensitivity.

**wobble meter**  A type of positive displacement flow meter that measures fluid volume flow rate by the rate of fluid displaced within alternating known volumes created by oscillating disks.

**word**  A collection of bits used to represent a number.

**zero drift**  A shift away from a zero output value under a zero input value condition.

**zero-order system**  A system whose behavior is independent of time-dependent characteristics, such as storage or inertia.

**zero-order uncertainty**  Random uncertainty estimate based only on a measurement system's resolution or the finest increment of measure.

# INDEX

*Note*: Page numbers in *italics* refer to illustrations